Mathematical Sciences Research Institute Publications

22

Mathematical Sciences Research Institute Publications

Volume 1	Freed and Uhlenbeck: Instantons and Four-Manifolds Second Edition
Volume 2	Chern (ed.): Seminar on Nonlinear Partial Differential Equations
Volume 3	Lepowsky, Mandelstam, and Singer (eds.): Vertex Operators in Mathematics and Physics
Volume 4	Kac (ed.): Infinite Dimensional Groups with Applications
Volume 5	Blackadar: K-Theory for Operator Algebras
Volume 6	Moore (ed.): Group Representations, Ergodic Theory, Operator Algebras, and Mathematical Physics
Volume 7	Chorin and Majda (eds.): Wave Motion: Theory, Modelling, and Computation
Volume 8	Gersten (ed.): Essays in Group Theory
Volume 9	Moore and Schochet: Global Analysis on Foliated Spaces
Volume 10	Drasin, Earle, Gehring, Kra, and Marden (eds.): Holomorphic Functions and Moduli I
Volume 11	Drasin, Earle, Gehring, Kra, and Marden (eds.): Holomorphic Functions and Moduli II
Volume 12	Ni, Peletier, and Serrin (eds.): Nonlinear Diffusion Equations and their Equilibrium States I
Volume 13	Ni, Peletier, and Serrin (eds.): Nonlinear Diffusion Equations and their Equilibrium States II
Volume 14	Goodman, de la Harpe, and Jones: Coxeter Graphs and Towers of Algebras
Volume 15	Hochster, Huneke, and Sally (eds.): Commutative Algebra
Volume 16	Ihara, Ribet, and Serre (eds.): Galois Groups over $\mathbb{Q}$
Volume 17	Concus, Finn, and Hoffman (eds.): Geometric Analysis and Computer Graphics
Volume 18	Bryant, Chern, Gardner, Goldschmidt, and Griffiths: Exterior Differential Systems
Volume 19	Alperin (ed.): Arboreal Group Theory
Volume 20	Dazord and Weinstein (eds.): Symplectic Geometry, Groupoids, and Integrable Systems
Volume 21	Moschovakis (ed.): Logic from Computer Science
Volume 22	Ratiu (ed.): The Geometry of Hamiltonian Systems

Tudor Ratiu
Editor

The Geometry of Hamiltonian Systems

Proceedings of a Workshop Held
June 5-16, 1989

With 126 Illustrations

Springer-Verlag
New York Berlin Heidelberg London
Paris Tokyo Hong Kong Barcelona

Tudor Ratiu
Department of Mathematics
University of California
Santa Cruz, CA 95064
USA

Mathematical Sciences Research Institute
1000 Centennial Drive
Berkeley, CA 94720
USA

The Mathematical Sciences Research Institute wishes to acknowledge support by the National Science Foundation.

Mathematical Subject Classifications: 58F05, 70F15, 58F07, 58F10, 93C10

Library of Congress Cataloging-in-Publication Data
Geometry of hamiltonian systems : proceedings of a workshop held June 5-16, 1989 / Tudor Ratiu, editor.
p. cm. – (Mathematical Sciences Research Institute publications : 22)
Includes bibliographical references.
ISBN-13: 978-1-4613-9727-4
1. Hamiltonian systems–Congresses. I. Ratiu, Tudor S. II. Series.
QA614.83.G46 1991 91-15922
514'.74–dc20

Printed on acid-free paper.

Camera-ready copy prepared by the Mathematical Sciences Research Institute using AMS-TEX.

9 8 7 6 5 4 3 2 1

ISBN-13: 978-1-4613-9727-4 e-ISBN-13: 978-1-4613-9725-0
DOI: 10.1007/978-1-4613-9725-0

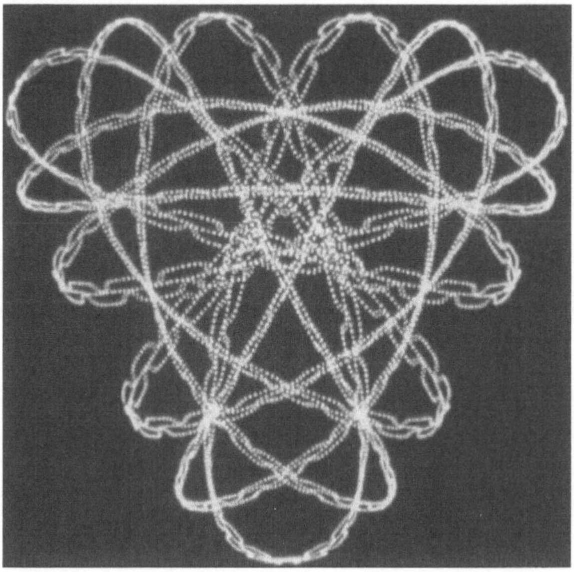

14,000 Iterations of a fourth-order symplectic integrator for the Hénon-Heiles Hamiltonian System. Photo by Clint Scovel and Paul Channell.

Preface

The papers in this volume are an outgrowth of the lectures and informal discussions that took place during the workshop on "The Geometry of Hamiltonian Systems" which was held at MSRI from June 5 to 16, 1989. It was, in some sense, the last major event of the year-long program on Symplectic Geometry and Mechanics. The emphasis of all the talks was on Hamiltonian dynamics and its relationship to several aspects of symplectic geometry and topology, mechanics, and dynamical systems in general. The organizers of the conference were R. Devaney (co-chairman), H. Flaschka (co-chairman), K. Meyer, and T. Ratiu.

The entire meeting was built around two mini-courses of five lectures each and a series of two expository lectures. The first of the mini-courses was given by A.T. Fomenko, who presented the work of his group at Moscow University on the classification of integrable systems. The second mini-course was given by J. Marsden of UC Berkeley, who spoke about several applications of symplectic and Poisson reduction to problems in stability, normal forms, and symmetric Hamiltonian bifurcation theory. Finally, the two expository talks were given by A. Fathi of the University of Florida who concentrated on the links between symplectic geometry, dynamical systems, and Teichmüller theory.

Due to the large number of participants, not everyone was able to speak at the conference. However, several special sessions were organized informally by some of the participants. Everyone was invited to participate, encouraged to speak, and submit a contribution to these proceedings. All papers that were received went through the refereeing process typical of a mathematics research journal and those that were accepted form the present volume. Unfortunately, there were several invited speakers who decided not to submit their contributions, so the present collection of papers does not accurately reflect all aspects of Hamiltonian dynamics that were discussed at the workshop.

As usual, getting a volume of proceedings of this magnitude off the ground entails a lot of work on the part of a number of people. First, thanks are due to the workshop co-chairmen, I. Kaplansky, the Director of MSRI, and V. Guillemin and A. Weinstein, the principal organizers of the entire special year, all of whom were of great help during the preparation of the workshop itself and beyond. The staff at MSRI has been particularly helpful, insuring that the workshop, one of the largest MSRI ever organized, did not collapse of its own weight. A. Baxter, the Manager of Business and Finance of MSRI, deserves the gratitude of all contributors for the excellent and difficult work she has done in shepherding the entire volume from its beginning right up to the last production stages. And last but not least, the participants at the conference must be thanked for offering everyone a two-week-long intense and highly interesting mathematical experience.

Tudor Ratiu

The Geometry of Hamiltonian Systems

Table of Contents

Heisenberg Algebras, Grassmannians and Isospectral Curves

MALCOLM R. ADAMS AND MAARTEN BERGVELT

Abstract. The connection between Heisenberg algebras and the geometry of branched covers of P^1 by Riemann surfaces is discussed using infinite Grassmannians.

§1. Introduction.

In recent years, several geometric and algebraic settings have been introduced for studying nonlinear partial differential evolution equations that display some properties common to Liouville completely integrable systems of ordinary differential equations. It is interesting to investigate the relations between these various approaches since this richness of structure allows one to use results that may be straightforward in one setting to obtain a deeper understanding of another setting.

For example, the Korteweg-de Vries equation (KdV) has been cast in the following three forms. The first is representation theoretic (see e.g., Date, Jimbo, Kashiwara, Miwa [**1**] or Kac, Raina [**2**]): the Plücker equations describing the group orbit through the highest weight vector of the basic representation $L(\Lambda_0)$ of the affine Kac-Moody algebra $A_1^{(1)}$ give the Hirota bilinear equations for the KdV hierarchy in the principal construction of $L(\Lambda_0)$. This construction uses the flows of the principal Heisenberg subalgebra of $A_1^{(1)}$ which correspond to the time variables of the KdV hierarchy. The next framework is more geometric (see Segal, Wilson [**3**]):the principal Heisenberg flows are realized as loop group flows on the infinite Grassmannian $Gr^{(2)}$. Finally, there is an algebro-geometric construction using linear flows of line bundles over Riemann surfaces (see Krichever [**4**]). (There is a fourth framework using Hamiltonian flows on the dual of a loop algebra (see e.g., Flaschka, Newell, Ratiu [**5**]), but we will not discuss this approach in this paper.)

Other hierarchies of equations can be found by choosing other Kac-Moody algebras and/or other constructions of the basic representation using other Heisenberg algebras. For example, the AKNS-Toda system may be studied using the homogeneous Heisenberg algebra of $A_1^{(1)}$. In fact, it is known in the theory of representations of affine Kac-Moody algebras that each conjugacy class of maximal Heisenberg subalgebras yields a construction of the basic module and a corresponding hierarchy of "integrable" equations in Hirota bilinear form. These hierarchies can be realized geometrically as the commutativity conditions for pairs of flows from the positive part of the Heisenberg algebra acting on an appropriate infinite Grassmannian. In this paper we want to describe the relation of the Heisenberg algebras in $A_n^{(1)}$ with the geometry of the Riemann surfaces. The details will appear in a later publication.

The authors would like to thank Robert Varley for many stimulating discussions.

§2. Heisenberg Algebras.

Let $\Lambda\mathrm{gl}(n,\mathbf{C})$ be the algebra of real analytic maps from the circle S^1 to $\mathrm{gl}(n,\mathbf{C})$. It is well known that the maximal Abelian subalgebras of $\Lambda\mathrm{gl}(n,\mathbf{C})$ (invariant under the Cartan involution) are, up to conjugacy, parametrized by the partitions of n (for a more precise statement and proof see ten Kroode [**6**]).

More explicitly, let $\underline{n} = (n_1 \geq n_2 \ldots \geq n_k > 0)$ be a partition of n, and let $\mathcal{H}^{\underline{n}}$ be the subalgebra of $\Lambda\mathrm{gl}(n,\mathbf{C})$ consisting of block matrices of the form

$$\begin{array}{c} \\ n_1 \\ n_2 \\ \vdots \\ n_k \end{array} \begin{array}{c} \begin{array}{cccc} n_1 & n_2 & \ldots & n_k \end{array} \\ \begin{pmatrix} B_1 & 0 & \ldots & 0 \\ 0 & B_2 & \ldots & 0 \\ \vdots & \vdots & \ddots & \vdots \\ 0 & 0 & \ldots & B_k \end{pmatrix} \end{array},$$

where every block B_i of size $n_i \times n_i$ is a power series in

$$n_i \overbrace{\begin{pmatrix} 0 & 0 & \ldots & 0 & z \\ 1 & 0 & \ldots & 0 & 0 \\ 0 & 1 & \ldots & 0 & 0 \\ \vdots & \vdots & \ddots & \vdots & \vdots \\ 0 & 0 & \ldots & 1 & 0 \end{pmatrix}}^{n_i}.$$

(Here $z = e^{i\theta}$ is the loop parameter on the circle). Then the $\mathcal{H}^{\underline{n}}$ are representatives for all the conjugacy classes of maximal abelian subalgebras if $\underline{n}$ runs over all partitions of n.

In the central extension of $\Lambda\mathrm{gl}(n,\mathbf{C})$ the $\mathcal{H}^{\underline{n}}$ are Heisenberg algebras and for simplicity we will also refer to $\mathcal{H}^{\underline{n}}$ (without central extension) as Heisenberg algebras.

§3. Grassmannians and vector bundles over P^1.

We need the following notation:

$$\Lambda\mathrm{Gl}(\mathrm{n},\mathbf{C}) = \{g : S^1 \to \mathrm{Gl}(n,\mathbf{C}), \text{real analytic}\},$$
$$\mathrm{H}^{(n)} = L^2(S^1,\mathbf{C}^n),$$
$$\mathrm{H}_+^{(n)} = \left\{ \begin{array}{r} f \in \mathrm{H}^{(n)} \mid f \text{ extends to an holomorphic} \\ \text{map from the disk to } \mathbf{C}^n \end{array} \right\}.$$

Define then

$$Gr^{(n)} = \{W \subset \mathrm{H}^{(n)} \mid W = g\mathrm{H}_+^{(n)}, g \in \Lambda\mathrm{Gl}(\mathrm{n},\mathbf{C})\}.$$

Note that

$$Gr^{(n)} \simeq \Lambda\mathrm{Gl}(\mathrm{n},\mathbf{C})/\Lambda_+\mathrm{Gl}(\mathrm{n},\mathbf{C}),$$

where

$$\Lambda_+\mathrm{Gl}(\mathrm{n},\mathbf{C}) = \left\{ \begin{array}{r} g \in \Lambda\mathrm{Gl}(\mathrm{n},\mathbf{C}) \mid g \text{ extends to a holomorphic map} \\ \text{from the disk to } \mathrm{Gl}(n,\mathbf{C}) \end{array} \right\}.$$

There is a geometric interpretation of $Gr^{(n)}$(see Wilson [7]). For $g \in \Lambda\mathrm{Gl}(\mathrm{n},\mathbf{C})$ we will define a holomorphic rank n bundle over P^1 as follows. Let $\{U_0, U_\infty\}$ be a cover of P^1 such that $S^1 \subset U_0 \cap U_\infty$ and let $\{e_i^{(0)}\}_{i=1}^n$, $\{e_i^{(\infty)}\}_{i=1}^n$ be frames for the trivial bundles over U_0, U_∞. Then the bundle E_g is defined by gluing with g: we put in $U_0 \cap U_\infty$

$$e_i^{(\infty)} = g e_i^{(0)}.$$

This gives us a correspondence

$$g \leftrightarrow \{\mathcal{E}_g, \{e_i^{(0)}\}_{i=1}^n, \{e_i^{(\infty)}\}_{i=1}^n\},$$

where $\mathcal{E}_g$ is the sheaf of germs of holomorphic sections of E_g. Forgetting about the trivialisation $\{e_i^{(0)}\}_{i=1}^n$ corresponds to quotienting by $\Lambda_+\mathrm{Gl}(\mathrm{n},\mathbf{C})$, so we have the following correspondences:

$$W_g = g\mathrm{H}_+^{(n)} \leftrightarrow g \mod \Lambda_+\mathrm{Gl}(\mathrm{n},\mathbf{C}) \leftrightarrow \{\mathcal{E}_g, \{e_i^{(\infty)}\}_{i=1}^n\}.$$

§4. Line bundles over Curves.

We consider the following geometric data:

(1) $f : X \to P^1$, an n-fold cover of P^1 by some algebraic curve X.
(2) $L \to X$, a holomorphic line bundle over X, with sheaf of germs of holomorphic sections $\mathcal{L}$,
(3) ϕ, a trivialisation of $\mathcal{L}$ over $f^{-1}(U_\infty)$,
(4) local coordinates z_i centered at ∞_i such that $f^*z = z_i^{n_i}$, where $D = \sum n_i \infty_i$ is the divisor $f^{-1}(\infty)$ and z^{-1} is a local coordinate at infinity on P^1.

REMARK: Since $D = \sum_{i=1}^k n_i \infty_i$ has degree n we have $\sum_{i=1}^k n_i = n$. So we have naturally associated to the geometric datum (1) a partition of n, and hence a Heisenberg algebra $\mathcal{H}^{\underline{n}}$.

Now we use the covering map f to to get a sheaf $f_*\mathcal{L}$ on P^1, which is the sheaf of germs of holomorphic sections of a rank n bundle on P^1. Similarly we get, from the trivialisation ϕ and the log coordinates z_i, a trivialisation of $f_*\mathcal{L}$. We will denote this by $f_*\phi$. So we have a map

$$\{f : X \to P^1, \mathcal{L}, \phi\} \mapsto \{f_*\mathcal{L}, f_*\phi\} \leftrightarrow W \in \mathrm{Gr}^{(n)}.$$

(More generally, if $n = mk$ we can consider a m-fold cover of P^1 by some curve X and a rank k bundle over X to obtain a point of $\mathrm{Gr}^{(n)}$ by pushing forward to P^1).

§5. Reconstruction of the Curve.

We generalize in this section the method of Segal and Wilson [**3**, sect. 6]. Suppose the point $W \in \mathrm{Gr}^{(n)}$ comes from the geometric data $\{f : X \to P^1, \mathcal{L}, \phi\}$ as in the previous section . We want to reconstruct the data from W alone. Note that

$$\begin{aligned} W &= f_*\mathcal{L}(U_0)|_{S^1}, \\ &= \mathcal{L}(f^{-1}(U_0))|_{f^{-1}(S^1)} \end{aligned}$$

Define

$$W^{\mathrm{alg}} = \{h \in W | h = \sum_{-\infty}^{N} h_i z^i, h_i \in C^n\}.$$

This can be identified with meromorphic sections of L which have only poles at the divisor $\bar{D} = \sum \infty_i$. Define then the *stabilizer algebra* of W to be

$$\mathcal{S}_W = \{\psi \in \Lambda\mathrm{gl}(n, \mathbf{C})^{\mathrm{alg}} \mid \psi W^{\mathrm{alg}} \subset W^{\mathrm{alg}}\}.$$

It is clear that $\mathcal{S}_W$ contains the pushforward of all meromorphic functions on X with only poles at $\bar{D}$. In fact if we define $\mathcal{S}_W^{\underline{n}} = \mathcal{S}_W \cap \mathcal{H}^{\underline{n}}$ we have

PROPOSITION: $f_* H^0(X - \bar{D}, \mathcal{O}_X) = \mathcal{S}_W^{\underline{n}}$.

So we have found a commutative subring $\mathcal{S}_W^{\underline{n}}$ of the noncommutative ring $\mathcal{S}_W$ and we are in a position to do algebraic geometry.

PROPOSITION: $\mathcal{S}_W^{\underline{n}}$ admits a filtration

$$\mathcal{S}_0 \subseteq \mathcal{S}_1 \subseteq \ldots \subseteq \mathcal{S}_W^{\underline{n}},$$

such that we can reconstruct the curve X from the associated graded rings

$$\mathcal{R}_X = \bigoplus_{i=0}^{\infty} \mathcal{S}_i, \quad \mathcal{R}_D = \bigoplus_{i=0}^{\infty} \mathcal{S}_i/\mathcal{S}_{i-1},$$

by

$$X = \mathrm{Proj}\mathcal{R}_X, \quad D = \mathrm{Proj}\mathcal{R}_D.$$

Here Proj$R$ denotes the scheme associated to a graded ring R; the point set of this scheme consists of the homogeneous prime ideals not containing the irrelevant ideal (see e.g., Hartshorne [**8**]).

§6. Flows on the Grassmannian and Isospectral Curves.

In general one can define a flow on the Grassmannian by

$$W \in \mathrm{Gr}^{(n)} \mapsto W(t) = e^{tp}W, p \in \Lambda\mathrm{gl}(n, \mathbf{C}).$$

Under these flows the stabilizer algebra transforms by conjugation: $\mathcal{S}_W^{\underline{n}}(t) = e^{tp}\mathcal{S}_W^{\underline{n}}e^{-tp}$. So if we want to fix $\mathcal{S}_W^{\underline{n}}$ for fixed partition $\underline{n}$ and all $W \in \mathrm{Gr}^{(n)}$ we have to take $p \in \mathcal{H}^{\underline{n}}$. If W comes from the geometric data $\{f : X \to P^1, \mathcal{L}, \phi\}$ as before the curve X will be "isospectral". More generally for generators $p_i^{\underline{n}}$ of the positive part of $\mathcal{H}^{\underline{n}}$ we can define $W(t_1, t_2, \ldots) = e^{\sum t_i p_i^{\underline{n}}}W$ and these commuting flows will preserve $f : X \to P^1$ and induce a linear flow on the Jacobian of X. As usual this can also be expressed in terms of matricial Lax equations, the solution of which can be expressed in terms of θ-functions. See e.g., Adams, Harnad, Hurtubise [**9**].

§7. Remarks and questions.

a. In representation theory not only the Heisenberg algebras $\mathcal{H}^{\underline{n}}$ but also the so–called translation group $\mathcal{T}^{\underline{n}}$ centralizing $\mathcal{H}^{\underline{n}}$ is important. $\mathcal{T}^{\underline{n}}$ is the discrete group generated by matrices $T = \mathrm{diag}(z^{l_1} 1_{n_1}, \dots, z^{l_k} 1_{n_k})$, with $\sum l_i = 0$. (Note that for the principal partition $\underline{n} = n$ the group $\mathcal{T}^{\underline{n}}$ is trivial). This group acts on the Grassmannian and preserves $\mathcal{S}^{\underline{n}}_W$. The action on the geometric data $\{f : X \to P^1, \mathcal{L}, \phi\}$ is simple to describe: the trivialisation and the covering are fixed and the sheaf $\mathcal{L}$ maps to $T \cdot \mathcal{L} = \mathcal{L} \otimes \mathcal{O}(\sum l_i \infty_i)$. In the simplest non trivial case of $n = 2$ and partition $\underline{2} = (1,1)$ the translation group leads to the Toda lattice (see Flaschka [**10**], Bergvelt, ten Kroode [**11**]).

b. One can start with an arbitrary $W \in \mathrm{Gr}^{(n)}$ and consider for all partitions $\underline{n}$ the various commutative rings $\mathcal{S}^{\underline{n}}_W$ and the corresponding schemes $X_{\underline{n}}$. It is unclear what the relation between these is. Even more ambitiously one can ask the question if it is possible to associate to the whole non-commutative ring $\mathcal{S}_W$ some geometric object (in some non commutative geometry).

References

1. E. Date, M. Jimbo, M. Kashiwara, T. Miwa, *Transformation groups for soliton equations*, in "Proceedings of RIMS symposium," editors M. Jimbo, T. Miwa, World Scientific, Singapore, 1983, pp. 39–120.
2. V. G. Kac, A. K. Raina, "Bombay lectures on highest weight repesentations of infinite dimensional Lie algebras," World Scientific, Singapore, 1987.
3. G. B. Segal and G. Wilson, *Loop Groups and equations of KdV type*, Publ. Math. I.H.E.S. **61** (1985), 5–65.
4. I. M. Krichever, *Algebraic curves and commuting matrix differential polynomials*, Funct. Anal. and Appl. **10** (1976), 144–146.
5. H. Flaschka, A. C. Newell, T. Ratiu, *Kac Moody algebras and soliton equations, II Lax equation associated with* $A_1^{(1)}$, Physica **9D** (1983), 300.
6. A. P. E. ten Kroode, "Affine Lie algebras and integrable systems," Thesis, University of Amsterdam, 1988.
7. G. Wilson, *Algebraic curves and soliton equations*, in "Geometry today," editors E. Arbarello et al., Birkhäuser, Boston, 1985, pp. 303–329.
8. R. Hartshorne, "Algebraic geometry," Springer, New York, 1977.
9. M. R. Adams, J. Harnad, J. Hurtubise, *Isospectral Hamiltonian flows in finite and infinite dimensions, II Integration of flows*, Comm. Math. Phys. (to appear).
10. H. Flaschka, *Relation between infinite dimensional and finite dimensional isospectral equations*, in "Non-linear integrable systems-classical theory and quantum theory," edited by M. Jimbo and T. Miwa, World Scientific, Singapore, 1983, pp. 219–240.
11. M. J. Bergvelt and A. P. E. ten Kroode, *Differential difference AKNS equations and homogeneous Heisenberg algebras*, J. Math. Phys. **28** (1987), 302–308; *τ functions and zero curvature equations of AKNS-Toda type*, J. Math. Phys **29** (1988), 1308–1320.

Department of Mathematics, University of Georgia, Athens, GA 30602, USA.
Research partially supported by U.S. Army grant DAA L03-87-k-0110

Coadjoint Orbits, Spectral Curves and Darboux Coordinates

M.R. ADAMS, J. HARNAD AND J. HURTUBISE

Abstract. For generic rational coadjoint orbits in the dual $\widetilde{gl}(r)^{+*}$ of the positive half of the loop algebra $\widetilde{gl}(r)$, the natural divisor coordinates associated to the eigenvector line bundles over the spectral curves project to Darboux coordinates on the $Gl(r)$-reduced space. The geometry of the embedding of these curves in an ambient ruled surface suggests an intrinsic definition of symplectic structure on the space of pairs (spectral curves, duals of eigenvector line bundles) based on Serre duality. It is shown that this coincides with the reduced Kostant-Kirillov structure. For all Hamiltonians generating isospectral flows, these Darboux coordinates allow one to deduce a completely separated Liouville generating function, with the corresponding canonical transformation to linearizing variables identified as the Abel map.

§0. Introduction.

Large classes of integrable Hamiltonian systems can be formulated as isospectral Hamiltonian flows with respect to the natural Lie-Poisson structure on the dual of a loop algebra ([**RS**], [**AHP**], [**AvM**], [**FNR**]). Identifying the algebra with its dual via an invariant bilinear form, one obtains equations of Lax type:

$$\dot{L}(\lambda) = [M(\lambda, L(\lambda)),\ L(\lambda)] \ . \tag{0.1}$$

In the "finite gap" case, essentially when $L(\lambda)$ is a polynomial or a rational function, these can be integrated using methods of algebraic geometry (see, e.g., [**RS**], [**K**], [**AvM**], [**vMM**]). Equation (0.1) corresponds to linear flows of line bundles in the Jacobian $J(S)$ of the spectral curve S of $L(\lambda)$, which is a compactification of the curve in $\mathbf{C}^2$ defined by the characteristic equation

$$\det(z1 - L(\lambda)) = 0 \ . \tag{0.2}$$

These methods lead to explicit solutions of (0.1) in terms of theta-functions on $J(S)$, but do not seem to be related to the Hamiltonian structure underlying such systems. It is the purpose of this note to put this integration process into a Hamiltonian framework and to explain that it is, in essence, just the Liouville generating function technique of canonical transformation theory. Earlier work of Dickey [**D**] gave a formulation based on the Liouville theorem in the context of differential algebras, but the relation between the algebro-geometric constructions and the coadjoint orbit structure in loop algebras was not developed. Here, we first examine the symplectic structure of rational coadjoint orbits (sec. 1), obtaining Darboux coordinates after a simple reduction. We then give a geometric interpretation of the orbital symplectic structure for loop algebras, in terms of the algebraic geometry of spectral curves (sec. 2). The symplectic structure can be naturally constructed using Serre duality. In a final section, we explain briefly the application of these coordinates to the integration of isospectral flows via the Liouville generating function technique.

In order to explain succinctly the essential features of the construction, we shall restrict ourselves to the simplest generic case. Let $\widetilde{gl}(r)$ denote the algebra of loops of the form

$$\sum_{i=-\infty}^{K} M_i\lambda^i \ , \qquad M_i \ \epsilon \ gl(r, \mathbf{C}) \ ,$$

and let $\widetilde{gl}(r)^+$ be the subalgebra of polynomial loops. One has the invariant pairing:

$$\langle A, B\rangle = tr[(AB)_0] \ ,$$

where the subscript 0 denotes the 0-th order term. Under this pairing, the dual of the Lie algebra $\widetilde{gl}(r)^+$ is identified with the subalgebra $\widetilde{gl}(r)_-$ of loops $\{\sum_{i=-\infty}^{0} M_i\lambda^i\}$. The Adler-Kostant-Symes theorem gives completely integrable Hamiltonian systems on coadjoint orbits in $(\widetilde{gl}(r)^+)^*$. Using the identification with $\widetilde{gl}(r)_-$, one obtains Lax equations of the form:

$$\dot{L}(\lambda) = [P(\lambda, L(\lambda))^+, \ L(\lambda)] \tag{0.3}$$

where $P(\lambda, z)$ is a polynomial and the superscript $+$ denotes the natural projection to $\widetilde{gl}(r)^+$. We consider here the orbits of elements

$$N(\lambda) = \lambda \sum_{i=1}^{n} \frac{R_i}{(\lambda - \alpha_i)} \tag{0.4}$$

viewed as series in λ^{-1}. (The α_i are chosen distinct.) The orbit of this (rational) element is of finite dimension, consisting of the set (cf. [**AHP**])

$$\left\{\lambda\sum_{i=1}^{n}\frac{g_iR_ig_i^{-1}}{\lambda-\alpha_i},\quad g_i\in Gl(r,\mathbf{C})\right\}.$$

The action of $Gl(r,\mathbf{C})$ on this orbit by conjugation is Hamiltonian, with an associated moment map that is just evaluation at $\lambda=\infty$. It is the symplectic reduction of the orbits by this action that will be of interest to us. This reduction consists of fixing the Jordan form of $N(\infty)$, and then quotienting by $Gl(r,\mathbf{C})$. For simplicity, we shall make the genericity assumption that the eigenvalues of $N(\infty)$ are distinct. We also simplify the discussion by assuming that the spectral curve of $N(\lambda)$ is smooth.

A more complete version of this work, including a discussion of the non-reduced orbits, a treatment of the case of singular curves and explicit examples, may be found in [**AHH2**], [**AHH3**].

§1. Darboux Coordinates.

Consider an element $N(\lambda)$ of the orbit discussed above. Let

(1.1) $$a(\lambda)=\Pi_{i=1}^{n}(\lambda-\alpha_i)$$

(1.2) $$L(\lambda)=\lambda^{-1}a(\lambda)N(\lambda)\ .$$

Thus, $L(\lambda)$ is a matricial polynomial of degree $(n-1)$. Let T be the total space of the line bundle $\mathcal{O}(n-1)$ over $\mathbf{P}_1(\mathbf{C})$, with two sets of coordinates (z,λ), $(z',\lambda')=(z/\lambda^{n-1},1/\lambda)$ defined respectively over $\lambda\neq0,\ \infty$. One can then embed the spectral curve S of $N(\lambda)$ (which is isomorphic to that of $L(\lambda)$) into T, where it has defining equation

(1.3) $$\det(z1-L(\lambda))=0\ .$$

The projection $\pi:T\to\mathbf{P}_1$ restricted to S gives an r-sheeted branched cover of $\mathbf{P}_1(\mathbf{C})$. Associated to $N(\lambda)$ we can define a sheaf E over T, supported over S, by the exact sequence

(1.4) $$0\to\mathcal{O}(-n-1)^{\oplus r}\overset{z1-L(\lambda)}{\longrightarrow}\mathcal{O}^{\oplus r}\to E\to0\ .$$

where $\mathcal{O}(j)$ is the pull-back to T from $\mathbf{P}_1(\mathbf{C})$ of the unique line bundle of degree j. Our genericity hypotheses tell us that E is a line bundle, the dual of the bundle of eigenvectors of $L^T(\lambda)$ with eigenvalue z at (z, λ), which has degree $(g+r-1)$, where

$$g \;=\; \text{genus}(S) \;= \tfrac{1}{2}[(r-1)(r(n-1)-2)] \tag{1.5}$$

and also that

$$H^0(S,\ E(-1)) = 0\ . \tag{1.6}$$

In [**AHH1**], it is shown how such an E determines $L(\lambda)$ up to global conjugation.

If one defines the weighted degree δ of a monomial $\lambda^j z^k$ by the rule

$$\delta(\lambda^j z^k) = j + k(n-1)\ , \tag{1.7}$$

and extends this in the natural way to polynomials, one can show that, in the natural trivialisation, the sections of the line bundle $\mathcal{O}(j)$ are the polynomials $p(\lambda, z)$ with $\delta(p) \leq j$. The defining equation $f(\lambda, z) = 0$ of a spectral curve then satisfies $\delta(f) = r(n-1)$. Let $\mathcal{C}$ be the set of smooth compact curves S' in the linear system $|\mathcal{O}(r(n-1))|$ of S (i.e. cut out by polynomials f' with same degree δ), which coincide with S on the $r(n+1)$ points over $\lambda = \alpha_i,\ \infty$:

$$\begin{aligned}(S'|_{\lambda=\alpha_i}) &= (S|_{\lambda=\alpha_i})\\ (S'|_{\lambda=\infty}) &= (S|_{\lambda=\infty})\end{aligned} \tag{1.8}$$

The reduced coadjoint orbit has an open dense set isomorphic to the variety V of pairs

$$V = \Big\{\ (S', E') \text{ with } S' \in \mathcal{C} \text{ and } E' \text{ a line bundle of degree } g+r-1 \text{ over } S' \text{ satisfying (1.6)}\Big\}\ .$$

V is $2g$-dimensional, and $\mathcal{C}$ is g-dimensional (see [**AHH1**], [**B**]).

Let $N(\infty) = \text{diag}(\mu_i)$, μ_i distinct. By (1.4), the sections of E are just projections of sections of the trivial bundle $\mathcal{O}^{\oplus r}$. The section $(1, 0, \cdots, 0)^T$ of $\mathcal{O}^{\oplus r}$ thus gives a section s of E that has $(r-1)$ zeroes over $\lambda = \infty$ and

g zeroes over the finite portion of the curve, with coordinates (λ_ν, z_ν), $\nu = 1, \cdots, g$. This procedure defines $2g$ functions (λ_ν, z_ν) on the reduced orbit, in a vicinity of $N(\lambda)$. Let

$$\zeta_\nu \equiv z_\nu / a(\lambda_\nu) \; . \tag{1.9}$$

THEOREM 1.1. *The functions $(\lambda_\nu, \zeta_\nu)_{\nu=1\dots g}$ project to a set of Darboux coordinates for the reduced Kostant-Kirillov form K on the reduced coadjoint orbit:*

$$K = \sum_\nu d\lambda_\nu \wedge d\zeta_\nu \; . \tag{1.10}$$

SKETCH OF PROOF: Setting

$$\zeta \equiv z/a(\lambda) \; ,$$

we have

$$\det\,(\zeta 1 - \lambda^{-1} N(\lambda)) = 0 \iff \det(z1 - L(\lambda)) = 0 \; .$$

The section s vanishes at (λ_ν, z_ν) if and only if $(1, 0, \dots, 0)^T$ is in the image of

$$M(\lambda, \zeta) = (\zeta 1 - \lambda^{-1} N(\lambda))$$

at (λ_ν, ζ_ν). If $\widetilde{M}$ is the matrix of cofactors of M, so that $M\widetilde{M} = \det(M) \cdot 1$, this is equivalent, under the genericity assumptions, to the requirement that

$$\widetilde{M}(\lambda_\nu, \zeta_\nu) \cdot (1, 0, \dots, 0)^T = 0 \; , \tag{1.11}$$

i.e., the first column $M_{j1}(\lambda_\nu, \zeta_\nu)$ must vanish.

The Poisson brackets of the functions defined by evaluating matrix elements of N at points (λ, σ) of $\mathbf{C}$ are given by

$$\{N(\lambda)_{ij}, N(\sigma)_{kl}\} = \frac{\sigma N(\lambda)_{il} - \lambda N(\sigma)_{il}}{\lambda - \sigma} \delta_{kj} - \frac{\sigma N(\lambda)_{kj} - \lambda N(\sigma)_{kj}}{\lambda - \sigma} \delta_{il} \; , \tag{1.12}$$

Setting $\widetilde{M}^\lambda_{ij} = \widetilde{M}_{ij}(\lambda, \zeta)$, $\widetilde{M}^\sigma_{kl} = \widetilde{M}_{kl}(\sigma, y)$, etc., one obtains:

$$\{\widetilde{M}^\lambda_{i1}, \widetilde{M}^\sigma_{k1}\} = \frac{1}{(\lambda - \sigma)} \left[\begin{array}{c} \frac{1}{det M^\sigma} \left((\widetilde{M}^\sigma \widetilde{M}^\lambda)_{i1} \, \widetilde{M}^\sigma_{k1} - (\widetilde{M}^\sigma \widetilde{M}^\lambda)_{k1} \widetilde{M}^\sigma_{i1} \right) \\ + \frac{1}{det M^\lambda} \left((\widetilde{M}^\lambda \widetilde{M}^\sigma)_{k1} \, \widetilde{M}^\lambda_{i1} - (\widetilde{M}^\lambda \widetilde{M}^\sigma)_{i1} \, \widetilde{M}^\lambda_{k1} \right) \end{array} \right] . \tag{1.13}$$

If $\Lambda^2\widetilde{M}$ denotes the natural action of $\widetilde{M}$ on $\Lambda^2\mathbf{C}^r$, the r.h.s. of (1.13) is

$$\frac{1}{(\lambda-\sigma)}\left[\frac{1}{det M^{\sigma}}\left(\Lambda^2\widetilde{M}^{\sigma}(\widetilde{M}^{\lambda}(e_1)\wedge e_1)\right)-\frac{1}{det M^{\lambda}}\left(\Lambda^2\widetilde{M}^{\lambda}(\widetilde{M}^{\sigma}(e_1)\wedge e_1)\right)\right]_{ik}$$

where $e_1 = (1,0,\ldots,0)^T$. Since $\widetilde{M}$ is of rank 1 over the spectral curve, $\Lambda^2\widetilde{M}$ vanishes, and the above remains well defined over S.

The points (λ_ν, ζ_ν) are defined by the conditions $\widetilde{M}_{k1}(\lambda_\nu,\zeta_\nu) = 0$ for all k. Generically, these points are cut out by only two of these equations, let us say $\widetilde{M}_{21} = \widetilde{M}_{11} = 0$, in the sense that the matrix

$$F_\nu \equiv \begin{pmatrix} \frac{\partial \widetilde{M}_{11}}{\partial\lambda} & \frac{\partial \widetilde{M}_{11}}{\partial\zeta} \\ \frac{\partial \widetilde{M}_{21}}{\partial\lambda} & \frac{\partial \widetilde{M}_{21}}{\partial\zeta} \end{pmatrix}(\lambda_\nu,\zeta_\nu) \tag{1.14}$$

is invertible. By implicit differentiation, the Poisson brackets of the functions (λ_ν, ζ_ν) are then:

$$\begin{pmatrix} \{\lambda_\nu,\lambda_\mu\} & \{\lambda_\nu,\zeta_\mu\} \\ \{\zeta_\nu,\lambda_\mu\} & \{\zeta_\nu,\zeta_\mu\} \end{pmatrix} \tag{1.15}$$

$$= (F_\nu)^{-1}\begin{pmatrix} \{\widetilde{M}_{11}^{\lambda_\nu}, \widetilde{M}_{11}^{\lambda_\mu}\} & \{\widetilde{M}_{11}^{\lambda_\nu}, \widetilde{M}_{21}^{\lambda_\mu}\} \\ \{\widetilde{M}_{21}^{\lambda_\nu}, \widetilde{M}_{11}^{\lambda_\mu}\} & \{\widetilde{M}_{21}^{\lambda_\nu}, \widetilde{M}_{21}^{\lambda_\mu}\} \end{pmatrix}(F_\mu)^{T-1}.$$

Computing, one obtains

$$\{\lambda_\nu,\ \lambda_\mu\} = \{\zeta_\nu\ ,\zeta_\mu\} = 0, \qquad \{\lambda_\nu,\zeta_\mu\} = \delta_{\nu\mu}\,. \tag{1.16}$$

These brackets are computed on the unreduced orbit. To reduce, one imposes the constraints $N_{ij}(\infty) - \delta_{ij}\mu_j = 0$, and then quotients by the isotropy group. The functions (λ_ν, z_ν) do not necessarily commute with all the constraints, although they are invariant under those that generate the stabilizer of $N(\infty)$; i.e., the "first class" constraints. To compute the Poisson brackets on the reduced manifold, we may modify these functions by adding multiples of the constraints, so that the resulting set does commute with the constraints, at least on the constrained submanifold. It is sufficient to do this for the $\widetilde{M}_{ij}^{\lambda}$. One finds that the modification

$$\widetilde{M}_{i1}^{\lambda} \mapsto \widetilde{M}_{i1}^{\lambda} + \sum_{k\neq l} \widetilde{M}_{il}^{\lambda}\delta_{k1}(\mu_1-\mu_l)^{-1}N_{lk}(\infty) \tag{1.17}$$

is the correct one. This modification is shown by direct computation to leave the Poisson brackets $\{\widetilde{M}_{k1}^{\lambda_\nu}, \widetilde{M}_{l1}^{\lambda_\mu}\}$ unchanged. Since the (λ_ν, ζ_ν) are invariant under the stabiliser of $N(\infty)$, one can pass to the quotient with the brackets unchanged, showing that the $2g$ functions (λ_ν,ζ_ν) obtained by projection to the $2g$ dimensional reduced symplectic manifold satisfy the correct brackets to be a Darboux coordinate system.

§2. Geometric Interpretation.

In the quotiented orbit R, one has a $2g$-dimensional open dense subvariety V of pairs (smooth curves S, line bundles E over S) mapping to a g-dimensional space $\mathcal{C}$ of curves: $\pi : V \longrightarrow \mathcal{C}$. Fixing $S \in \mathcal{C}$, the fiber $\pi^{-1}(S)$ is naturally isomorphic to an open subvariety of $Pic^{g-1}(S)$, which is essentially the complement of the theta-divisor. Infinitesimal deformations of these line bundles are parametrised by $H^1(S, \mathcal{O})$. On the other hand, infinitesimal deformations of the curve S in T are given by elements of $H^0(S, N)$, where N is the normal bundle. These sections of N, however, are constrained to be zero at $\lambda = \alpha_i, \infty$ and so the deformations of S correspond to elements of $H^0(S, N(-n-1))$. By the adjunction formula [**GH**], this is isomorphic to $H^0(S, K_S)$ where K_S is the canonical bundle of S. Thus, at a point (S, E) of V, we have

$$0 \to H^1(S, \mathcal{O}) \to TV \to H^0(S, K_S) \to 0 . \tag{2.1}$$

Since $H^1(S, \mathcal{O})$ and $H^0(S, K_S)$ are Serre duals, given any splitting of (2.1), with elements of TV written as pairs $\{(a, v),\ a \in H^1(S, \mathcal{O}),\ b \in H^0(S, K_S)\}$, we can define a natural skew form W on TV by

$$W((a, v), (b, u)) = a(u) - b(v) . \tag{2.2}$$

Such a splitting is tantamount to finding a way of fixing the line bundle and varying the curve, at least infinitesimally. One way of doing this is to extend the line bundle over S to a neighbourhood of S in T. One then has a well defined "base" line bundle to restrict to any variation of S. It is sufficient to take this extension to the first formal neighbourhood $S^{(1)}$ of S. There are many such extensions; for our purposes however, they are all equivalent.

Suppose then that we have chosen an extension of E to $S^{(1)}$, and so defined a splitting of (2.1) and a form W. Let $T(\lambda, z)$ represent the transition function for the line bundle E extended to $S^{(1)}$, from the open set $U_0 = \{\lambda \neq \infty\}$ to the open set $U_1 = \{\lambda \neq 0\}$. The section s of E over S that cuts out the divisor (λ_ν, z_ν) on S is represented by functions $s_i(\lambda, z)$ on U_i, such that on $U_0 \cap U_1$

$$s_0 = T \cdot s_1 \cdot (1 + rf)$$

where $f(\lambda, z) = 0$ is the defining equation of the curve, and r is some function. We now have:

THEOREM 2.1. *Let S be the spectral curve of an element of the quotiented orbit R. Suppose that S is smooth, with r distinct points over $\lambda = \infty$. Then the skew form W is the (reduced) Kostant-Kirillov form.*

PROOF: Choose a two parameter family $(S(x,t), E(x,t))$ in our space. Let $S(x,t)$ be defined by equations $f(\lambda, z, x, t) = 0$, and let the family of sections of $E(x,t)$ be given over U_0 by the functions $s_0(\lambda, z, x, t)$. We will evaluate the symplectic 2-form on the tangent vectors $\frac{\partial}{\partial t}, \frac{\partial}{\partial x}$ to R at $(x,t) = (0,0)$.

To do this, we use the description given by Theorem 1.1. If $\sum_i(\lambda_\nu, z_\nu)$ is the divisor of s away from $\lambda = \infty$, so that the points (λ_ν, z_ν) are given by the simultaneous vanishing of f and s_0, then the Kostant-Kirillov form evaluated on $\frac{\partial}{\partial t}, \frac{\partial}{\partial x}$ is

$$K\left(\frac{\partial}{\partial t}, \frac{\partial}{\partial x}\right) = \sum_\nu \frac{(\lambda_\nu)_t(z_\nu)_x - (\lambda_\nu)_x(z_\nu)_t}{a(\lambda_\nu)} . \tag{2.3}$$

where the subscripts x, t denote differentiation. Now define new variables $\hat{\lambda}_\nu, \hat{z}_\nu$ by

$$\begin{aligned} \hat{\lambda}_\nu(0,0) &= \lambda_\nu(0,0) \\ f(\hat{\lambda}_\nu(x,t), z(\hat{\lambda}_\nu(x,t)), 0, 0) &= 0 \\ s_0(\hat{\lambda}_\nu(x,t), z(\hat{\lambda}_\nu(x,t)), x, t) &= 0 \\ \hat{z}_\nu(0,0) &= z_\nu(0,0) \\ f(\lambda_\nu(0,0), \hat{z}_\nu(x,t), x, t) &= 0 . \end{aligned}$$

$\hat{\lambda}_\nu(x,t)$ is the λ-coordinate of the point cut out on $S(0,0)$ by the function $s_0(\lambda, z, x, t)$; $\hat{z}_\nu$ represents the variation of the z-coordinate of $S(x,t)$ over $\lambda = \lambda_\nu(0,0)$. It is then straightforward to show that at $(x,t) = (0,0)$,

$$(\lambda_\nu)_t(z_\nu)_x - (\lambda_\nu)_x(z_\nu)_t = (\hat{\lambda}_\nu)_t(\hat{z}_\nu)_x - (\hat{\lambda}_\nu)_x(\hat{z}_\nu)_t .$$

Let d denote exterior differentiation along the curve $S(x,t), (x,t)$ fixed. One has for (x,t) small,

$$\hat{\lambda}_\nu = \frac{1}{2\pi i} \oint_{C_\nu} \lambda \, d \ln s_0 .$$

for some suitable contour C_ν, and so

$$(\hat{\lambda}_\nu)_{t \text{ or } x} = \frac{1}{2\pi i}\oint_{C_\nu} (d\, ln\, s_0)_{t \text{ or } x} \,.$$

At $(x,t) = (0,0)$

$$(\hat{z}_\nu)_{t \text{ or } x} = -\frac{f_{t \text{ or } x}(\lambda_\nu, z_\nu, 0, 0)}{f_z(\lambda_\nu, z_\nu, 0, 0)} \,.$$

We choose a base point λ_0 and cut open the Riemann surface $S(0,0)$ into a $4g$-gon ($g = \text{genus}(S)$) in the standard fashion, so that $\int_{\lambda_0}^{\lambda} \frac{f_{t \text{ or } x}}{a(\lambda)\cdot f_z}\, d\lambda$ is well defined. If we set

$$F = \left[\left(\int_{\lambda_0}^{\lambda} \frac{f_x}{a(\lambda) f_z}\, d\lambda\right)(d\, ln\, s_0)_t\right] - [x \leftrightarrow t]$$

then at (0,0), we have

$$\frac{(\lambda_\nu)_t(z_\nu)_x - (\lambda_\nu)_x(z_\nu)_t}{a(\lambda_\nu)} = \frac{1}{2\pi i}\oint_{C_\nu} F \,.$$

Thus, for the Kostant Kirillov symplectic form K, evaluated on $\frac{\partial}{\partial t}, \frac{\partial}{\partial x}$:

$$\begin{aligned} K\left(\frac{\partial}{\partial t}, \frac{\partial}{\partial x}\right) &= \frac{1}{2\pi i}\sum_\nu \oint_{C_\nu} F \\ &= \frac{1}{2\pi i}\left(\oint_{\text{edge of 4g-gon}} F - \sum_j \oint_{D_j} F\right) \end{aligned} \tag{2.4}$$

where the D_j are contours around the points p_j over $\lambda = \infty$. It is easy to see that the contribution of the edge is zero. Also, the 1-form $\frac{f_x}{a(\lambda) f_z}\, d\lambda$ is holomorphic at $\lambda = \infty$. We then write the transition function for the line bundle $E(x,t)$, to first order in (x,t) as $T(\lambda, z)\, e^{t\beta_t + x\beta_x}$, so that

$$s_0(\lambda, z, x, t) = s_1(\lambda, z, x, t)\, T(\lambda, z)\, e^{t\beta_t + x\beta_x}\, (1 + r(\lambda, z, x, t) f(\lambda, z, x, t))$$

for some function r on $U_0 \cap U_1$. Substituting into (2.4), we get

$$\begin{aligned} &K\left(\frac{\partial}{\partial t}, \frac{\partial}{\partial x}\right) \\ &= \frac{1}{2\pi i}\sum_j \left(\oint_{D_j}\left(\int_{\lambda_0}^{\lambda}\frac{f_x d\lambda}{a(\lambda) f_z}\right)\left(d\beta_t + (dln s_1)_t + dln(1 + rf)_t\right)\right) - (x \leftrightarrow t) \end{aligned}$$

Of the three terms in the integrand, the last two give zero when integrated and antisymmetrized ($x \leftrightarrow t$), giving :

$$K\left(\frac{\partial}{\partial t},\frac{\partial}{\partial x}\right)=\frac{1}{2\pi i}\sum_j \oint_{D_j}\left(\int_{\lambda_0}^{\lambda}\frac{f_x d\lambda}{a(\lambda)f_z}\right)d\beta_t \quad - \quad (x\leftrightarrow t)$$
$$=\frac{1}{2\pi i}\sum_j \oint_{D_j}\frac{f_x}{a(\lambda)f_z}d\lambda\cdot\beta_t \quad - \quad (x\leftrightarrow t)$$

To show that $K\left(\frac{\partial}{\partial t},\frac{\partial}{\partial x}\right)=W\left(\frac{\partial}{\partial t},\frac{\partial}{\partial x}\right)$ we note that:

1) $\frac{f_x}{a(\lambda)f_z},\frac{f_t}{a(\lambda)f_z}$ are simply the elements of $H^0(S,K_S)$ corresponding to the variations of the curves in the x,t directions, under the identification $N(-n-1)\simeq K_S$.

2) β_t,β_x are representative cocycles for the elements of $H^1(S,O)$ representing the variations of the line bundles $E(x,t)$

3) The sum of the contour integrals around the D_j is the explicit representation of the Serre duality pairing, when the cocycles are chosen with respect to the U_0,U_1 covering.

§3. Liouville Integration.

Recall from section 1, the definition (1.7) of the weighted degree $\delta(p)$ of a polynomial $p(\lambda,z)$. The defining equation $f(\lambda,z)=0$ of a spectral curve then satisfies $\delta(f)=r(n-1)$, and the set $\mathcal{C}$ of curves can be described as an open dense subset of the set of curves with equations

$$f_P(\lambda,z)=f(\lambda,z)+a(\lambda)P(\lambda,z)=0 \tag{3.1}$$

where $\delta(P)\le r(n-1)-n-1$. The set of such P's is a g-dimensional linear space; the coefficients $P_1,\ldots,P_g$ of $P(\lambda,z)$ then give us a set of coordinates on $\mathcal{C}$.

Typically, the Hamiltonians generating the Lax pairs (0.3) are functions $h(P_1,\cdots P_g)$ of the P_j's. From the geometric description of the symplectic structure, one has that the fibers of the projection $\pi:V\to S$ form a Lagrangian foliation of V, and that $P_1,\cdots,P_g$ give a complete set of integrals of motion. To complete the integration process, one simply needs to find the complementary Darboux coordinates $Q_1, \ldots, Q_g$. Following the method of the Liouville theorem, this is determined by a generating function, starting from our Darboux coordinates (λ_ν, ζ_ν).

Let $\zeta(\lambda, P_1, \cdots, P_g)$ be the algebraic function defined implicitly by $f_P(\lambda, a(\lambda)\zeta) = 0$. The generating function G is defined by

$$G(\lambda_\nu, P_\mu) = \sum_{i=1}^{g} \int_{\lambda_0}^{\lambda_\nu} \zeta(\lambda, P_1, \cdots, P_g) d\lambda \,. \tag{3.2}$$

One has $\frac{\partial G}{\partial \lambda_\nu} = \zeta_\nu$, and so, if P_μ is the coefficient of $\lambda^\alpha z^\beta$ in $P(\lambda, z)$, setting

$$Q_\mu = \frac{\partial G}{\partial P_\mu} = \sum_{i=1}^{g} \int_{\lambda_0}^{\lambda_\nu} \frac{\lambda^\alpha z^\beta}{(\partial f_P / \partial z)} d\lambda \tag{3.3}$$

gives us the required coordinates. The flows corresponding to the Hamiltonian $h(P_1, \ldots, P_g)$ are then linear when expressed in the coordinates Q_μ.

We then remark that the integrands in (3.3), as μ varies, form a basis for the holomorphic differentials on the curve. This is a simple consequence of the Poincaré residue formula [**GH**]. Up to a change of basis, the vector $Q = (Q_1, \ldots, Q_g)$ is thus the expression in coordinates of the Abel map of the g fold symmetric product of the curve into its Jacobian. The inversion of this map, expressing the coordinates λ_ν in terms of Q, can be determined in the usual way, in terms of theta-functions. Invariantly, $Q(t)$ describes a linear flow of line bundles on the curve. The Liouville integration process gives us an explicit "classical" derivation of the Krichever-Novikov method, in Hamiltonian terms.

As a final remark, the application of this method to the determination of the matrix $L(\lambda, t)$ requires an extension of these results, since one must determine the flow directly on the coadjoint orbits, rather than on the reduced space. This can, in fact, be done. One defines an extended Darboux coordinate system on a suitable $2(g + r - 1)$ dimensional symplectic submanifold Z of the orbit; essentially, that obtained by constraining the off-diagonal elements of $N(\infty)$ to zero. Again, a geometric interpretation can be given; Z has an open dense subset V' that may be identified with the space of pairs consisting of (smooth curves S in T, line bundles over S with a trivialisation over $\lambda = \infty$, modulo an overall scaling), with the curves now unconstrained over $\lambda = \infty$. The relevant infinitesimal deformation spaces are now $H^0(S, K_S(1))$, $H^1(S, \mathcal{O}(-1))$, which are again Serre duals, and, after a small modification, one has a similar interpretation of the symplectic form. The Liouville generating function technique in this case leads to singular abelian differentials with poles at $\lambda = \infty$. Full details, together with the modifications for dealing with singular spectral curves, and some applications may be found in [**AHH2**], [**AHH3**].

References

[**AHH1**] Adams, M., Harnad, J. and Hurtubise, J., *Isospectral Hamiltonian Flows in Finite and Infinite Dimensions II. Integration of Flows*, Commun. Math. Phys. (1990, in press).

[**AHH2**] ______________, *Darboux Coordinates and Liouville-Arnold Integration in Loop Algebras*, CRM preprint (1990).

[**AHH3**] ______________, *Liouville Generating Function for Isospectral Flow in Loop Algebras*, CRM preprint **1652** (1990). to appear in "Integrable and Superintegrable Systems" (ed. B. Kupershmidt), World Scientific, Singapore (1990).

[**AHP**] Adams, M., Harnad, J. and Previato, E., *Isospectral Hamiltonian Flows in Finite and Infinite Dimensions I. Generalised Moser Systems and Moment Maps into Loop Algebras*, Commun. Math. Phys. **117** (1988), 451–500.

[**AvM**] Adler, M. and van Moerbeke, P., *Completely Integrable Systems, Euclidean Lie Algebras, and Curves*, Adv. Math. **38** (1980), 267–317.

[**B**] Beauville, A., *Jacobiennes des courbes spectrales et systèmes Hamiltoniens complètement intégrables*, Preprint (1989).

[**D**] Dickey, L.A., *Integrable Nonlinear Equations and Liouville's Theorem I, II*, Commun. Math. Phys. **82**, 345-360;; ibid. **82** (1981), 360–375.

[**FNR**] Flaschka, H., Newell, A.C. and Ratiu, T., *Kac-Moody Algebras and Soliton Equations II. Lax Equations Associated to* $A_1^{(1)}$, Physica **9D** (1983), p. 300.

[**GH**] Griffiths, P. and Harris, J., "Principles of Algebraic Geometry," Wiley, New York, 1978.

[**K**] Krichever, I.M., *Methods of Algebraic Geometry in the Theory of Nonlinear Equations*, Russ. Math. Surveys **32** (1980), 53–79.

[**vMM**] van Moerbeke, P. and Mumford, D., *The Spectrum of Difference Operators and Algebraic Curves*, Acta Math. **143** (1979), 93-154.

[**RS**] Reiman, A.G., and Semenov-Tian-Shansky, M.A., *Reduction of Hamiltonian systems, Affine Lie algebras and Lax Equations I, II*, Invent. Math. **54** (1979), 81–100; ibid. **63** (1981), 423–432.

M.R. ADAMS: Department of Mathematics, University of Georgia, Athens, GA 30602, USA

J. HARNAD: Department of Mathematics, Concordia University, Montréal, Qué., and Centre de Recherches Mathématiques, Université de Montréal, C.P. 6128-A, Montréal, Qué., Canada, H3C 3J7

J. HURTUBISE: Department of Mathematics, McGill University, Montréal, Qué., Canada, H3A 2K6

Research supported in part by the Natural Sciences and Engineering Research Council of Canada and by U.S. Army grant DAA L03-87-k-0110

Hamiltonian Systems on the Jacobi Varieties

SOLOMON J. ALBER
September 25, 1989

Abstract. New sequences of series of integrable Hamiltonian systems are obtained as the systems defined on the Jacobi varieties. Equations of Korteweg-de Vries (KdV) and inverse KdV type as well as sine-Gordon and nonlinear Schrödinger equations together with the corresponding Hamiltonian systems are only initial members of the found sequences. We also investigate discrete systems associated with the integrable continuous systems.

§1. Introduction.

In many papers deep link was established between the theory of nonlinear integrable systems and algebraic geometry. In accordance with this approach the periodic and finite-gap problems are considered as problems on some Jacobi or Prym varieties [**1-8**].

In 1985 there were found [**9**] infinite sequences of series of nonlinear integrable equations with corresponding canonically conjugate variables and Hamiltonian structures. Equations of KdV and inverse KdV type, sine-Gordon and nonlinear Schrödinger equations are only initial members of the found sequences. Finite-gap and periodic problems for these systems are represented in the form of meromorphic Hamiltonian systems defined on the corresponding Jacobi varieties.

In the present paper we try to find all possible completely integrable Hamiltonian systems defined on the Jacobi varieties of the hyperelliptic curves. As a result different odd and even classes of infinite sequences of series of integrable Hamiltonian systems are obtained. Then the importance of these classes is demonstrated using some well-known problems as examples. We also investigate the Jacobi-Neumann problems [**13**] corresponding to these systems. Note that all systems mentioned above are also initial members of the found new sequences.

In conclusion discrete integrable systems with the corresponding Hamiltonians are determined as systems associated with some continuous integrable problems.

§2. Basic classes of the Hamiltonian systems on the Jacobi varieties.

Here we describe integrable Hamiltonian systems defined on the corresponding Jacobi varieties.

Therefore, let $\Gamma: W^2 + C_N(\lambda) = 0, C_N(\lambda) = C_0\lambda^N + \cdots + C_N$ be the Riemann surface of genus n.

DEFINITION 1.1: Hamiltonian system belongs to the odd class if the degree of the polynomial $C_N(\lambda)$ is an odd number $N = 2n+1$. Otherwise it belongs to the even class $N = 2n + 2$.

We consider the Jacobi variety T^n of the Riemann curve Γ with the coordinates $\lambda_1, \ldots, \lambda_n$ and corresponding cotangent bundle T^*T^n with coordinates $(\lambda_1, \ldots, \lambda_n, W_1, \ldots, W_n)$.

In order to define Hamiltonian functions for the desired classes of systems we introduce a polynomial

$$G(\lambda) = \prod_{j=1}^{n}(\lambda - \lambda_j) = \lambda^n + b_1\lambda^{n-1} + \cdots + b_n \tag{2.1}$$

and a series of polynomials

$$G_l(\lambda) = \lambda^l + b_1\lambda^{l-1} + \cdots + b_l \tag{2.2}$$

$$G^*_m(\lambda) = b_{n-m}\lambda^m + b_{n-m-1}\lambda^{m-1} + \cdots + b_n \tag{2.3}$$

Let also $\alpha \in C^n$ and $\beta \in C^n$ be n-dimensional vectors $\alpha = (\alpha_0, \ldots, \alpha_{n-1})$ and $\beta = (\beta_0, \ldots, \beta_{n-1})$.

THEOREM 2.1. *For every rational function of the form*

$$M(\lambda; \alpha, \beta) = \sum \alpha_l G_l(\lambda) + \sum \beta_m \frac{G^*_m(\lambda)}{\lambda^{m+1}} \tag{2.4}$$

there exists a completely integrable Hamiltonian system with the following Hamiltonian

$$H = \sum_{j=1}^{n} \frac{W_j^2 + C_N(\lambda_j)}{\prod_{r\neq j}(\lambda_j - \lambda_r)} M(\lambda_j; \alpha, \beta) \tag{2.5}$$

It is equivalent to the Jacobi system of inversion

(2.6)

$$\frac{1}{2}\sum \int_{\lambda_j^0}^{\lambda_j} \frac{\lambda_j^{k-1} d\lambda_j}{\sqrt{-P(\lambda_j)}} = (\sum_{l=0}^{n-1} \alpha_l \delta^k_{n-l} - \sum_{m=0}^{n-1} \beta_m \delta^k_{m+1})(x - x^0) + D_k(x^0) =$$

$$m_{\alpha,\beta,k}(x - x^0) + D_k(x^0)$$

with the polynomial $P(\lambda) = C_N(\lambda)$.

PROOF: The proof is based on the introduction of the action-angle variables and is similar to the general case to be proved in the next section.

If N is an odd number systems (2.5) are systems of the Korteweg-de Vries (KdV) hierarchy. But in case of the even degree $N = 2n + 2$ of the polynomial $C_N(\lambda)$, Hamiltonians in the same form (2.5) are Hamiltonians of the two-potential systems with the generating equation

$$\frac{\partial U}{\partial t} = -\frac{B'''}{2} + 2B'U + BU' \tag{2.7}$$

and potential

$$U = \lambda^2 + u\lambda + v \tag{2.8}$$

First term in this hierarchy (analogous to the KdV equation) is the two-potential system defined by the following equations on the potentials u and v

$$\begin{cases} \frac{\partial u}{\partial t} = -\frac{3}{2}u\frac{\partial u}{\partial x} + \frac{\partial v}{\partial x} \\ \frac{\partial v}{\partial t} = \frac{1}{4}\frac{\partial^3 u}{\partial x^3} - v\frac{\partial u}{\partial x} - \frac{1}{2}u\frac{\partial v}{\partial x} \end{cases} \tag{2.9}$$

Different Hamiltonian systems can be obtained as systems defined on the Riemann surfaces of the following type

$$W^2 + \frac{C_N(\lambda)}{\lambda} = 0 \tag{2.10}$$

THEOREM 2.2. *For every polynomial* $M(\lambda; \alpha, \beta)$ *the Hamiltonian system with Hamiltonian*

$$H = \sum_{j=1}^{n} \frac{W_j^2 + \frac{C_N(\lambda_j)}{\lambda_j}}{\prod_{s\neq j}(\lambda_j - \lambda_s)} \lambda_j M(\lambda_j; \alpha, \beta) \tag{2.11}$$

is completely integrable and can be reduced to the Jacobi system of inversion (2.6) with the polynomial in the form $P(\lambda) = \lambda C_N(\lambda)$.

Proof also follows from the general case.

The sine-Gordon equation is the most important example of integrable systems of such kind. In case of the sine-Gordon hierarchy (see for details [**13**]) the order of the polynomial $C_N(\lambda)$ is equal to $N = 2n$ and the polynomial M is equal to

$$M(\lambda; \alpha, \beta) = 1 - \frac{b_n}{\lambda} \tag{2.12}$$

At last, we consider one more class of Integrable Problems arised from the earlier investigation of the Jacobi problem of geodesics on quadrics and the inverse KdV equation. The latest equation was shown [**13, 14**] to be an integrable model for the Jacobi problem. At the same time a direct isomorphism between the space of stationary solutions of this equation and the space of solutions of the Jacobi problem was established.

THEOREM 2.3. *If the polynomial $M(\lambda;\alpha,\beta)$ in the series of Hamiltonians*

$$H = \sum_{j=1}^{n} \frac{W_j^2 + \frac{C_N(\lambda_j)}{\lambda_j}}{\prod_{s\neq j}(\lambda_j - \lambda_s)} M(\lambda_j;\alpha,\beta) \tag{2.13}$$

is constant then every Hamiltonian system (2.10) is completely integrable and can be reduced to the Jacobi system of inversion with the polynomial $P(\lambda) = \lambda C_N(\lambda)$.

Notice that following the method of **[13,14]** one obtains Jacobi system of inversion containing meromorphic Abelian differentials. This difficulty is resolved using a special change of the independent variable.

§3. General case.

After considering two different types of the integrable systems over Jacobi varieties now we can investigate the general case of the completely integrable systems over the hyperelliptic curves.

Assume that the polynomial $C_N(\lambda)$ is factorized into the product of two polynomials

$$C_N(\lambda) = R_L(\lambda) S_{N-L}(\lambda) \tag{3.1}$$

here

$$S_{N-L}(\lambda) = s_0\lambda^{N-L} + \cdots + s_{N-L}$$

with additional condition $L \leq (N+1)/2$. We consider the system with the meromorphic Hamiltonian

$$H = \sum_{j=1}^{n} \frac{R_L(\lambda_j)W_j^2 + S_{N-L}(\lambda_j)}{\prod_{r\neq j}(\lambda_j - \lambda_r)} M(\lambda_j;\alpha,\beta) \tag{3.2}$$

defined on the Jacobi variety of Riemann surface

$$W^2 + \frac{S_{N-L}(\lambda)}{R_L(\lambda)} = 0 \tag{3.3}$$

We represent the main result in the form of the theorem.

THEOREM 3.1. *For every polynomial $M(\lambda_j;\alpha,\beta)$ Hamiltonian system with Hamiltonian (3.2) is completely integrable and can be reduced to the Jacobi system (2.6) with the polynomial $P(\lambda) = C_N(\lambda)$.*

PROOF: The first part of the Hamiltonian system (3.2) has the following form

$$\frac{d\lambda_j}{dx} = \frac{2W_j R_L(\lambda_j)}{\prod_{r\neq j}(\lambda_j - \lambda_r)} M(\lambda_j;\alpha,\beta) \tag{3.4}$$

Integrating the second part one gets a system of first integrals

$$W_j^2 + \frac{S_{N-L}(\lambda_j)}{R_L(\lambda_j)} = 0 \tag{3.5}$$

We introduce action-angle variables and then reduce the problem to the Jacobi problem of inversion as follows.

Let S be the action-function

$$S(\lambda_1, \ldots, \lambda_n; H_1, \ldots, H_n) = \sum_{j=1}^{n} \int_{\lambda_j^0}^{\lambda_j} \sqrt{\frac{-S_{N-L}(\lambda_j)}{R_L(\lambda_j)}} d\lambda_j \tag{3.6}$$

Here $H_k = s_{N-L-n+k}$. Then the angle variables conjugate to the H_k can be defined as the partial derivatives of the action-function S

$$W_k^H = -\frac{\partial S}{\partial H_k} = \frac{1}{2} \sum_{j=1}^{n} \int_{\lambda_j^0}^{\lambda_j} \frac{\lambda_j^{n-k}}{\sqrt{-C_N(\lambda_j)}} d\lambda_j \tag{3.7}$$

Using Hamiltonian equations (3.4) and (3.5) we see that

$$\begin{gathered} \frac{dW_k^H}{dx} = \frac{1}{2} \sum_{j=1}^{n} \frac{\lambda_j^{n-k}}{\sqrt{-C_N(\lambda_j)}} \lambda_j = \\ \sum_{j=1}^{n} \frac{\lambda_j^{n-k}}{\prod_{r \neq j}(\lambda_j - \lambda_r)} M(\lambda_j; \alpha, \beta) = \sum_{l=0}^{n-1} \alpha_k \delta_{n-l}^k - \sum_{m=0}^{n-1} \beta_m \delta_{m+1}^k \end{gathered} \tag{3.8}$$

Integrating this equations we get the Jacobi system of inversion on the Jacobi variety of the Riemann surface (3.3).

Notice that all systems with the same $C_N(\lambda)$ have common angle variables but different action variables.

Now we give several examples. In case when $N = 2n+1$, $L = n+1$ and polynomial $M(\lambda_j; \alpha, \beta) = 1$ our integrable system (3.2) is equivalent to the C. Neumann problem of the motion of particles on an n-dimensional sphere in the field of the quadratic potential. In terms of the canonical variables (q_j, P_j) Hamiltonian of this problem can be expressed in the form

$$H = \frac{1}{2}\{\sum l_j q_j^2 + \sum P_k^2 \sum q_k^2 - (\sum P_k q_k)^2\} \tag{3.9}$$

In case when $N = 2n+2$, $L = n+1$ and $M(\lambda_j; \alpha, \beta) = 1$ we get a new system of Neumann type (of even class) with the Hamiltonian

$$H = \frac{1}{2}\{-\sum l_j q_j^2 + \sum P_k^2 \sum q_k^2 - (\sum P_k q_k)^2 + (\sum a_j q_j^2)^2\} \tag{3.10}$$

Here $l_j = a_j^2 + a_j \sum_{k=1}^{n+1}(a_k - l)$, $l = const.$

The last example is defined by the following Hamiltonian

$$H = \sum_{j=1}^{n} \frac{R_L(\lambda_j)W_j^2 + S_{N-L}(\lambda_j)\lambda_j}{\lambda_j \prod_{r\neq j}(\lambda_j - \lambda_r)} M(\lambda_j;\alpha,\beta) \tag{3.11}$$

here polynomial M is constant. It is a completely integrable system which can be reduced to Jacobi system of inversion.

If the number N is even ($N = 2n$) then this system is equivalent to classical Jacobi problem on geodesics.

But in the case of odd number ($N = 2n + 1$) we obtain new integrable system.

§4. Periodic problem.

In 1976 H. Flaschka and D. McLaughlin investigated [4] the periodic problem for the KdV equation and Toda lattice and introduced new canonically conjugate variables and Hamiltonians.

In our case if $N = 2n$ is an even number and if the polynomial $C_N(\lambda)$ is represented in the form

$$C_N(\lambda) = \Delta^2(\lambda) - 1 \tag{4.1}$$

then we can introduce Flaschka-McLaughlin momenta and Hamiltonians. As a result we obtain a theorem.

THEOREM 4.1. *For every polynomial $M(\lambda_j;\alpha,\beta) = 1$ a Hamiltonian system with the Hamiltonian*

$$H = \sum_{j=1}^{n} \frac{(-1)^{n+j+1}cosh(\frac{P_j}{2}) + \Delta(\lambda_j)}{\prod_{s\neq j}(\lambda_j - \lambda_s)} M(\lambda_j;\alpha,\beta) \tag{4.2}$$

can be determined on the Riemann surface of the function

$$cosh(\frac{P}{2}) = (-1)^{n+j}\Delta(\lambda) \tag{4.3}$$

It is completely integrable and it can be reduced to the Jacobi problem of inversion.

Proof is similar to the general case.

Note that the space of solutions of this periodic problem is usually a half-dimensional subspace of the space of solutions of the finite-gap problem.

Flaschka-McLaughlin representation can be also used for the relativistic Toda lattices. This new class of integrable discrete systems has been recently introduced [**16**] by Ruijsenaars and Schneider.

Hamiltonians of these systems in the case of finite-gap solutions have the form [**12**]

$$H = \sum_{j=1}^{n} \frac{W_j^2 + C_N(\lambda_j)}{\lambda_j \prod_{s \neq j}(\lambda_j - \lambda_s)} M(\lambda_j; \alpha, \beta) \tag{4.4}$$

Therefore, Hamiltonians of the periodic problems can be found in the form

$$H = \sum_{j=1}^{n} \frac{(-1)^{n+j+1} \lambda_j^{\frac{n}{2}} cosh(\frac{W_j}{2}) + \Delta(\lambda_j)}{\lambda_j \prod_{s \neq j}(\lambda_j - \lambda_s)} M(\lambda_j; \alpha, \beta) \tag{4.5}$$

In both cases nonstandard Poisson structure is introduced by the skew - symmetric matrix

$$\begin{cases} P_{k,n+j} = \delta_j^k \lambda_j & \text{for } 1 \leq j \leq n; \\ P_{n+k,j} = -\delta_j^k \lambda_j & \text{for } 1 \leq j \leq n. \end{cases} \tag{4.6}$$

§5. Integrable systems.

Let Γ be any Riemann surface (3.3) described above. Let the polynomials

$$M_1, \dots, M_h$$

and corresponding Hamiltonians

$$H_1, \dots, H_h$$

be chosen from the same sequence with the same $R_L(\lambda)$ and $S_{N-L}(\lambda)$.

DEFINITION 5.1: The set of Hamiltonian systems with Hamiltonians

$$H_r(x_r, M_r), \qquad r = 1, \dots, h$$

will be called as integrable system of the type

$$(\Gamma, x_r, M_r, H_r). \tag{5.1}$$

Notice that corresponding Hamiltonian flows are commutative.

Theorem 5.1. *Integrable system of type (5.1) is equivalent to h Jacobi systems of inversion*

$$\frac{1}{2}\sum\int_{\lambda_j^0}^{\lambda_j}\frac{\lambda_j^{k-1}d\lambda_j}{\sqrt{-P(\lambda_j)}}=m_{\alpha_r,\beta_r,k}(x_r-x_r^0)+D_k(x_r^0) \tag{5.2}$$

Definition 5.2: We replace first q continuous systems of inversion (5.2) by the following discrete systems of inversion

$$\frac{1}{2}\sum\int_{\lambda_{j,l}}^{\lambda_{j,l+1}}\frac{\lambda_j^{k-1}d\lambda_j}{\sqrt{-P(\lambda_j)}}=m_{\alpha_r,\beta_r,k}\Delta x_r, \qquad r=1,\ldots,q. \tag{5.3}$$

Then obtained discrete system is called as the discrete system associated with the continuous system (5.2) and vice versa.

Discrete transformations defined by the discrete equations (5.3) are canonical. They preserve the Hamiltonian structure of the remaining continuous part.

In conclusion notice that in [**12**] continuous integrable systems associated with the hierarchies of Toda lattices, relativistic Toda lattices and Volterra lattices are found.

§6. Associated discrete and continuous integrable systems.

There were a lot of attempts to find a limit continuous approximation for the discrete integrable systems. In particular the Boussinesq equation and Korteweg-de Vries (KdV) equation were obtained as a limiting systems for the usual Toda Lattice [**17**].

In accordance with our approach the following Definition plays a central role.

Definition 6.1: The discrete and continuous integrable systems are associated with each other

1) If they have the same spectrum (Riemann surface Γ) as well as the same Hamiltonian for the continuous part of each system

2) If the discrete Jacobi system of inversion is in fact a lattice-discretization of the continuous Jacobi system.

Three hierarchies of discrete systems, namely Toda lattices, relativistic Toda lattices and Volterra lattices can be described using similar stationary and dynamical equations.

The stationary equations are equivalent to the discrete Jacobi system of inversion

$$\frac{1}{2}\sum \int\limits_{\lambda_{j,k}}^{\lambda_{j,k+1}} \frac{\lambda_j^{n-m} d\lambda_j}{\sqrt{-C_N(\lambda_j)}} = U_m \tag{6.1}$$

Here U_m does not depend on k.

The continuous part of these systems is a Hamiltonian system with the Hamiltonian

$$H_D = \sum_{j=1}^{n} \frac{W_{j,k}^2 + C_N(\lambda_{j,k})}{\lambda_{j,k}^{\epsilon} \prod_{s \neq j}(\lambda_{j,k} - \lambda_{s,k})} M(\lambda_{j,k}; \alpha, \beta) \tag{6.2}$$

Here $\epsilon = 0$ for Toda lattices and $\epsilon = 0$ for relativistic Toda lattices.

Comparing spectrums, Hamiltonians and Jacobi systems of inversion for the Toda lattices and two-potential KdV systems (2.7)–(2.9) we see that two-potential systems are associated with the hierarchy of the Toda lattices.

Spectrum, discrete and continuous Jacobi systems of inversion for the relativistic Toda lattices coincide with those in case of usual Toda lattices. Therefore, associated continous system can be also determined as a two-potential systems defined by the same generating equation (2.7). But Poisson structure, Hamiltonians and potentials in the case of relativistic Toda lattices are quite different.

Notice that using suggested method two classes of Volterra lattices were shown to be subclasses of the usual and relativistic Toda lattices [**12**].

REMARK: If the number $(n+k)$ of the variables $\lambda_1, \ldots, \lambda_{n+k}$ is greater than genus n of the Riemann surface (3.3), then the corresponding Hamiltonian system is defined on the noncompact generalized Jacobi variety. This Hamiltonian system is also completely integrable and is equivalent to the Jacobi system of inversion with holomorphic and meromorphic differentials.

References

1. P. Lax, *Periodic solutions of the KdV equation*, American Math. Soc., Lectures in App. Math. **5** (1974), 85–96.
2. S. Novikov, *A periodic problem for the Korteweg-de Vries equation*, Funct. Anal. **8** (1974), 54–66.
3. H. McKean, *Integrable systems and algebraic curves*, Lecture Notes in Mathematics (1979), Springer.
4. H. Flaschka and D. McLaughlin, *Canonically conjugate variables for the Korteweg-de Vries equation and the Toda lattice with periodic boundary conditions*, Prog. Theor. Phys. **55** 2 (1976), 438–456.
5. S. Novikov, B. Dubrovin and V. Matveev, *Nonlinear equations of Korteweg-de Vries type, finite-zone linear operators and Abelian manifolds*, Uspekhi Mat. Nauk, Moscow **31** (1976), 55–136.
6. S. Alber, *Investigation of equations of Korteweg-de Vries type by the method of recurrence relations*, Institute of Chemical Physics, USSR Academy of Sciences (1976); J. London Math. Soc. (2) (1979), 467–480.
7. I. Krichever, *Methods of algebraic geometry in the theory of nonlinear equations*, Uspekhi Mat. Nauk, Moscow **32** (1977), 183–208.
8. D. Mumford and P. van Moerbeke, *The spectrum of difference operators and algebraic curves*, Acta Math. **143** (1979), 93–154.
9. G. Segal and G. Wilson, *Loop groups and equation of KdV type*, Publ. Math. I.H.E.S. **61** (1985), 3–64.
10. S. Alber and M. Alber, *Hamiltonian formalism for finite-zone solutions of integrable equations*, C.R. Acad. Sci. Paris 301, Ser.1 **16** (1985), 777–780.
11. A. Veselov, *Integrable systems with discrete time and difference operators*, Funct. Anal. Appl. **22** (1988), 1–13.
12. S.Alber, *Associated integrable systems*, (1989). (in print).
13. S. Alber and M. Alber, *Hamiltonian formalism for nonlinear Schrödinger and sine-Gordon equations*, J. London Math. Soc. (2) **36** (1987), 176–192.
14. S. Alber and M. Alber, *Stationary problems for equations of KdV type and geodesics on quadrics*, Institute of Chemical Physics, USSR Academy of Sciences (1984).
15. S. Alber, *On stationary problems for equations of Korteweg-de Vries type*, Comm. Pure Appl. Math. **34** (1981), 259–272.
16. S. Ruijsenaars and H. Schneider, *A new class of integrable systems and its relation to solitons*, Ann. Phys. (NY) **170** (1986), 370–405.
17. M. Toda, "Theory of Nonlinear Lattices," Springer-Verlag, Berlin, New York, 1989.

Department of Mathematics
University of Pennsylvania
Philadelphia, PA 19104-6395

A Universal Reduction Procedure for Hamiltonian Group Actions

JUDITH M. ARMS, RICHARD H. CUSHMAN, AND MARK J. GOTAY

ABSTRACT. We give a universal method of inducing a Poisson structure on a singular reduced space from the Poisson structure on the orbit space for the group action. For proper actions we show that this reduced Poisson structure is nondegenerate. Furthermore, in cases where the Marsden-Weinstein reduction is well-defined, the action is proper, and the preimage of a coadjoint orbit under the momentum mapping is closed, we show that universal reduction and Marsden-Weinstein reduction coincide. As an example, we explicitly construct the reduced spaces and their Poisson algebras for the spherical pendulum.

1. INTRODUCTION

Reduction of the order of mechanical systems with symmetry is a venerable topic dating back to Jacobi, Routh and Poincaré. Although reduction in the regular case is well understood [1, 2], interesting and important applications continue to arise [3, 4]. (See also [5] for a comprehensive exposition.) The singular case has received much less attention, despite the growing realization that it is the rule rather than the exception. (See for instance [6–11].) For example, the solution spaces of classical field theories are invariably singular and the ramifications of this have only recently begun to be explored [12, 13]. Singularities also play an increasingly critical role in understanding the behavior of simple mechanical systems, such as the photon [14], the Lagrange top [15], coupled rigid bodies [16], particles with zero angular momentum [17] and even homogeneous Yang-Mills fields [18].

The central ingredient in the reduction of a mechanical system with symmetry is the construction of a symplectic reduced space of invariant states. This reduction process can be formulated abstractly as follows. Let (M, ω) be a symplectic manifold and G a Lie group with Lie algebra $\mathcal{G}$. Suppose that there is a Hamiltonian action of G on (M, ω) with an Ad^*-equivariant momentum mapping $J : M \longrightarrow \mathcal{G}^*$. We consider the problem of inducing symplectic structures on the reduced spaces

$$M_\mu = J^{-1}(\mu)/G_\mu$$

as μ ranges over $J(M) \subseteq \mathcal{G}^*$. Here G_μ is the isotropy group of μ under the coadjoint action of G on $\mathcal{G}^*$. Marsden and Weinstein [19] solved this problem in "regular" cases. (See also [20].) Specifically they showed that if μ is a weakly regular value of J and if G_μ acts freely and properly on $J^{-1}(\mu)$, then M_μ is a symplectic quotient manifold of $J^{-1}(\mu)$ in a natural way.

This Marsden-Weinstein reduction breaks down in singular situations. In this context various reduction techniques were developed and studied in [21] and [22]. Unfortunately these alternative reductions are not always applicable, and so it is sometimes necessary to switch from one reduction technique to another as μ varies in $J(M)$. Obviously this would cause havoc in the context of perturbation theory. Even worse, when these reduction procedures apply, they do not necessarily agree. Other techniques give reduced Poisson algebras which, however, are not always function algebras on M_μ. We note that the reduction procedure of Śniatycki and Weinstein [23] suffers this difficulty.

The purpose of this paper is to present a new reduction procedure which assigns a Poisson structure to each reduced space M_μ, even in the presence of singularities. The salient features of this reduction are that:

(1) it always works;

(2) it is natural;

(3) it can be applied uniformly to every M_μ for every $\mu \in J(M)$.

One can regard it as a universal method of reduction. As such it has advantages over the other singular reduction procedures discussed above.

As this paper was being written, conversations with P. Dazord [24] indicated that he has independently discovered ideas similar to the notion of universal reduction treated here. However, his techniques and results differ from ours in several respects.

The notation and terminology are explained in the appendix.

2. Universal reduction

Let (M, ω), G and J be as in §1. Our reduction procedure is motivated by the following commutative diagram:

$$\begin{array}{ccc} J^{-1}(\mu) & \hookrightarrow & M \\ \pi_\mu \downarrow & & \downarrow \pi \\ M_\mu & \xrightarrow{i_\mu} & M/G . \end{array} \tag{1}$$

Here π and π_μ are the G- and G_μ-orbit projections. The map i_μ is defined by associating, to each point $m_\mu \in M_\mu$, the G-orbit in which $\pi_\mu^{-1}(m_\mu)$ lies. Consequently $i_\mu \circ \pi_\mu = \pi | J^{-1}(\mu)$. An easy argument using the Ad^*-equivariance of J shows that i_μ is one to one. In fact i_μ is a homeomorphism of M_μ onto $\pi(J^{-1}(\mu))$. In what follows we will identify M_μ with its image under i_μ. Note that M_μ is also $\pi(J^{-1}(\mathcal{O}_\mu))$ where $\mathcal{O}_\mu$ is the G-coadjoint orbit through μ. Observe that we make *no* assumptions regarding the regularity of J or the smoothness of either M_μ or M/G.

The key observation is that M/G is a Poisson variety. The Poisson bracket $\{\,,\}_{M/G}$ on $C^\infty(M/G)$ is given by

$$\{f,h\}_{M/G}(\pi(m)) = \{f \circ \pi, h \circ \pi\}_M(m) \tag{2}$$

where $\{\,,\}_M$ is the Poisson bracket on M associated with ω. This definition makes sense since $C^\infty(M/G) = C^\infty(M)^G$ is a Lie subalgebra of $C^\infty(M)$. This allows the possibility of inducing a Poisson structure on each reduced space M_μ by restricting the Poisson structure on M/G. Our main result is the observation that reduction by restriction *always* works.

Theorem 1 *For each $\mu \in J(M)$, M_μ inherits the structure of a Poisson variety from M/G.*

Proof: For each pair of Whitney smooth functions $f_\mu, h_\mu \in W^\infty(M_\mu)$, set

$$\{f_\mu, h_\mu\}_\mu = \{f,h\}_{M/G} | M_\mu \tag{3}$$

where f,h are any smooth extensions of f_μ, g_μ to M/G. Equation (3) defines a Poisson bracket $\{\,,\}_\mu$ on $W^\infty(M_\mu)$, provided we can show that its right hand side is independent of the choice of extensions f,h. This amounts to showing that the ideal $\mathcal{I}(M_\mu)$ of smooth functions vanishing on M_μ is a *Poisson ideal* in $C^\infty(M/G)$, that is,

$$\{C^\infty(M/G), \mathcal{I}(M_\mu)\}_{M/G} \subseteq \mathcal{I}(M_\mu). \tag{4}$$

Since $C^\infty(M/G) = C^\infty(M)^G$ and $M_\mu = \pi(J^{-1}(\mathcal{O}_\mu))$, equation (4) is equivalent to

$$\{C^\infty(M)^G, \mathcal{I}(J^{-1}(\mathcal{O}_\mu))^G\}_M \subseteq \mathcal{I}(J^{-1}(\mathcal{O}_\mu))^G, \tag{5}$$

where $\mathcal{I}(J^{-1}(\mathcal{O}_\mu))^G$ is the ideal of smooth G-invariant functions which vanish on $J^{-1}(\mathcal{O}_\mu)$. Let $F, H \in C^\infty(M)^G$ with $H|J^{-1}(\mathcal{O}_\mu) = 0$. We must show that for every $m \in J^{-1}(\mathcal{O}_\mu)$,

$$(6) \qquad \{F, H\}_M(m) = -X_F\, H(m) = 0.$$

Now for every $\xi \in \mathcal{G}$,

$$\{J^\xi, F\}_M = -\xi_M F = 0,$$

where $J^\xi(m) = J(m)(\xi)$. The first equality follows from the definition of the momentum mapping J, while the second follows from $F \in C^\infty(M)^G$. Consequently for every $\xi \in \mathcal{G}$, J^ξ is constant along the integral curves of X_F. Hence the integral curve $t \longrightarrow \varphi_t^F(m)$ lies in $J^{-1}(J(m))$. This proves

$$(7) \qquad H(\varphi_t^F(m)) = 0.$$

Differentiating (7) with respect to t and evaluating at $t = 0$ gives (6). □

In this fashion each orbit space M_μ is naturally equipped with a Poisson structure. The corresponding reduced Poisson algebra is $(W^\infty(M_\mu), \{\,,\}_\mu)$. It follows from the definitions that the underlying function space can be represented as

$$\begin{aligned} W^\infty(M_\mu) &= C^\infty(M/G)/\mathcal{I}(M_\mu) \\ &= C^\infty(M)^G/\mathcal{I}(J^{-1}(\mathcal{O}_\mu))^G \\ &= C^\infty(M)^G/\mathcal{I}(J^{-1}(\mu))^G \end{aligned}$$

where $\mathcal{I}(J^{-1}(\mu))^G$ is the ideal of smooth G-invariant functions on M which vanish on $J^{-1}(\mu)$. In the last equality we have used the fact that $J^{-1}(\mathcal{O}_\mu) = G \cdot J^{-1}(\mu)$, which is a consequence of the Ad^*-equivariance of J.

Formally we see that reduction in singular cases works exactly as it does in regular cases. We caution, however, that the results of singular reduction may seem a bit strange. For example, in the regular case, one knows that the reduced Poisson algebra is *nondegenerate* in the sense that it contains no nontrivial elements which Poisson commute with everything, that is, there are no nontrivial Casimirs. This is not necessarily so in the singular case without additional assumptions. Likewise in the regular case M_μ is a finite union of symplectic leaves of M/G. This also is no longer true if the reduced algebra has nontrivial Casimirs. We illustrate these remarks with the following example. Lift the irrational flow on T^2 to an $\mathbb{R}$-action on T^*T^2. Here each M_μ may be identified with $T^2/\mathbb{R} \times \mathbb{R}$. The Poisson algebra consists entirely of Casimirs, some nontrivial. Certainly M_μ is not "symplectic". Note that the symplectic leaves of $T^*T^2/\mathbb{R}$ are just points. (See [21, example 3.4] for more details.) These kinds of behavior cannot occur if the action is proper. When this is the case, we will show in §3 that the reduced Poisson bracket is nondegenerate.

If G is noncompact, we encounter another phenomenon which can arise even in the regular case. For example, let $M = T^*G$ with G acting on M by the lift of left translation. Let $J : M \longrightarrow \mathcal{G}^*$ be the momentum map given by $J(\alpha_g) = R_g^*\alpha_g$ for $\alpha_g \in M$. Every $\mu \in \mathcal{G}$ is a regular value of J and $M_\mu = J^{-1}(\mu)/G_\mu$ is symplectically diffeomorphic to $\mathcal{O}_\mu$ with its Kostant-Kirillov symplectic structure [1]. On the other hand, T^*G/G is isomorphic as a Poisson manifold to $\mathcal{G}^*$ with its Lie-Poisson structure [4]. Thus the basic commutative diagram (1) becomes

$$\begin{array}{ccc} J^{-1}(\mu) & \hookrightarrow & T^*G \\ \pi_\mu \downarrow & & \downarrow \pi \\ \mathcal{O}_\mu & \overset{i_\mu}{\hookrightarrow} & \mathcal{G}^* \end{array}$$

where i_μ is the inclusion mapping. Now suppose that $\mathcal{O}_\mu$ is not closed in $\mathcal{G}^*$. Then the Marsden-Weinstein reduction, which views $\mathcal{O}_\mu$ abstractly as a quotient space, differs slightly from the universal reduction, which views $\mathcal{O}_\mu$ as a subset of $\mathcal{G}^*$. This distinction is apparent on the level of function algebras. The Marsden-Weinstein reduced function algebra is $C^\infty(\mathcal{O}_\mu)$, whereas the corresponding function algebra for the universal reduction is $W^\infty(i_\mu(\mathcal{O}_\mu))$. In applications there are plausible arguments for the appropriateness of both function spaces. In general relativity and other field theories, the momentum for the gauge group is constrained to vanish for all physically admissible states. (See [25] for more details.) This situation strongly suggests using the Marsden-Weinstein reduced function algebra $C^\infty(\mathcal{O}_\mu)$: the extension of functions to nearby, physically unrealizable values of momentum is irrelevant. In mechanics, however, momentum usually is a parameter which may be varied. In such cases, Hamiltonians should be smoothly extendable to nearby values, and therefore should lie in the universal reduced function algebra $W^\infty(i_\mu(\mathcal{O}_\mu))$.

A concrete example of the above situation is obtained by taking G to be the subgroup of $Sl(2,\mathbb{R})$ consisting of matrices of the form

$$\begin{pmatrix} a & b \\ 0 & a^{-1} \end{pmatrix}, \quad a > 0.$$

Let

$$\xi_1 = \begin{pmatrix} 1 & 0 \\ 0 & -1 \end{pmatrix} \text{ and } \xi_2 = \begin{pmatrix} 0 & 1 \\ 0 & 0 \end{pmatrix}$$

be the standard basis for $\mathcal{G}$. Viewing (ξ_1, ξ_2) as coordinates on $\mathcal{G}^*$, the coadjoint orbits can be pictured as

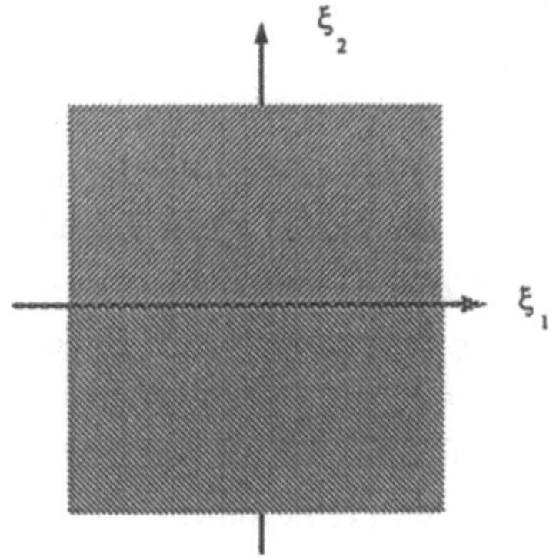

Figure 1. The coadjoint orbits of G on $\mathcal{G}^*$.

There are two classes of orbits: (1) the open upper and lower half planes, where $G_\mu = \{e\}$, and (2) points along the ξ_1-axis where $G_\mu = G$.

3. Universal reduction for proper actions

In this section we give conditions which eliminate the unusual features in the reduction process illustrated by the examples in §2. We will show that if the action of G is proper, then the universal reduced Poisson algebras $(W^\infty(M_\mu), \{\ ,\ \}_\mu)$ are nondegenerate and hence give "symplectic structures" on the reduced spaces M_μ. If in addition to having a proper action we know that $J^{-1}(\mathcal{O}_\mu)$ is closed in M and that the Marsden-Weinstein reduction procedure applies, then the Marsden-Weinstein and universal reductions agree. To obtain these results requires some machinery. As a byproduct we will show that the singularities in $J^{-1}(\mu)$ are quadratic.

We first recall some definitions. Throughout this section we will assume that the action of G on M is *proper*. This means that

$$\Phi : G \times M \longrightarrow M \times M : (g, m) \longrightarrow (g \cdot m, m)$$

is a proper mapping, that is, the preimage of any compact set is compact. For $m \in M$, let G_m be the isotropy group of m. A (smooth) *slice* at m for the G-action is a submanifold $S_m \subseteq M$ such that

(*i*) $G \cdot S_m$ is an open neighborhood of the orbit $\mathcal{O}_m$; and

(*ii*) there is a smooth equivariant retraction

$$r : G \cdot S_m \longrightarrow G \cdot m$$

such that $S_m = r^{-1}(m)$.

From these properties one easily derives the following (cf. [26], propositions 2.1.2 and 2.1.4):

(a) S_m is closed in $G \cdot S_m$, which we call the *sweep* of S_m.

(b) m belongs to S_m.

(c) S_m is invariant under G_m.

(d) For $g \notin G_m$, the sets S_m and $g \cdot S_m$ are disjoint.

(e) Let $\sigma : \mathcal{U} \longrightarrow G$ be a local cross-section of G/G_m. Then

$$F : \mathcal{U} \times S_m \longrightarrow M : (u, s) \longrightarrow \sigma(u) \cdot s$$

is a diffeomorphism onto an open subset of M.

In the literature a slice is sometimes defined by requiring (a-e) or (i) and (a-d); these definitions are equivalent to that given above.

Proposition 1 *There is a slice for the G-action at each $m \in M$. Each isotropy group G_m is compact, and M/G is Hausdorff.*

Proof: The first statement follows immediately from proposition 2.2.2 and remark 2.2.3 of [26]. The fact that G_m is compact is a direct consequence of the properness of the action, since $G_m = \rho(\Phi^{-1}(m, m))$ where $\rho(g, m) = g$. Finally, since Φ is proper, it is a closed mapping. Now apply proposition 4.1.19 of [1] to conclude that M/G is Hausdorff. □

Remark: Palais' definition of proper action [26] differs from ours, but a straightforward exercise in point set topology shows that the two definitions are equivalent.

The existence of slices for the G-action has several important consequences. One is that it guarantees that there are enough smooth G-invariant functions to separate orbits. This is a corollary of the following proposition.

Proposition 2 *Let N be any G-invariant subset of M and suppose that $f \in W^\infty(N)$ is constant on each G-orbit. Then there is a smooth G-invariant extension F of f with $F \in C^\infty(M)^G$ and $F|N = f$.*

Proof: The main ideas of this proof appear in Palais' argument showing the existence of an invariant Riemannian metric for a smooth proper action ([26], theorem 4.3.1). We will pick a set of local slices whose sweeps are a locally finite open cover of M, and then construct a G-invariant partition of unity subordinated to this cover. The properties of slices allow us to extend f as desired on each sweep. Then we patch these local extensions together using the partition of unity.

As M/G is locally compact, σ-compact, and Hausdorff, we can choose a sequence of points m_i and slices S_i at these points such that $\{\pi(S_i)\}$ is a locally finite cover of M/G. Then $\{G \cdot S_i\}$ is a locally finite cover of M. Furthermore M/G is normal, so we may assume that the slices were chosen so that each m_i has a relatively compact neighborhood $N_i \subseteq S_i$ such that $\{\pi(N_i)\}$ also covers M/G. On each slice S_i construct a function $\tilde{h}_i$ which is positive on N_i and has compact support in S_i. As G_{m_i} is compact, we may average over its orbits to get an invariant function on S_i. (Here we have used both proposition 1 and property (c) of the slice.) By properties (d) and (i) of the slice, we may extend $\tilde{h}_i$ to the sweep and then to all of M by requiring it to be G-invariant and to vanish outside of the sweep. Now by the local finiteness of the covering by sweeps, the $\tilde{h}_i$ can be normalized to a G-invariant partition of unity h_i. That is, let $h_i = \tilde{h}_i / \sum_j \tilde{h}_j$, so that $\sum_j h_j = 1$ at all points of M.

To extend f, we first define a function F_i on each sweep $G \cdot S_i$. If $N \cap S_i$ is empty, set $F_i = 0$. Otherwise, we can extend $f|S_i \cap N$ smoothly to S_i because f belongs to $W^\infty(N)$ and S_i is a submanifold of M. As above we can average over the orbits of the isotropy group on S_i and extend to a G-invariant function F_i on the sweep. Now define F belonging to $C^\infty(M)^G$ by setting $F = \sum_i h_i \cdot F_i$. It is clear that F has the desired properties. □

Corollary 1 *$C^\infty(M)^G$ separates G-orbits in M.*

Proof: Let $N = \mathcal{O}_n \cup \mathcal{O}_m$ and let $f = 1$ on $\mathcal{O}_m$ and $f = 0$ on $\mathcal{O}_n$; such a function exists and belongs to $W^\infty(N)$ since these orbits are closed, embedded submanifolds of M (because the action is proper). Thus by proposition 2, f has an invariant extension F which separates $\mathcal{O}_m$ and $\mathcal{O}_n$. □

Another consequence of the existence of slices is that the singularities of J must have a particularly simple form.

Proposition 3 *For each $\mu \in J(M)$, the singularities of $J^{-1}(\mu)$ are quadratic.*

Proof: We first reduce to the case $\mu = 0$ by the following standard construction (see theorem 4.1 of [13]). Define an action of G on $M \times \mathcal{O}_\mu$ by $(g, (m, \nu)) \longrightarrow (g \cdot m, Ad^*_{g^{-1}}\nu)$. This action is proper because the original action on M is. Then give $M \times \mathcal{O}_\mu$ the symplectic structure $\Theta = \pi_1^*\omega - \pi_2^*\Omega$ where ω is the symplectic structure on M, Ω is the Kostant-Kirillov symplectic structure on $\mathcal{O}_\mu$, and π_i is the projection of $M \times \mathcal{O}_\mu$ on the ith factor.

Note that on $(M \times \mathcal{O}_\mu, \Theta)$ the G-action is Hamiltonian with Ad^*-equivariant momentum mapping $\mathcal{J} : M \times \mathcal{O}_\mu \longrightarrow \mathcal{G}^*$ given by $\mathcal{J}(m,\nu)(\xi) = J^\xi(m) - \nu(\xi)$ for all $\xi \in \mathcal{G}$.

The level sets $J^{-1}(\mu)$ and $\mathcal{J}^{-1}(0)$ are closely related. In fact we can construct a local diffeomorphism

$$\alpha : M \times \mathcal{U} \longrightarrow M \times \mathcal{U},$$

where $\mathcal{U}$ is a neighborhood of $\mu \in \mathcal{O}_\mu$, such that

$$\alpha(J^{-1}(\mu) \times \mathcal{U}) = \mathcal{J}^{-1}(0) \cap (M \times \mathcal{U}).$$

To do this use a local section for G/G_μ to construct a map $\beta : \mathcal{U} \longrightarrow G$ such that $\beta(\mu)$ is the identity element of G and $Ad^*{}_{\beta(\nu)^{-1}}\mu = \nu$. Then define $\alpha(m,\nu) = (\beta(\nu) \cdot m, \nu)$. It follows easily that α is a local diffeomorphism. The equivariance of the action implies that $J(m) = \mu$ if and only if $\mathcal{J}(\alpha(m,\nu)) = 0$ for all $\nu \in \mathcal{U}$. Thus $J^{-1}(\mu)$ has a quadratic singularity if and only if $\mathcal{J}^{-1}(0)$ has.

For the case $\mu = 0$, the result follows immediately from [8]; we only need to confirm that the hypotheses of that paper are satisfied. By theorem 4.3.1 of [26], a smooth proper action admits an invariant Riemannian metric. The usual polarization construction allows us to assume that the metric and symplectic structures are compatible in the sense of being related by means of a G-invariant almost complex structure (cf. p.8 of [27]). By proposition 1, there is a slice at each point. Also by proposition 1, the isotropy group of each point is compact, so we can construct a metric on the dual Lie algebra which is invariant under the coadjoint action of the isotropy group. Thus by theorem 5 of [8] the singularities of $J^{-1}(0)$ are quadratic. □

Since the singularities in $J^{-1}(\mu)$ are quadratic, we have

Corollary 2 *Each $J^{-1}(\mu)$ is locally path connected.*

In fact proposition 3 shows that nearby points of $J^{-1}(\mu)$ can be connected by piecewise smooth curves in M lying in $J^{-1}(\mu)$. With these results in hand, we are ready to prove the main results of this section.

Theorem 2 *Let G act properly on M. Then for each $\mu \in J(M)$ the reduced Poisson algebra $(W^\infty(M_\mu), \{\ ,\ \}_\mu)$ is nondegenerate.*

In other words, each Casimir in $W^\infty(M_\mu)$ is locally constant. Thus in the context of universal reduction we are able to answer a question of Weinstein [4, p.429].

Proof: Suppose that $c_\mu \in W^\infty(M_\mu)$ is a Casimir. This means that for all $H \in C^\infty(M)^G$

$$\{C,H\}_M | J^{-1}(\mathcal{O}_\mu) = 0,$$

where C is any smooth G-invariant function M with $C|J^{-1}(\mathcal{O}_\mu) = c_\mu \circ \pi$. Since by Ad^*-equivariance $J^{-1}(\mathcal{O}_\mu) = G \cdot J^{-1}(\mu)$, this is equivalent to

$$\{C,H\}_M | J^{-1}(\mu) = -(X_C\, H)|J^{-1}(\mu) = 0. \tag{8}$$

By corollary 1 the G-invariant smooth functions separate G-orbits on M. Thus (8) implies that $X_C|J^{-1}(\mu)$ is tangent to the G-orbits in $J^{-1}(\mu)$, that is, at each $m \in J^{-1}(\mu)$ there is $\xi \in \mathcal{G}$ such that $X_C(m) = \xi_M(m)$. Equivalently,

$$dC(m) = dJ^\xi(m). \tag{9}$$

Since by corollary 2 $J^{-1}(\mu)$ is locally path connected, equation (9) implies that C is locally constant on $J^{-1}(\mu)$ and hence that c_μ is locally constant on M_μ. □

Theorem 3 *Suppose that G acts properly on M, $J^{-1}(\mathcal{O}_\mu)$ is closed in M, and the Marsden-Weinstein reduction is well-defined. Then the Marsden-Weinstein and universal reductions are the same.*

Proof: The reduced spaces are the same and both Poisson brackets descend from that on M, so the only question is whether the reduced function algebras are the same. For the Marsden-Weinstein reduction, the function algebra is

$$C^\infty(M_\mu) = C^\infty(J^{-1}(\mu))^{G_\mu} = C^\infty(J^{-1}(\mathcal{O}_\mu))^G.$$

As $J^{-1}(\mathcal{O}_\mu)$ is closed, any smooth function on $J^{-1}(\mathcal{O}_\mu)$ has a smooth extension, and by proposition 2 we may assume that the extension is G-invariant. Thus

$$C^\infty(J^{-1}(\mathcal{O}_\mu))^G = C^\infty(M)^G / \mathcal{I}(J^{-1}(\mathcal{O}_\mu))^G.$$

But the left hand side of the preceding equation is $C^\infty(M_\mu)$ while the right hand side is $W^\infty(M_\mu)$, so the Marsden-Weinstein and universal reduced function algebras coincide. □

The results of this section demonstrate that proper actions are rather well behaved as far as reduction is concerned. In particular theorem 2 shows that the orbit space M/G for such an action is *always* partitioned into symplectic leaves, these being the connected components of the reduced spaces M_μ.

4. Universal reduction for compact groups

In this section we briefly discuss universal reduction when G is a compact Lie group. This is the nicest of all possible situations, because theorems 2 and 3 apply.

According to a recent result of Gotay and Tuynman [28], every symplectic manifold of finite type which admits a Hamiltonian action of a compact (connected) Lie group G can be obtained by equivariant reduction from a linear symplectic action of G on $\mathbb{R}^{2n}$ with its standard symplectic structure ω. Therefore, without loss of generality, we may suppose that $M = \mathbb{R}^{2n}$ with its standard symplectic form ω and that G is a subgroup of the group of linear symplectic mappings of $\mathbb{R}^{2n}$ onto itself. This symplectic G-action is Hamiltonian because the mapping $J : \mathbb{R}^{2n} \longrightarrow \mathcal{G}^*$ given by $J(m)\xi = \frac{1}{2}\omega(\xi_{\mathbb{R}^{2n}}(m), m)$ for every $\xi \in \mathcal{G}$ is an Ad^*-equivariant momentum mapping [29].

To construct the G-orbit space $\mathbb{R}^{2n}/G$ we use invariant theory. Since G is compact and acts linearly on $\mathbb{R}^{2n}$, the algebra of G-invariant polynomials is finitely generated [30]. Let $\sigma_1, \ldots, \sigma_k$ be a set of generators. Define the Hilbert map for the G-action by

$$\sigma : \mathbb{R}^{2n} \longrightarrow \mathbb{R}^k : m \longrightarrow (\sigma_1(m), \ldots, \sigma_k(m)) .$$

Since G is compact, σ separates G-orbits [31]. Therefore $\sigma(\mathbb{R}^{2n})$ is the G-orbit space $\mathbb{R}^{2n}/G$. Because σ is a polynomial mapping, $\sigma(\mathbb{R}^{2n})$ is a semialgebraic subset of $\mathbb{R}^k$, by the Tarski-Seidenberg theorem [32].

According to Mather's refinement of Schwarz's theorem [33], the mapping $\sigma^* : C^\infty(\mathbb{R}^k) \longrightarrow C^\infty(\mathbb{R}^{2n})^G : f \longrightarrow f \circ \sigma$ is split surjective. An easy diagram chase shows that the space of smooth functions on $\mathbb{R}^{2n}/G$ is isomorphic as a Fréchet space to the space of Whitney smooth functions on $\mathbb{R}^{2n}/G \subseteq \mathbb{R}^k$. We know that the space of smooth functions on the orbit space has a Poisson bracket $\{\ ,\ \}_{\mathbb{R}^{2n}/G}$. The invariant theoretic construction of $\mathbb{R}^{2n}/G$ gives rise to the question: is there a Poisson bracket $\{\ ,\ \}_{\mathbb{R}^k}$ on $C^\infty(\mathbb{R}^k)$ such that $C^\infty(\mathbb{R}^{2n}/G)$ is a Poisson subalgebra? In all known examples, the answer is yes.

We can identify the reduced space M_μ with $\sigma(J^{-1}(\mathcal{O}_\mu))$. Moreover, $J^{-1}(\mathcal{O}_\mu)$ is an algebraic variety, since J is a polynomial mapping and $\mathcal{O}_\mu \subseteq \mathcal{G}^*$ is an algebraic variety. Consequently, $M_\mu = \sigma(J^{-1}(\mathcal{O}_\mu))$ is a semialgebraic variety, which is a subvariety of the orbit space $\mathbb{R}^{2n}/G$. By theorem 1 we know that we have a Poisson bracket on the space of Whitney smooth functions on M_μ. Comparing this with [21], in particular propositions 5.5, 5.6, and 5.7, we see that in the present case universal reduction agrees with geometric reduction and also Dirac reduction, when the latter applies. It then follows from either theorem 2 or proposition 5.9 of [21] that the Poisson structure on M_μ is nondegenerate. In [21] one can find an example which shows that universal reduction can differ from the Śniatycki-Weinstein reduction even in the case of a compact linear group action.

5. The spherical pendulum

To make this somewhat abstract discussion more concrete, we use invariant theory to construct the reduced spaces and Poisson algebras occurring in the spherical pendulum. For other treatments see [34, 35].

Consider the linear symplectic action of S^1 on $(T\mathbb{R}^3, \omega)$ defined by

$$\Phi : S^1 \times T\mathbb{R}^3 \longrightarrow T\mathbb{R}^3 : (t, (x, y)) \longrightarrow (R_t\, x, R_t\, y)$$

where

$$R_t = \begin{pmatrix} \cos t & \sin t & 0 \\ -\sin t & \cos t & 0 \\ 0 & 0 & 1 \end{pmatrix}.$$

Φ leaves the subspace defined by $\{x_1 = x_2 = y_1 = y_2 = 0\}$ fixed and is the diagonal action of $SO(2, \mathbb{R})$ on $\mathbb{R}^2 \times \mathbb{R}^2$, which is the subspace defined by $\{x_3 = y_3 = 0\}$. A theorem of Weyl [36] states that the algebra of S^1-invariant polynomials is generated by

$$\begin{array}{lll} \tau_1 = x_3 & \tau_3 = y_1^2 + y_2^2 & \tau_5 = x_1^2 + x_2^2 \\ \tau_2 = y_3 & \tau_4 = x_1 y_1 + x_2 y_2 & \tau_6 = x_1 y_2 - x_2 y_1. \end{array}$$

Later calculations are simplified by using the equivalent set of generators

$$\begin{array}{lll} \sigma_1 = x_3 & \sigma_3 = y_1^2 + y_2^2 + y_3^2 & \sigma_5 = x_1^2 + x_2^2 \\ \sigma_2 = y_3 & \sigma_4 = x_1 y_1 + x_2 y_2 & \sigma_6 = x_1 y_2 - x_2 y_1. \end{array}$$

The Hilbert map for the S^1-action is

$$\sigma : T\mathbb{R}^3 \longrightarrow \mathbb{R}^6 : m \longrightarrow (\sigma_1(m), \ldots, \sigma_6(m)).$$

Here the σ_i satisfy the relation

$$\sigma_4^2 + \sigma_6^2 = \sigma_5(\sigma_3 - \sigma_2^2) \tag{10}$$

together with

$$\sigma_3 \geq 0 \;\&\; \sigma_5 \geq 0. \tag{11}$$

As is shown in [36] these are the only relations. Thus the S^1-orbit space $T\mathbb{R}^3/S^1$ is the semialgebraic variety $\sigma(T\mathbb{R}^3)$ defined by (10) and (11).

Now observe that

$$TS^2 = \{(x,y) \in T\mathbb{R}^3 \,|\, x_1^2 + x_2^2 + x_3^2 = 1 \ \&\ x_1y_1 + x_2y_2 + x_3y_3 = 0\}$$

is invariant under Φ. Moreover $\omega|TS^2$ is a symplectic form on TS^2. Therefore the S^1-orbit space TS^2/S^1 of Φ is the semialgebraic variety $\sigma(TS^2)$ defined by (10), (11) and

$$\begin{aligned} \sigma_5 + \sigma_1^2 &= 1 \\ \sigma_4 + \sigma_1\sigma_2 &= 0. \end{aligned} \tag{12}$$

Solving (12) for σ_5 and σ_4, and substituting the result into (10) gives

$$\begin{aligned} \sigma(TS^2) &= \{(\sigma_1,\sigma_2,\sigma_3,\sigma_6) \in \mathbb{R}^4 | \\ &\qquad \sigma_1^2\sigma_2^2 + \sigma_6^2 = (1-\sigma_1^2)(\sigma_3 - \sigma_2^2),\ |\sigma_1| \le 1 \ \&\ \sigma_3 \ge 0\}. \end{aligned}$$

The S^1 action Φ has the angular momentum mapping

$$L : T\mathbb{R}^3 \longrightarrow \mathbb{R} : (x,y) \longrightarrow x_1y_2 - x_2y_1 = \sigma_6.$$

Set $J = L|TS^2$. Then the reduced space

$$M_\ell = \sigma(J^{-1}(\ell)) = \sigma(L^{-1}(\ell) \cap TS^2)$$

is the semialgebraic subvariety of $\sigma(TS^2)$ given by

$$\sigma_6 = \ell. \tag{13}$$

Equivalently M_ℓ is the semialgebraic variety in $\mathbb{R}^3$ defined by

$$(1-\sigma_1^2)\sigma_3 = \sigma_2^2 + \ell^2$$

with $|\sigma_1| \le 1$ & $\sigma_3 \ge 0$. When $\ell \ne 0$, M_ℓ is diffeomorphic to $\mathbb{R}^2$, being the graph of the function

$$\sigma_3 = \frac{\sigma_2^2 + \ell^2}{1-\sigma_1^2}, \qquad |\sigma_1| < 1.$$

When $\ell = 0$, M_0 is not the graph of a function, because it contains the vertical lines $\{(\pm 1, 0, \sigma_3) \in \mathbb{R}^3 | \sigma_3 \ge 0\}$ (see figure 2). However, this singular space is still homeomorphic to $\mathbb{R}^2$.

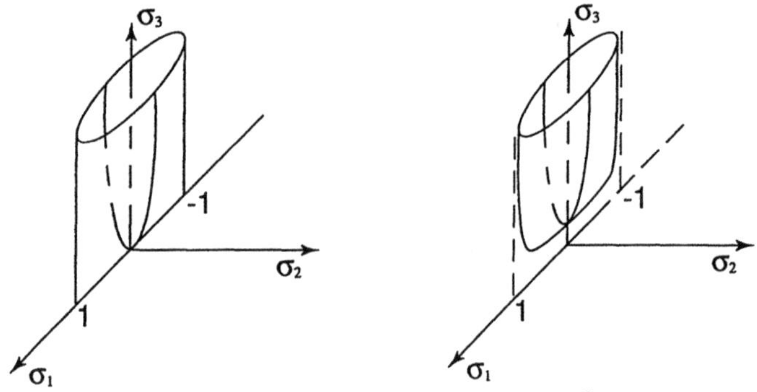

Figure 2. The reduced spaces M_ℓ.

Next we compute the Poisson structure on TS^2/S^1. We could do this using (2) and the symplectic form on TS^2. However, we proceed in a slightly different (but equivalent) way. A straightforward calculation shows that $C^\infty(\mathbb{R}^6)$ with coordinates (σ_i) has a Poisson bracket whose structure matrix is the skew symmetric matrix, half of which is given in table 1.

$\{A,B\}_{\mathbb{R}^6}$	σ_1	σ_2	σ_3	σ_4	σ_5	σ_6	B
σ_1	0	1	$2\sigma_2$	0	0	0	
σ_2		0	0	0	0	0	
σ_3			0	$-2(\sigma_3-\sigma_2^2)$	$-4\sigma_4$	0	
σ_4				0	$-2\sigma_5$	0	
σ_5					0	0	
σ_6						0	
A							

Table 1. Structure matrix for the Poisson bracket on $\mathbb{R}^6$.

Another calculation shows that $C_1 = \sigma_4^2 + \sigma_6^2 - \sigma_5(\sigma_3 - \sigma_2^2)$ and $C_2 = \sigma_6$ are Casimirs. Since $\sigma(T\mathbb{R}^3) \subseteq \mathbb{R}^6$ is defined by $C_1 = 0$, the Poisson bracket on $T\mathbb{R}^3/S^1$ has the same structure matrix as the Poisson bracket on $\mathbb{R}^6$. Because

$$\{\sigma_5 + \sigma_1^2, \sigma_4 + \sigma_1\sigma_2\}_{\sigma(T\mathbb{R}^3)} | \sigma(TS^2) = 2(\sigma_5 + \sigma_1^2)|\sigma(TS^2) = 2,$$

$\sigma(TS^2)$ is a cosymplectic subvariety of $\sigma(T\mathbb{R}^3)$. Consequently, the Poisson bracket on $\sigma(TS^2) = TS^2/S^1$ may be computed using the Dirac process [37]. The structure matrix of $\{\ ,\ \}_{TS^2/S^1}$ is given in table 2.

$\{A,B\}_{TS^2/S^1}$	σ_1	σ_2	σ_3	σ_6	B
σ_1	0	$1-\sigma_1^2$	$2\sigma_2$	0	
σ_2		0	$-2\sigma_1\sigma_3$	0	
σ_3			0	0	
σ_6				0	
A					

Table 2. Structure matrix for the Poisson bracket on TS^2/S^1.

Since the reduced space $M_\ell \subseteq \sigma(TS^2)$ is defined by $C_2 = \ell$, the Poisson bracket on M_ℓ has the same structure matrix as given in table 2 (with the last row and column deleted). If we set

(14) $$\psi(\sigma_1, \sigma_2, \sigma_3) = \sigma_3(1-\sigma_1^2) - \sigma_2^2 - \ell^2,$$

then a careful look at table 2 shows that

(15) $$\{\sigma_i, \sigma_j\}_\ell = \sum_k \epsilon_{ijk} \frac{\partial \psi}{\partial \sigma_k}.$$

From (15) we deduce that

(16) $$\{f_\ell, g_\ell\}_\ell = (\nabla f_\ell \times \nabla g_\ell) \cdot \nabla \psi$$

via the calculation:

$$\begin{aligned}
\{f_\ell, g_\ell\}_\ell &= df_\ell \cdot X_{g_\ell} = \sum_j \frac{\partial f_\ell}{\partial \sigma_j} d\sigma_j \cdot X_{g_\ell} \\
&= \sum_j \frac{\partial f_\ell}{\partial \sigma_j} \{\sigma_j, g_\ell\}_\ell = -\sum_j \frac{\partial f_\ell}{\partial \sigma_j} dg_\ell \cdot X_{\sigma_j} \\
&= \sum_{j,k,i} \frac{\partial f_\ell}{\partial \sigma_j} \frac{\partial g_\ell}{\partial \sigma_k} \epsilon_{jki} \frac{\partial \psi}{\partial \sigma_i} \\
&= (\nabla f_\ell \times \nabla g_\ell) \cdot \nabla \psi.
\end{aligned}$$

Therefore Hamilton's equations for $h_\ell \in W^\infty(M_\ell)$ are

$$\dot{\sigma} = \nabla h_\ell \times \nabla \psi$$

where $\times$ is the vector product on $\mathbb{R}^3$, because

$$\begin{aligned}
\dot{\sigma_i} &= \{\sigma_i, h_\ell\}_\ell \\
&= (\nabla \sigma_i \times \nabla h_\ell) \cdot \nabla \psi \\
&= (\nabla h_\ell \times \nabla \psi) \cdot \nabla \sigma_i \\
&= (\nabla h_\ell \times \nabla \psi)_i.
\end{aligned}$$

So far we have only treated the S^1 symmetry of the spherical pendulum. To treat the dynamics, we consider the Hamiltonian

$$H : T\mathbb{R}^3 \longrightarrow \mathbb{R} : (x, y) \longrightarrow \frac{1}{2}(y_1^2 + y_2^2 + y_3^2) + x_3.$$

When restricted to TS^2, H is the Hamiltonian of the spherical pendulum. Since H is S^1-invariant, it is a function of the invariants, namely

$$H = \frac{1}{2}\sigma_3 + \sigma_1. \tag{17}$$

The reduced Hamiltonian on M_ℓ is h_ℓ, given also by (17). A short calculation shows that the critical points of h_ℓ occur only when $\sigma_2 = 0$. Geometrically they occur in the σ_1, σ_3-plane when the line defined by $h_\ell = e$ is tangent to the fold curve of the projection of M_ℓ on the σ_1, σ_3-plane (see figure 3).

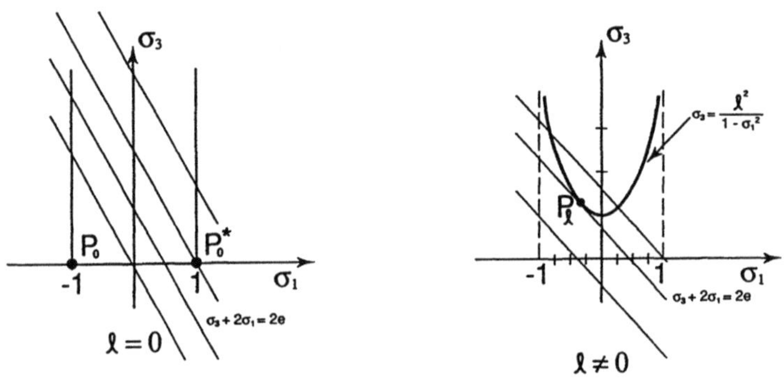

Figure 3. The darkened curve is the fold curve of the projection of M_ℓ on the $\{\sigma_2 = 0\}$ plane. The dots are the critical points of h_ℓ on $M_\ell \cap \{\sigma_2 = 0\}$.

Over every point on a line in figure 3 lie two points of M_ℓ except when (*i*) the line intersects the darkened fold curve where there is only one point of M_ℓ or (*ii*) outside the darkened curve where there are no points of M_ℓ. Hence P_ℓ, P_0 are stable equilibrium points of X_{h_ℓ}, and correspond to stable periodic orbits of X_H. On the other hand, P_0^* is an unstable equilibrium point of X_H.

Acknowledgments

The authors would like to thank Gene Lerman. Geoff Mess and Alan Weinstein for very useful discussions while this paper was being conceived. The authors received partial support from the MSRI while attending its program on symplectic geometry.

6. Appendix: Notation and terminology

We deal with varieties in a generalized sense. For us a variety consists of a topological space X together with a choice of a set of "smooth" functions $C^\infty(X) \subseteq C^0(X)$. This choice defines a differential structure on X. If $\pi : X \longrightarrow Y$ is a surjection of a variety X onto a space Y, we make Y a quotient variety of X by putting the quotient topology on Y and taking

$$C^\infty(Y) = \{f \in C^0(Y) | f \circ \pi = F \;\; \textit{for some } F \in C^\infty(X)\}.$$

Similarly if Y is a subset of a variety X, we make Y into a subvariety of X by putting the relative topology on Y and taking $C^\infty(Y) = W^\infty(Y)$, where

$$W^\infty(Y) = \{f \in C^0(Y) | f = F|Y \;\; \textit{for some } F \in C^\infty(X)\}$$

are the Whitney smooth functions on Y. A map $f : X \longrightarrow Y$ between varieties is smooth provided it is continuous and $f^* C^\infty(Y) \subseteq C^\infty(X)$. If $Y \subseteq X$ is a subvariety, then the ideal $\mathcal{I}(Y) \subseteq C^\infty(X)$ of Y consists of all smooth functions on X which vanish when restricted to Y.

Let G be a Lie group, which acts smoothly on a variety X. Let $Y = X/G$ be the corresponding orbit space. From the definitions it follows that $C^\infty(X/G) = C^\infty(X)^G$, the space of G-invariant smooth functions on X.

References

1. Abraham, R. and Marsden, J., "Foundations of Mechanics", 2nd edition, Benjamin/Cummings, Reading, 1978.
2. Arnol'd, V.I., Kozlov, V.V., and Neishtadt, A.I., *Mathematical Aspects of Classical and Celestial Mechanics*, in: "Dynamical Systems III", ed. V.I. Arnol'd, Springer-Verlag, New York, 1988.
3. Marsden, J., "Lectures on Geometric Methods in Mathematical Physics", CBMS Reg. Conf. Ser. Math., **37**, SIAM, Philadelphia, 1981.
4. Weinstein, A., *Poisson structures and Lie algebras*, in: "The Mathematical Heritage of Élie Cartan", Astérisque (1985), Numéro Hors Sèrie, pp. 421–434.
5. Marsden, J. and Ratiu, T., *Mechanics and Symmetry*, (to appear).
6. Cushman, R., *Reduction, Brouwer's Hamiltonian and the critical inclination*, Celest. Mech., **31** (1983), 401–429.
7. Cushman, R., *A survey of normalization techniques applied to perturbed Keplerian systems*, (to appear in Dynamics Reported).
8. Arms, J., Marsden, J., and Moncrief, V., *Symmetry and bifurcations of momentum mappings*, Comm. Math. Phys., **78**, (1981), 455–478.
9. van der Meer, J.C., "The Hamiltonian Hopf bifurcation", Lect. Notes in Math., **1160**, Springer Verlag, New York, 1985.
10. Kummer, M., *On resonant Hamiltonian systems with finitely many degrees of freedom*, Lect. Notes in Phys., **252**, (1986), 19–31.
11. Kummer, M., *Realizations of the reduced phase space of a Hamiltonian system with symmetry*, Lect. Notes in Phys., **252**, (1986), 32–39.
12. Marsden, J., *Spaces of solutions of relativistic field theories with constraints*, Lect. Notes in Math., **987**, (1982), 29–43.
13. Arms, J., *Symmetry and solution set singularities in Hamiltonian field theories*, Acta Phys. Polon., **B17**, (1986), 499–523
14. Gotay, M.J., *Poisson reduction and quantization for the (n+1)-photon*, J. Math. Phys., **25**, (1984), 2154–2159.
15. Cushman, R. and Knörrer, H., *The momentum mapping of the Lagrange top*, Lect. Notes in Math., **1139**, (1985), 12–24.
16. Patrick, G., *The dynamics of two coupled rigid bodies in three space*, Contemporary Mathematics, **97** (1989), 315–335.
17. Gotay, M.J. and Bos, L., *Singular angular momentum mappings*, J. Diff. Geom., **24**, (1986), 181–203.
18. Gotay, M.J., *Reduction of homogeneous Yang-Mills fields*, J. Geom. Phys., **6**, (1989), 349–365.
19. Marsden, J. and Weinstein, A., *Reduction of symplectic manifolds with symmetry*, Rep. Math. Phys., **5**, (1974), 121–130.
20. Meyer, K., *Symmetries and integrals in mechanics*, in: "Dynamical Systems", ed. M. Peixoto, pp. 259–272, Academic Press, New York, 1973.
21. Arms, J., Gotay, M.J., and Jennings, G., *Geometric and algebraic reduction for singular momentum mappings*, Adv. in Math., **74**, (1990), 43–103.
22. Śniatycki, J., *Constraints and quantization*, Lect. Notes in Math., **1037**, (1983), 301–344.
23. Śniatycki, J. and Weinstein, A., *Reduction and quantization for singular momentum mappings*, Lett. Math. Phys., **7**, (1983), 155–161.
24. Dazord, P., *private communication*, May 1989.
25. Gotay, M.J., Isenberg, J., Marsden, J., Montgomery, R., Śniatycki, J., and Yasskin, P., "Momentum Mappings and the Hamiltonian Structure of Classical Field Theories with Constraints", (to appear).

26. Palais, R., *On the existence of slices for actions of non-compact Lie groups*, Ann. Math., **73**, (1961), 295–323.
27. Weinstein, A., "Lectures on Symplectic Manifolds", CBMS Ref. Conf. Ser. Math. **29**, American Mathematical Society, Providence, 1977.
28. Gotay, M.J. and Tuynman, G.M., *A symplectic analogue of the Mostow-Palais theorem*, (to appear in Proceedings of the Séminaire Sud-Rhodanien de Géométrie à Berkeley).
29. Cushman, R., *The momentum mapping of the harmonic oscillator*, Symposia Math., **14**, (1975), 323–342.
30. Kempf, G., *Computing invariants*, Lect. Notes in Math., **1278**, (1987), 62–80.
31. Poènaru, V., "Singularités C^∞ en présence de symétrie", Lect. Notes in Math., **510**, Springer Verlag, New York, 1976.
32. Collins, G., *Quantifier elimination for real closed fields by cylindrical algebraic decomposition*, Lect. Notes in Comp. Sci., **33**, (1974), 134–183.
33. Mather, J., *Differentiable invariants*, Topology, **16**, (1977), 145–156.
34. Duistermaat, J.J., *On global action angle coordinates*, Comm. Pure. Appl. Math., **33**, (1980), 687–706.
35. Cushman, R., *Geometry of the energy momentum mapping of the spherical pendulum*, Centrum voor Wiskunde en Informatica Newsletter, **1**, (1983), 4–18.
36. Weyl, H., "The classical groups", 2nd ed., Princeton University Press, Princeton, 1946.
37. Dirac, P.A.M., "Lectures on Quantum Mechanics", Academic Press, New York, 1964.

J. M. Arms: Mathematics Department, University of Washington, Seattle, WA 98195, USA.

R. H. Cushman: Mathematics Institute, Rijksuniversiteit Utrecht, 3508 TA Utrecht, The Netherlands.

M. J. Gotay: Mathematics Department, United States Naval Academy, Annapolis, MD 21402, USA.

Supported in part as a Ford Foundation Fellow and by NSF grant DMS-8805699.

Linear Stability of a Periodic Orbit in the System of Falling Balls

JIAN CHENG
MACIEJ P. WOJTKOWSKI
September 10, 1989

Abstract. We study linear stability of a periodic orbit in the hamiltonian system with many degrees of freedom introduced in [**W1**].

§0. Introduction.

Hamiltonian systems with many degrees of freedom are likely to exhibit strong mixing behavior produced by exponential divergence of nearby orbits. It is of great interest to establish to what extent this is so. We know that symmetries (hidden or explicit) bring about integrability and the rigid behavior associated with it. Although strict integrability is easily destroyed by perturbations the KAM theory guarantees its survival on some exotic subsets. Numerical experiments reveal the emergence of instability zones. This interplay of integrability and nonintegrability is still a great challenge for the theory. There are systems where the mixing behavior is present in all of the phase space. These should be the most accessible to rigorous study. A new class of such systems was recently introduced in [**W1**] and [**W2**]. Their important features are that they are one dimensional (interacting particles in a line) and that we are free to increase the size of the system (number of particles) arbitrarily without making it intractable. This 'approachability' is further confirmed in the present paper (we do here things which one would not expect to be doable in a typical system).

Our system consists of n impenetrable particles in the half line falling with constant acceleration to the floor and colliding elastically with each other and the floor. If the masses of the particles are equal the system is completely integrable. Since the system is not continuous (because of multiple collisions) the KAM theory does not restrict its behavior and indeed even a small change in the masses can destroy all invariant tori of the integrable case. It was established in [**W1**] that if higher particles have smaller masses the system has strong mixing properties in all of the phase space.

The method is such that one establishes in particular that all periodic orbits are linearly unstable (have Floquet exponents off the unit circle). One of our goals in this paper was to find out if the condition of decreasing masses is necessary for the results of [**W1**]. The system possesses a highly symmetric periodic orbit with a fairly short period. The study of (in)stability properties of this periodic orbit played some role in the proofs in [**W1**]. We show here that if the masses increase (as we go up) this periodic orbit becomes in general linearly stable. In working on this problem we realized that in the theory of Hamiltonian systems there are more criteria for instability than for stability of a periodic orbit. Could it be a manifestation of the fact that stability is more fragile after all than instability ? In the case of an equilibrium there is at least the simple case of a minimum of the potential energy which guarantees linear (and nonlinear) stability but it has no counterpart for periodic orbits.

The content of the paper is as follows. In Section 1 we describe the system in detail and construct the periodic orbit. In Section 2 we compute explicitly (!) the nonlinear Poincaré section map for our periodic orbit. The formulas resulting from our choice of coordinates are summarized in Proposition 1. They are remarkably simple and we believe that they constitute a good candidate for numerical study. One attractive goal would be to establish the presence of Arnold diffusion – since we can take many particles it could probably be speeded up to the point where one could distinguish it from numerical effects. In Section 3 we compute the characteristic polynomial of the monodromy matrix of our periodic orbit (the derivative of the Poincaré section map). Its special form (Proposition 3) allows to count the roots which lie on the unit circle. We do it in Sections 4 and 5. The picture is roughly the following. For equal masses the eigenvalues are roots of unity of order $n-1$ and they all have multiplicity two. If there are some increases in the sequence of the masses the double eigenvalues split apart staying on the unit circle but they go off the unit circle if the masses decrease. In Theorem 1 we consider the case of almost equal masses. One gap in the masses may be sufficient to produce linear stability as demonstrated in Theorem 3, where no smallness assumptions are necessary. The cases covered there are those of one heavier particle at the top and one lighter particle at the bottom. It turns out that if the groups of neighboring particles with equal masses are of sizes which are not relatively prime then some of the double eigenvalues do not move at all and what is more the periodic orbit is linearly unstable (because of the presence of Jordan blocks associated with them). That is the content of Theorem 2.

Now what we do not do in the paper (and should). We could not establish the nondegeneracy of the Birkhoff normal form of our periodic orbit. The straightforward computation in the highdimensional case looks hopeless (we do not even have the normal modes explicitly). It seems that there are no working sufficient conditions in the literature of the subject. Hence strictly speaking the linear stability of the orbit does not restrict the actual instability or even strong mixing behavior of the system in all of the phase space. It is though very unlikely that the periodic orbit is so highly degenerate.

§1. Description of the system and the periodic orbit.

We consider a system of n point particles in the half line $q \geq 0$ with masses $m_1, \ldots, m_n$. They collide elastically with each other and the bottom particle collides with the floor $q = 0$. They are all subjected to constant negative acceleration i.e., they fall down. Between collisions the motion of the particles is governed by the Hamiltonian

$$H = \sum_{i=1}^{n} \left(\frac{p_i^2}{2m_i} + m_i q_i \right)$$

where q_i are the positions and $p_i = m_i v_i$ the momenta of the particles, $i = 1, \ldots, n$. We assume that the particles are impenetrable i.e., $0 \leq q_1 \leq q_2 \leq \cdots \leq q_n$ and at the moment of a collision $q_i = q_{i+1}$, $1 \leq i \leq n-1$, of i-th and $i+1$ particles there is an instantaneous change of momenta

$$\begin{aligned} p_i^+ &= \gamma_i p_i^- + (1+\gamma_i) p_{i+1}^- \\ p_{i+1}^+ &= (1-\gamma_i) p_i^- - \gamma_i p_{i+1}^-, \end{aligned} \tag{1}$$

where $\gamma_i = \frac{m_i - m_{i+1}}{m_i + m_{i+1}}$, the superscript $^-$ denotes the momenta before the collision and the superscript $^+$ the momenta after the collision. At the collision $q_1 = 0$ of the bottom particle with the floor

$$p_1^+ = -p_1^-. \tag{2}$$

Independent of the number of particles or their masses the system possesses a simple periodic orbit with a fairly short period. In particular on this periodic orbit the individual energies of the particles $h_i = \frac{p_i^2}{2m_i} + m_i q_i$, $i = 1, \dots, n$, are constant. The description of the orbit is slightly different in the case of even and the case of odd number of particles. We will discuss in detail the case of even $n = 2k$ and we will indicate when necessary what has to be changed in the case of odd $n = 2k + 1$.

At time $t = 0$ the first particle collides with the second, the third collides with the fourth, etc., so that

$$q_{2l-1} = q_{2l}, \quad l = 1, \dots, k. \tag{3}$$

Moreover in each of the colliding pairs the center of mass is at rest, i.e.,

$$p_{2l-1}^- + p_{2l}^- = 0, \quad l = 1, \dots, k. \tag{4}$$

By (1) in such collisions the momenta are reversed

$$p_i^+ = -p_i^-, \quad i = 1, \dots, n.$$

Further the initial positions q_i and momenta $p_i, i = 1, \dots, n$ are such that at time $t = t_0$ the bottom particle collides with the floor, the second collides with the third, the fourth collides with the fifth, etc., so that

$$\begin{aligned} q_1 &= \frac{p_1^-}{m_1} t_0 + \frac{1}{2} t_0^2, \\ q_{2l} - q_{2l+1} &= \left(\frac{p_{2l}^-}{m_{2l}} - \frac{p_{2l+1}^-}{m_{2l+1}} \right) t_0, \quad l = 1, \dots, k-1. \end{aligned} \tag{5}$$

Again colliding pairs have their centers of mass at rest and the top particle assumes zero momentum

$$\begin{aligned} p_{2l}^- + p_{2l+1}^- &= -(m_{2l} + m_{2l+1}) t_0, \quad l = 1, \dots, k-1, \\ p_n^- &= -m_n t_0. \end{aligned} \tag{6}$$

The system of equations (3), (4), (5) and (6) has a unique solution depending only on t_0 and the masses $m_1, \dots, m_n$. Actually we get from these equations that

$$\begin{aligned} p_{2l-1}^- &= (m_{2l} + m_{2l+1} + \cdots + m_n) t_0, \\ p_{2l}^- &= -(m_{2l} + m_{2l+1} + \cdots + m_n) t_0, \quad l = 1, \dots, k. \end{aligned} \tag{7}$$

In view of the reversibility of our dynamical system the orbit with such initial conditions will return to the initial state at time $t = 2t_0$, i.e., the orbit is periodic with period $T = 2t_0$. Changing t_0 moves the orbit to another energy level $H = const$. On two energy levels our system differs only by appropriate scaling: if $(q(t), p(t)), -\infty < t < +\infty$ is a trajectory of our system, so is $(\alpha^2 q(\frac{t}{\alpha}), \alpha p(\frac{t}{\alpha})), -\infty < t < +\infty$ for any $\alpha > 0$.

In the case of odd number of particles $n = 2k+1$ the top particle has zero initial momentum so that $p_n^- = -m_n t_0$ is replaced by $p_n^- = 0$ and in (5) and (6) l runs from 1 through k. In particular (7) holds unchanged.

§2. The Poincaré section map.

We are going to compute the Poincaré section map for our periodic orbit. Towards that end we consider two local manifolds transversal to our periodic orbit in its energy level at time $t = -\frac{t_0}{2}$ and $t = \frac{t_0}{2}$. We will use the canonical coordinate system furnished by individual energies h_i and velocities v_i of the particles

$$h_i = m_i q_i + \frac{p_i^2}{2m_i},$$
$$v_i = \frac{p_i}{m_i}, \qquad i = 1, \dots, n.$$

Let $(\bar{h}, \bar{v}(t)), -\infty < t < +\infty$ denote our periodic orbit (the individual energies $h_i, i = 1, \dots, n$ are constant on the orbit). In the vicinity of $(\bar{h}, \bar{v}(\pm\frac{t_0}{2}))$ we introduce linear coordinates

$$\delta h = h - \bar{h},$$
$$\delta v = v - \bar{v}(\pm\frac{t_0}{2}).$$

We let the local transversal sections at time $t = -\frac{t_0}{2}$ and $t = \frac{t_0}{2}$ be $\mathcal{T}_-$ and $\mathcal{T}_+$ respectively, where

$$\mathcal{T}_\pm = \{(\delta h, \delta v) | \sum_{i=1}^n \delta h_i = 0, \sum_{i=1}^n m_i \delta v_i = 0\}.$$

Dynamics in the neighborhood of our periodic orbit is described by two return maps

$$F_1 : \mathcal{V}_1 \to \mathcal{T}_+$$

and

$$F_2 : \mathcal{V}_2 \to \mathcal{T}_-,$$

where $\mathcal{V}_1 \subset \mathcal{T}_-$ and $\mathcal{V}_2 \subset \mathcal{T}_+$ are sufficiently small neighborhoods of the origin. The Poincaré section map F is equal to the composition of the two maps $F = F_2F_1$.

The formulas for F_1 and F_2 can be obtained by straightforward computations. An important step is to describe a collision $q_i = q_{i+1}$ in (h, v) coordinates by

$$\begin{aligned} h_i^+ &= \gamma_i h_i^- + (1+\gamma_i)h_{i+1}^- - \delta_i(v_{i+1}^- - v_i^-)^2, \\ h_{i+1}^+ &= (1-\gamma_i)h_i^- - \gamma_i h_{i+1}^- + \delta_i(v_{i+1}^- - v_i^-)^2, \\ v_i^+ &= \gamma_i v_i^- + (1-\gamma_i)v_{i+1}^-, \\ v_{i+1}^+ &= (1+\gamma_i)v_i^- - \gamma_i v_{i+1}^-, \end{aligned} \tag{8}$$

where $\delta_i = \gamma_i(m_i^{-1} + m_{i+1}^{-1})^{-1}$. (8) follows from (1) . Strictly speaking (8) is an extension of the collision map (1) to all of $R^n \times R^n$ which changes q_i and q_{i+1} unless $q_i = q_{i+1}$.

The result is the following. Let us define for any real γ,α and δ an auxiliary map $C : R^2 \times R^2 \to R^2 \times R^2$ by the following formula

$$C(x_1, x_2, y_1, y_2; \gamma, \alpha, \delta) = (\hat{x}_1, \hat{x}_2, \hat{y}_1, \hat{y}_2),$$

$$\begin{aligned} \hat{x}_1 &= \gamma x_1 + (1+\gamma)x_2 - \alpha(y_2 - y_1) - \delta(y_2 - y_1)^2, \\ \hat{x}_2 &= (1-\gamma)x_1 - \gamma x_2 + \alpha(y_2 - y_1) + \delta(y_2 - y_1)^2, \\ \hat{y}_1 &= \gamma y_1 + (1-\gamma)y_2, \\ \hat{y}_2 &= (1+\gamma)y_1 - \gamma y_2. \end{aligned}$$

We put $(\hat{\delta h}, \hat{\delta v}) = F_r(\delta h, \delta v)$ for $r = 1, 2$. Now F_1 is given by

$$\begin{aligned} &(\hat{\delta h}_{2l-1}, \hat{\delta h}_{2l}, \hat{\delta v}_{2l-1}, \hat{\delta v}_{2l}) \\ &\qquad = C(\delta h_{2l-1}, \delta h_{2l}, \delta v_{2l-1}, \delta v_{2l}; \gamma_{2l-1}, \alpha_{2l-1}, \delta_{2l-1}), \end{aligned} \tag{9}$$

for $l = 1, \ldots, k$.

F_2 is given by

$$\begin{aligned} \hat{\delta h}_1 &= \delta h_1, \\ \hat{\delta v}_1 + \delta t &= \delta v + f(-\delta h_1), \end{aligned} \tag{10}$$

$$(\hat{\delta h}_{2l}, \hat{\delta h}_{2l+1}, \hat{\delta v}_{2l} + \delta t, \hat{\delta v}_{2l+1} + \delta t) = C(\delta h_{2l}, \delta h_{2l+1}, \delta v_{2l}, \delta v_{2l+1}; \gamma_{2l}, \alpha_{2l}, \delta_{2l}),$$

for $l = 1, \ldots, k-1$. In the above formulas for $i = 1, \ldots, n-1$

$$\gamma_i = \frac{m_i - m_{i+1}}{m_i + m_{i+1}},$$

$$\delta_i = \gamma_i (m_i^{-1} + m_{i+1}^{-1})^{-1} = \frac{m_i m_{i+1}(m_i - m_{i+1})}{(m_i + m_{i+1})^2},$$

$$\alpha_i = 2\delta_i \left(\bar{v}_{i+1}(\pm\frac{t_0}{2}) - \bar{v}_i(\pm\frac{t_0}{2}) \right) = 2\gamma_i (m_{i+1} + \cdots + m_n) t_0,$$

$$f(u) = 2\left(\sqrt{\frac{t_0^2 M^2}{m_1^2} - \frac{2u}{m_1}} - \frac{t_0 M}{m_1} \right), \quad M = m_1 + \cdots + m_n,$$

and δt is defined by the condition that $\sum_{i=1}^n m_i \hat{\delta v}_i = 0$.

The formulas (9) and (10) can be further simplified by introducing yet another linear canonical coordinate system (ξ, η) in $\mathcal{T}_\pm$. We define these coordinates by

$$\delta h_1 = -\xi_1,$$

$$\delta h_i = \xi_{i-1} - \xi_i, \text{ for } i = 2, \ldots, n-1,$$

$$\delta h_n = \xi_{n-1}$$

and

$$\eta_i = \delta v_{i+1} - \delta v_i, \text{ for } i = 1, \ldots, n-1.$$

We summarize the result of this last change of coordinates in the following

PROPOSITION 1. *In the (ξ, η) coordinates if we put $(\hat{\xi}, \hat{\eta}) = F_r(\xi, \eta), r = 1, 2$, and $\xi_0 = \xi_n = 0, \eta_0 = \eta_n = 0$ we get for F_1*

$$\hat{\xi}_i = \begin{cases} (1-\gamma_i)\xi_{i-1} - \xi_i + (1+\gamma_i)\xi_{i+1} - \alpha_i \eta_i + \delta_i \eta_i^2, & \text{for odd } i \\ \xi_i, & \text{for even } i, \end{cases}$$

$$\hat{\eta}_i = \begin{cases} -\eta_i, & \text{for odd } i \\ (1+\gamma_{i-1})\eta_{i-1} + \eta_i + (1-\gamma_{i+1})\eta_{i+1}, & \text{for even } i. \end{cases}$$

For F_2 we get

$$\hat{\xi}_i = \begin{cases} \xi_i, & \text{for odd } i \\ (1-\gamma_i)\xi_{i-1} - \xi_i + (1+\gamma_i)\xi_{i+1} - \alpha_i \eta_i + \delta_i \eta_i^2, & \text{for even } i, \end{cases}$$

$$\hat{\eta}_1 = \eta_1 + (1-\gamma_2)\eta_2 - f(\xi_1),$$

$$\hat{\eta}_i = \begin{cases} -\eta_i, & \text{for even } i \\ (1+\gamma_{i-1})\eta_{i-1} + \eta_i + (1-\gamma_{i+1})\eta_{i+1}, & \text{for odd } i \geq 3. \end{cases}$$

□

Whatever changes are required in the case of odd $n = 2k+1$ in the formulas (9) and (10) the formulas in Proposition 1 are valid without any modifications.

§3. The derivative of the Poincaré section map (the monodromy matrix) and its characteristic polynomial.

The derivative of $F = F_2F_1$ at the origin, $P = D_0F$ is equal to the composition of $P_2 = D_0F_2$ and $P_1 = D_0F_1$. In (ξ, η) coordinates we obtain for $P_r, r = 1, 2$,

$$P_r = \begin{pmatrix} A_r & B_r \\ C_r & A_r^* \end{pmatrix},$$

where the $(n-1) \times (n-1)$ blocks are as follows.

$$B_1 = diag(-\alpha_1, 0, -\alpha_3, 0, \dots)$$

is the diagonal matrix with odd entries on the diagonal equal to $-\alpha_i$ and zero even entries.

$$B_2 = diag(0, -\alpha_2, 0, -\alpha_4, \dots)$$

is the diagonal matrix with zero odd entries on the diagonal and even entries equal to $-\alpha_i$. C_1 is zero matrix and

$$C_2 = diag(\frac{2}{t_0 M}, 0, 0, \dots)$$

is the diagonal matrix with only first diagonal entry different from zero.

$A_r, r = 1, 2$, are tridiagonal matrices. To describe them we introduce matrices $\mathcal{D}_i, i = 1, \dots, n-1$,

$$\mathcal{D}_1 = \begin{pmatrix} -1 & 1+\gamma_1 & & \\ 0 & 1 & & \\ & & \ddots & \\ & & & 1 \end{pmatrix}, \mathcal{D}_{n-1} = \begin{pmatrix} 1 & & & \\ & \ddots & & \\ & & 1 & 0 \\ & & 1-\gamma_{n-1} & -1 \end{pmatrix},$$

and for $2 \le i \le n-2$

$$\mathcal{D}_i = \begin{pmatrix} 1 & & & & & \\ & \ddots & & & & \\ & & 1 & 0 & 0 & \\ & & 1-\gamma_i & -1 & 1+\gamma_i & \\ & & 0 & 0 & 1 & \\ & & & & & \ddots & \\ & & & & & & 1 \end{pmatrix}$$

is the matrix which differs from the identity matrix in the i-th row only. Now

$$A_1 = \prod_{odd\ i}^{n-1} \mathcal{D}_i, \quad \text{and} \quad A_2 = \prod_{even\ i}^{n-1} \mathcal{D}_i.$$

We have $\mathcal{D}_i^2 = I$ and $\mathcal{D}_i$ commutes with $\mathcal{D}_j$ unless $|i-j| = 1$. Moreover all $\mathcal{D}_i$'s are isometries with respect to the scalar product given by

$$(11) \quad \|\xi\|^2 = \sum_{i=1}^{n} \frac{(\xi_i - \xi_{i-1})^2}{m_i} = \sum_{i=1}^{n-1} \left(\frac{1}{m_i} + \frac{1}{m_{i+1}}\right) \xi_i^2 - \sum_{i=2}^{n-1} \frac{2}{m_i} \xi_{i-1}\xi_i,$$

where we again introduced $\xi_0 = \xi_n = 0$.

Another description of $A_r, r = 1, 2$, is the following: the i-th odd (even) row of $A_1(A_2)$ is the same as that of $\mathcal{D}_i$ and the even (odd) rows are the same as in the identity matrix.

Let

$$\Upsilon(\lambda) = \det(P - \lambda I) = \det(P_2 P_1 - \lambda I)$$

be the characteristic polynomial of P. Since both P_1 and P_2 are symplectic their determinants are equal to 1 and hence

$$\Upsilon(\lambda) = \det(P_2 - \lambda P_1^{-1}).$$

But

$$P_1^{-1} = \begin{pmatrix} A_1 & -B_1 \\ 0 & A_1^* \end{pmatrix}$$

so that

$$(12) \qquad \Upsilon(\lambda) = \det \begin{pmatrix} A_2 - \lambda A_1 & B_2 + \lambda B_1 \\ C_2 & (A_2 - \lambda A_1)^* \end{pmatrix}.$$

Let

$$\mathcal{Z}_{n-1}(\lambda) = \mathcal{Z}_{n-1}(\lambda; \gamma_1, \ldots, \gamma_{n-1}) = \det(A_2 - \lambda A_1).$$

It is useful for the following to describe the matrix $A_2 - \lambda A_1$ in more detail. We have

$$(13) \quad A_2 - \lambda A_1 = \begin{pmatrix} 1+\lambda & -\lambda(1+\gamma_1) & 0 & & \\ 1-\gamma_2 & -(1+\lambda) & 1+\gamma_2 & 0 & \\ 0 & -\lambda(1-\gamma_3) & 1+\lambda & -\lambda(1+\gamma_3) & 0 \\ & & \ddots & \ddots & \ddots \end{pmatrix}.$$

We do not put the last row in the matrix (13) because it looks different in the case of even or odd n.

PROPOSITION 2.

(1) $\mathcal{Z}_{n-1}$ *is a symmetric polynomial in the following sense*

$$\lambda^{n-1}\mathcal{Z}_{n-1}(\frac{1}{\lambda}) = \mathcal{Z}_{n-1}(\lambda),$$

(2) *If* $-1 < \gamma_i < 1, i = 1, \ldots, n-1$ *then all roots of* $\mathcal{Z}_{n-1}(\lambda)$ *are simple and lie on the unit circle.*

PROOF: We have $\det \mathcal{D}_i = -1, i = 1, \ldots, n-1$, and $A_r^2 = I, r = 1, 2$. Hence

$$\begin{aligned}\mathcal{Z}_{n-1}(\lambda) &= \det(A_2A_1 - \lambda I)\det A_1 = \det(A_1 - \lambda A_2)\det(A_2A_1)\\ &= (-1)^{n-1}\lambda^{n-1}\det(A_2 - \frac{1}{\lambda}A_1)\det(A_2A_1) = \lambda^{n-1}\mathcal{Z}_{n-1}(\frac{1}{\lambda}),\end{aligned}$$

which proves the first part of the proposition.

Further we see above that $\mathcal{Z}_{n-1}(\lambda)$ is (up to a sign) the characteristic polynomial of the matrix A_2A_1 which is an isometry with respect to the scalar product (11). Hence all of its roots lie on the unit circle. Suppose now that it has a double root λ_0. Since A_2A_1 is an isometry it must then have two (complex) linearly independent eigenvectors $\xi^{(1)}$ and $\xi^{(2)}$

$$(A_2A_1 - \lambda_0 I)\xi^{(r)} = 0, r = 1, 2.$$

This is equivalent to

$$(A_2 - \lambda_0 A_1)A_1\xi^{(r)} = 0, r = 1, 2.$$

By (13) this system is tridiagonal with no zero entries on the three diagonals so it cannot have two linearly independent solutions. □

The next Proposition is crucial in our goal to locate the roots of $\Upsilon(\lambda)$.

PROPOSITION 3.

$$\begin{aligned}\Upsilon(\lambda) &= \mathcal{Z}_{n-1}^2(\lambda; \gamma_1, \ldots, \gamma_{n-1})\\ &+ \sum_{i=1}^{n-1}\sigma_{i+1}\gamma_i\prod_{j=1}^{i-1}(1+\gamma_j)\lambda^i\mathcal{Z}_{n-i-1}^2(\lambda; \gamma_{i+1}, \ldots, \gamma_{n-1}),\end{aligned}$$

where

$$\sigma_i = \frac{2\alpha_{i-1}}{\gamma_{i-1}Mt_0} = 4\frac{m_i + \cdots + m_n}{m_1 + \cdots + m_n}, \text{ for } i = 2, \ldots, n-1,$$

and $\mathcal{Z}_0(\lambda) \equiv 1$.

Proof: The basis for the computation of $\Upsilon(\lambda)$ is the formula (12) in which the off diagonal blocks are $B_2 + \lambda B_1 = diag(-\lambda\alpha_1, -\alpha_2, -\lambda\alpha_3, \dots)$ and $C_2 = diag(\frac{2}{t_0 M}, 0, 0, \dots)$. We use the Laplace theorem to expand the determinant (12) into $(n-1)\times(n-1)$ minors, i.e., we use the formula

$$\begin{aligned}\det &= d\xi_1 \wedge \cdots \wedge d\xi_{n-1} \wedge d\eta_1 \wedge \cdots \wedge d\eta_{n-1} \\ &= (d\xi_1 \wedge \cdots \wedge d\xi_{n-1}) \wedge (d\eta_1 \wedge \cdots \wedge d\eta_{n-1}).\end{aligned}$$

Since the off diagonal blocks contain mostly zeroes, after a moment of reflection one sees that there are only n terms in the expansion

$$\Upsilon(\lambda) = \det(A_2 - \lambda A_1)^2 + \sum_{i=1}^{n-1} \sigma_{i+1}\gamma_i \lambda^{\nu(i)} \det(A_2 - \lambda A_1)^2_{i,1} ,$$

where

$$\nu(i) = \begin{cases} 0, & \text{for even } i \\ 1, & \text{for odd } i, \end{cases}$$

and $(A_2 - \lambda A_1)_{i,1}$ is the matrix obtained from $A_2 - \lambda A_1$ by the removal of the i-th row and the first column.

Careful inspection of the matrix (13) brings out the fact that

$$(A_2 - \lambda A_1)_{i,1} = \begin{pmatrix} T_i & 0 \\ & Z_i \end{pmatrix},$$

where T_i is an $(i-1)\times(i-1)$ lower triangular matrix with elements on the diagonal equal to $(-\lambda(1+\gamma_1), (1+\gamma_2), -\lambda(1+\gamma_3), \dots)$. The matrix Z_i is the $(n-i)\times(n-i)$ lower-right diagonal block of $A_2 - \lambda A_1$ and as such it has the structure similar to that of $A_2 - \lambda A_1$ itself. In particular one can easily see that

$$\det Z_i = \pm \mathcal{Z}_{n-i-1}(\lambda; \gamma_{i+1}, \dots, \gamma_{n-1}).$$

This gives us the Proposition. □

§4. Linear stability.

We allow neighboring particles to have equal masses so that many γ_i's may be equal to zero. The following property of the set of indices $\{i|\gamma_i \neq 0\}$ will play an important role.

DEFINITION. *A subset G of $\{1, \ldots, n-1\}$ is called resonant of order $k \geq 2$ if the greatest common divisor of the numbers in $G \cup \{n\}$ is equal to k and nonresonant otherwise, i.e., if the numbers in $G \cup \{n\}$ are relatively prime.*

THEOREM 1. *For any nonresonant subset G of $\{1, 2, \ldots, n-1\}$ there is an ϵ, $0 < \epsilon \leq 1$, such that if $\gamma_i = 0$, for $i \notin G$ and $-\epsilon < \gamma_i < 0$, for $i \in G$ then all roots of $\Upsilon(\lambda)$ are simple and lie on the unit circle.*

In particular the derivative P of the Poincaré section map F is strongly stable and hence our periodic orbit is linearly stable.

We begin with the computation of $\mathcal{Z}_i$ at $\gamma_1 = \gamma_2 = \cdots = \gamma_i = 0$.

PROPOSITION 4. *At $\gamma_1 = \gamma_2 = \cdots = \gamma_i = 0$*

$$\mathcal{Z}_i(\lambda) = \pm \frac{\lambda^{i+1} - 1}{\lambda - 1}. \tag{14}$$

PROOF: By straightforward expansion of the determinant of (13) with respect to the last row we obtain for any γ's

$$\begin{aligned} \mathcal{Z}_i(\lambda; \gamma_1, \ldots, \gamma_i) = &\pm (1+\lambda)\mathcal{Z}_{i-1}(\lambda, \gamma_1, \ldots, \gamma_{i-1}) \\ &+ \lambda(1-\gamma_{i-2})(1+\gamma_{i-1})\mathcal{Z}_{i-2}(\lambda; \gamma_1, \ldots, \gamma_{i-2}), \end{aligned}$$

where we take $+$ if i is odd and $-$ if i is even.

Hence if $\gamma_1 = \cdots = \gamma_{2k+1} = 0$ we have for $s = 1, \ldots, k$

$$\begin{pmatrix} \mathcal{Z}_{2s} \\ \mathcal{Z}_{2s+1} \end{pmatrix} = \begin{pmatrix} \lambda & -(1+\lambda) \\ \lambda(1+\lambda) & -1-\lambda-\lambda^2 \end{pmatrix} \begin{pmatrix} \mathcal{Z}_{2s-2} \\ \mathcal{Z}_{2s-1} \end{pmatrix}. \tag{15}$$

But

(16)

$$\begin{pmatrix} \lambda & -(1+\lambda) \\ \lambda(1+\lambda) & -1-\lambda-\lambda^2 \end{pmatrix} = \frac{1}{\lambda-1} \begin{pmatrix} 1 & 1 \\ 1 & \lambda \end{pmatrix} \begin{pmatrix} -1 & 0 \\ 0 & -\lambda^2 \end{pmatrix} \begin{pmatrix} \lambda & -1 \\ -1 & 1 \end{pmatrix}.$$

(15) and (16) yields

(17)

$$\begin{pmatrix} \mathcal{Z}_{2s} \\ \mathcal{Z}_{2s+1} \end{pmatrix} = \frac{1}{\lambda-1} \begin{pmatrix} 1 & 1 \\ 1 & \lambda \end{pmatrix} \begin{pmatrix} (-1)^s & 0 \\ 0 & (-1)^s\lambda^{2s} \end{pmatrix} \begin{pmatrix} \lambda & -1 \\ -1 & 1 \end{pmatrix} \begin{pmatrix} \mathcal{Z}_0 \\ \mathcal{Z}_1 \end{pmatrix}.$$

Since $\mathcal{Z}_0 \equiv 1$ and $\mathcal{Z}_1 = 1 + \lambda$ (17) gives us the formula (14) . □

PROOF OF THEOREM 1: By Proposition 2 for any γ_i's we have

$$\frac{\mathcal{Z}_i^2(\lambda)}{\lambda^i} = |\mathcal{Z}_i(\lambda)|^2, \quad \text{for } |\lambda| = 1. \tag{18}$$

So that by Proposition 3

(19)
$$\frac{\Upsilon(\lambda)}{\lambda^{n-1}} = |\mathcal{Z}_{n-1}(\lambda; \gamma_1, \ldots, \gamma_{n-1})|^2 + \sum_{i \in G} \sigma_{i+1}\gamma_i \prod_{j=1}^{i-1}(1+\gamma_j)|\mathcal{Z}_{n-i-1}(\lambda; \gamma_{i+1}, \ldots, \gamma_{n-1})|^2,$$
for $|\lambda| = 1$.

At $\gamma_1 = \cdots = \gamma_{n-1} = 0$ the formula (14) shows that if G is nonresonant then $\mathcal{Z}_{n-1}(\lambda)$ and $\mathcal{Z}_{n-i-1}(\lambda), i \in G$, do not have any common roots. This property will be preserved also for sufficiently small negative $\gamma_i, i \in G$, and then by (19) every double root of $|\mathcal{Z}_{n-1}(\lambda)|^2$ bifurcates into a pair of distinct roots of $\Upsilon(\lambda)$ on the unit circle. Clearly by making (if necessary) the γ_i's, $i \in G$, even smaller we can guarantee that $\Upsilon(\lambda)$ has $2(n-1)$ distinct roots on the unit circle. □

It was proven in [**W1**,proof of Proposition 2] (by a completely different method) that for nonnegative γ_i's and nonresonant $G = \{i | \gamma_i \neq 0\}$ the matrix P has no eigenvalues on the unit circle. In view of (19) it means that $\mathcal{Z}_{n-1}(\lambda)$ and $\mathcal{Z}_{n-i-1}(\lambda), i \in G$, do not have any common roots if only $0 < \gamma_i < 1, i \in G$, and G is nonresonant. The formula (19) by itself shows clearly that if only $\gamma_{n-1} > 0$ and $\gamma_i \geq 0, i = 1, \ldots, n-2$, then $\Upsilon(\lambda)$ has no roots on the unit circle.

The condition that G be nonresonant cannot be removed in view of the following

THEOREM 2. *If $G = \{i | \gamma_i \neq 0\}$ is resonant of order $k \geq 2$ then $\left(\frac{\lambda^k - 1}{\lambda - 1}\right)^2$ divides $\Upsilon(\lambda)$ and the matrix P has a 2×2 Jordan block associated with each of the $k-1$ double roots. In particular P is unstable even if all of its eigenvalues are on the unit circle.*

We first prove the following algebraic

Proposition 5. *If $G = \{i | \gamma_i \neq 0\}$ is resonant of order $k \geq 2$ then $\frac{\lambda^k - 1}{\lambda - 1}$ divides $\mathcal{Z}_{n-1}(\lambda)$.*

Proof: We use induction on the size of G. For empty G our Proposition follows from (14). Let $G = \{i_1, \ldots, i_l\}, l \geq 1$. Expanding the determinant of (13) with respect to i_1 row we obtain

$$\begin{aligned}\mathcal{Z}_{n-1}(\lambda; \gamma_1, \ldots, \gamma_{n-1}) = & \\ & \pm(\lambda + 1)\mathcal{Z}_{i_1-1}(\lambda; 0, \ldots, 0)\mathcal{Z}_{n-i_1-1}(\lambda; \gamma_{i_1+1}, \ldots, \gamma_{n-1}) \\ & - \lambda(1 - \gamma_{i_1})\mathcal{Z}_{i_1-2}(\lambda; 0, \ldots, 0)\mathcal{Z}_{n-i_1-1}(\lambda; \gamma_{i_1+1}, \ldots, \gamma_{n-1}) \\ & - \lambda(1 + \gamma_{i_1})\mathcal{Z}_{i_1-1}(\lambda; 0, \ldots, 0)\mathcal{Z}_{n-i_1-2}(\lambda; \gamma_{i_1+2}, \ldots, \gamma_{n-1}).\end{aligned}$$

Since k divides i_1 we have that $\frac{\lambda^k - 1}{\lambda - 1}$ divides $\mathcal{Z}_{i_1-1}(\lambda)$ by Proposition 4 and $\mathcal{Z}_{n-i_1-1}(\lambda)$ by the inductive assumption. Thus the inductive step is proven. □

Proof of Theorem 2: The first part of the Theorem follows from Proposition 3 and Proposition 5. To establish the presence of Jordan blocks we abandon the algebraic approach and go back to the original dynamical system.

Under our assumption of resonant $G = \{i_1, \ldots, i_l\}$ we have $l + 1$ groups of neighboring particles containing

$$k_1 = i_1, k_2 = i_2 - i_1, \ldots, k_{l+1} = n - i_l$$

particles respectively, with equal masses, $m_1, \ldots, m_{l+1}$. Clearly k is the greatest common divisor of $k_1, \ldots, k_{l+1}$.

Let the period of our periodic orbit be T. By the construction of our periodic orbit two colliding particles with equal masses have the same energies. Hence the positions and velocities of particles in one group are positions and velocities of one freely falling (and colliding with the floor) particle at different times. We exercise the option of allowing particles with equal masses in these groups to pass freely through each other. Mathematically it amounts to introducing a finite extension of our original dynamical system. The period of the periodic orbit in this extension is equal to NkT and the monodromy matrix is just P^{Nk}, where $N = \frac{k_1 \ldots k_{l+1}}{k^{l+1}}$.

The particles in the extension split naturally into k groups, in such a way that all collisions take place between particles in one group. Another description of the splitting is the following: two particles with equal masses belong to the same group if after time kT (or its multiple) one of them will assume the position and velocity of the other. Since the particles in different groups do not interact on our periodic orbit the dynamical system in its vicinity is the cartesian product of k copies of the dynamical system describing the motion of particles in one group. For example if all particles have equal masses (empty G) then the dynamics in the vicinity of the periodic orbit in the extension is the cartesian product of n copies of one particle system.

Let us consider the derivative of the flow after the period (at any point on the periodic orbit, original or extended) in the full phase space, i.e., without restricting the total energy of the system. This derivative has eigenvalue 1 (because of energy preservation) with multiplicity at least two (because of the hamiltonian character of the system). In view of the previously described scaling properties of our system the period of our orbit varies with the total energy level H, which immediately implies that this derivative has a 2×2 Jordan block corresponding to the eigenvalue 1.

Returning to the cartesian product of k copies we see that the derivative of the flow after the period has k 2×2 Jordan blocks with eigenvalue 1 (the total energy in each group is preserved). In linear approximation restricting the value of the total energy in the whole system and taking the Poincaré section amounts to restricting the derivative of the flow to a codimension one subspace and taking the quotient by the one dimensional subspace spanned by the velocity vector of our flow – the resulting linear operator being equal to P^{Nk}. In doing this we could reduce the number of Jordan blocks by at most two. It is not hard to see that actually P^{Nk} has $k-1$ 2×2 Jordan blocks with eigenvalue 1. Taking a power of a linear map can neither create nor destroy a Jordan block so that we obtain that P has $k-1$ 2×2 Jordan blocks with eigenvalues which are roots of unity of order Nk. We still have to show that they are actually roots of $\frac{\lambda^k-1}{\lambda-1}$.

Particles in one group form a non resonant system which is an N-to-1 covering of the system with all particles colliding. Thus if we decide that particles with equal masses do collide if only they belong to one and the same of the k groups then we obtain a system which is still the cartesian product so that the above argument applies and the derivative of the Poincare section map of the periodic orbit is equal to P^k. Now since P^k has $k-1$ 2×2 Jordan blocks with eigenvalue 1 then P must have $k-1$ Jordan blocks of the same size with eigenvalues which are roots of unity of order

k. As we mentioned before in the nonresonant case if $0 < \gamma_i < 1, i \in G$, P has no eigenvalues on the unit circle. Under the same assumption in the resonant case we have that P^k has eigenvalue 1 with multiplicity equal to exactly $2(k-1)$. Hence in this case P has $k-1$ Jordan blocks with eigenvalues which are roots of $\frac{\lambda^k-1}{\lambda-1}$. Finally by continuity that must be also the case in general. □

§5. Special cases.

Our fairly explicit formulas allow us to go beyond the case of small γ_i's treated in previous section. We will do it in two special and quite interesting cases

(1) $\gamma_1 = \cdots = \gamma_{n-2} = 0, \gamma_{n-1} = \gamma < 0,$
(2) $\gamma_1 = \gamma < 0, \gamma_2 = \cdots = \gamma_{n-1} = 0.$

THEOREM 3. *Both in case (1) and case (2) all roots of $\Upsilon(\lambda)$ are simple and lie on the unit circle. In particular the periodic orbit is linearly stable.*

It is quite plausible that also in general no requirements on the smallness of γ_i's are necessary, i.e., Theorem 1 holds with $\epsilon = 1$. Using the explicit criteria of stability in dimension 4, which one can find for example in [**H-M**], we checked that this is indeed the case for three particles.

PROOF: By straightforward expansion of the determinant of (13) with respect to the last(first) row we obtain

$$\mathcal{Z}_{n-1}(\lambda) = \begin{cases} \pm(1+\lambda)\mathcal{Z}_{n-2}(\lambda) + \lambda(1-\gamma)\mathcal{Z}_{n-3}(\lambda), & \text{in case (1)} \\ \pm(1+\lambda)\mathcal{Z}_{n-2}(\lambda) + \lambda(1+\gamma)\mathcal{Z}_{n-3}(\lambda), & \text{in case (2).} \end{cases}$$

Hence using Proposition 4 we get

$$\mathcal{Z}_{n-1}(\lambda) = \pm\left(\frac{\lambda^n-1}{\lambda-1} \pm \gamma\Big(\frac{\lambda^{n-1}-1}{\lambda-1} - 1\Big)\right), \tag{20}$$

where we take $+\gamma$ in case (1) and $-\gamma$ in case (2). Further by formula (19) we have for $|\lambda| = 1$

$$\frac{\Upsilon(\lambda)}{\lambda^{n-1}} = \begin{cases} |\mathcal{Z}_{n-1}(\lambda)|^2 + \gamma\sigma_n, & \text{in case (1)} \\ |\mathcal{Z}_{n-1}(\lambda)|^2 + \gamma\sigma_2|\mathcal{Z}_{n-2}(\lambda)|^2, & \text{in case (2).} \end{cases} \tag{21}$$

Our goal is achieved in the same way as in the proof of Theorem 1: by counting the changes of sign of $\frac{\Upsilon(\lambda)}{\lambda^{n-1}}$ on the unit circle using (21) . In case (2) this is especially simple. First of all by Proposition 2 and the formula (20) $\mathcal{Z}_{n-1}(\lambda)$ and

$$\mathcal{Z}_{n-2}(\lambda) = \pm\frac{\lambda^{n-1} - 1}{\lambda - 1}$$

have all simple roots on the unit circle and do not have any common roots if only $\gamma > -1$. In particular the relative arrangement of the roots does not change as γ changes from 0 to -1. We see that at $\gamma = 0$ the roots of $\mathcal{Z}_{n-1}(\lambda)$ and of $\lambda^{n-1} - 1$ intertwine. Hence they intertwine for all $-1 < \gamma < 0$ and they divide the unit circle into $2(n-1)$ arcs on which by (21) $\frac{\Upsilon(\lambda)}{\lambda^{n-1}}$ changes sign. Indeed the only doubtful case is that of an arc bounded by 1. In that case we need only to check that $\Upsilon(1) > 0$. But

$$\Upsilon(1) = \frac{(m_1 + (n-1)m_n)^2}{(m_1 + m_n)^2} - \frac{(n-1)^3 m_n(m_n - m_1)}{(m_1 + m_n)(m_1 + (n-1)m_n)},$$

which is clearly positive.

In case (1) the roots of

$$|\mathcal{Z}_{n-2}(\lambda)|^2 = \frac{|\lambda^n + \gamma\lambda^{n-1} - \gamma\lambda - 1|^2}{|\lambda - 1|^2}$$

divide the unit circle into $(n-1)$ arcs. In view of (21) it is sufficient to show that the local maximum of $|\mathcal{Z}_{n-2}(\lambda)|^2$ in any of these intervals exceeds $-\gamma\sigma_n$. Putting $\lambda = e^{i\varphi}$ we get by straightforward computations

$$|\mathcal{Z}_{n-2}(\lambda)|^2 = \frac{(1 - \cos(n-2)\varphi)\gamma^2 + 2(\cos\varphi - \cos(n-1)\varphi)\gamma + (1 - \cos n\varphi)}{1 - \cos\varphi}.$$

Since this quadratic function of γ is nonnegative it is most likely a complete square. Indeed after further transformations we get

$$\begin{aligned}
|\mathcal{Z}_{n-2}(\lambda)|^2 &= \frac{2}{1-\cos\varphi}\left(\gamma\sin(n-2)\frac{\varphi}{2} + \sin n\frac{\varphi}{2}\right)^2 \\
&= \frac{2}{1-\cos\varphi}\left((\gamma\cos\varphi + 1)\sin n\frac{\varphi}{2} - \gamma\sin\varphi\cos n\frac{\varphi}{2}\right)^2 \\
&= \frac{2(\gamma^2 + 2\gamma\cos\varphi + 1)}{1-\cos\varphi}\sin^2(n\frac{\varphi}{2} + \psi),
\end{aligned}$$

where ψ depends continuously on φ and γ and is between $-\frac{\pi}{2}$ and $\frac{\pi}{2}$. Hence clearly the local maxima of $|\mathcal{Z}_{n-2}(\lambda)|^2$ exceed the minimum of the function

$$\frac{2(\gamma^2 + 2\gamma\cos\varphi + 1)}{1 - \cos\varphi},$$

which is equal to $(1-\gamma)^2$. Finally it is easy to check that $(1-\gamma)^2 > -\gamma\sigma_n$. $\square$

References

[H-M] J.E. Howard and R.S. MacKay, *Linear stability of symplectic maps*, J. Math. Phys **28** (1987), 1036–1051.

[W1] M.P. Wojtkowski, *A system of one dimensional balls with gravity*, Comm. Math. Phys. **126** (1990), 507–533.

[W2] M.P.Wojtkowski, *The system of one dimensional balls in an external field II*, Comm. Math. Phys. **127** (1990), 425–432.

Department of Mathematics
University of Arizona
Tucson AZ 85721

The second author acknowledges partial support by the NSF Grant DMS–8807077 and the Sloan Foundation. Research support from the Arizona Center for Mathematical Sciences, sponsored by AFOSR Contract F49620–86–C0130 with the University Research Initiative Program at the University of Arizona, is gratefully acknowledged.

Birth and Death of Invariant Tori for Hamiltonian Systems

LUIGI CHIERCHIA

Abstract. The stability of invariant tori with (highly) irrational periods in the context of nearly integrable Hamiltonian systems and symplectic diffeomorphysms is considered. A theorem, based on computer-assisted implementations of recent KAM techniques, establishing the persistence of "golden-mean-tori" for "large" values of the nonlinearity parameter ϵ in two paradigmatic models is presented. Numerical investigations on the distribution of complex singularities in the parameter ϵ indicates that the method of proof, which involves the explicit construction of approximating surfaces, is optimal at least in the models considered here.

The problems I shall address here concern the destiny of certain compact invariant surfaces that arise naturally in the theory of perturbed Hamiltonian systems.

To simplify the exposition and to fix the ideas, let us take as (generalized) phase space $\mathbf{M}^{2d+1} \equiv \mathsf{R}^d \times \mathsf{T}^{d+1}$ ($\mathsf{T} \equiv \mathsf{R}/2\pi\mathsf{Z}$), endowed with the standard symplectic form $\sum_{i=1}^{d} dy_i \wedge dx_i$, $(y,x) \in \mathsf{R}^d \times \mathsf{T}^d$, and consider on $\mathbf{M}^{2d+1}$ the Hamiltonian

$$H = \frac{1}{2} y \cdot y + \epsilon V(x,t) \tag{1}$$

where dot denotes the standard inner product in R^d, V is a real-analytic function and ϵ a (possibly *complex*) nonlinearity-parameter.

The equations of motion read

$$\ddot{x} + \epsilon V_x(x,t) = 0 \qquad (y = \dot{x}) \tag{2}$$

and in the integrable case ($\epsilon = 0$) the dynamics is immediately described: $\mathbf{M}^{2d+1}$ is foliated by *invariant tori* $\mathcal{T}_y \equiv \{y\} \times \{(x,t) \in \mathsf{T}^{d+1}\}$ on which the flow, ϕ^s, is linear: $(y,x,t) \xrightarrow{\phi^s} (y, x + ys, t + s)$.

The basic content of the celebrated KAM theory ([**Ko**], [**A**], [**Mo1**], [**Mo2**], [**Bo**] for a review and [**G**] for an elementary exposition) is that many of the above invariant tori are preserved under perturbation provided ϵ is small enough.

Let me immediately mention here that the smallness requirement dictated by classical KAM techniques or by simple reorganization of them (cfr. [**G**] Problems 7 and 12, pag. 519, 520) are so stringent to be of little, if any, "practical" relevance. Furthermore, even from a purely theoretical point of view the gap between "observations" and KAM predictions might lead to the opinion that perturbation techniques are not adequate for a full explanation of non-integrable phenomena.

One of the tasks of this lecture will be to show how, at least in simple models, one can not only prescribe "realistic" smallness requirement but also shed some light on crucial and delicate questions regarding the "disappearance" of invariant tori when ϵ is increased.

Before going into more details, let me formulate a modern version of the KAM result for the case we are considering (cfr. [**Po**], [**C-G**], [**C-Z**], [**C-1989**]).

THEOREM 1. *For any real-analytic Hamiltonian* $H = \frac{1}{2}y\cdot y + \epsilon V(x,t)$ *on* $\mathbf{M}^{2d+1}$ *and for any integer* $p \geq 1$, *one can construct a triple* (S,Γ,h) *where (i)* $S : (\eta,\theta,t) \in \mathbf{M}^{2d+1} \to (y,x,t) \in \mathbf{M}^{2d+1}$ *is a* C^∞ *symplectic diffeomorphysm (i.e.* $dy\wedge dx = d\eta\wedge d\theta$*)* C^p*-close to the identity, (ii)* $\Gamma\subset \mathbf{R}^d$ *is a Cantor set of positive Lebesgue measure, (iii)* h *is a* C^∞ *function on* Γ C^p*-close to* $\frac{1}{2}\eta\cdot\eta$. *The triple* (S,Γ,h) *is related to* H *by the equation*

$$H\circ S(\eta,\theta,t) = h(\eta), \qquad \forall \eta\in\Gamma.$$

Furthermore:

$$\lim_{R\uparrow\infty} vol\,(\{|\,y\,|\geq R\}\setminus\Gamma) = \lim_{\epsilon\downarrow 0} vol\,(\{|\,y\,|\leq R\}\setminus\Gamma) = 0$$

and each "frequency" $\omega \equiv \frac{\partial h}{\partial\eta}(\eta), \eta\in\Gamma$ *satisfies the Diophantine inequalities*

$$|\,\omega\cdot n + m\,| \geq \frac{1}{\gamma\,|\,n\,|^\tau} \qquad \forall n\in\mathbf{Z}^d\setminus 0, \quad m\in\mathbf{Z}$$

with suitable positive constants γ *and* τ.

It follows that $S(\Gamma\times\mathbf{T}^{d+1})$ form a set of H-invariant tori $\mathcal{T}_\omega$ on which the flow, ϕ^s, is linear:

$$\phi^s\circ S(\eta,\theta,t) = S(\eta,\theta+\omega s, t+s), \quad (\omega = \frac{\partial h}{\partial\eta}(\eta), \eta\in\Gamma).$$

Now, informally speaking, these tori $\mathcal{T}_\omega$ depend continuously on ϵ (actually, they are analytic near $\epsilon = 0$, [**Mo2**]) and as $|\,\epsilon\,|$ is increased above a critical value they disintegrate.

In order to give a precise mathematical description of this phenomenon, I will restrict my attention to low-dimensional systems, namely to Hamiltonians on $\mathbf{M}^3 \equiv \mathbf{R} \times \mathbf{T}^2$ $(d = 1$ in (2)) and to area-preserving twist diffeomorphisms of $\mathbf{M}^2 \equiv \mathbf{R} \times \mathbf{T}$ of the form

$$(3) \qquad F : (y, x) \in \mathbf{M}^2 \to (y', x') = (y - \epsilon f_x(x), x + y')$$

where f is a real-analytic 2π-periodic function. The dynamics of such maps is described by the following discrete analogous of eqn. (2):

$$(4) \qquad x_{n+1} - x_n + x_{n-1} + \epsilon f_x(x_n) = 0$$

where (y_n, x_n) is the nth iterate of an initial point (y_0, x_0), $(y_n, x_n) \equiv F^n(y_0, x_0)$, and $y_n = x_n - x_{n-1}$. Famous examples are the Chirikov-Greene ([**Ch**], [**Gr**]) "standard map" ($f = \cos x$) and, for the Hamiltonian case, the "Escande-Doveil pendulum" ([**E-D**]) [$V = \cos x + \cos(x - t)$].

In 1982 J. Mather ([**Ma0**]) proved that if ϵ is *big* enough there do not exist *any* (homotopically non-trivial) invariant one-dimensional tori for the standard map and the same hold true for maps as in (3). On the other hand, by the so-called Aubry-Mather theory ([**A-LD**], [**Ma1**], [**Ma2**], [**Ma3**]) F admits for *any* irrational ω a closed, topologically transitive, invariant set $\mathcal{M}_\omega$, which is (possibly strictly) contained in the graph of a Lipschitz function and such that $F \restriction \mathcal{M}_\omega$ is (semi-) conjugated to a rotation by ω. Furthermore, $\mathcal{M}_\omega$ is either a (topological) circle or a Cantor set. In the first case $\mathcal{M}_\omega \equiv \mathcal{T}_\omega$ and is unique, while in the second case there may be (uncountably) many $\mathcal{M}_\omega$'s with the same rotation number ω (see [**Ma2**], [**C-1990**]).

If one replaces the standard map by the Escande-Doveil pendulum and $\mathcal{M}_\omega$ by $\mathcal{M}_\omega \cap \{t = 0\}$, what said for maps extends to one-dimensional, time-dependent Hamiltonian systems (see [**MK**] for non-existence in $\mathbf{M}^3 \cap \{| y | \leq 1\}$, and [**Mo3**], [**D**], [**Ba**] for the generalization of Aubry-Mather theory).

Thus, when ω is Diophantine and $| \epsilon |$ is small, $\mathcal{T}_\omega \equiv \mathcal{M}_\omega$ but as ϵ is increased $\mathcal{T}_\omega$ disintegrates and bifurcates into possibly many Cantor sets $\mathcal{M}_\omega$.

In this low-dimensional setting it is possible to give a proper definition of the critical value ϵ_ω immediately above which $\mathcal{T}_\omega$ breaks down:

$$\epsilon_\omega \equiv \sup\{\epsilon_0 : \mathcal{M}_\omega \text{is a graph of a continuous function } \forall \epsilon \in [0, \epsilon_0]\};$$

(in the Hamiltonian case, substitute $\mathcal{M}_\omega$ by $\mathcal{M}_\omega \cap \{t=0\}$). Notice that, in principle, it may happen that $\mathcal{M}_\omega$ is a Cantor set lying in an invariant circle ("Denjoy's counterexample"), but, in this case, such an invariant circle is not conjugated to a rotation and therefore is not one of the objects that we are considering here.

Two fundamental questions are:

Q 1: It is possible to give "realistic" estimates of ϵ_ω and to actually construct $\mathcal{T}_\omega(\epsilon)$ for $|\epsilon| \tilde{<} \epsilon_\omega$?

Q 2: How can one mathematically explain the bifurcation of smooth (or at least continuous) tori $\mathcal{T}_\omega$ into Cantor sets or, as I.Percival [**P**] would put it, into Cantori?

An answer to the first question, for the standard map [(3) with $f = \cos x$] and the Escande-Doveil pendulum [(1) with $V = \cos x + \cos(x-t)$], is contained in the following computer-assisted theorem (for description and discussions of computer-assisted techniques see [**La1**], [**La2**], [**E-W**], [**F-Ll**], [**C-C1**], [**C-C2**]).

THEOREM 2. *Let $\omega = (\sqrt{5}-1)\pi$ [respectively $\omega = (\sqrt{5}-1)/2$] and let ϵ be a complex number with $|\epsilon| \leq \rho \equiv 0.66$ [resp. $\rho \equiv 0.018$]. Then the standard map [resp. the Escande-Doveil pendulum] admits a unique invariant torus $\mathcal{T}_\omega$ with rotation number ω. More explicitely, $\mathcal{T}_\omega$ is given by*

$$\mathcal{T}_\omega \equiv \{(y,x) = \gamma(\theta) \equiv (\omega + u(\theta) - u(\theta-\omega), \theta + u(\theta)) \mid \theta \in \mathrm{T}\},$$

[*resp.*

$$\mathcal{T}_\omega \equiv \{(y,x,t) = \gamma(\theta,t) \equiv (\omega + \omega u_\theta + u_t, \theta + u, t) \mid (\theta,t) \in \mathrm{T}^2\}],$$

where $u(\theta,\epsilon)$ [resp. $u(\theta,t,\epsilon)$] is a real-analytic function 2π-periodic in θ [resp. (θ,t)] satisfying

$$\min |1 + u_\theta| > 0\ ,$$

the minimum being taken over $|\epsilon| \leq \rho$ and $\theta \in \mathrm{T}$ [resp. $(\theta,t) \in \mathrm{T}^2$]. Furthermore, one can construct a polinomial approximation v of u,

$$v \equiv \sum_{k=1}^{40} \sum_{\substack{n\in\mathbf{Z}\\ |n|\leq k}} \tilde{v}_{k,n}\epsilon^k \sin(n\theta)$$

[*resp.*

$$v \equiv \sum_{k=1}^{40} \sum_{\substack{n \in \mathbf{Z}^2 \\ |n| \leq k}} \tilde{v}_{k,n} \epsilon^k \sin(n_1\theta + n_2 t)]$$

such that

$$\sup_{|\epsilon| \leq \rho} \| u - v \| \leq 8.510^{-4}, \text{ [resp. } \quad 9.310^{-4}]$$

where $\| \cdot \|$ *denotes the* $C^1(\mathrm{T})$ [*resp.* $C^1(\mathrm{T}^2)$] *norm and the* $\tilde{v}_{k,n}$ *are rational numbers with a finite binary expansion.*

This theorem is proved in [**C-C2**] and is a strengthened version of Theorem 1 and 2 of [**C-C1**] (see also [**C-C3**] for a short description of the main ideas involved in the proof). The above value of ρ for which existence is established should be compared with numerical experiments, which predict the disintegration of $\mathcal{T}_\omega$ when ϵ reaches the value of 0.971 [resp. 0.0276] (cfr. [**Gr**], [**G-P**], [**E**], [**Fa**]).

Naturally, Q 2 is much more delicate and even though the "renormalization theory" of [**E-D**] and [**MK**] (see also [**E**] and [**R**]) provides an extremely interesting context of ideas, a mathematically rigorous explanation is still missing. Here is however a new indication ([**B-C**]).

Imagine that one fixes ("randomly") $\theta_0 \in \mathrm{T}$ and looks at the ϵ-expansion of the function u of Theorem 2 (in the case of the standard map). This is a honest power-series with its singularities and its radius of convergence, $r_\omega(\theta_0)$. Obviously $\epsilon_\omega \geq \inf_{\theta_0 \in \mathrm{T}} r_\omega(\theta_0)$. Now, let

$$R_N \equiv \frac{P_N(\epsilon;\theta_0)}{Q_N(\epsilon;\theta_0)}$$

be the *Padè diagonal approximant* of order N. That is, P_N and Q_N are ϵ-polynomials of degree N such that the first $2N$ terms of the Taylor expansion at 0 of u and R_N coincide (see [**Bak**]). Then one expects (at least in "non-degenerate situations") that as $N \to \infty$ the zeroes of Q_N tend to the singularities of u.

The figure below has been obtained in [**B-C**]; the +'s indicate the (numerically founded) zeroes of $\epsilon \to Q_{90}(\epsilon;1)$ plotted in the complex ϵ-plane (the thick circle has radius 0.971 $\sim$ believed break-down threshold; different θ_0's yield analogous pictures) .

What the figure strongly suggests is that $\epsilon \to u(\epsilon;\theta)$ has (for "typical" values of θ) a *natural boundary of singularities* formed by a *disk* whose radius coincide with the break-down value ϵ_ω.

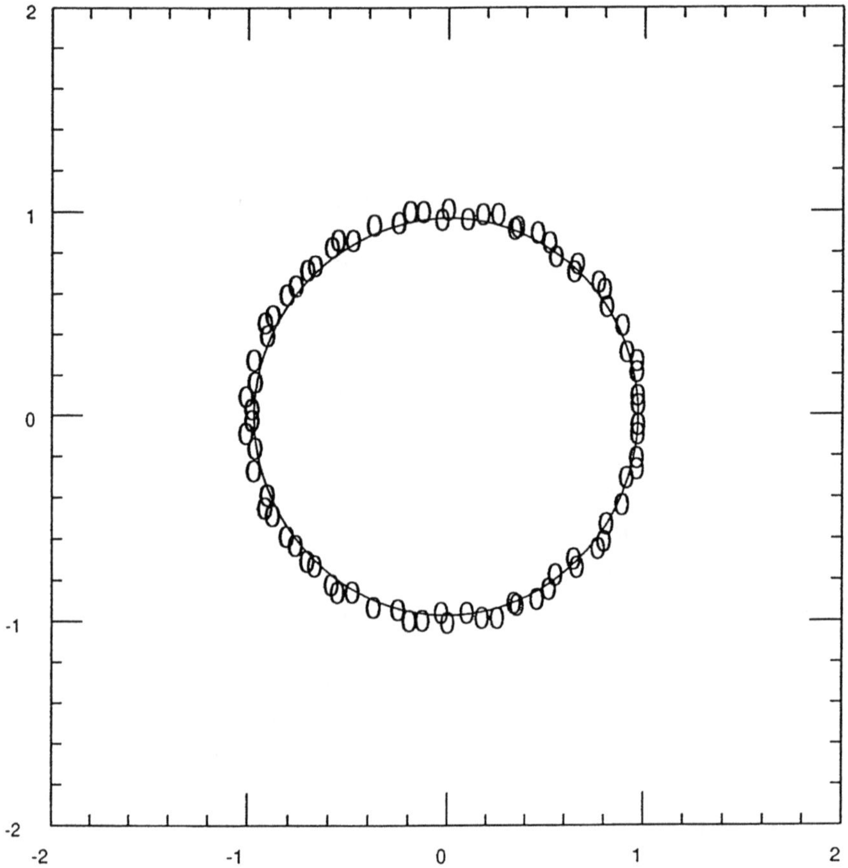
2
1
0
-1
-2
-2
-1
0
1
2

REFERENCES

[**A-LD**] Aubry S., Le Daeron P.Y., *The Discrete Frenkel-Kantorova Model and its Extensions I*, Physica D **8** (1983), p. 381.

[**A**] Arnold V.I, *Proof of a theorem of A.N. Kolmogorov on the invariance of quasiperiodic motions under small perturbations of the Hamiltonian*, Russ. Math. Surv. **18** (1963), p. 9.

[**Bak**] Baker G., "Essentials of Padé Approximants," Academic Press.

[**Ba**] Bangert, V., *Mather sets for twist mappings and geodesics on tori*, Dynamics Reported **1** (1987), Wiley and Teubner.

[**B-C**] Berretti A., Chierchia L., *On the complex analytic structure of the golden invariant curve for the standard map*, Nonlinearity **3** (1990), p. 39.

[**Bo**] Bost J.-B., *Tores invariantes des systémes dynamiques Hamiltoniens*, Astérisque **133-134** (1986).

[**C-C1**] Celletti A., Chierchia L., *Construction of Analytic KAM Surfaces and Effective Stability Bounds*, Commun. Math. Phys. **118** (1988), p. 119.

[**C-C2**] Celletti A., Chierchia L., *A computer-assisted approach to small-divisors problems arising in Hamiltonian mechanics*, to appear in the proceedings of the conference "Computer aided proofs in analysis", March 1989, Cincinnati.

[**C-C3**] Celletti A., Chierchia L., *On rigorous stability results for low-dimensional KAM surfaces*, Physics Lett. A **128** (1988), p. 166.

[**C-1989**] Chierchia L., *A direct method for constructing solutions of the Hamilton-Jacobi equation*, Preprint 1989, to appear in Meccanica.

[**C-1990**] Chierchia L., *On a result by J.N. Mather concerning invariant sets for area-preserving twist maps*, Preprint 1990, to appear in ZAMP.

[**C-G**] Chierchia L., Gallavotti G., *Smooth prime integrals for quasi-integrable Hamiltonian systems*, Il Nuovo Cimento **67 B** (1982), p. 277.

[**C-Z**] Chierchia L., Zehnder E., *Asymptotic expansion for quasi-periodic motions*, Ann. Scuola Norm. Sup. Pisa **16**, n. 2 (1989), p. 245.

[**Ch**] Chirikov B.V., *A universal instability of many dimensional oscillator systems*, Physics Reports **52** (1979), p. 263.

[**D**] Denzler J., *Mather sets for plane Hamiltonian systems*, J. Appl. Math. Physics (ZAMP) **38** (1987), p. 791.

[**E-W**] Eckmann J.-P., Wittwer P., *Computer methods and Borel summability applied to Feigenbaum's equation*, Lecture Notes Physics, **227** (1985), Springer, Berlin, Heidelberg, New York.

[**E**] Escande D.F., *Stochasticity in Classical Hamiltonian Systems: Universal Aspects*, Phys. Rep. **121** (1985), p. 165.

[**E-D**] Escande D.F., Doveil F., *Renormalization method for computing the thresholds of the large-scale instability in two degrees of freedom Hamiltonian systems*, J. Stat. Physics **26** (1981), p. 257.

[**Fa**] Falcolini C. Private communication.

[**F-Ll**] Fefferman C., de la Llave R., *Relativistic stability of matter I*, Rev. Mat. Iber. (2) (1986), p. 119.

[**G**] Gallavotti G., "The Elements of Mechanics," Springer-Verlag, New York, 1983.

[**Gr**] Greene J.M., *A Method for Determining a Stochastic Transition*, J. Math. Phys. **20** (1979), p. 1183.

[**G-P**] Greene J.M, Percival I.C., Physica D **3** (1981), p. 530.

[**Ko**] Kolmogorov A.N, *On the conservation of conditionally periodic motions under small pewrturbation of the Hamiltonian*, Dokl. Akad. Nauk SSR **98** (1954), p. 469.

[**La1**] Lanford O.E., *A computer-assisted proof of the Feigenbaum conjectures*, Bull. Amer. Math. Soc., New Series **6** (1982), p. 427.

[La2] Lanford O.E., *Computer-assisted proofs in analysis*, Phys. A **124** (1984), p. 465.

[MK] MacKay R.S., *A criterion for non-existence of invariant tori for Hamiltonian systems*, in "Nonlinear Dynamics," G. Turchetti Edt., World Scientific, 1989.

[Ma0] Mather J.N., *Non-existence of invariant circles*, Ergod. Th. Dyn. Syst. **4** (1984), p. 301.

[Ma1] Mather J.N., *Existence of Quasi-periodic Orbits for Twist Homeomorphisms of the Annulus*, Topology **21** (1982), p. 457.

[Ma2] Mather J.N., *Non-uniqueness of solutions of Percival's Euler-Lagrange equation*, Commun. Math. Physics **86** (1982), p. 465.

[Ma3] Mather J.N., *More Denjoy Minimal Sets for Area Preserving Diffeomorphisms*, Comment. Math. Helv. **60** (1985), p. 508.

[Mo1] Moser J., *On invariant curves for area-preserving mappings of the annulus*, Nach. Akad. Wiss. Göttingen, Math. Phys. K1. II **1** (1962).

[Mo2] Moser J., *Convergent series expansions for quasi-periodic motions*, Math. Ann. **169** (1967), p. 136.

[Mo3] Moser J., *Minimal solutions of variational problems on a torus*, Ann. Inst. Henri Poincaré **3** (1986), p. 229.

[P] Percival I.C., *Variational Principle for Invariant Tori and Cantori*, in "Non-linear Dynamics and the Beam-Beam Interaction," AIP Conference Proceedings **57**, eds. M. Month, J.C. Herrera, 1980, p. 302.

[Po] Pöschel J., *Integrability of Hamiltonian systems on Cantor sets*, Comm. Pure Appl. Math. **35** (1982), p. 220.

[R] Rand D.A., *Universality and Renormalization in Dynamical Systems*, to appear in *New Directions in Dynamical Systems*, eds. T. Bedford, J.W. Swift.

Permanent address: Dipartimento di Matematica, II Università di Roma "Tor Vergata", Via del Fontanile di Carcaricola 00133 Roma, Italy (CHIERCHIA@VAXTVM.INFN.IT).

Text of a lecture delivered at the *"Workshop on The Geometry of Hamiltonian Systems"* (M.S.R.I., Berkeley, June 1989).

Dynamical Aspects of the Bidiagonal Singular Value Decomposition

PERCY DEIFT, JAMES DEMMEL, LUEN-CHAU LI, CARLOS TOMEI

Abstract. In this paper we describe some striking stability properties of the singular value decomposition (SVD) of a bidiagonal matrix and of the Hamiltonian flow which interpolates the standard SVD algorithm at integer times.

The singular value decomposition (SVD) of a real $n \times n$ matrix A is the factorization $A = U\Sigma V^T$, where U and V are orthogonal matrices and $\Sigma = \text{diag}\{\sigma_1, \dots, \sigma_n\}$, and $\sigma_1 \geq \sigma_2 \geq \cdots \geq \sigma_n \geq 0$. The σ_i's are the *singular values* of A, the columns v_i of V the *right singular vectors* of A, and the columns u_i of U the *left singular vectors* of A. In this paper we will consider the SVD of a bidiagonal matrix B,

$$B = \begin{pmatrix} a_1 & b_1 & & \\ & \ddots & \ddots & O \\ & & & b_{n-1} \\ & O & & a_n \end{pmatrix}, \tag{1}$$

where we may assume without loss of generality that the a_i and b_i are positive. Such matrices arise as the final stage in the computation of the SVD of a general matrix A [**GK65**], [**GVL 83**], as well as in computing the eigendecomposition of a symmetric positive-definite tridiagonal matrix T [**BD 88**]. Both computations arise frequently in a wide variety of applications. It turns out that the SVD of a bidiagonal matrix B can be computed far more accurately than the SVD of a general matrix A, and the goal of this paper is to explain certain aspects of this phenomenon with the aid of a Hamiltonian differential equation underlying the SVD algorithm for B. We present no proofs: the reader is referred to [**DDLT89**] for details.

How accurately can the SVD of a general matrix A be computed? The standard approach is to bound the uncertainty δA in the initial data by its two-norm $\|\delta A\|$: we say that δA is an *absolute error of scale η in A* if $\|\delta A\|/\|A\| \leq \eta$. Regular perturbation theory then shows that if σ_i are the singular values of A and σ_i' are the singular values of $A + \delta A$, then

$$|\sigma_i - \sigma_i'| \leq \eta\|A\| \ . \tag{2}$$

This bound is useful for large singular values σ_j, $\sigma_j \sim \sigma_1 = \|A\|$, but for tiny singular values $\sigma_j << \sigma_1$, the bound may greatly exceed their value.

Now return to the case of a bidiagonal matrix B and introduce the *relative error of* δB

$$\eta_r \equiv (2n-1)\max_{i,j} \left|\log\left(\frac{B_{ij}+\delta B_{ij}}{B_{ij}}\right)\right| \tag{3}$$

instead of $\|\delta B\|$: when $\eta_r << 1$, this means that $\sum_{i,j} |\delta B_{ij}/B_{ij}|$ is approximately bounded by η_r. Using η_r as the measure of uncertainty in B, a perturbation argument now shows that

$$e^{-\eta_r} \leq \frac{\sigma_i'}{\sigma_i} \leq e^{\eta_r} \tag{4}$$

where σ_i, σ_i' are the singular values of B and $B + \delta B$ respectively. When $\eta_r << 1$, this implies

$$|\sigma_i' - \sigma_i| \leq \sigma_i \eta_r \ . \tag{5}$$

Thus small relative perturbations in the entries of B only cause small relative perturbations in the σ_i, independent of their magnitudes.

One can easily see that (4) is always at least about as strong as its counterpart (2). To show how much stronger it can be, consider making relative perturbations of size 10^{-10} in the entries of a 3×3 bidiagonal matrix with singular values $\sigma_1 = 1$, $\sigma_2 = 2 \cdot 10^{-20}$ and $\sigma_3 = 10^{-20}$. Since the norm of the perturbation $\|\delta A\|$ is about 10^{-10}, we may apply (2) to get the absolute error bound $10^{-10} >> \sigma_3$ for σ_3'. On the other hand, applying (4) we get a relative error bound $|(\sigma_3' - \sigma_3)/\sigma_3| \leq 5 \cdot 10^{-10}$. Thus we have at least 9 decimal digits of accuracy in σ_3', whereas (2) predicts changes 10^{20} times as large. Using η_r in place of η, there are similar improvements in the error bounds for singular vectors (see [**DDLT89**]).

Given the much greater accuracy to which singular values and singular vectors of bidiagonal matrices are determined by the data, it is desirable to have an algorithm which computes these quantities to their inherent accuracy. Such an algorithm was introduced in [**DK88**]: the algorithm is a hybrid of the standard, shifted SVD algorithm in [**GK65**], [**BDMS79**], and a new, stable implementation of SVD with a zero shift. The analysis in [**DK88**] and [**DDLT89**] shows that the hybrid algorithm is indeed optimally accurate in the desired sense, viz., suppose each entry of the original bidiagonal matrix B contains a small but unknown relative error (which could come from previous computation, or simply rounding the exact entry to fit in the machine). Then the singular values and singular vectors of B will have a certain inherent uncertainty due to their initial errors. Rounding and other errors introduced by the SVD algorithm contribute an extra uncertainty *but the analysis shows* that the contribution is of approximately the same size as this inherent uncertainty; thus the singular values and singular vectors are computed as accurately as the data permits. This accuracy can be much greater than that attained by the standard algorithm, and usually with no sacrifice in speed (see [**DK88**] and [**DDLT89**]).

It turns out that the striking stability properties of hybrid-SVD as outlined above, and in particular the properties of zero-shift SVD, rest heavily on the existence of an underlying Hamiltonian flow which interpolates the algorithm at integer times: as noted earlier our goal is to explain certain aspects of this phenomenon, and for this purpose we will focus our attention on four facts (see below) that were obtained during initial numerical experiments with the algorithm.

To proceed, we need to introduce some notation. QR iteration with a zero shift applied to a general invertible matrix A_0 produces a sequence of orthogonally similar matrices A_i as follows. Given A_i, compute its QR decomposition $A_i = QR$, where Q is orthogonal and R is upper triangular with positive diagonal. Then, $A_{i+1} = RQ = Q^T A_i Q$. It is well known that if A_0 has eigenvalues with distinct moduli, then A_i converges to a triangular matrix with the eigenvalues on the diagonal as $i \to \infty$. This algorithm may be applied to the bidiagonal singular value problem as follows [**GVL83**]. Let B_0 be our initial bidiagonal matrix. Given B_i, compute the QR decompositions $B_i B_i^T = Q_1 R_1$ and $B_i^T B_i = Q_2 R_2$. Then let $B_{i+1} = Q_1^T B_i Q_2$. Then B_i is bidiagonal for all i and converges as $i \to \infty$ to a diagonal matrix with the singular values on the diagonal. Observe that $B_{i+1} B_{i+1}^T = R_1 Q_1$ and $B_{i+1}^T B_{i+1} = R_2 Q_2$, so that the above *zero-shift* SVD *algorithm* implicitly applies the usual QR iteration to $B_i B_i^T$ and $B_i^T B_i$ simultaneously.

We think of the SVD algorithm as a mapping from $\mathbb{R}^{2n-1}$ (the entries of B_i) to $\mathbb{R}^{2n-1}$ (the entries of B_{i+1}). To understand how errors propagate through iterations of the SVD algorithm, it is natural to look at the Jacobian of this map, since the Jacobian describes how small perturbations in B_i affect B_{i+1}. However, since we are interested in the propagation of *relative* errors, we will look at a Jacobian which maps small relative perturbations in B_i to relative perturbations in B_{i+1}. To this end, we will work with the logarithms of the entries B_i and B_{i+1}, since small perturbations in the logarithms of the matrix entries are equivalent to small relative perturbations in the matrix entries themselves. Thus, we will think of a bidiagonal B as a point in $\mathbb{R}^{2n-1}$ through the identification (recall that the nontrivial entries of B are positive)

$$\begin{pmatrix} a_1 & b_1 & & \\ & \ddots & \ddots & \\ & & \ddots & b_{n-1} \\ & & & a_n \end{pmatrix} \Leftrightarrow \begin{pmatrix} \log b_1 \\ \vdots \\ \log b_{n-1} \\ \log a_1 \\ \vdots \\ \log a_n \end{pmatrix}$$

and think of one step of the SVD algorithm as a map F which takes vectors of logarithms of entries of B_i to vectors of logarithms of entries of B_{i+1}, $i = 0, 1, 2, \ldots$. Thus for $j > i$

$$F^{(j-i)}(B_i) = \underbrace{F \circ \cdots \circ F}_{j-i \text{ times}}(B_i) = B_j \; .$$

We will call its Jacobian $M(j,i)$, which by the chain rule is the product of the one step Jacobians $M(j,i) = M(j,j-1) \cdots M(i+1,i)$. It is $M(j,i)$ which describes how initial relative errors in B and roundoff errors committed during prior SVD iterations propagate during later SVD iterations.

The following four facts (see above) were observed during initial numerical experiments:

Fact 1: The eigenvalues of $M(j,i)$ appear in reciprocal pairs. In other words, if λ is an eigenvalue, so is $1/\lambda$.

Fact 2: Near convergence (i.e. for i large enough), the eigenvalues of $M(i+1,i)$ are simple, approach 1 and all lie on the unit circle.

Fact 3: As $i \to \infty$, $M(i+1,i)$ converges to the constant matrix

$$(3) \quad M_\infty = \begin{pmatrix} I_{n-1} & \Gamma_n \\ 0 & I_n \end{pmatrix} , \text{ where } \Gamma_n = \begin{pmatrix} -2 & 2 & & \\ & -2 & 2 & O \\ & \ddots & \ddots & \\ O & & -2 & 2 \end{pmatrix} .$$

Fact 4: $\|M(j,i)\|$ grows linearly in the number of SVD steps $j-i$. More precisely, we observed numerically for a large class of problems that $\|M(j,i)\|_\infty \leq 5.06n(j-i)$. This is the essential property of roundoff error propagation which lets us prove that the algorithm computes singular vectors to the desired accuracy. This linear growth is to be expected because near convergence we have

$$M(j,i) \approx M_\infty^{j-i} = \begin{pmatrix} I_{n-1} & (j-i)\Gamma_n \\ 0 & I_n \end{pmatrix} ,$$

which grows linearly in norm.

For a matrix A, let A_+, A_- denote its strictly upper, strictly lower triangular parts respectively, and set $\pi_0(A) = A_- - A_-^T$. To explain **Fact 1**, we begin with the following result.

THEOREM 4 (Interpolation at integer times). *Let $A(t)$ be the solution of*

$$\frac{dA}{dt} = A(\pi_0(\log A^T A)) - (\pi_0(\log AA^T))A , \tag{5}$$

where

$$A(t=0) = A_0 \textit{ is invertible.}$$

Then

$$A(k) = A_k , \quad k = 0, 1, 2, \dots . \tag{6}$$

where $A_0, A_1, A_2, \dots$ are the iterates of the zero shift SVD algorithm.

□

The above result is due essentially to Chu [**Chu 86**], and is modeled on related results, ([**Sym 80**], [**Sym 82**]) and [**DNT83**], for the symmetric eigenvalue problem. Note that if we set $T = T(A) = A^T A$, then (5) implies

$$\frac{dT}{dt} = [T, \pi_0(\log T)] , \tag{7}$$

which is the flow that interpolates the QR algorithm for symmetric matrices (see [**DNT 83**], [**DLNT 86**]).

For smooth functions $\phi : M_n(\mathbf{R}) \to \mathbf{R}$, set

$$D'\phi(A) \equiv (\nabla\phi(A))^T A \tag{8}$$

and

$$D\phi(A) \equiv A(\nabla\phi(A))^T \,, \tag{9}$$

where $\nabla\phi(A)$ is the matrix with entries $\dfrac{\partial\phi}{\partial A_{ij}}(A)$. The space $M_n(\mathbb{R})$ carries a natural (gl(n,$\mathbb{R}$)-ad) invariant pairing

$$(A, B) = \operatorname{tr} AB \,, \tag{10}$$

$$(A, [B, C]) = -([B, A], C) \,. \tag{11}$$

Introduce the linear map $R : M_n(\mathbb{R}) \to M_n(\mathbb{R})$,

$$R(A) \equiv A_+ - A_- \,. \tag{12}$$

Then we have the following result, due to Sklyanin (see [**Sem 84**]).

THEOREM 13. *For smooth functions ϕ, ψ,*

$$\{\phi, \psi\}_R(A) \equiv (R(D\phi(A)), D\psi(A)) - \big(R(D'\phi(A)) \,,\, D'\psi(A)\big) \tag{14}$$

defines a Poisson bracket on $M_n(\mathbb{R})$. □

REMARKS: The above bracket is an example of a so-called Sklyanin-Poisson structure. Also, the map R is an example of a *classical r-matrix* and solves the *modified Yang-Baxter equation*

$$[R(A), R(B)] - R([A, R(B)] + [R(A) \,, B]) = -[A, B] \tag{15}$$

(see [**Sem 84**]). Note that, as opposed to Lie-Poisson structures which are linear in A, the Sklyanin structure (14) is quadratic in A.

THEOREM 16. *The equation*

$$\frac{d}{dt}\phi(A(t)) = \{\phi, H_{SVD}\}_R(A(t)) \,, \quad A(t=0) = A_0 \,, \tag{17}$$

generated by the Hamiltonian $H_{SVD}(A) \equiv -\frac{1}{4}\operatorname{tr}(\log A^T A)^2$, is equivalent to (5). □

Assembling Theorem 4 and 16 we see that the SVD algorithm is the integer time evaluation of a Hamiltonian flow.

THEOREM 18. *The set B_Δ of bidiagonal matrices B with positive entries a_p, b_q and fixed determinant,*

$$\det B = \prod_{p=1}^{n} a_p = \Delta \ , \tag{19}$$

is a $(2n-2)$-dimensional symplectic leaf for the Sklyanin bracket $\{\cdot,\cdot\}_R$. Moreover $\mathcal{B}_\Delta$ has a global Darboux coordinate system given by

$$x_i \equiv \log b_i \ , \quad 1 \le i \le n-1 \ , \tag{20}$$

$$y_i \equiv \log \prod_{j=1}^{i} a_j \ , \quad 1 \le i \le n-1 \ , \tag{21}$$

$$\{x_i, x_j\}_R = 0 \ , \quad \{y_i, y_j\}_R = 0 \ , \quad \{x_i, y_j\}_R = \delta_{ij} \ . \tag{22}$$

□

Set $y_n \equiv \log \prod_{j=1}^{n} a_j$.

REMARK: The fact that $\mathcal{B}_\Delta$ is a natural phase space for the Hamiltonian version of SVD is in striking parallel to the fact that T_r, the tridiagonal matrices with nonzero off diagonal elements and prescribed trace r, provides a natural phase space for the Hamiltonian version of QR. The relevant Poisson structure for QR is given by the Lie-Poisson structure on the dual of the Lie algebra of the lower triangular group (see [**Kos79**], [**Adl79**]; see also [**DLNT86**]). However, the map $B \mapsto B^T B$ from $\mathcal{B}_\Delta$ to T_r is *not* symplectic and to understand the relationship between the SVD flow on B and the QR flow on $B^T B$ one must consider a so-called "second Poisson structure" on positive definite matrices. We will address this matter in a later publication.

We are now in a position to prove **Fact 1**. Thus, if $M(j,i)$ is the Jacobian of the iterated SVD map from B_i to B_j expressed in the variables $\log b_1, \ldots, \log b_{n-1}, \log a_1, \ldots, \log a_n$, we must show

$$\lambda \in \text{spec } M(j,i) \Leftrightarrow \lambda^{-1} \in \text{spec } M(j,i) \ .$$

For a bidiagonal matrix B set

$$\beta_i = \log b_i \ , \quad 1 \le i \le n-1 \ , \tag{23}$$

$$\alpha_i = \log a_i \ , \qquad 1 \le i \le n \ , \tag{24}$$

so that

$$\begin{pmatrix}\beta\\ \alpha\end{pmatrix} = \begin{pmatrix}1 & 0\\ 0 & N\end{pmatrix}\begin{pmatrix}x\\ y\end{pmatrix}, \tag{25}$$

where N is the $n \times n$ matrix

$$\begin{pmatrix}1 & & & O\\ -1 & 1 & & \\ & \ddots & \ddots & \\ O & & -1 & 1\end{pmatrix}.$$

Note

$$N^{-1} = \begin{pmatrix}1 & & & O\\ 1 & 1 & & \\ \vdots & & \ddots & \\ & & 1 & \\ 1 & 1 & 1 & 1\end{pmatrix}.$$

Let $\begin{pmatrix}\beta^i\\ \alpha^i\end{pmatrix} = \begin{pmatrix}1 & 0\\ 0 & N\end{pmatrix}\begin{pmatrix}x^i\\ y^i\end{pmatrix}$, $\begin{pmatrix}\beta^j\\ \alpha^j\end{pmatrix} = \begin{pmatrix}1 & 0\\ 0 & N\end{pmatrix}\begin{pmatrix}x^j\\ y^j\end{pmatrix}$ be the coordinates of the ith and j^{th} SVD iterates B_i and B_j, with $i < j$. Then

$$M(j,i) \equiv \frac{\partial(\beta^j,\alpha^j)}{\partial(\beta^i,\alpha^i)} = \begin{pmatrix}1 & 0\\ 0 & N\end{pmatrix}\frac{\partial(x^j,y^j)}{\partial(x^i,y^i)}\begin{pmatrix}1 & 0\\ 0 & N^{-1}\end{pmatrix}. \tag{26}$$

Under the iteration, $y^j_n = \log\ \det B^j = \log\ \det B^i = y^i_n$. Hence

$$\frac{\partial(x^j,y^j)}{\partial x^i,y^i)} = \begin{pmatrix} \frac{\partial x^j_1,\dots,x^j_{n-1},y^j_1,\dots,y^j_{n-1})}{\partial(x^i_1,\dots,x^i_{n-1},y^i_1,\dots,y^i_{n-1}} & \begin{matrix}\frac{\partial x^j_1}{\partial y^i_n}\\ \vdots\\ \frac{\partial y^j_{n-1}}{\partial y^i_n}\end{matrix}\\ 0\cdots 0 & 1\end{pmatrix}. \tag{27}$$

Now recall the following well known fact from Hamiltonian mechanics ([**Ar**]): let J denote the standard matrix $\begin{pmatrix}0 & I\\ -I & 0\end{pmatrix}$ and suppose that the curve $(x(t;x,y),y(t;x,y))$ is the solution of a Hamiltonian system of equations in $\mathbb{R}^{2n-2}$ in canonical form

(28)

$$\frac{d}{dt}\begin{pmatrix}x\\ y\end{pmatrix} = J\nabla H = \begin{pmatrix}0 & I\\ -I & 0\end{pmatrix}\begin{pmatrix}H_x\\ H_y\end{pmatrix}, \quad (x(0;x,y),y(0;x,y)) = (x,y),$$

for some Hamiltonian $H : \mathbb{R}^{2n-2} \to \mathbb{R}$. Then for any t, the Jacobian $D = \frac{\partial(x(t;x,y),y(t;x,y))}{\partial(x,y)}$ is symplectic,i.e. $D^T JD = J$. But $\det(D^T - \lambda) = \det(JD^{-1}J^{-1} - \lambda) = \det(D^{-1} - \lambda)$. Thus if D is symplectic,

$$\lambda \in \text{spec } D \Leftrightarrow \lambda^{-1} \in \text{spec } D \; . \tag{29}$$

Finally from Theorems 4, 16 and 18, $x_1^j, \dots, x_{n-1}^j, y_1^j, \dots, y_{n-1}^j$ is the time $t = j - i$ evaluation in canonical variables of a Hamiltonian flow with initial data $x_1^i, \dots, x_{n-1}^i$, $y_1^i, \dots, y_{n-1}^i$. Hence the matrix of partial derivatives $\frac{\partial(x_1^j,\dots,x_{n-1}^j,y_1^j,\dots,y_{n-1}^j)}{\partial(x_1^i,\dots,x_{n-1}^i,y_1^i,\dots,y_{n-1}^i)}$ is symplectic and (29) holds. **Fact 1** now follows from (26) and (27).

To summarize, **Fact 1** follows from the happy coincidence that relative errors correspond to logarithms of the entries of B, and thus are the Darboux coordinates for a symplectic leaf of the Sklyanin structure $\{\cdot,\cdot\}_R$ through which the Hamiltonian $H_{SVD}(A) = -\frac{1}{4}\text{tr}(\log A^T A)^2$ generates a flow which interpolates SVD at integer times.

To explain **Fact 4**, we take the gradient of the SVD flow

$$\frac{dx_i}{dt} = \frac{\partial H_{SVD}}{\partial y_i}(x_1, \dots, y_{n-1}; y_n) \; , \quad 1 \le i \le n-1 \; , \tag{30}$$

$$\frac{dy_i}{dt} = \frac{\partial H_{SVD}}{\partial x_i}(x_1, \dots, y_{n-1}; y_n) \; , \quad 1 \le i \le n-1 \; , \tag{31}$$

$$\frac{dy_n}{dt} = 0 \tag{32}$$

with respect to the initial data

$$x_i(0) = x_i \; , \quad y_i(0) = y_i \; , \quad 1 \le i \le n-1 \; , \quad y_n(0) = y_n(t) = y_n \; ,$$

to obtain the equation

$$\frac{dK_n}{dt} = \begin{pmatrix} J & \nabla_n^2 H_{SVD} \\ 0 & \cdots \quad 0 \end{pmatrix} K_n \; , \quad K_n(0) = I \tag{33}$$

for the full $(2n-1) \times (2n-1)$ Jacobian matrix

$$K_n(t) = \frac{\partial(x_1(t), \dots, x_{n-1}(t), y_1(t), \dots, y_n(t))}{\partial(x_1, \dots, x_{n-1}, y_1, \dots, y_n)},$$

where $\nabla_n^2 H_{SVD}$ is the Hessian matrix

$$\nabla_n^2 H_{SVD} = \left(\frac{\partial^2 H_{SVD}}{\partial z_i \partial z_j}\right)_{1\le i\le 2n-2,\, 1\le j\le 2n-1},$$

$$(z_1, \dots, z_{2n-1}) = (x_1, \dots, x_{n-1}, y_1, \dots, y_n).$$

Inserting the asymptotics

$$a_i(t) = \sigma_i + o(1) \tag{34}$$

$$b_i(t) \sim b_i^\infty (\sigma_{i+1}/\sigma_i)^{2t} \tag{35}$$

as $t \to \infty$, where $b_i^\infty > 0$, into (33) and evaluating the asymptotics of K_n using standard ODE techniques (see [**CL 55**]), we are led to the following result, Theorem 40 below.

For α_m, β_n as above, set

$$B_{21} \equiv \left(\frac{\partial \log \sigma_i}{\partial \beta_m}\right)_{1\le i\le n,\ 1\le m\le n-1} \tag{36}$$

and

$$B_{22} \equiv \left(\frac{\partial \log \sigma_i}{\partial \alpha_m}\right)_{1\le i\le n,\ 1\le m\le n}. \tag{37}$$

A straightforward computation shows that

$$\Gamma_n B_{21} = \left(\frac{\partial \log \sigma_{i+1}^2/\sigma_i^2}{\partial \beta_m}\right)_{1\le i\le n-1,\ 1\le m\le n-1} \tag{38}$$

and

$$\Gamma_n B_{22} = \left(\frac{\partial \log \sigma_{i+1}^2/\sigma_i^2}{\partial \alpha_m}\right)_{1\le i\le n-1,\ 1\le m\le n}. \tag{39}$$

THEOREM 40. *Under the bidiagonal SVD flow (30)–(32), the Jacobian matrix* $M(t,0) = \frac{\partial(\beta(t),\alpha(t))}{\partial(\beta,\alpha)}$ *satisfies*

$$M(t,0) = \begin{pmatrix} B_{11} + t\Gamma_n B_{21} & B_{12} + t\Gamma_n B_{22} \\ B_{21} & B_{22} \end{pmatrix} (1 + o(1)) \tag{40}$$

where the term $o(1)$ *is exponentially decreasing as* $t \to \infty$, *the matrices* B_{ij} *are constant, and* B_{21}, B_{22}, $\Gamma_n B_{21}$, $\Gamma_n B_{22}$ *satisfy (36)-(39) respectively.*

Also

$$\|M(t,0)\|_\infty \le (8n-4)t + O(1) \tag{41}$$

as $t \to \infty$. □

Fact 4 now follows easily from (the proof of) Theorem 40. Note that the constant in (41) is in good agreement with the experimentally obtained bound $\|M(j,i)\|_\infty \le 5.06n(j-i)$ noted above.

The asymptotics (34) and (35) are the analogs for the SVD flow of the well known asymptotics (see [**Mos 75**]) of the Toda flow

$$\frac{dT}{dt} = [T, \pi_0(T)] \tag{42}$$

(cf. (7)) on tridiagonal matrices $T = \begin{pmatrix} c_1 & d_1 & & O \\ d_1 & & \ddots & \\ & \ddots & \ddots & d_{n+1} \\ O & & d_{n-1} & c_n \end{pmatrix}$, for which

$$c_i(t) \to \lambda_i \ , \qquad 1 \le i \le n \ , \tag{43}$$

$$d_i(t) \sim d_i^\infty \, e^{-(\lambda_i - \lambda_{i+1})t} \ , \quad d_i^\infty > 0 \ , \quad 1 \le i \le n-1 \ , \tag{44}$$

as $t \to \infty$, where $\lambda_1 > \cdots > \lambda_n$ denotes the spectrum of T. Thus (34) and (35) correspond to setting $\lambda_i \to \sigma_i^2$ and $e^{-(\lambda_i - \lambda_{i+1})t} \to e^{-(\log \lambda_i - \log \lambda_{i+1})t} = (\sigma_{i+1}/\sigma_i)^{2t}$ in (43) and (44). Under Flaschka's change of variables (see e.g. [**Mos 75**])

$$y_i = -2c_i \ , \qquad 1 \le i \le n \ , \tag{45}$$

$$e^{(x_i - x_{i+1})} = 4d_i^2 \ , \quad 1 \le i \le n-1 \ , \tag{46}$$

equation (42) becomes a Hamiltonian flow for n particles on the line with Hamiltonian

$$\frac{1}{2}\sum_{k=1}^{n} y_k^2 + \sum_{k=1}^{n-1} e^{(x_k - x_{k+1})}$$

in the standard Poisson structure $(\sum_{i=1}^{n} dx_i \wedge dy_i,\ \mathbb{R}^{2n})$. The asymptotics (43), (44) now show that, as time goes on, the particles $x_1, \dots, x_n$ scatter

$$x_i - x_{i+1} \to -\infty$$

at a linear rate, and eventually move at fixed velocities

$$y_i \to -2\lambda_i \ .$$

Thus the asymptotics (34) and (35) describe the scattering of an underlying system of n particles on the line, *and the content of Theorem 40* (see in particular formula (40)), *and hence* **Fact 4**, *is that these asymptotics can be differentiated with respect to the initial data.*

The explanation of **Fact 2** and en route, **Fact 3**, follows from a scaling analysis of the Jacobian K of the SVD map $B \to B'$.

$$K = \frac{\partial(x_1', \dots, x_{n-1}', y_1', \dots, y_{n-1}')}{\partial(x_1, \dots, x_{n-1}, y_1, \dots, y_{n-1})}$$

(recall (26)), and Krein's theory of strongly stable Hamiltonian systems ([**Kre50**], [**Kre55**]).

One starts by showing that for $b^2 \equiv \sum_{i=1}^{n-1} b_i^2$ small, K can be written in the form

$$K = I + \begin{pmatrix} E_1 \delta^2 & E_2 + O_0(b^2) \\ \delta(I + \delta E_3 \delta)\delta & \delta^2 E_4 \end{pmatrix} , \tag{47}$$

where

$$E_2 = \begin{pmatrix} -4 & 2 & & & & \\ 2 & -4 & 2 & & O & \\ & 2 & \ddots & \ddots & & \\ & & \ddots & & & \\ O & & & & 2 & \\ & & & 2 & -4 \end{pmatrix} , \quad O_0(b^2) \text{ is lower Hessenberg,}^{(1)} \tag{48}$$

(49) E_1, E_4 are lower Hessenberg with entries $(E_1)_{i,i+1}$, $(E_4)_{i,i+1}$ of the form (positive constant $+\,O(b^2)$) ,

(1) A matrix (A_{ij}) is lower Hessenberg if $A_{ij} = 0$ for $j > i + 1$.

(50) E_3 is strictly lower triangular,

and the *scaling matrix* δ has the form

(51) $$\delta = \operatorname{diag}(p_1 b_1, \dots, p_{n-1} b_{n-1}) ,$$

with

(52) $$p_i = \sqrt{\frac{1}{\sigma_i^2} + \frac{\sigma_{i+1}^2}{\sigma_i^4} + O(b^2)} , \quad 1 \le i \le n-1 .$$

Letting $t = k \to \infty$, and hence $b_j(k) \to 0$ by (35), the proof of **Fact 3** is immediate: indeed

$$K \to K_\infty = \begin{pmatrix} I & E_2 \\ 0 & I \end{pmatrix} ,$$

and the result follows from (26).

The proof of **Fact 2** is in 3 steps:

Step 1. One shows that the eigenvalues λ_j, $1 \le j \le 2n-2$, of $K = K(t, t+1)$ eventually lie in Gershgorin-type disks contained in a fixed cone with vertex $\lambda = 1$, symmetric about $1 + i\mathbf{R}$, and with aperture $2\theta < \pi$. More precisely, if $\lambda_j = 1 + \nu - j$, then eventually ν_j must lie in one of the $2(n-1)$ Gershgorin disks $D_j^{\pm}$, $1 \le j \le n-1$,

(53) $$\{z : |4 + (z/p_j b_j)^2| < \rho_n\} ,$$

where $\rho_n = \frac{1}{2}(4\cos\frac{\pi}{n} + 4) < 4$. In particular $\lambda_j \to 1$ as $t \to \infty$.

Step 2. Using (47), Step 1 and the lower Hessenberg properties of the E_j's, one shows that the eigenvalues of K are eventually geometrically simple.

Step 3. Again using (47) and Step 1, one shows that

(54) $$\operatorname{sgn}(\bar{h}, iJh) = \operatorname{sgn}(\operatorname{Im} \lambda) \quad \text{for } \lambda \notin \mathbf{R} ,$$

where $h \ne 0$ is the eigenvector corresponding to λ, $Kh = \lambda h$, and $(\cdot,\cdot)$ denotes the real Euclidean inner product in $\mathbf{R}^{2n-2}$.

Fact 2 now follows using observations essentially due to Krein. Suppose $\lambda \ne \bar{\lambda}^{-1}$ is an eigenvalue of K, $Kh = \lambda h$. By Step 1, $\lambda \notin \mathbf{R}$ and hence by Step 3, $(\bar{h}, iJh) \ne 0$. But as K is symplectic,

$$(\bar{h}, iJh) = (K\bar{h}, iJKh) = \bar{\lambda}\lambda(\bar{h}, iJh) ,$$

which implies $(\bar{h}, iJh) = 0$, a contradiction. Thus $\lambda = \bar{\lambda}^{-1}$ and spec $K(t, t+1) \subset \{z : |z| = 1\}$, for t large. Finally, suppose λ corresponds to a nontrivial Jordan block,

$$K = U \begin{pmatrix} \lambda & 1 & & \\ 0 & \lambda & & \\ \vdots & \ddots & \ddots & \\ & & & \end{pmatrix} U^{-1} ,$$

$$K^T = U^{-T} \begin{pmatrix} \lambda & 0 & \cdots & \\ 1 & \lambda & & \\ & \ddots & \ddots & \\ & & & \end{pmatrix} U^T ,$$

for some invertible matrix U. Set $h = Ue_1 \neq 0$; then $Kh = \lambda h$. But $K^T J\bar{h} = JK^{-1}\bar{h} = \bar{\lambda}^{-1} J\bar{h}$, so that $J\bar{h} = \sum_{j\neq 1} c_j U^{-T} e_j$, for suitable constants c_j, where $j = 1$ is excluded as $U^{-T}e_1$ is *not* an eigenvector of K^T. Thus

$$(\bar{h}, iJh) = -(h, iJ\bar{h}) = -\sum_{j\neq 1} c_j (Ue_1, U^{-T}e_j) = 0 ,$$

contradicting Step 3. Thus K is diagonalizable and λ has algebraic multiplicity 1, by Step 2. This completes the proof of **Fact 2.**

REMARK: Some critical ideas in the above proof of **Fact 2** were suggested to the authors by Gene Wayne (see also [**dlLW89**]).

In conclusion:

In describing the application of Hamiltonian mechanics to the problem of the computation of the singular values of a bidiagonal matrix, the authors hope to have drawn the attention of researches in dynamics and Hamiltonian mechanics to a wide variety of problems in numerical analysis, in which the ideas of ode's play a central role. The interested reader should consult [**DDLT 89**] and the references therein; for other directions, see [**BS89**] and [**B89**].

REFERENCES

[Adl79] M. Adler, *On a trace functional for formal pseudodifferential operators and the symplectic structure of the Korteweg-de Vries type equations*, Inv. Math. **50** (1979), 219–248.

[BD88] J. Barlow and J. Demmel, *Computing Accurate Eigensystems of Scaled Diagonally Dominant Matrices*, Computer Science Dept. Technical Report 421, Courant Institute, New York, NY, December 1988. submitted to SIAM J. Num. Anal.

[BDMS79] J. Bunch, J. Dongarra, C. Moler, and G.W. Stewart, "LINPACK User's Guide," SIAM, Philadelphia, PA, 1979.

[B89] S. Batterson, *Convergence of the Shifted QR Algorithm on* 3×3 *Normal Matrices*, Emory University, preprint.

[BS89] S. Batterson and J. Smillie, *The dynamics of Rayleigh quotient iteration*, SIAM J. Num. Anal. **26, n. 3** (June 1989).

[Chu86] M. Chu, *A differential equation approach to the singular value decomposition of bidiagonal matrices*, Lin. Alg. Appl. **80** (1986), 71–80.

[CL55] E.A. Coddington and N. Levinson, "Theory of ordinary differential equations," McGraw-Hill, New York, 1955.

[DDLT89] P. Deift, J. Demmel, L.-C. Li and C. Tomei, *The bidiagonal singular value decomposition and Hamiltonian mechanics*, Comp. Sci. Dept. Technical Report 458 (July 1989), Courant Inst., New York, NY. To appear in SIAM J. Num. Anal., 1991.

[DK88] J. Demmel and W. Kahan, *Accurate Singular Values of Bidiagonal Matrices*, Comp. Sci. Dept. Technical Report 326 (March 1988), Courant Institute, New York, NY. To appear in SIAM J. Sci. Stat. Comp.

[dlLW89] R. de la Llave and C.E. Wayne, *Whiskered and low dimensional tori in nearly integrable Hamiltonian systems*, preprint (1989).

[DLNT86] P. Deift, L.C. Li, T. Nanda, and C. Tomei, *The Toda flow on a generic orbit is integrable*, Comm. Pure Appl. Math. **39** (1986), 183–232.

[DNT83] P. Deift, T. Nanda, and C. Tomei, *Differential equations for the symmetric eigenvalue problem*, SIAM J. Num. Anal. **20** (1983), 1–22.

[GK65] G. Golub and W. Kahan, *Calculating the singular values and pseudo-inverse of a matrix*, SIAM J. Num. Anal. (Series B) **2**(2) (1965), 205–224.

[GVL83] G. Golub and C. Van Loan, "Matrix Computations," Johns Hopkins University Press, Baltimore, MD, 1983.

[Kos79] B. Kostant, *The solution to a generalized Toda lattice and representation theory*, Adv. in Math. **34** (1979), 195–338.

[Kre50] M.G. Krein, *A generalization of several investigations of A.M. Lyaponuv,*, Dokl. Akad. Nauk **73** (1950), 445–448.

[Kre55] M.G. Krein, *The basic propositions of the theory of* λ*-zones of stability of a canonical system of linear differential equations with periodic coefficients, Pamyati A A. Androvna,*, Izvestia Akad. Nauk (1955), 413–498.

[Mos75] J. Moser, *Finitely many mass points on the line under the influence of an exponential potential—an integrable system*, in "Dynamical Systems Theory and Applications," Springer-Verlag, New York, Berlin, Heidelberg, 1975.

[Sem84] M. Semenov-Tyan-Shanskii, *What is a classical matrix?*, Funct. Anal. Appl. (1984), 250–272.

[Sym80] W.W. Symes, *Hamiltonian group actions and integrable systems*, Physica **1D** (1980), 339–374.

[Sym82] W.W. Symes, *The QR algorithm for the finite nonperiodic Toda lattice*, Physica **4D** (1982), 275–280.

P. DEIFT, J. DEMMEL:
Courant Institute
New York, NY 10012

L.-C. LI:
Mathematics Department
Pennsylvania State University
University Park, PA 16802

C. TOMEI:
Departamento de Matemática
PUC-Rio, Rio de Janeiro, Brazil

Percy Deift and Luen-Chau Li acknowledge the support of NSF grants DMS–8802305 and DMS–8704097. James Demmel acknowledges the support of NSF grants DCR–8552474 and ASC–8715728. Carlos Tomei thanks CNPq, Brazil and the Department of Mathematics of Yale University for their hospitality during Spring 1989, when this research was completed.

The Rhomboidal Four Body Problem. Global Flow on the Total Collision Manifold

J. Delgado-Fernández and E. Pérez-Chavela

Abstract. In this work, we have considered a particular case of the planar four-body problem, obtained when the masses form a rhomboidal configuration. If we take the ratio of the masses α as a parameter, this problem is a one parameter family of non-integrable Hamiltonian systems with two degrees of freedom. We use the blow up method introduced by McGhee to study total collision. This singularity is replaced by an invariant two-dimensional manifold, called the total collision manifold. Using numerical methods we prove first that there are two equilibrium points for the flow on this manifold, and second, that there are only two values of α for which there is a connection between the invariant submanifolds of the equilibrium points. For these values of α the problem is not regularizable.

§1. Introduction.

We consider four masses m_1, m_2, m_3, m_4 in the plane, where $m_1 = m_2$ and $m_3 = m_4$, at the vertices of a rhombus. We give initial symmetric conditions (in positions and velocities) with respect to the axes in the plane, such that they always keep a symmetric rhomboidal configuration when moving under the newtonian law of attraction. If the center of mass is fixed at the origin, the system has two degrees of freedom. The goal of this paper is to study the total collision singularity. By means of a change of coordinates and rescalings it is possible to extend the vectorfield to a limiting total collision manifold. The importance of the study of this flow is that it gives information about orbits which pass close to total collision.

§2. Description of the Problem.

According to the figure (1). Let x be the semidistance between the particles of mass $m_1 = m_2$, and y the semidistance between the particles of mass $m_3 = m_4$. Let α be the ratio of masses $\alpha = m_3/m_1$, then we can suposse $m_1 = m_2 = 1$ and $m_3 = m_4 = \alpha$.

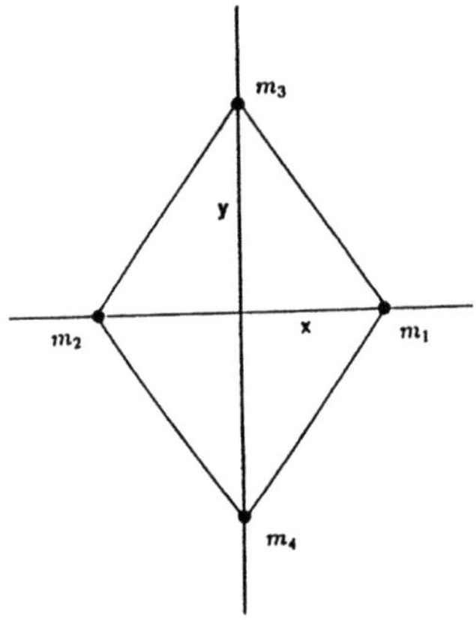

Figure 1. The rhomboidal 4-body problem.

In these coordinates, the equations of motion may be written as

(1)
$$\ddot{x} = -\frac{1}{4x^2} - \frac{2\alpha x}{(x^2+y^2)^{3/2}}$$
$$\ddot{y} = -\frac{\alpha}{4y^2} - \frac{2y}{(x^2+y^2)^{3/2}}.$$

If we make $q = \binom{x}{y}$, $M = \begin{pmatrix} 2 & 0 \\ 0 & 2\alpha \end{pmatrix}$, $p = M\dot{q}$, then the equations of motion (1) may be written as a first order system of differential equations in Hamiltonian form,

(2)
$$\dot{q} = \frac{\partial H}{\partial p}$$
$$\dot{p} = -\frac{\partial H}{\partial q},$$

where $H(q,p) = \frac{1}{2}p^t M^{-1} p - U(q)$ and

$$U(q) = \frac{1}{2x} + \frac{\alpha^2}{2y} + \frac{4\alpha}{\sqrt{x^2+y^2}}.$$

Since M^{-1} appears in the Hamiltonian and we are interested in studying the cases when $\alpha \to 0$ and $\alpha \to \infty$ we will work with an equivalent system, where the mass matrix does not depend on α.

Let $x_1 = x, \quad x_2 = \sqrt{\alpha}\, y$, the equations of motion are

$$\begin{aligned} \ddot{x}_1 &= -\frac{1}{4x_1^2} - \frac{2\alpha^{5/2}x_1}{(\alpha x_1^2 + x_2^2)^{3/2}} \\ \ddot{x}_2 &= -\frac{\alpha^{5/2}}{4x_2^2} - \frac{2\alpha^{3/2}x_2}{(\alpha x_1^2 + x_2^2)^{3/2}}, \end{aligned} \tag{3}$$

in this case we let $q = \binom{x_1}{x_2}$, $M = \begin{pmatrix} 2 & 0 \\ 0 & 2 \end{pmatrix}$, $p = M\dot{q}$. System (1) keeps the Hamiltonian form

$$\begin{aligned} \dot{q} &= \frac{\partial H}{\partial p} \\ \dot{p} &= -\frac{\partial H}{\partial q}, \end{aligned} \tag{4}$$

where

$$H(q,p) = \frac{1}{2}p^t M^{-1} p - U(q) \tag{5}$$

and

$$U(q) = \frac{1}{2x_1} + \frac{\alpha^{5/2}}{2x_2} + \frac{4\alpha^{3/2}}{\sqrt{\alpha x_1^2 + x_2^2}}. \tag{6}$$

The system (3) has the symmetry

$$(x_1, x_2, \dot{x}_1, \dot{x}_2, \alpha) \to (x_2, x_1, \dot{x}_2, \dot{x}_1, \alpha^{-1}), \tag{7}$$

if we allow a reparametrization of time $t = \alpha^{5/4}t'$ it is sufficient to study the problem when $0 < \alpha \leq 1$.

§3 The total collision manifold.

Here we use McGhee's coordinates [4].
Let

$$
\begin{aligned}
r &= (q^t M q)^{1/2} \\
\underline{s} &= r^{-1} q = M^{-1/2} \begin{pmatrix} \cos\theta \\ \sin\theta \end{pmatrix} \\
v &= r^{1/2} (\underline{s} \cdot p) \qquad\qquad (8) \\
\underline{u} &= r^{1/2} (M^{-1} p - (\underline{s} \cdot p)\underline{s}) = u M^{-1/2} \begin{pmatrix} -\sin\theta \\ \cos\theta \end{pmatrix} \\
w &= \frac{\sin 2\theta}{2\sqrt{W(\theta)}} u,
\end{aligned}
$$

where $0 \leq \theta \leq \pi/2$, and

$$
U(\theta) = \frac{1}{\sqrt{2}\cos\theta} + \frac{\alpha^{5/2}}{\sqrt{2}\sin\theta} + \frac{4\sqrt{2}\alpha^{3/2}}{\sqrt{\alpha\cos^2\theta + \sin^2\theta}} \tag{9}
$$

$$
2W(\theta) = U(\theta)\sin(2\theta). \tag{10}
$$

We also scale the time variable of the system by

$$
\frac{dt}{d\tau} = \frac{r^{3/2}\sin(2\theta)}{2\sqrt{W(\theta)}},
$$

so that (3) becomes

$$
\begin{aligned}
\frac{dr}{d\tau} &= \frac{\sin(2\theta) r v}{2\sqrt{W(\theta)}} \\
\frac{dv}{d\tau} &= \sqrt{W(\theta)}(1 - \frac{\sin(2\theta)}{4W(\theta)}(v^2 - 4rh)) \\
\frac{d\theta}{d\tau} &= w \qquad\qquad (11) \\
\frac{dw}{d\tau} &= \cos(2\theta)(1 + \frac{\sin(2\theta)}{2W(\theta)}(2rh - v^2)) - \frac{1}{4}\frac{vw\sin(2\theta)}{\sqrt{W(\theta)}} \\
&\quad + \frac{\dot{W}(\theta)}{2W(\theta)}(\sin(2\theta) - w^2),
\end{aligned}
$$

where $\dot{W}$ means derivation with respect to the variable θ.

The energy relation goes over to

$$\frac{w^2}{\sin(2\theta)} - 1 = \frac{\sin(2\theta)}{2W(\theta)}(rh - \frac{v^2}{2}), \tag{12}$$

and the total collision manifold is given by

$$\Lambda = \{(r, v, \theta, w) \quad / \quad r = 0, \quad w^2 + \frac{v^2 \sin^2(2\theta)}{4W(\theta)} = \sin(2\theta)\}. \tag{13}$$

Λ is invariant under the flow, since $\dot{r} = 0$ when $r = 0$, also Λ is independent of h; so each energy surface has the same total collision manifold. See figure (2).

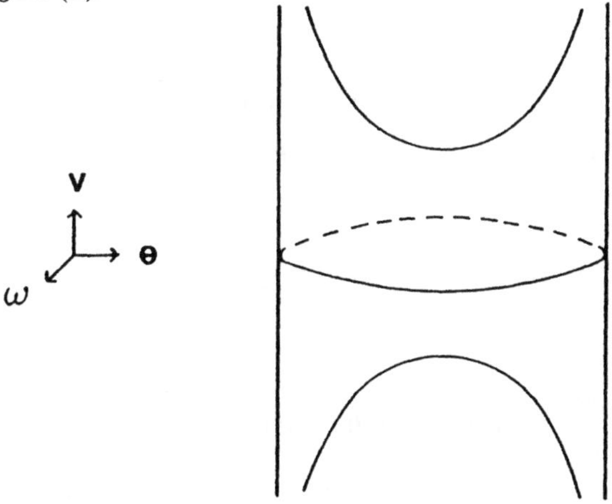

Figure 2. The regularized collision manifold.

§4. The flow on the total collision manifold.

Writing (11) with $r = 0$, we obtain that the flow on Λ is given by:

$$
\begin{aligned}
\frac{dv}{d\tau} &= \sqrt{W(\theta)}(1 - \frac{\sin(2\theta)}{4W(\theta)}v^2) = \sqrt{W(\theta)}(\frac{w^2}{\sin(2\theta)}) \\
\frac{d\theta}{d\tau} &= w \qquad\qquad (14) \\
\frac{dw}{d\tau} &= \cos(2\theta)(1 - v^2\frac{\sin(2\theta)}{2W(\theta)}) - \frac{1}{4}\frac{vw\sin(2\theta)}{\sqrt{W(\theta)}} \\
&+ \frac{\dot{W}(\theta)}{2W(\theta)}(\sin(2\theta) - w^2).
\end{aligned}
$$

Since $dv/d\tau > 0$ if $w > 0$, and $w = 0$ corresponds to an equilibrium point we have the following result.

PROPOSITION 1. *The flow on Λ is gradient-like with respect to the coordinate v.*

From (14) and (10) we obtain:

PROPOSITION 2. *If (v_0, θ_0, w_0) is an equilibrium point for the flow on Λ, then $w_0 = 0, \dot{U}(\theta_0) = 0$ and $v_0^2 = 2U(\theta_0)$.*

Now $\dot{U}(\theta) = 0$ if and only if

$$\tan\theta\sin\theta - \alpha^{5/2}\frac{\cos\theta}{\tan\theta} - \frac{2(1-\alpha)\alpha^{3/2}\sin^2(2\theta)}{(\alpha\cos^2\theta + \sin^2\theta)^{3/2}} = 0,$$

Let $\tan\theta = a$, then $\cos\theta = 1/\sqrt{1+a^2}$, and $\sin\theta = a/\sqrt{1+a^2}$, therefore the last equation is equivalent to

$$0 = (a^3 - \alpha^{5/2})(\alpha + a^2)^{3/2} - 8(1-\alpha)\alpha^{3/2}a^3,$$

or, if we divide by a^3

$$(1 - \frac{\alpha^{5/2}}{a^3})(\alpha + a^2)^{3/2} = 8(1-\alpha)\alpha^{3/2}.$$

Noting that the left hand side is increasing when $0 \leq \theta \leq \pi/2$, we obtain the following result

PROPOSITION 3. *The potential $U(\theta)$ has only one critical point for each value of α.*

By proposition (**2**) we obtain that the flow on Λ has exactly two equilibrium points.

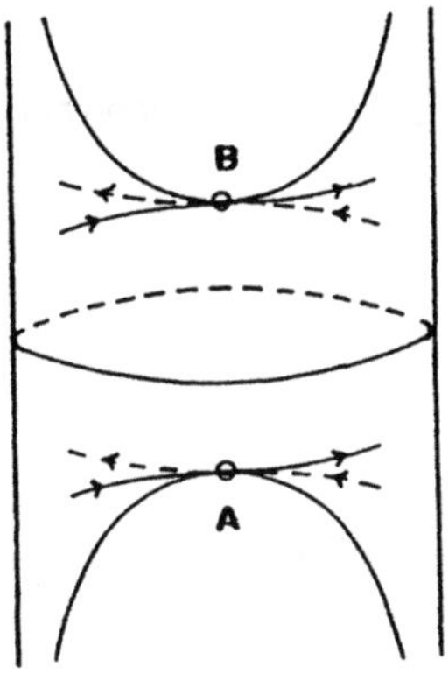

Figure 3. Saddle points on Λ

The two equilibrium points of the flow are hyperbolic, in fact saddle points, so that they have one dimensional invariant submanifolds. We denote by A and B the equilibrium points which correspond to $v = -\sqrt{2U(\theta_0)}$ and $v = \sqrt{2U(\theta_0)}$. See figure (3).

In table (1), we show the values found for θ and v at the critical point A for some values of α, they were found using the Newton-Raphson method.

Table 1.

α	v	θ
1.0	−3.9132733739	$7.8539816340 \times 10^{-1}$
1.0×10^{-1}	−1.5053851535	$4.8545169947 \times 10^{-1}$
1.0×10^{-2}	−1.2217868339	$1.7089843252 \times 10^{-1}$
1.0×10^{-3}	−1.1924761819	$5.4697977384 \times 10^{-2}$
1.0×10^{-4}	−1.1895341344	$1.7318154855 \times 10^{-2}$
1.0×10^{-5}	−1.1892398181	$5.4771511460 \times 10^{-3}$
1.0×10^{-6}	−1.1892103853	$1.7320484539 \times 10^{-3}$
1.0×10^{-7}	−1.1892074420	$5.4772248307 \times 10^{-4}$
1.0×10^{-8}	−1.1892071477	$1.7320507840 \times 10^{-4}$
1.0×10^{-9}	−1.1892071183	$5.4772255676 \times 10^{-5}$
1.0×10^{-10}	−1.1892071153	$1.7320508073 \times 10^{-5}$

Note that

$$\lim_{\alpha \to 0} v = -2^{1/4} = -1.1892071153\cdots. \tag{17}$$

§5 Projection on the $\theta - v$ plane.

From (12) and (14) we obtain

$$\frac{w^2}{\sin(2\theta)} = 1 - \frac{\sin(2\theta)}{4W(\theta)}v^2$$

then

$$\frac{w^2}{\sin(2\theta)}\sqrt{W(\theta)} = \sqrt{W(\theta)}(1 - \frac{\sin(2\theta)}{4W(\theta)}v^2) = v',$$

so that

$$w = \pm\frac{\sqrt{v' \sin(2\theta)}}{(W(\theta))^{1/4}}. \tag{18}$$

Using (10), (14) and (18), we finally obtain

$$\begin{aligned} \frac{dv}{d\theta} = \frac{v'}{\omega} &= \pm\frac{\sqrt{v'}(W(\theta))^{1/4}}{\sqrt{\sin 2\theta}} \\ &= \pm\frac{W(\theta)^{1/4}(1 - v^2\frac{\sin(2\theta)}{4W(\theta)})^{1/2}}{(\sin 2\theta)^{1/2}} \\ &= \pm\frac{\sqrt{2U(\theta) - v^2}}{2}. \end{aligned} \tag{19}$$

Observe that (19) has singularities for $\theta = 0$ and $\theta = \pi/2$ and the sign $+(-)$ corresponds to $w > 0 (w < 0)$.

For studying the possible connection between the invariant submanifolds of the equilibrium points we introduce a new change of independent variable, see figure (4).

$$\theta = J(\gamma) = \frac{\pi}{4}(1 + \sin(2\gamma)), \tag{20}$$

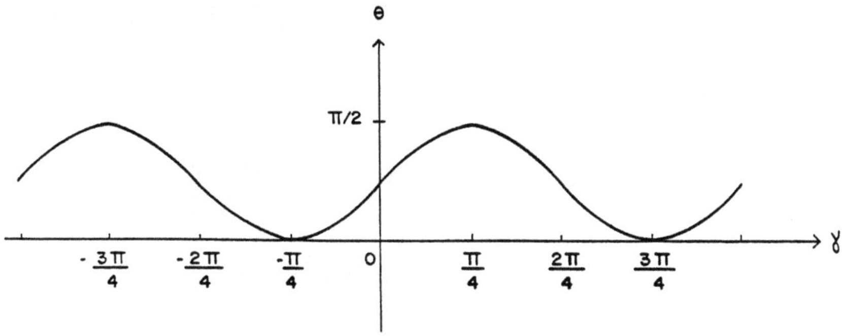

Figure 4. Relation between θ and the new independent variable γ

and study the differential equation

$$\frac{dv}{d\gamma} = \pm\frac{\pi}{4}\sqrt{2U(\theta) - v^2} \quad \cos(2\gamma). \tag{21}$$

We denote the critical point of $U(\theta)$ by θ_0 and γ_0 the corresponding value in $[-\pi/4, \pi/4]$.

For the unstable branch of the point A where $\theta_0 \leq \theta < \pi/2$ or $\gamma_0 \leq \gamma < \pi/4$ we have

$$\frac{dv}{d\theta} = +\sqrt{2U(\theta) - v^2} \qquad \text{and} \qquad \cos(2\gamma) > 0.$$

Therefore,

$$\frac{dv}{d\gamma} = \frac{\pi}{4}\sqrt{2U(\theta) - v^2}\,|\cos(2\gamma)| = \frac{\pi}{4}\sqrt{2U(\theta)\cos^2(2\gamma) - v^2\cos^2(2\gamma)}.$$

For the part of this branch corresponding to $\omega < 0$, this means that θ decreases from $\pi/2$ to 0, we have $\pi/4 < \gamma < 3\pi/4$ and $\cos(2\gamma) < 0$.

Therefore,

$$\begin{aligned}\frac{dv}{d\gamma} &= -\frac{\pi}{4}\sqrt{2U(\theta) - v^2}\cos(2\gamma) = \frac{\pi}{4}\sqrt{2U(\theta) - v^2}\,|\cos(2\gamma)| \\ &= \frac{\pi}{4}\sqrt{2U(\theta)\cos^2(2\gamma) - v^2\cos^2(2\gamma)}.\end{aligned}$$

Then in general we find that the ambiguity of sign disappears giving us

$$(22)\quad \frac{dv}{d\gamma} = \pm\frac{\pi}{4}\sqrt{2U(\theta) - v^2}\cos(2\gamma) = \frac{\pi}{4}\sqrt{2U(\theta)\cos^2(2\gamma) - v^2\cos^2(2\gamma)}.$$

Observe that

$$2U(\theta)\cos^2(2\gamma) = \frac{\sqrt{2}\cos^2(2\gamma)}{\cos(J(\gamma))} + \frac{\sqrt{2}\alpha^{5/2}\cos^2(2\gamma)}{\sin(J(\gamma))} + \frac{8\sqrt{2}\alpha^{3/2}\cos^2(2\gamma)}{\sqrt{\alpha\cos^2(J(\gamma)) + \sin^2(J(\gamma))}}$$

still has singularities for $\gamma = \ldots - \pi/4, \pi/4, 3\pi/4, \ldots$, however these singularities are removable, since

$$\lim_{\gamma\to\frac{\pi}{4}} \frac{\sqrt{2}\cos^2(2\gamma)}{\cos(J(\gamma))} = \frac{8\sqrt{2}}{\pi} \quad \text{and} \quad \lim_{\gamma\to\frac{3\pi}{4}} \frac{\sqrt{2}\alpha^{5/2}\cos^2(2\gamma)}{\sin(J(\gamma))} = \frac{8\sqrt{2}\alpha^{5/2}}{\pi}.$$

In this way we have regularized the singularities due to double collisions.

Since the original Hamiltonian is quadratic in the momenta we have the following symmetry

$$(23)\qquad (r, v, \theta, w, \tau) \xrightarrow{L} (r, -v, \theta, -w, -\tau).$$

We denote by

$W_A^{u,+(-)}$ the branch of the unstable submanifold in A which corresponds to $w > 0 (w < 0)$.

$W_B^{s,+(-)}$ the branch of the stable submanifold in B which corresponds to $w > 0 (w < 0)$.

From the symmetry of the invariant manifolds,

$$(24)\qquad L(W_A^{u,+}) = W_B^{s,-} \quad \text{and} \quad L(W_A^{u,-}) = W_B^{s,+},$$

we have:

Proposition 4. *$W_A^{u,+}$ coincides with $W_B^{s,-}$ if and only if $W_A^{u,+}$ arrives to the plane $v = 0$ with a value of $\theta = 0$ or $\theta = \pi/2$, or a value of γ which is an odd multiple of $\pi/4$.*

A similar result applies for the coincidence of $W_A^{u,-}$ with $W_B^{s,+}$. Let us observe that because of the symmetry (24) we also have

PROPOSITION 5. $W_A^{u,+} = W_B^{s,+}$ *if and only if* $W_A^{u,-} = W_B^{s,-}$.

Then by the arguments described previously it is enough to study the behavior of $W_A^{u,+}$ and $W_A^{u,-}$ until they reach the value $v = 0$. We denote with $x(+)$ ($x(-)$) the value of γ for which $W_A^{u,+}$ ($W_A^{u,-}$) reaches the value $v = 0$. The computations have been done using a RKF-78 routine. The table (2) shows some results obtained.

Table 2.

α	$x(+)$	$x(-)$
1.0	2.100017	−2.100017
0.2230995423	1.0778347432	$-2.3561945 = -3\pi/4$
1.0×10^{-1}	1.021096	−2.537965
1.0×10^{-2}	8.357815×10^{-1}	−2.381411
1.0×10^{-3}	7.966514×10^{-1}	−2.360311
1.0×10^{-4}	7.875486×10^{-1}	−2..56926
1.0×10^{-5}	7.857881×10^{-1}	−2.356325
1.0×10^{-6}	7.854679×10^{-1}	−2.356218
1.0×10^{-7}	7.854106×10^{-1}	−2.356199
1.0×10^{-8}	7.854004×10^{-1}	−2.356195
1.0×10^{-9}	7.853985×10^{-1}	−2.356195
1.0×10^{-10}	7.853982×10^{-1}	−2.356195

From this numerical computations, we conclude the following theorem.

THEOREM 1. *For* $\alpha = \alpha_1 = 0.2230995423\cdots$ *we have* $W_A^{u,-} = W_B^{s,+}$.

The figure (5) shows the behavior of the invariant submanifolds of the equilibrium point A, for some values of α.

For $\alpha = 0$ we have a limiting case where $\theta_0 = 0$ and $\gamma_0 = -\pi/4$. From (9) we have

$$U(\theta) = \frac{1}{\sqrt{2}\cos\theta}$$

and (19) has the form

$$\frac{dv}{d\theta} = \pm\frac{1}{2}\sqrt{\sqrt{2}/\cos\theta - v^2},$$

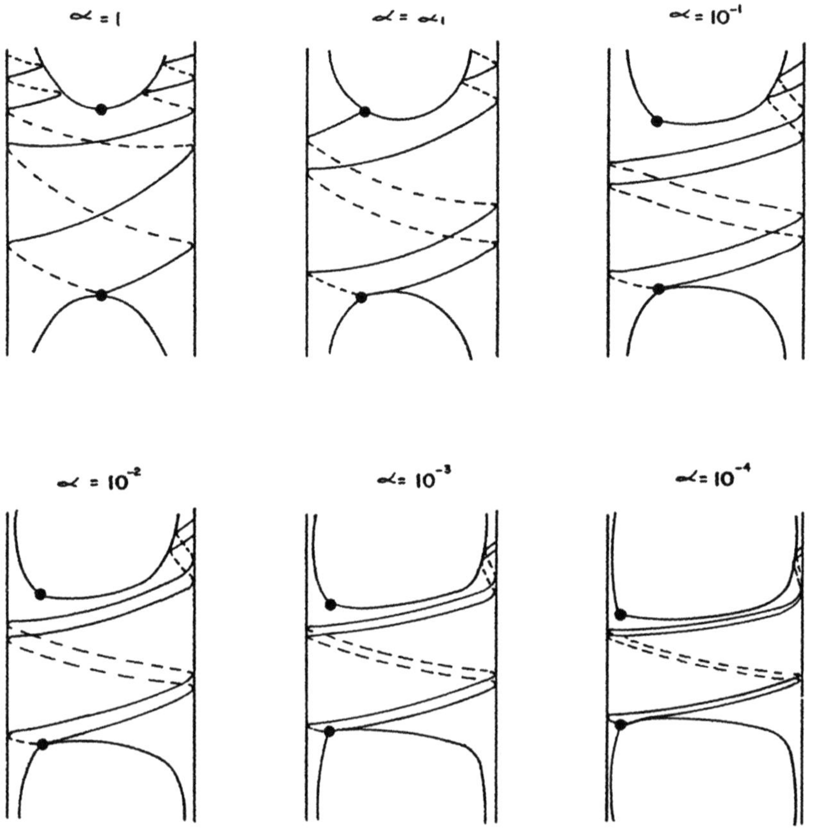

Figure 5. Behavior of the invariants manifolds on Λ.

solving this equation with $v(0) = \lim_{\theta_0 \to 0}(-\sqrt{2U(\theta_0)}) = -2^{1/4}$ (see (17)), we have

$$v(\theta) = -2^{1/4}\sqrt{\cos\theta}$$

therefore $v = 0$ if $\theta = \pi/2$.

We have now the following result:

PROPOSITION 6. *For α small enough and positive, there is no intersection between invariant submanifolds. An intersection occurs in the limiting case $\alpha = 0$.*

§6 Conclusion.

From the symmetry introduced in (7), and by the previous results we can conclude.

THEOREM 2. *In the Rhomboidal Four-Body problem there are only two values of the mass ratio α, one of them is $\alpha_1 < 1$, and other one is $\alpha_2 = \alpha_1^{-1} > 1$ for which there are connections between the invariant submanifolds of the equilibrium points.*

Now for the value α_1 we have $W_A^{u,-} = W_B^{s,+}$ but $W_A^{u,+} \neq W_B^{s,-}$ (see table (2)) and for the value α_2 we have $W_A^{u,+} = W_B^{s,-}$, but $W_A^{u,-} \neq W_B^{s,+}$. Hence, from [6] we see that the problem is not regularizable.

§7 Acknowledgements.

The authors wish to express their gratitude to Dr. Ernesto Lacomba by his valuable advice and useful criticisms in the preparation of this paper. Thanks also to Mrs. Beatriz Arce for her fine work of typesetting the manuscript.

References

1. Devaney, R., *Triple Collision in the Planar Isosceles Three Body Problem*, Inventiones Math. **60** (1980), 249–267.
2. Lacomba, E. Simó, C., *Boundary Manifolds for Energy Surfaces in Celestial Mechanics*, Cel. Mechanics **28** (1982), 37–48.
3. Lacomba, E. Simó, C., *Analysis of some degenerate quadruple collisions*, Cel. Mechanics **28** (1982), 49–62.
4. McGehee, R., *Triple Collision in the Collinear Three-Body Problem*, Inventiones Math. **27** (1974), 191–227.
5. Simó, C. Martínez, R., *Qualitative study of the Planar isosceles Three-Body Problem*, Cel. Mechanics **41** (1988), 179–251.
6. Simó, C., *Necessary and Sufficient Conditions for the Geometrical Regularization of Singularities*, in "Proc. IV Congreso Ec. Dif. y Aplic., Sevilla, Spain," 1981, pp. 193–202.

Departamento de Matemáticas, Universidad Autónoma Metropolitana-Iztapalapa. Apdo. Postal 55–534, C.P. 09340 México, D.F.

Toward a Topological Classification of Integrable PDE's

NICHOLAS M. ERCOLANI AND DAVID W. MCLAUGHLIN

Abstract. We model Fomenko's topological classification of 2 degree of freedom integrable stratifications in an infinite dimensional soliton system. Specifically, the analyticity of the Floquet discriminant $\Delta(q, \lambda)$ in both of its arguments provides a transparent realization of a Bott function and of the remaining building blocks of the stratification; in this manner, Fomenko's structure theorems are expressed through the inverse spectral transform. Thus, soliton equations are shown to provide natural representatives of the classification in the context of PDE's.

§I. Introduction.

Integrable phase space stratifications are remarkably rigid. Two examples have been discussed in this workshop:

(i) Fomenko's topological classification of isoenergetic 3-manifolds for integrable hamiltonian systems with two degrees of freedom;

(ii) our detailed description of critical level sets in function space for integrable soliton PDE's.

Such detailed stratifications can be useful for understanding qualitative behavior of solutions of the integrable equation itself, as well as the behavior of solutions of nearby perturbed equations. In one degree of freedom Hamiltonian systems, phase plane structure is visible at a glance: invariant critical sets are typically points, saddles or centers, and these completely organize the topology of orbits. In higher dimensions, the situation is more complicated; the invariant critical sets may be manifolds of quite varied structure. Nevertheless, for integrable 2 degree of freedom systems Fomenko has shown that possibilities for critical sets are quite tractable: each is constructed from basic building blocks, of which *there are only five types*! Moreover, these critical sets completely determine the geometry of an integrable stratification.

In (i), the integrable stratification begins with a second constant of the motion whose critical sets determine the degenerate strata which, in turn, yields a decomposition of the global stratification into canonical building blocks with discrete invariants. Any two integrable hamiltonian systems with the same invariants will be topologically equivalent as stratified spaces.

In (ii), the stratification is determined by the spectral theory of that linear operator which integrates the soliton equation. Central to this spectral theory is a function called the Floquet discriminant Δ:

$$\Delta \colon \mathcal{F} \times \mathbb{C} \to \mathbb{C},$$

where $\mathcal{F}$ denotes a fixed function space of periodic functions. The discriminant $\Delta(q, \lambda)$ is analytic in both of its arguments. Its multiple spectral points (in λ) label the degenerate strata (in q). This Floquet discriminant provides the basic analytic tool for soliton equations which applies to an arbitrary, even infinite, number of degrees of freedom.

For integrable two degree of freedom Hamiltonian systems, Fomenko's topological classification is very general. It is applicable to integrable systems which may not be algebraically integrable. It does not require analytic functions or Jacobians of Riemann surfaces. However, this method, while an advance over one degree of freedom phase space analysis, is likely to encounter formidable if not impassable difficulties when generalized to higher, let alone infinite, dimensions. Our hope is that the analytic tools from spectral theory which have been developed for soliton equations will provide important models for the topological intricacies of higher dimensional stratifications. One may further hope that these analytic models realize all of the fundamental types of integrable stratifications up to topological equivalence.

In this spirit we present a comparison of the two stratifications on the common ground of two-phase, even solutions of the cubic nonlinear Schroedinger (NLS) equation. In this regime (i) and (ii) overlap: inside the function space of NLS is an invariant 4 dimensional submanifold on which Fomenko's classification can be implemented with analytic tools from spectral theory. This implementation is accomplished in Section IV, after a summary of the two classifications of critical level sets. In Section V we conclude with some general remarks concerning extensions to infinite dimensions and to the sine-Gordon problem.

§II. Fomenko's classification: Critical Level Sets.

Fomenko [**1,2**] studies two degree of freedom hamiltonian systems which possess two integrals (H and F) in involution. He restricts attention to a compact level set Q: $H = h$ for a "regular value" h so that Q is a smooth nonsingular 3-manifold on which H has no critical points. There is a proviso that Q be "nonresonant"; i.e., those (H, F) Liouville tori on Q whose H orbits have an irrational dense winding should form a dense set in Q. F may have critical points on Q; in fact, these critical points cannot be isolated and must form manifolds of dimension one or greater. The classification of these critical sets is the basis of the theory. Again there is a proviso. F must be "Bottean"; i.e., its critical sets are smooth submanifolds with Hessian nondegenerate in normal directions. The last condition is central to the classification. Away from critical values of F, the connected components of the (H, F) level sets are 2-tori: the product of two circles. At a critical value, each connected component of the level set is a disjoint union of manifolds. There are two categories in the classification: (1) the critical set is properly contained in its level set or (2) the critical set equals the entire level set.

In *Category (1)*, F does *not* attain its maximum or minimum on the level set, and the critical set consists of a disjoint union of circles, called "saddles" for F. (For a generic system there is only one circle.) The remainder of the level set consists of two dimensional manifolds, each of which is homeomorphic to a product of a circle with a line. The product may be trivial, or it may be twisted, a so-called "Seifert product". (However, this case is always the 2:1 quotient of a trivial product in a system with a Z_2 symmetry.) The two dimensional pieces of the level set are clutched to one another along the critical saddles (circles). In the generic case there is only one saddle, and the level set is either the (homoclinic) circle fibration over the oriented graph:

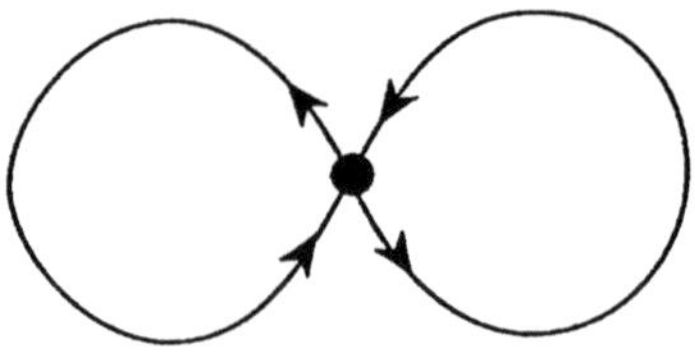

Figure 1a

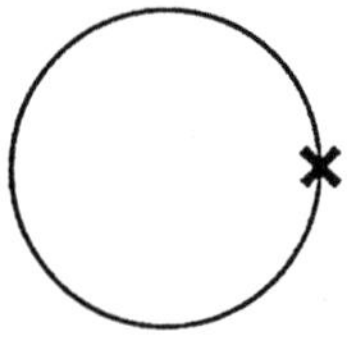

Figure 1b

or, the level set is the (Seifert) circle fibration over the unoriented graph. The double cover of "1b" is "1a".

In *Category (2)*, the critical set *equals* the entire level set. On it, F attains a maximum or a minimum. There are three cases here: the critical (level) set is a circle or a 2-torus or a Klein bottle.

To recapitulate, critical level sets are constructed from five types of building blocks:

$$\underset{\text{Category 1}}{\{\text{Homoclinic, Seifert}\}} \text{ or } \underset{\text{Category 2}}{\left\{S^1, T^2, \text{Klein Bottle}\right\}}$$

In the first two types (Category 1), the critical set is properly contained in its level set and the integral F does *not* attain a max or a min on the critical set. In the last three types (Category 2), the critical set equals the entire level set and F attains a max (or min) on the critical set. Next, Fomenko embeds the critical set locally into a 3-manifold, and by gluing the local pieces together, reconstructs the isoenergetic manifold Q. In the NLS setting, only two types (homoclinic or S^1) will arise. Figure 2 illustrates this gluing procedure for these two cases, in an almost self-explanatory fashion.

§III. Critical Level Sets for the NLS Equation.

The nonlinear Schroedinger (NLS) equation is a typical example of soliton equations. In this section we describe its critical level sets in terms of the nonlinear spectral transform.

NLS under periodic boundary conditions,

$$2iq_t = q_{xx} + \frac{1}{2}qq^*q \tag{3.1}$$

$$q(x+L,t) = q(x,t) \tag{3.1b, c}$$

is a Hamiltonian system with Hamiltonian $\mathcal{H}\colon \mathcal{F} \to \mathbf{R}$,

$$\mathcal{H} = \int_0^L \left[q_x q_x^* - \frac{1}{4}(qq^*)^2 \right] dx. \tag{3.2}$$

The complete integrability of NLS is established through a nonlinear transform which is defined in terms of the following linear spectral problem:

$$\begin{aligned} &\widehat{L}\vec{\psi} = \lambda\vec{\psi}, \\ &\widehat{L} = i\begin{pmatrix} 1 & 0 \\ 0 & -1 \end{pmatrix}\frac{d}{dx} + \begin{pmatrix} 0 & -q \\ q^* & 0 \end{pmatrix} \end{aligned} \tag{3.3}$$

Indeed, NLS defines an "isospectral flow" for this linear problem, the spectrum of which provides sufficient invariants to establish the integrability of NLS.

To be precise, the operator $\widehat{L}$ is viewed as an operator on the Hilbert space $\mathbf{L} = L^2(\mathbf{R})$:

$$\widehat{L}\colon \mathcal{D} \subset L^2(\mathbf{R}) \to L^2(\mathbf{R}),$$

where

$$\mathbf{L}^2(\mathbf{R}) \equiv \left\{ \vec{\varphi}\colon \mathbf{R} \to \mathbf{C}^2 \mid \int_{-\infty}^{\infty} \vec{\varphi}^* \cdot \vec{\varphi}\, dx < \infty \right\},$$

and $\mathcal{D}$ is the (dense) domain of $\widehat{L}$. In this setting $\widehat{L}$ has only continuous spectrum $\sigma(\widehat{L})$. Since $\widehat{L}$ is *not* self adjoint, its spectrum need not be real. These complex curves of spectrum are invariant under the NLS flow.

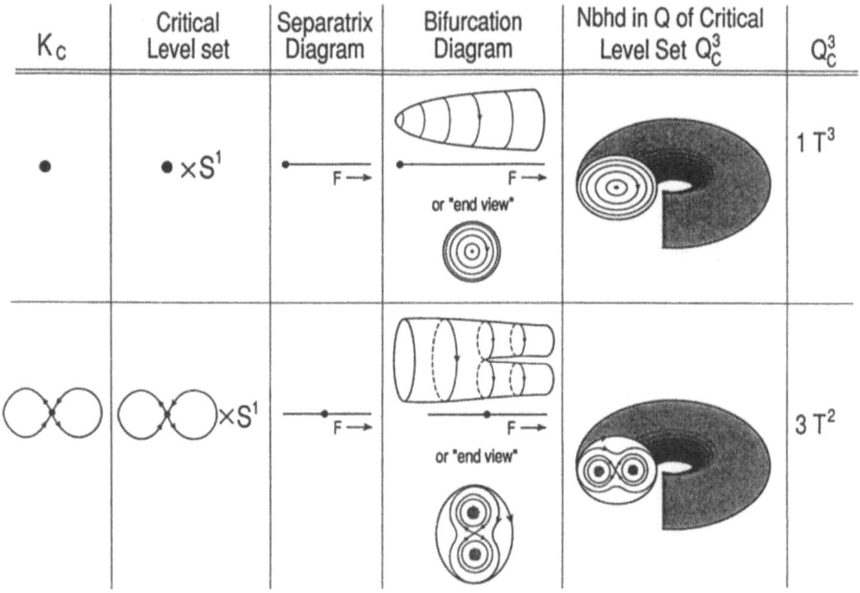

Figure 2
The gluing procedure for a Q-neighborhood of the critical level sets.
Two types of critical level sets are depicted: S^1 and Homoclinic.
In both cases we show (i) the critical graph K_c; (ii) the critical level set; (iii) the "separatrix-diagram" which sketches the values of the Bott function F near its critical value; (iv) the "bifurcation diagram" as a surface over F values; (v) the Q-neighborhood of the critical level set; (vi) a boundary of this neighborhood.

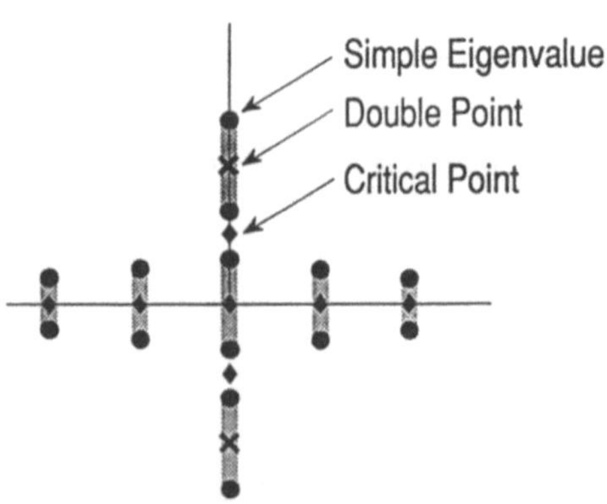

Figure 3
A double point is formed when two simple eigenvalues coalesce, necessarily at a critical point.

The spectrum $\sigma(\widehat{L})$ can be characterized through an analytic function $\Delta(q, q^*; \lambda)$, called the Floquet discriminant, which is defined as follows: One begins with the fundamental matrix solution M,

$$\widehat{L}M = \lambda M \tag{3.4}$$
$$M(x = y, y; \lambda) = \begin{pmatrix} 1 & 0 \\ 0 & 1 \end{pmatrix},$$

and then defines Δ through its trace,

$$\Delta(q, q^*; \lambda) \equiv \operatorname{tr} M(y + L, y; q, q^*; \lambda). \tag{3.5}$$

The Floquet discriminant $\Delta(q, q^*; \lambda)$ is invariant under the NLS flow, and the spectrum σ is then characterized in terms of this constant of the motion,

$$\sigma(\widehat{L}(q, q^*)) = \{\lambda \in \mathbb{C} \mid \Delta(q, q^*; \lambda) \text{ is real and } -2 \leq \Delta \leq 2\}. \tag{3.6}$$

Two other spectral quantities are important, (i) critical points and (ii) multiple points:

(3.7a, b)

$$\textit{critical points:} \quad \left\{\lambda \in \mathbb{C} \mid \frac{d\Delta}{d\lambda} = 0\right\}$$
$$\textit{multiple points:} \quad \{\lambda \in \mathbb{C} \mid \lambda \text{ is a critical point and } \Delta(\lambda) = \pm 2\}$$

The curves of spectrum terminate at *periodic or antiperiodic eigenvalues* $\{\lambda_j\}$,

$$\Delta(q, q^*; \lambda_j) = \pm 2. \tag{3.8}$$

Two curves of spectrum can join or touch at a multiple point, creating an eigenvalue of higher multiplicity (see Figure 3).

We now describe the critical level sets in terms of these spectral ingredients. The Floquet discriminant $\Delta(q;\lambda)$ is analytic in both of its arguments, the real functions $u(x), v(x)[q(x) = u(x) + iv(x)]$ and the complex parameter λ. The description proceeds by interplaying these two dependencies. First, one uses the Floquet discriminant to define the "isospectral" class of q",

$$\mathcal{M}_q \equiv \{r \in \mathcal{F} \mid \Delta(r;\lambda) = \Delta(q;\lambda) \forall \lambda\}. \tag{3.9}$$

Since, for every complex λ, the Floquet discriminant $\Delta(q;\lambda)$ is invariant for the NLS equation, Δ provides an infinite set of constants of the motion which are labeled by λ. Moreover, these integrals can be shown to be in involution and to be complete. Thus, the isospectral class $\mathcal{M}$ consists of "invariant manifolds" of NLS.

By completeness, the generic level sets of the map

$$\begin{aligned} \Delta\colon \mathcal{F} &\to \{\text{analytic functions of } \lambda\} \\ q &\to \Delta(q;\lambda) \end{aligned} \tag{3.10}$$

are (possibly infinite dimensional) Liouville tori. The critical values of Δ, viewed as a functional of q, occur at functions q_{cr} for which the spectrum $\sigma(\widehat{L}(q_{cr}))$ contains multiple points. (Generically such multiple points are double, and that is the only case that we treat.) Note that, in this fashion, the critical behavior of Δ, viewed as a functional of q, is related to its multiple behavior in the complex parameter λ.

Our arguments in reference [**3**] for the sine-Gordon equation can be readily adapted to this NLS case. These methods show that the configuration of double points for $\Delta(\lambda)$ completely determines the structure of the critical level set. This result may be stated concisely as

THEOREM 1. *Let $q(x)$ be a potential for which $\sigma(\widehat{L}(q))$ has 2N simple eigenvalues, while it has 2M double points off of the real axis, with the remaining eigenvalues double and real. The associated critical set is a disjoint union of manifolds diffeomorphic to*

$$T^N \times (S^1 \times \mathbb{R}^1)^j \tag{3.11}$$

where T^N is a real N-torus, S^1 is the circle, and all values $0 \le j \le M$ occur.

The potentials Q in the strata of (3.11) can be *explicitly* constructed by Bäcklund transformations from the quasi-periodic potentials, q, in T^N:

$$Q = q + 4(\nu - \nu^*)\frac{\varphi_1\varphi_2^*}{\varphi_1\varphi_1^* + \varphi_2\varphi_2^*} \tag{3.12a}$$

where ν is a nonreal double point, and

$$\vec{\varphi} = \alpha\vec{\psi}^{(+)} + \beta\vec{\psi}^{(-)} \tag{3.12b}$$

is a general solution (eigenfunction) of linear problem (3.3) at $\lambda = \nu$. (In (3.12b), $\{\vec{\psi}^{(+)}, \vec{\psi}^{(-)}\}$ denotes the Floquet basis of eigenfunctions for (3.3).) Thus, the Bäcklund transformation formula (3.12a) provides a family of periodic NLS solutions, parametrized by α/β, which is isospectral to q. Iteration generates the entire singular stratum (3.11).

An alternative parametrization of the level sets, critical or not, will also be useful. This second parametrization is closely related to action-angle coordinates of the Liouville tori. The role of all action variables is played simultaneously by the analytic (in λ) function $\Delta(q;\lambda)$; this invariant selects a specific level set, which is in turn coordinatized by "angle variables". In the setting of soliton equations, such angle information is carried by the "auxiliary spectrum", $\{\mu_j\}$. This auxiliary spectrum is defined through the fundamental matrix, equation (3.4):

$$M_{12}(x' + L, x'; \lambda; q) \mid_{\lambda=\mu_j(x';q)} = 0. \tag{3.13}$$

While the Floquet spectrum $\sigma(\widehat{L}(q))$ remains invariant under the NLS flow, the auxiliary eigenvalues evolve in time and represent the dynamics of the system as "particle" motion in the spectral plane. Because (3.3) is *not* self-adjoint, the μ_j do not travel on fixed trajectories in the complex plane. However, trajectories (which may be constructed using the higher commuting NLS flows) do traverse closed paths on the Riemann surface which are homologous to fixed cycles. These cycles are called "γ cycles". For more details we refer the reader to [**3,4,5**].

§IV. Spectral Theory and the Topological Classification.

In this section we realize the topological classification of Fomenko through the spectral transform. We work in a regime where the two theories overlap:

the NLS system with at most six simple eigenvalues whose coefficient $q(x)$ is even in space. This is a two degree of freedom Hamiltonian system and Fomenko's classification will apply.

In this regime there are two spectral configurations: all six of the eigenvalues reside on the imaginary axis, or else two are on the imaginary axis and four occur in complex conjugate pairs. (Because of the even symmetry of q(x), if λ is a simple eigenvalue, so is $-\lambda$.) These spectral configurations are depicted in Figure 4, together with two cases of multiple points in which a double point resides on either the real or the imaginary axis.

Thus, in our case the phase space consists of all smooth functions $q = u + iv$ which are even with fixed spatial period and which have at most six simple eigenvalues. This phase space has four real dimensions and is coordinatized by two constants of the motion and two angle variables. For the angles we use μ_0 and μ_1, where μ_0 is the "extra" μ variable of the NLS system [**6**], and where μ_1 denotes that Dirichlet eigenvalue which is associated to the "first" critical spectral point $\lambda_c = \lambda_c(q)$ in the closed first quadrant i.e.,$\Delta'(q, \lambda_c) = 0$. (Because of the even symmetry, this critical point λ_c is either purely imaginary or real.) The generic level sets are two-tori.

We find it convenient to fix the simple eigenvalue λ_0 (instead of the energy H). Let Q denote a level set for λ_0. For the second invariant, we choose the Floquet discriminant *evaluated at the critical point* $\lambda_c(q)$,

$$F(q) = \Delta(q; \lambda_c). \tag{4.1}$$

In our arguments we will continually use an interplay between the λ dependence and the q dependence of $\Delta(q; \lambda)$. It will be crucial that critical (and degenerate) behavior in λ (as labeled by λ_c (and λ_d)) be kept distinct from critical behavior of F(q) in q, (as labeled by q_d). In addition, critical sets of $F(q)$ in function space must be kept distinct from (the sometimes larger) critical <u>level</u> sets of $\Delta(q, .)$ on which they reside.

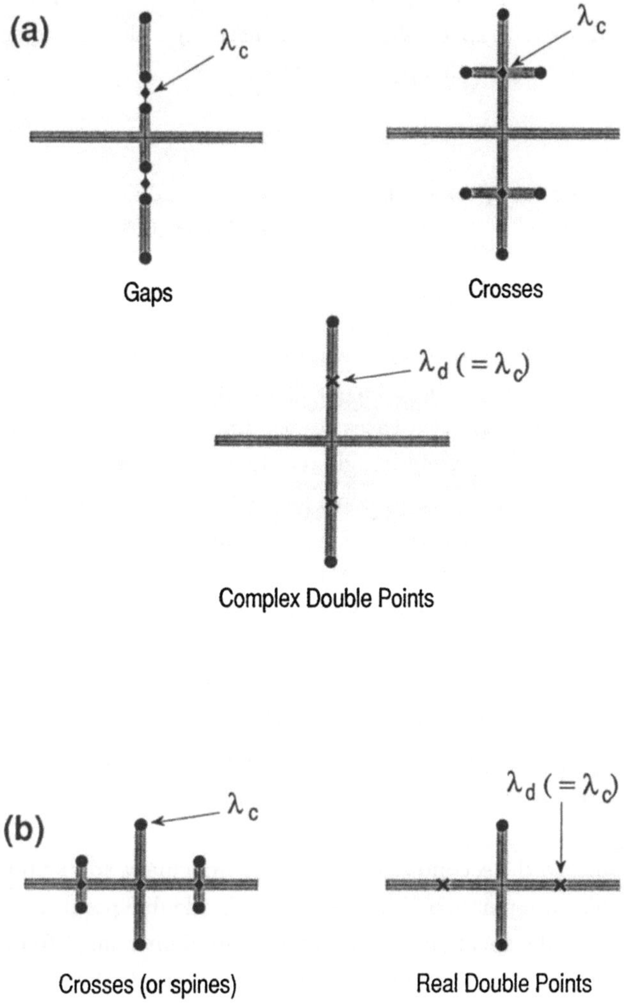

Figure 4
Six simple eigenvalues in "cross" or "gap" configurations.
(a) Complex critical and double points; (b) Real critical and double points.
The "cuts" denote Floquet spectrum.

The critical sets of F(q) in the phase space are characterized by the condition that the critical spectral point λ_c is actually a double eigenvalue λ_d *with the Dirichlet eigenvalue* μ_1 *locked to it*,

$$\lambda_d = \lambda_c = \mu_1.$$

These critical sets are (homeomorphic to) circles and are coordinatized by μ_0. At these critical circles, the Floquet discriminant $\Delta = 2$ (or -2);

$$F(q) = \Delta(q; \lambda_c) = 2(or - 2). \tag{4.2}$$

The behavior of F(q) for q in a neighborhood of a point q_d on a critical circle in phase space can be inferred from the behavior of $\Delta(q;\lambda)$ as a function of λ. There are two cases, depending upon whether the double point $\lambda_d = \lambda_c$ is real or purely imaginary. In the first case of a real double point, the behavior is assessed by considering $\Delta(q, \lambda)$ for real λ. In this case "reality" $[q = u + iv, u = u^*, v = v^*, \lambda = \lambda^*]$ implies that the fundamental matrix M, and hence the transfer matrix, is unitary for all real λ. This in turn implies that the entire real λ axis is always spectrum (see Figure 4b), and that only spines (*no gaps!*) of spectrum can emerge from perturbations of a real double point. This means that in this real case, [or $\Delta(q, \lambda_c(q)) \geq -2$] with a maximum of 2 [or minimum of -2] attained at $q_d, \lambda_c(q_d) = \lambda_d$. Thus, in this first case of a real double point, F is a maximum (or minimum); the critical circle is the entire critical level set. This situation is depicted in Figure 5a.

Next, we turn to the second case of a purely imaginary double point (Figure 4a). In this case perturbations can open the double point into either a spine or a gap in the spectrum. This can be seen numerically [**6**] or perturbatively [**7**]. Thus, at a complex critical point, $\Delta(q; \lambda_c(q))$ can take values above or below 2 [or -2, which we omit for ease of presentation]. In this second case, F attains neither a max nor a min on the critical set; the critical circle is of saddle type. In its neighborhood, F can increase producing a spectral configuration with a "gap", or it can decrease, producing a spectral configuration with a "cross". The critical circle itself is coordinatized by the single variable μ_0, with the variable μ_1, locked to the double point $\mu_1 = \lambda_d$. On the other hand, for the critical level set, the variable μ is not locked to the double point. Rather, it is free to move on a curve homoclinic to the complex double point λ_d. In this manner one sees that the entire level set is the disjoint union of three pieces: two cylinders ($= S^1 \times \mathbb{R}$), each of which is homoclinic to the third piece, which is a critical circle. This situation is depicted in Figure 5b. We note that Figure 5 really contains the whole story.

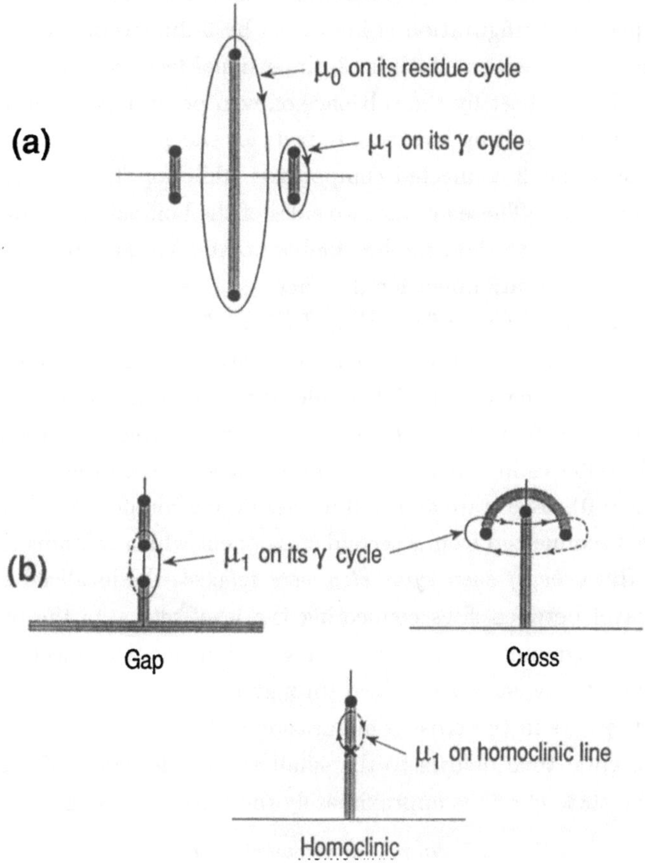

Figure 5a,b
γ-cycles. (a) In a neighborhood of real double points;
(b) in neighborhoods of complex double points.
Note, in particular, the homoclinic case.

Finally, we turn to connectivity as a sample of the type of information that can be extracted by examining the critical sets and their neighborhoods. Without the even symmetry, the full NLS level set associated to the above spectral configuration is known to be 3-dimensional and *connected* [8]. Under evenness constraints, 2-dimensional level sets arise as slices of the full NLS level set by the subspace of even potentials. This *slicing* may disconnect the resulting level set. In fact, for the gap configuration (Figure 5b) the slice has 2 connected components while for the cross (Figure 5b) there is only one. These are the two sides of the homoclinic saddle (see Figure 2) associated to the complex double points. We give both an intuitive and an analytical argument for this fact:

Intuitively, the gap configuration of Figure 5a can be deformed, without crossing any critical values, to the near-breather configuration of Figure 6a. In this even setting the spatial profile of the large breather can either be centered at $x = 0$ or at $x = L$. The spatially localized excitation at $x = 0$ ($x = L$) oscillates in time, but it cannot travel to the other spatial site at $x = L$ ($x = 0$). The state at $x = 0$ resides in one component of the level set which is disconnected from a second component which contains the state at $x = L$. However, *if even symmetry were relaxed*, the localized excitations could travel between sites connecting the level sets. On the other hand, under even symmetry, the cross configuration has a connected level set. This connectivity can also be seen through a deformation argument. First, replace the cuts in the cross configuration of Figure 6b with an equivalent choice of cuts; then deform to the small amplitude region of Figure 6b in which the state $q(x,t)$ is approximately the linear standing wave

$$q(x,t) \simeq \cos \omega t \cos kx,$$

for k, ω constant. As time t increases, the maximum of the wave changes its location between $x = 0$ and $x = L$ periodically in t, always maintaining even symmetry. This periodic exchange reflects the connectivity of the level set.

The analytical explanation of this bifurcation is best given by complexifying the canonical angle coordinate associated to μ_1:

$$\theta_1(t) = \int^{\mu_1^2(t)} \frac{dz}{i\sqrt{z(z-\lambda_0^2)(z-\lambda_1^2)(z-\lambda_2^2)}} = \omega_1 t + \theta_1^0 \tag{4.3}$$

If one permits μ_1 to take arbitrary values in the λ-plane, θ_1 as a function of complex t is doubly periodic with period basis $\{\omega_1, \omega_2\}$, where ω_2 is complex while ω_1 is real. The image is a 1-dimensional complex torus whose period parallelogram for the cross and gap configurations are shown respectively in Figure 7a and 7b. For the gap, ω_2 is pure

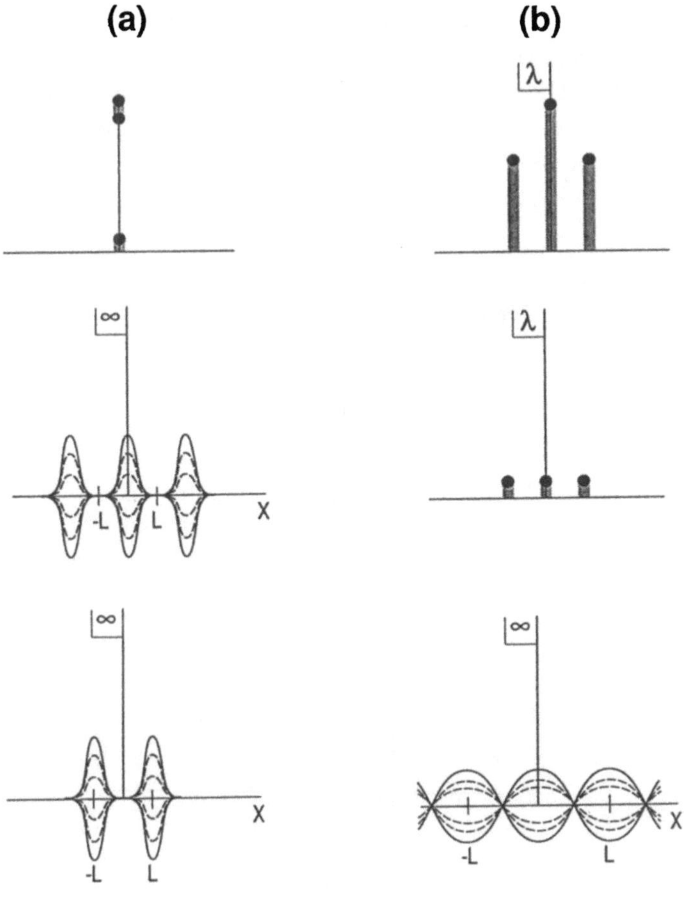

Figure 6
Spectral deformations and spatial profiles. (a) The "gap configuration" has two disconnected level sets, with two distinct spatial profiles – one centered at $x = 0$ and the second at $x = L$. (b) The "cross configuration", its spectral deformation, and a spatial profile. The cross-configuration has a single connected component.

imaginary, while for the cross ω_2 has non-zero real part. In both cases the "real" points corresponding to real t flows appear as two vertical segments separated by a 1/2-period. In Figure 8a these are disjoint circles while in Figure 8b the two segments reconnect modulo periods so there is only one real component. This scenario is completely analogous to the topological distinction between kink trains and breather trains in periodic sine-Gordon [**5**]. We remark that these considerations about connectivity are closely related to recent numerical experiments about chaos in near-integrable NLS equations [**9,10,11**], where the perturbation can induce an irregular jumping between the two disconnected components of the unperturbed phase space saddle structure (Figure 2).

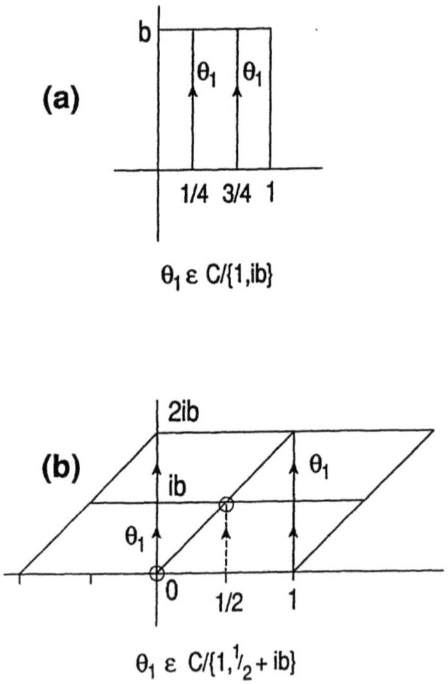

Figure 7
Period parallelograms for the (a) cross and (b) gap configurations.

§V. Conclusion.

In this note we have compared our spectral theoretic classification of critical level sets of [**3**] with the more topological classification of [**2**]. Our main result is to show that the Floquet discriminant, *evaluated at multiple spectral points*, provides a natural and intuitive "Bott function". *The max, min, or saddle structure of this Bott function, realized through spectral theory, is completely transparent with the multiple behavior in* Δ *as a function of* λ *mirroring perfectly its critical behavior as a functional of* q.

Beyond this, the structural ingredients of the stratification–critical graphs, separatrix diagrams, splitting of connected components–each has a natural analogue in spectral theory–singular γ cycles, critical points of Δ, and the transition from "crosses" to "gaps".

We emphasize that spectral theory naturally extends beyond four to higher (even infinite) dimensions. In general, there is a one to one correspondence between critical points and Dirichlet $\{\mu_j\}$ eigenvalues; thus, one can define a Bott function for each critical point, and hence for each angle variable. Specifically,

$$F_j(q) = \Delta(q; \dot{\lambda}_j),$$

where λ_j denotes the jth critical point. Given the analyticity of Δ in both q and λ, the Floquet discriminant provides a powerful tool for studying topological questions in these soliton models. We have begun this study in the cases of the Toda lattice, the discrete NLS systems and the sine-Gordon system.

Even in the two degree of freedom case, some interesting questions remain. For example, in the sine-Gordon setting examples exist in three topological classes rather than just the two which arise in NLS. In this case, the relationship between "charge", the integer "m" of reference [**3**], and the possibility of "twisted products" remains to be sorted out.

We close with a remark concerning an application of the topological classification of integrable systems. Perturbations of integrable systems can possess chaotic solutions. In our studies of one such example, a damped driven NLS equation, we have introduced [**9,10**] a low dimensional mode truncation of the exact dynamics as a model problem. Although the truncation is a severe one which retains only the first two Fourier modes, we have argued that the truncation provides a useful model because it is "faithful to the geometry of the unperturbed system". A more precise statement is that the unperturbed bifurcations in both the truncated model and the full equation belong to the same topological class; hence, one expects qualitatively similar behavior in both systems under perturbations. In fact, the chaos seen in these equations is closely related to the connectivity discussed in Section 4 [**9,11,12**]. In this manner the topological classification can provide a rationale for truncated model systems. Moreover, in this setting of near-integrable pde's the Bott integral $\Delta(q, \dot{\lambda}_j(q))$ is the most natural integral for the generation of a Melnikov function and its associated distance measurements.

REFERENCES

1. A.T. Fomenko and H. Zieschang, *Topological Classification of Integrable Hamiltonian Systems*, Preprint France (1988), IHES; *Criterion of Topological Equivalence of Integrable Hamiltonian Systems with Two Degrees of Freedom*, Izvestiya Akad. Nauk SSSR (1990).
2. A.T. Fomenko, and H. Zieschang, *Symplectic Topology of Completely Integrable Hamiltonian Systems*, Uspechi Matem. Nauk **44,** No. 1 (1989), 145–173.
3. N. Ercolani, M.G. Forest, and D.W. McLaughlin, *Geometry of the Modulational Instability*, Memoirs of the AMS (to appear); Physica D **43** (1990), 349–384.
4. M.G. Forest and D.W. McLaughlin, *Modulations of Sinh-Gordon and Sine-Gordon Wavetrains*, Studies in Appl. Math. **68** (1983), 11–59.
5. N. Ercolani and M.G. Forest, *The Geometry of Real Sine-Gordon Wavetrains*, Commun. Math. Phys. **99** (1985), 1–49.
6. E. Overman and C. Schober. Private communications (1989).
7. M.S. Ablowitz and B.M. Herbst, *On Homoclinic Boundaries in the Nonlinear Schröedinger Equation*, Proc. CRM Workshop on Hamiltonian Systems, ed. by J. Harnad and J.E. Marsden, CRM Publication, Univ. Montreal (1989).
8. E. Previato, *Hyperelliptic Quasi-periodic and Soliton Solutions of the Nonlinear-Schroedinger Equation*, Duke Math. J. **52** (1985), p. 329.
9. A. Bishop, M.G. Forest, D.W. McLaughlin, and E.A. Overman, *Model Representations of Chaotic Attractors for the Driven Damped Pendulum Chain*, Phys. Lett. **A144** (1990), 17–25.
10. A.R. Bishop, D.W. McLaughlin, M.G. Forest, and E.A. Overman II, *Quasi-periodic Route to Chaos in a Near Integrable PDE: Homoclinic Crossings*, Phys. Lett. **A127** (1988), 335–340.
11. H.T. Moon, *Homoclinic Crossings and Pattern Selection*, Phys. Rev. Lett. **64** (1990), 412–414.
12. G. Kovacic and S. Wiggins, *Orbits Homoclinic to Resonances: Chaos in a Model of the Forced and Damped Sine-Gordon Equation*, Preprint, Cal. Inst. Tech. (1989).

N. ERCOLANI: Department of Mathematics
Univesity of Arizona
Tucson, Arizona 85721

D. MCLAUGHLIN: Department of Mathematics
Program in Applied and Computational Mathematics
Princeton University
Princeton, New Jersey 08544

Support is gratefully acknowledged from the National Science Foundation under grant # DNS8703397 and from the United States Air Force under grant #AFOSRU F49620–86–C0130. We wish to take this opportunity to thank R. Devaney, H. Flaschka, K. Meyer, and T. Ratiu for organizing the workshop at the Mathematical Sciences Research Institute where we first learned of Professor Fomenko's work through his lectures at that workshop. Finally, one of us (DWM) acknowledges several extremely useful conversations with P. Deift.

Topological Classification of All Integrable Hamiltonian Differential Equations of General Type With Two Degrees of Freedom

A. T. FOMENKO

This paper is dedicated to Professor Stephen Smale

CONTENTS

CHAPTER 1. THE BASIC NOTIONS AND THE CLASSIFICATION THEOREMS

1. Introduction.

This paper is based on the series of lectures which were delivered by the author in 1989 at the Mathematical Sciences Research Institute in Berkeley. As part of the year-long 1988–1989 program in Symplectic Geometry and Mechanics, the Berkeley Mathematical Sciences Research Institute hosted a two-week workshop on the Geometry of Hamiltonian Systems on the period June 5 to June 16, 1989.

The author expresses his thanks to MSRI and its Director, Professor Irving Kaplansky, for the invitation to take part in the work of the conference in *Symplectic Geometry and Mechanics*. The author thanks Professor J. Marsden and Professor T. Ratiu who suggested that I give this course on *Geometry and Topology of Integrable Hamiltonian Systems*.

The author thanks Professor J. Marsden, Professor T. Ratiu and Professor A. Weinstein for humerous useful discussions. I want to mention here the deep impression from the workshop, in particular from the excellent lectures of Professor J. Marsden and Professor A. Weinstein. I thank Professor T. Ratiu for the help in the organization of the lectures.

The author thanks Professor M. Adler, Professor J. Arms, Professor R. Cushman, Professor H. Flaschka, Professor D. Rod, Professor C. Scovel, Professor D. McLaughlin, Professor P. van Moerbeke, Professor L. Haine and Professor H. Ito for useful conversations on the problems of Hamiltonian mechanics.

The paper contains the new theory of topological classification of integrable Hamiltonian systems of differential equations with two degrees of freedom. This theory was created by the author and then was developed in collaboration with my colleagues, in particular with H. Zieschang, S. V. Matveev, A. V. Bolsinov, A. A. Oshemkov, A. V. Brailov and V. V. Sharko. Many interesting results were then obtained by Nguen Tyen Zung, G. G. Okuneva, L. S. Polyakova, E.N. Selivanova, B. B. Kruglikov, V. V. Kalashnikov (junior).

The theory develops some important ideas of I. M. Gel'fand, S. P. Novikov, V. I. Arnold, S. Smale, R. Bott, J. Marsden, T. Ratiu, A. Weinstein, J. Moser, V. V. Kozlov, M. P. Charlamov and F. Waldhausen. It should be mentioned that the origin of this general theory can be deduced from the famous paper of S. Smale *Topology and Mechanics* [**94**]. The author of the present paper was deeply impressed by this excellent work and here I want to express my admiration toward the right ideas of Smale concerning the topology of Hamiltonian systems.

The basis of the theory is contained in the following main publications:

A. T. Fomenko, *The topology of surfaces of constant energy in integrable Hamiltonian systems, and obstructions to integrability*, Math. USSR Izvestiya., Vol. 29, No. 3 (1987), 629–658 (see [**29**]).

A. T. Fomenko, "Integrability and Non-integrability in Geometry and Mechanics", Kluwer Acad. Publ., 1988 (see [**33**]).

A. T. Fomenko, *Topological invariants of Hamiltonian systems integrable in Liouville sense*, Func. Anal. and its Appl., Vol. 22, No. 4 (1988), 38–51 (see [**30**]).

A. T. Fomenko, *Symplectic topology of completely integrable Hamiltonian systems*, Uspekhi Matem. Nauk., Vol. 44, No. 1(265) (1989), 145–173 (see [**31**]).

The further development of the theory can be seen in:

A. V. Bolsinov, A. T. Fomenko and S. V. Matveev, *Topological classification and arrangement with respect to complexity of integrable Hamiltonian systems of differential equations with two degrees of freedom. The list of all integrable systems of low complexity*, Uspekhi Matem. Nauk., Vol. 45, No. 2 (1990), 49–77 (see [**15**]).

A. T. Fomenko and H. Zieschang, *Criterion of topological equivalence of integrable Hamiltonian systems with two degrees of freedom*, Izvestiya Akad. Nauk. SSSR, Vol. 54, No. 3, (1990), 546–575 (see [**43**]).

See also:

A. V. Brailov and A. T. Fomenko [**19**], A. T. Fomenko [**28**]–[**36**], A. T. Fomenko and V. V. Sharko [**38**], A. T. Fomenko and H. Zieschang [**41**]–[**43**], S. V. Matveev, A. T. Fomenko and V. V. Sharko [**70**], S. V. Matveev and A. T. Fomenko [**71**], [**72**].

In the present paper we consider Hamiltonian systems of differential equations which are integrable in Liouville sense and are systems "of general type", or "in general position" (see details below). We will consider the integrable systems up to topological equivalence, namely two systems are topologically equivalent iff there exists some diffeomorphism which transforms the set of Liouville tori of the first system to the set of Liouville tori of the second system. Let us recall here several basic problems:

1) Let us consider two integrable Hamiltonian systems "of general type" v_1 and v_2. How does one decide whether they are topologically equivalent or not?

2) Classify integrable Hamiltonian systems "of general type" up to topological equivalence.

3) Does there exist some topological invariant which can classify integrable systems?

4) Find some new topological obstructions to integrability.

5) What is the "complexity" of an integrable system?

6) How does one describe all integrable Hamiltonian systems (of general type) which have low complexity?

It happens that all above questions have a positive answer (in some strong sense as we shall see below).

Let us present here a rough outline of the whole paper. We list the main topics of the work.

1) Liouville integrability of Hamiltonian systems. Liouville tori and Liouville foliation of isoenergy surfaces.

2) Bott integrals. Non-resonance integrable systems. The systems of "general type" = the systems "in general position". In the non-resonance case the whole theory does not depend on the choice of a Bott integral for a given integrable system.

3) Topological equivalence of integrable systems.

4) Formal brief description of the topological classification of integrable systems.

5) Geometrical construction of new topological invariants for integrable systems.

6) The invariant $I(H,Q)$ and the marked invariant $I(H,Q)^*$.

7) Main theorem: the integrable Hamiltonian systems of general type are topologically equivalent iff their marked invariants coincide, i.e. $I(H_1,Q_1)^* = I(H_2,Q_2)^*$.

8) Classification of integrable Hamiltonian systems of general type with two degrees of freedom.

9) Classification of three-dimensional isoenergy surfaces of integrable Hamiltonian systems. Canonical representation of constant-energy surfaces of integrable equations.

10) New Morse type theory for the Bott integrals on the isoenergy surfaces.

11) Classification of all bifurcations of Liouville tori inside the isoenergy surfaces of integrable systems.

12) The main examples from mathematical physics and mechanics.

13) The non-singular constant-energy surfaces of integrable Hamiltonian systems possess specific properties which distinguish them among all smooth three-dimensional manifolds. Not all three-manifolds are constant-energy surfaces for some integrable system.

14) Multidimensional case; the theory of integrable systems with arbitrary number of degrees of freedom.

15) New topological obstacles to integrability of Hamiltonian systems in the class of Bott integrals. If the isoenergy manifold does not belong to some remarkable class (H), then the Hamiltonian system is non-integrable on Q.

16) Application to the theory of three-dimensional closed hyperbolic manifolds. Calculation of the volume of about 1000 closed compact hyperbolic three-manifolds.

17) The estimation of the number of periodic solutions of integrable Hamiltonian systems in terms of homology groups of corresponding isoenergy surfaces (three-dimensional case).

18) The notion of complexity for integrable systems with two degrees of freedom. The connection with the deep properties of three-dimensional manifolds.

19) The algorithm of enumeration of all integrable systems (up to topological equivalence).

20) The list of all integrable systems of low complexity.

21) The code of integrable systems can be represented as some "molecule" which is constructed from "atoms". In this sense the set of all integrable systems is precisely the set of all "molecules". The list of all integrable systems is similar to the table of all chemical elements.

23) Connection with the theory of Morse–Smale flows. The Hamiltonian mechanics and the theory of "round Morse functions".

24) Computers in Hamiltonian mechanics and symplectic topology. Our computer experiments and results.

25) The values of new topological invariant for famous physical systems (Kovalevskaya case, Euler case, Lagrange case, Toda lattice and so on).

26) The "physical zone" inside the list of all integrable Hamiltonian systems. This zone is formed by all the systems which are discovered today in concrete problems of physics and mechanics. It is very interesting to "draw" this zone inside the table of all possible integrable systems (most of them of course have formal character and are not yet discovered in physics).

This program is realized in the present paper. Of course the lack of place prevents us from giving many details, but we will try to present here all the main ideas and results. (Most recent results see in Appendix 3).

2. Basic notions of the theory.

2.1. Symplectic manifolds and Hamiltonian systems.

DEFINITION. *A smooth even-dimensional manifold M^{2n} is called symplectic if on this manifold there is an exterior differential two-form (that is, a form of degree 2)*

$$\omega = \sum_{i<j} \omega_{ij} dx_i \wedge dx_j$$

which possesses the following properties:

1) This form is non-degenerate, that is, the matrix of its coefficients $\Omega(x) = (\omega_{ij}(x))$ is non-degenerate at each point of the manifold.

2) This form is closed, that is, its exterior differential is equal to zero: $d\omega = 0$. Such a form is called a symplectic structure on the manifold M^{2n}.

Here $x_1, \dots, x_{2n}$ are local regular coordinates on the manifold.

DEFINITION. *Let f be a smooth function on M^{2n} and ω a symplectic structure on M. A smooth vector field on M which is uniquely defined by the relation $\omega(v, \operatorname{sgrad} f) = v(f)$, where v runs through the set of all smooth vector fields on M and $v(f)$ is the value of the field v on the function f (that is, the derivative of the function f in the direction of the field v), is called the skew-symmetric gradient* sgrad f *of the function f (or, simply, a skew gradient).*

We see that the definition of sgrad f copies the definition of the vector field grad f. The only difference is that, instead of the *symmetric* tensor g_{ij} of Riemannian metric, one considers the *skew-symmetric* tensor ω_{ij}.

Let us recall the famous Darboux theorem. Let ω_{ij} be a symplectic structure on M^{2n}. Then, for any point $x \in M$, there always exists an open neighborhood with local regular coordinates $p_1, \dots, p_n, q_1, \dots, q_n$, such that the form ω is written in the canonical form $\Sigma dp_i \wedge dq_i$, that is, at each point of this neighborhood the matrix (ω_{ij}) has the simple form $\begin{pmatrix} o & E \\ -E & o \end{pmatrix}$.

DEFINITION. *Local coordinates* $p_1, \dots, p_n, q_1, \dots, q_n$ *on a symplectic manifold are called* ***symplectic*** *if in these coordinates the form* ω *is written in canonical form.*

How will a vector field sgrad f be written in these symplectic coordinates? Since the matrix of the form ω looks like $\begin{pmatrix} o & E \\ -E & o \end{pmatrix}$, and since the gradient is given by the formula:

$$\operatorname{grad} f = (\partial f / \partial p_1, \dots, \partial f / \partial q_n),$$

it follows that the skew gradient is written as

$$\operatorname{sgrad} f = (\partial f / \partial q_1, \dots, \partial f / \partial q_n, -\partial f / \partial p_1, \dots, -\partial f / \partial p_n).$$

Consequently each symplectic structure can be locally reduced to the form $\omega = \Sigma dp_i \wedge dq_i$.

DEFINITION. *A smooth vector field* v *on a symplectic manifold* M*, which has the form* $v = \operatorname{sgrad} H$*, is called a* ***Hamiltonian field*** *on* M *if the smooth function* H *is defined on the entire manifold* M*. The function* H *is then called* ***Hamiltonian*** *(or energy function).*

Each Hamiltonian vector field can be considered as a Hamiltonian system of differential equations on M and conversely. Such a system can be written (in the corresponding symplectic coordinates) in the form:

$$dp_i/dt = \partial H / \partial q_i, \quad dq_i/dt = -\partial H / \partial p_i.$$

The function $\{f, g\} = \omega(\operatorname{sgrad} f, \operatorname{sgrad} g)$ is called the *Poisson bracket* of smooth functions f and g on a symplectic manifold M with a form ω. If we denote by ω^{ij} the coefficients of the inverse matrix to (ω_{ij}), then the Poisson bracket can be written in local coordinates $x_1, \dots, x_{2n}$ as

$$\{f, g\} = \sum \omega^{ij} \partial f / \partial x_i \, \partial g / \partial x_j.$$

The Poisson bracket has a simple interpretation. There exists an important relation: $\{f, g\} = (\operatorname{sgrad} f) g$, that is, *the Poisson bracket coincides with the derivative of the function* g *in the direction of the vector field* $\operatorname{sgrad} f$.

2.2. The Liouville theorem. Liouville tori.

We will say that *two functions f and g are in involution* on a symplectic manifold M if their Poisson bracket is exactly zero. It turns out that for Hamiltonian systems it suffices to find only n independent integrals in involution, in order to completely describe the integral trajectories. We will present the known Liouville theorem which solves this integration problem **[6]**.

THEOREM 2.1. *Let a set of n smooth functions $f_1, \dots, f_n$ in involution be given on a symplectic manifold M^{2n}. For example let $f_1, \dots, f_n$ be the independent involutive integrals of some Hamiltonian system $v = \operatorname{sgrad} H$, where $H = f_1$ is a given smooth Hamiltonian on M. We will denote by M_ξ the common level surface given by the system of equations*

$$f_1(x) = \xi_1, \dots, f_n(x) = \xi_n.$$

Suppose that on this surface all the functions f_i are functionally independent. Then:

1) The level surface M_ξ is a smooth n-dimensional submanifold invariant with respect to each vector field $v_i = \operatorname{sgrad} f_i$, that is, all these fields are tangent to the level surface M_ξ.

2) If the level surface M_ξ is connected and compact, it is diffeomorphic to an n-dimensional torus T^n. In the non-compact case M_ξ is diffeomorphic to a factor of the Euclidean space R^n by a certain lattice ("cylinder").

3) If the level surface is a torus T^n, then in a certain open neighborhood of this surface, one may introduce regular curvilinear coordinates $s_1, \dots, s_n$; $\varphi_1, \dots, \varphi_n$ called "action-angle variables", which have the following properties:

3a. The functions $s_1(x), \dots, s_n(x)$ define coordinates in the directions transversal to the torus T^n and are functionally expressed through the integrals $f_1, \dots, f_n$. In these coordinates, the equation of the torus is given as $s_1 = \dots = s_n = 0$.

3b. The vector field v has the simplest possible form on the torus T^n in the coordinates $\varphi_1, \dots, \varphi_n$: its components are all constant, and its integral trajectories are a rectilinear winding on the torus, that is, almost periodic motion on the torus.

For other details of the Liouville theorem see, for example, **[6]** or **[33]**. Thus, if the integrals $f_1, \dots, f_n$ are independent, then *non-singular compact* *common level surfaces are unions of tori.* On the torus itself, the Hamiltonian system of differential equations $v = \operatorname{sgrad} H$ which we now investigate is constructed extremely simple. In the coordinates $\varphi_1, \ldots, \varphi_n$, it becomes a field with *constant* components, i.e. it is completely defined by setting the velocity vector at a certain point of the torus. All other velocity vectors are obtained from it by parallel transport in the coordinates $\varphi_1, \ldots, \varphi_n$. We will say that a Hamiltonian system of differential equations $v = \operatorname{sgrad} H$ *is completely Liouville-integrable (or admits a complete commutative integration)* if for this system there exists a set of n independent functions $f_1(= H), f_2, \ldots, f_n$ which are in involution (i.e. satisfying the conditions of the Liouville theorem). The independence of the functions f_i we consider as *almost everywhere* on the manifold or in the neighborhood of some fixed isoenergy surface. The tori T^n in the Liouville theorem are called *Liouville tori.* These tori define *the Liouville foliation* (with singularities) of the phase space M^{2n}.

Hamiltonian differential equations of "general position" are, as a rule, non-integrable. Since of particular importance for mechanical applications are the cases where the equations are integrable in Liouville sense, it becomes clear how difficult it is to find such rare integrable systems in the vast ocean of systems, the "majority" of which are non-integrable. It is natural to start with the following question: *How is the property of integrability of a Hamiltonian system of equations associated with the topology of the phase (or configuration) manifold?*

Complete Liouville integrability of a Hamiltonian equation is traditionally assumed to provide a more or less detailed qualitative description of the behavior of integral trajectories. In principle, this is certainly the case. But one often ignores the fact that for such description one should effectively find the action-angle variables. Embeddings of Liouville tori into the phase manifold M may be rather complicated. It is therefore relevant to ask the following questions. How are the Liouville tori arranged in the phase space? How do they adjoin one another, fill open domains, how are they transformed in the neighborhood of critical surfaces, etc.? In other words, what is the way of constructing a qualitative theory of topological arrangements and interactions of Liouville tori, for instance, on a constant-energy surface of a system.

The famous Kolmogorov–Arnold–Moser theorem states that small perturbations of integrable systems are not ergodic and have an invariant subset of positive measure. More precisely, if a non-perturbed Hamiltonian system is non-degenerate (or isoenergy-non-degenerate), then for sufficiently small Hamiltonian perturbations, most of the non-resonant Liouville tori do not disappear. They slightly deform and are covered by almost periodic integral trajectories. These tori are "typical" in the following sense: the measure of the complement to their union is small when the perturbation is small. If a Hamiltonian system is close to an integrable system (and is non-degenerate on isoenergy surfaces), the invariant tori cover most of the manifold of constant energy. Thus, all Hamiltonian systems which are close to integrable ones, are characterized by the existence of a "rich" set of invariant tori on isoenergy surfaces. A small perturbation of an initially integrable Hamiltonian system induces some small diffeomorphism of its isoenergy surface. As a result we obtain the isoenergy surface of the non-integrable system (in the case of perturbation of general type). Consequently, the information on the topology of isoenergy surfaces of integrable systems gives us the information about the topology of isoenergy surfaces of all non-integrable systems close to integrable ones. The study of this topology is the aim of our work.

2.3. Momentum mapping for integrable systems.

Let M^{2n} be a smooth symplectic manifold (compact or non-compact) and $v = \operatorname{sgrad} H$ be a Hamiltonian system (vector field) on M with a smooth single-valued Hamiltonian H. Let the system v be Liouville integrable, i.e. v has a complete set of commuting integrals: $f_1 = H, f_2, \ldots, f_n$ such that $\{f_i, f_j\} = 0$ and all functions are independent (almost everywhere) on M.

By $F : M^{2n} \to R^n$ we denote the *momentum mapping (the mapping of the moment)* of an integrable system, defined by

$$F(x) = (f_1(x), \ldots, f_n(x)).$$

The point $x \in M$ is called *a critical point* for the mapping F if the $\operatorname{rank} dF(x) < n$. The set of all critical points for the mapping F is denoted by N. Then its image, i.e. the set $\Sigma = F(N) \subset R^n$ is called the *bifurcation diagram* of the momentum mapping. The points $c \in \Sigma$ are called *the critical values of the momentum mapping*, while the points $a \in R^n \backslash \Sigma$ are called *the non-critical values*. Let a be a non-critical value. Then, in view of the Liouville theorem, each connected compact component of the complete inverse image $F^{-1}(a)$ is homeomorphic to an n-dimensional torus T^n. If the point a slides along a smooth curve, the Liouville torus $F^{-1}(a)$ is deformed inside the manifold M. These tori are somehow rearranged when the point a "pierces" the bifurcation set Σ. Later we will describe all such rearrangements in general position and classify them.

For the sake of simplicity, we begin with the four-dimensional case, although we have proved almost all the results formulated in the general multidimensional case as well.

2.4. Isoenergy surface of integrable systems, i.e. constant-energy surface.

Let M^4 be a four-dimensional (compact or noncompact) symplectic manifold, and $v = \operatorname{sgrad} H$ be a Hamiltonian system with the Hamiltonian H, having a second supplementary independent smooth integral f, such that $\{H, f\} = 0$. We will study the integrability of the system on a single isolated constant-energy surface $Q^3 = \{H = \text{const}\}$.

We should emphasize that, in our opinion, special attention should be given to *integrability on a separate fixed constant-energy surface.* The reason is that in mechanics and physics one often has to deal with a Hamiltonian system integrable only on *one isoenergy surface* and non-integrable on the others. Analytic systems are usually integrable either simultaneously on all regular level surfaces of Hamiltonian or on none of them [**52**], [**43**], [**120**], [**121**] (V. V. Kozlov, S. L. Ziglin). See also the work of G. G. Okuneva in Appendix 3.

For this reason we think it to be instructive to consider *smooth systems and smooth integrals* which admit a simultaneous presence of both "integrable" and "non-integrable" constant-energy surfaces inside M.

In this connection, the following general problem arises. Let a Hamiltonian system with a Hamiltonian H be given. One should establish whether among the constant-energy surfaces of this system there exists at least one on which the system is integrable.

It is natural to hypothesize that many Hamiltonian systems (non-integrable globally) do have such a surface. Clearly, a smooth (or analytic) integral f, which integrates a system $v = \operatorname{sgrad} H$ only on one Hamiltonian level surface, satisfies an equation weaker that the ordinary equation

$\{H, f\} = 0$. Namely, this holds on the integral surface itself, whereas outside this surface, it may fail. In the simplest form, this condition can be written as follows: $\{H, f\} = \lambda(H)$, where the function $\lambda(H)$ is such that $\lambda(0) = 0$ and $\lambda'(0) = 0$. We assume that the isolated Hamiltonian level surface of interest is given by the equation $H = 0$.

Thus, the general equation $\{H, f\} = \lambda(H)$ deserves a most thorough examination. On a Hamiltonian level surface, consider two vector fields $\operatorname{sgrad} H$ and $\operatorname{sgrad} f$. For these vector fields to commute on a level surface, it suffices that the gradient of the function $\{H, f\}$ vanish on the surface $H = 0$. Indeed, we know that $[\operatorname{sgrad} H, \operatorname{sgrad} f] = \operatorname{sgrad}\{H, f\}$. Thus, the function $\lambda(H)$ must be quadratic in H in the neighborhood of the value $H = 0$. In particular, it should be possible first to investigate the general properties of the equation $\{H, f\} = \varepsilon H^2$, where ε is a non-zero constant.

From the Liouville theorem it follows that all non-singular two-dimensional compact level surfaces of a second integral f (on a constant-energy manifold) are unions of tori. It turns out that the *structure of singular level surfaces of the integral f can also be completely described.*

Since H is an integral of the system v, it follows that the field v may be restricted to an invariant three-dimensional isoenergy surface Q, that is, $Q = \{x \in M : H(x) = \text{const}\}$. Being a symplectic manifold, M is orientable, and *therefore the isoenergy manifold Q is also orientable.* Consider a non-critical level surface Q, that is, such a surface on which $\operatorname{grad} H \neq 0$. Then $\operatorname{sgrad} H \neq 0$ in all points $x \in Q$, i.e. $v \neq 0$ on Q. Let us consider the second independent integral f (see above). We restrict it to the surface Q and obtain a smooth function (we denote this function again by f). As has already been mentioned, we will consider integrability of the system only *on one separate isoenergy surface.* G. G. Okuneva proved that any smooth integrable system can be perturbed in such a way, that the integrability remains only on precisely one isoenergy surface.

2.5. Bott integrals.

DEFINITION (see [**28**], [**29**]). *We will call a smooth integral f a Bott integral on an isoenergy surface Q if the critical points of the function f form on Q non-degenerate critical smooth submanifolds.*

The general properties of such functions were studied in the well-known papers of R. Bott (see, for example [**16**]). Based on this, it makes sense to call such functions *Bott functions.*

Recall that a critical submanifold of a function f is called *non-degenerate* if the Hessian $d^2 f$ of the function f is non-degenerate on normal planes to the submanifold. In other words, the integral f can be considered as a Morse function in all *normal directions* to the critical submanifold (Figure 1). In our case, non-degenerate critical submanifolds for the integral f on Q may be *one-dimensional and two-dimensional.* We do not have *zero-dimensional* critical submanifolds because $v \neq 0$ on Q and f is constant on all integral trajectories of the field v. The field v does not have isolated points as integral trajectories.

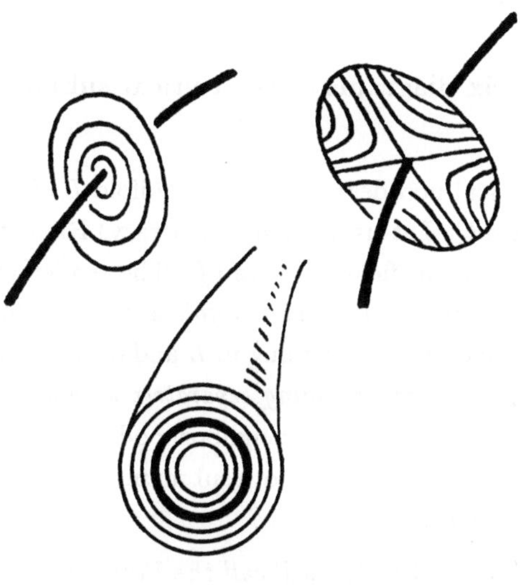

Figure 1

Important "experimental fact": the investigation of concrete integrable systems (see, for instance, the works of M. P. Charlamov [**21**]–[**23**], T. I. Pogosyan [**89**], A. A. Oshemkov [**84**]–[**87**], A. T. Fomenko, A. V. Bolsinov, L. S. Polyakova and M. R. Vovchenko [**15**], V. V. Kozlov [**52**] etc.) *has shown that the overwhelming majority of the discovered integrals in concrete mechanical and physical systems are Bott integrals on almost all regular isoenergy surfaces Q^3 in M^4.* See also the papers of E. N. Selivanova and B. S. Kruglikov.

Consequently, the class of Bott integrals introduced in our paper, is sufficiently large.

CONJECTURE. *Bott integrals are the integrals "in general position" (of general type) in integrable Hamiltonian mechanics.*

This conjecture is not yet proved because we need here the strong notion of "general position". It would be very interesting to describe the set of integrable systems having Bott integrals on almost all isoenergy surfaces.

DEFINITION. *The subdivision of the phase space M^4 and the isoenergy surface Q^3 into the union of the Liouville tori and the connected components of the critical surfaces of the Bott integral f will be called Liouville foliaton on M and Q respectively.*

2.6. Separatrix diagrams of the critical submanifolds of Bott integrals.

Consider the critical non-degenerate submanifold L of the integral f on an isoenergy surface Q. Let us fix some (arbitrary) Riemannian metric on Q and consider the vector field $w = \operatorname{grad} f$. The reader will recall that a *separatrix diagram of the critical submanifold* L is the union of all integral trajectories of the field $\operatorname{grad} f$ incoming in L and outgoing from L. In view of this, we shall speak of an *incoming separatrix diagram* and an *outgoing separatrix diagram.* In a small open neighborhood of a critical submanifold L, both in- and out-diagrams are smooth submanifolds. They may be either *orientable* or *non-orientable.*

DEFINITION (see [**28**], [**29**]). *We will call the Bott integral f orientable on the isoenergy surface Q if all separatrix diagrams of its critical submanifolds are orientable. If at least one of its separatrix diagrams is non-orientable, we say that the integral f is non-orientable.*

2.7. Classification of critical submanifolds of Bott integrals.

It is evident that a smooth Bott integral f cannot have isolated critical points on a non-singular isoenergy surface Q. The proof follows from the fact that Q contains no critical points of the function H. Therefore, from each critical point x_o of the function f on Q, there goes a non-degenerate integral trajectory of the field v which consists entirely of the critical points of the function f.

LEMMA 2.1. *The critical connected submanifolds L of a smooth Bott integral on a compact non-singular isoenergy surface Q^3 are either circles, two-dimensional tori, or Klein bottles.*

If f is a Bott integral of the system v and L is its critical submanifold, then L admits a non-degenerate tangent vector field, namely the restriction of the field v to L. Thus, its Euler characteristic is equal to zero.

It turns out that in some strong sense the critical Klein bottles are "non-interesting submanifolds".

STATEMENT 2.1 (see [**29**]). *Let Q^3 be non-singular compact isoenergy surface in M^4 and f a Bott integral on Q having critical Klein bottles among its critical submanifolds. Let $U(Q)$ be a sufficiently small tubular neighborhood of the surface Q in M. Then there exists a two-sheeted (two-fold) covering*

$$\pi : (U'(Q'), H', f') \to (U(Q), H, f)$$

(with a fibre Z_2), where $U'(Q')$ is a symplectic manifold with a new Hamiltonian system $v' = \operatorname{sgrad} H'$ with the new Hamiltonian H' which is of the form $H' = \pi^*(H)$. This new system v' is integrable on the manifold $Q' = \pi^{-1}(Q)$ by means of a Bott integral $f' = \pi^*(f)$. All critical submanifolds of the integral f' are orientable (i.e. circles and tori). In this case, all the critical Klein bottles in Q "unfold" into critical tori T^2 in Q'. The manifold $U'(Q')$ is a tubular neighborhood of the manifold Q'.

Consequently, if we consider the integrable systems up to two-fold coverings of their isoenergy surfaces, we can assume that all critical submanifolds of the Bott integral f are orientable (no Klein bottles!). It should be mentioned here that our experience in the investigation of concrete mechanical systems shows that practically in all cases we do not see critical Klein bottles in real physical situations. Thus, every integrable system with Bott integral can be covered by some new integrable system, where the new Bott integral does not have critical Klein bottles.

From this it follows that if f is a Bott integral (on Q) with critical Klein bottles, then $\pi_1(Q) \neq 0$ and the group $\pi_1(Q)$ contains a subgroup of index two. Here $\pi_1(Q)$ denotes the fundamental group of the manifold Q. If, for instance, the isoenergy surface Q is homeomorphic to the three-sphere S^3 (an important case in mechanics), then *any Bott integral f on the sphere S^3 does not have critical Klein bottles.*

2.8. The bifurcations of Liouville tori. Simple and complicated integrals.

The basic idea of the papers [**28**]–[**30**] is as follows. We investigate the motion of Liouville tori inside the manifold Q, induced by the change in value of the integral f. The tori can transform, disappear, and appear. This process is a particular case of the general phenomena of dependence of the solutions of differential equations of the initial data (see details below). It should be mentioned here that today there exist powerful computer programs which allow us to see on the computer screen the transformations of Liouville tori in phase space (see, for example the excellent paper of P. J. Channell and C. Scovel [**20**]). The general theory of transformations of Liouville tori and their classification was constructed by the author in [**28**]–[**30**].

Let us describe the transformations of Liouville tori when they pass through the critical value c of the integral f. Let us denote by R the *connected* component of the set $f^{-1}(c)$ containing the critical points of f. Two possibilities can occur: a) The critical submanifold L in R is *connected.* We call the integral f in this case *simple* (on R). b) The critical submanifold L in R consists of several connected components (more than 1). In this case we call the integral f *complicated* (on R). We will call the integral f *simple (complicated)* on Q if f is *simple (complicated)* on all sets $f^{-1}(c)$ (on at least one set $f^{-1}(c)$). Problem: Is it possible to describe the bifurcations of the Liouville tori inside Q? The answer is positive.

THEOREM 2.2 (see [**29**]). *Let us suppose that the integral f is simple on the isoenergy surface Q. Then only the following cases are possible:*

a) *L is a maximal circle (the local maximum of the integral f). Then there exists the one-parameter family $\{T_\varepsilon \subset f^{-1}(c-\varepsilon), \varepsilon > 0\}$ of the Liouville tori T_ε which decreases when ε decreases and which coincides with the circle when $\varepsilon = 0$. (See Figure 2–a, where the formal picture is represented; the real picture is obtained by multiplying on the circle S^1; see Figure 2–b). In the case of the minimal circle L (the local minimum of the integral f) the Liouville tori transform in a similar way: the critical circle transforms (blows) into a one-parameter family of Liouville tori. Hence $L = R$.*

b) *L is the maximal or minimal two-dimensional torus (the local maximum or minimum of f). Then there exist two one-parameter families of Liouville tori, which come together (when the value of the integral tends*

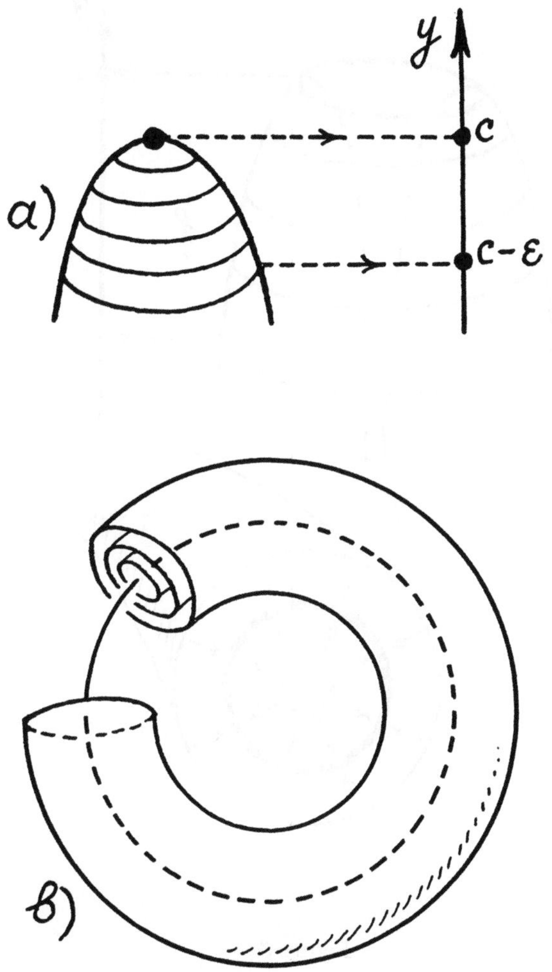

Figure 2

to the critical one) and finally stick together in the torus L (Figure 3). (It is important that from the topological point of view we do not have any transformation of a single Liouville torus. Consequently, in further considerations we include the minimal and maximal tori in the set of "usual" Liouville tori). Here we also have $L = R$.

c) L is the maximal or minimal Klein bottle (the local maximum or minimum of f). Then L is the limit of a one-parameter family of Liouville

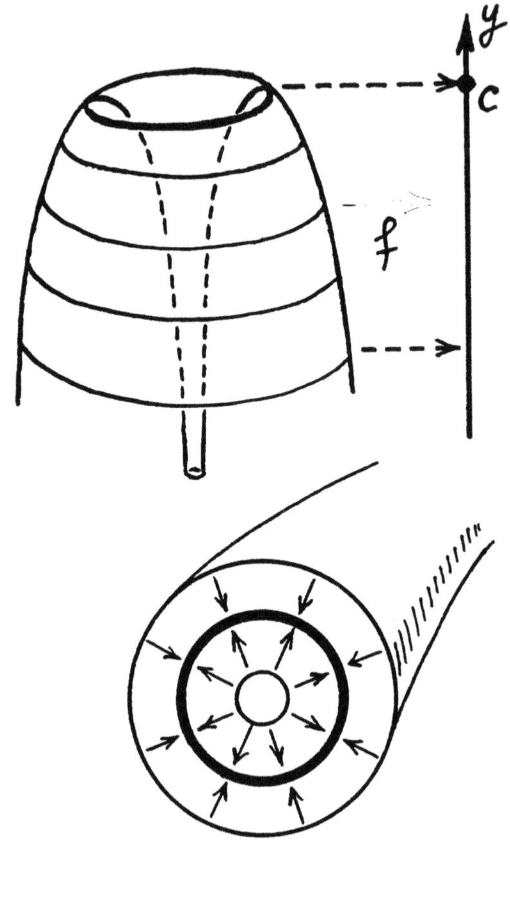

Figure 3

tori. (As we mentioned above, in real physical systems critical Klein bottles practically never appear. Thus usually we will suppose that our integral does not have Klein bottles. Of course, the general theory in **[29]** *was constructed in the case of arbitrary Bott integral). Here* $L = R$.

d) *The connected component* R *of the set* $f^{-1}(c)$ *contains precisely one hyperbolic critical circle* L. *In that case* R *is the compact piecewise smooth two-dimensional polyhedron which has singularities of multiplicity four: the singular curve can be locally represented as the transversal intersection of* *two planes (Figure 4). Each such singular curve is closed and coincides with one of the hyperbolic (saddle) critical circles of the integral f. Let X be the "cross"—the bouquet of 4 segments with common end. The neighborhood of this circle L in R is obtained from the direct product $X \times I$ of X on the segment I, when we identify two crosses on the boundary using either the identity mapping or the symmetry with respect to the vertex of the cross (Figure 5–a in the first, orientable, cases and Figure 5–b in the second, non-orientable, case). In the first case we will speak about an orientable saddle, in the second case, about a non-orientable saddle.*

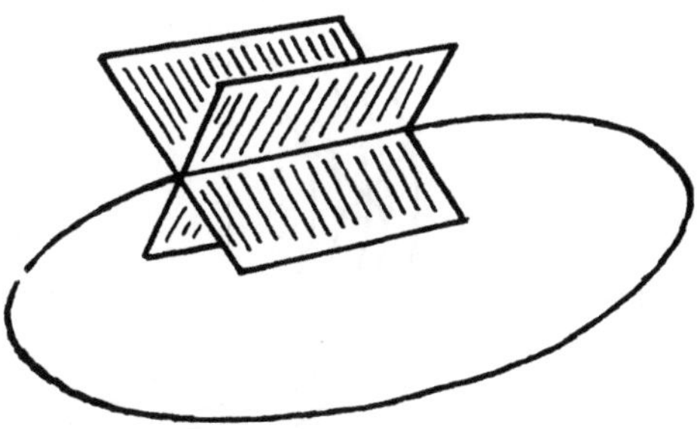

Figure 4

In case d) the separatrix diagram of the critical circle can be orientable and non-orientable. It is clear that on Figure 5–a the diagram is orientable and in Figure 5–b, non-orientable.

THEOREM 2.2* (see [**29**]). *In the case of an arbitrary Bott integral f on Q (in particular, in the case of a complicated integral) every bifurcation of Liouville tori is the composition of elementary bifurcations of the types* a)–d) *described in Theorem* 2.2.

In this sense the five types of elementary bifurcations (namely, a, b, c, $d1$, $d2$) form the *basis* in the space of all possible bifurcations of Liouville tori. The decomposition of arbitrary bifurcation in the sequence of elementary bifurcations is non-unique. Our theory gives also the classification of all complicated bifurcations of Liouville tori. This classification is the classification of all *letters–atoms*, which will be described below.

The natural question is whether it is possible to perturb an arbitrary *complicated* integral f of some integrable Hamiltonian system v to obtain a

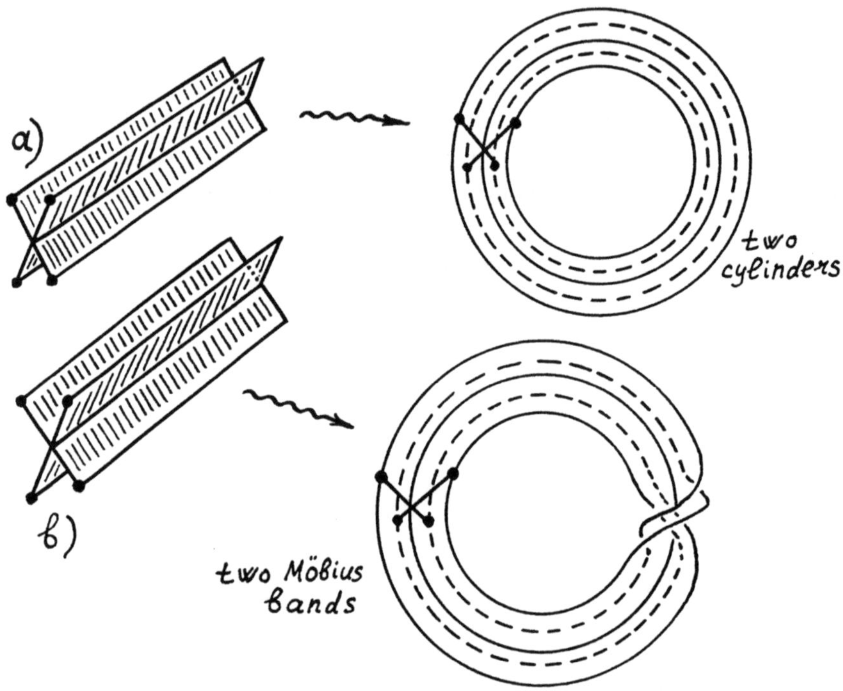

Figure 5

simple integral f'. It is easy to see that if we perturb only f (without changing H or ω), then the answer is negative. In fact, the following theorem of Nguen Tyen Zung is valid.

Let us consider the symplectic manifold M^4 and an integrable system v on some isoenergy surface Q. Then there exists small perturbations $f \to f'$ and $H \to H'$ (or $f \to f'$ and $\omega \to \omega'$) such that the new Hamiltonian system v' on M is integrable and the new integral f' is simple on Q.

Consequently, small deformation of the integral f and H (or ω) can transform a complicated integral into a simple one. But in the case of real physical systems it is interesting to classify all bifurcations of Liouville tori without the perturbation of the Hamiltonian (which is given to us by God). It turns out that this classification exists and is given by our theory based on the new topological invariant $I(H,Q)^*$.

2.9. Non-resonant integrable Hamiltonian systems.

Let us consider an integrable system of arbitrary dimension and the Liouville foliation of an isoenergy surface, i.e. the set of Liouville tori T^n (the manifold M^{2n} has dimension $2n$).

DEFINITION. *Let us call the Hamiltonian H and the system $v = \operatorname{sgrad} H$ non-resonant on a given isoenergy surface ($H = \text{const}$) if on this surface there are everywhere dense Liouville tori on which the integral trajectories of the system v form a dense irrational winding.*

In the non-resonant case the closure of the integral trajectory in general position is, consequently, the Liouville torus containing this trajectory (or initial point of the trajectory).

The experience of the study of concrete known physical systems on M^4 shows that on four-dimensional manifolds Hamiltonians are mainly non-resonant ones on almost all surfaces Q^3 (Oshemkov, Charlamov, Fomenko, Bolsinov, Kozlov and Tatarinov). In the multidimensional case, i.e., on M^{2n} where $n > 2$, interesting examples are known where the Hamiltonian H is "essentially" resonant on almost all surfaces Q. This happens, in particular, when a system is integrable in a non-commutative sense, i.e. has a complete set of integrals which form a *non-commutative* Lie algebra. For an analysis of the main cases of non-commutative integrability (starting from the works of E. Cartan) see the books [**31**], [**33**] and also the review by Trofimov and Fomenko in [**39**], [**105**]. In the case of non-commutative integrability the system is "degenerate" (resonant) in the sense that its trajectories are everywhere dense only on "low-dimensional tori" (dim $< n$). They are organized into "large" Liouville tori (in a compact case). Consequently, an initial Hamiltonian system is irrational only on small tori, but not so on "large tori". In the four-dimensional case, however, the non-resonant nature of the system is a typical situation, because here we do not have real non-commutative integrability: every semi-simple Lie algebra of dimension *two* is *commutative.*

In the non-resonant case the closure of the integral trajectory, passing from the point x, is the Liouville torus T^n which we denote T^n_x. Thus, we have the correspondence $x \to T^n_x$. Changing x, we change the torus T^n_x. These tori can be transformed (by some bifurcations) for some special values of x (Figure 6).

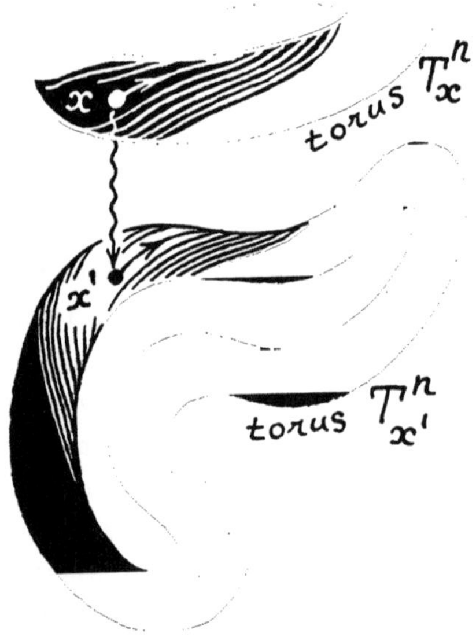

Figure 6

Consequently, the deformations and transformations of the Liouville tori

T_x^n as a function of x, show the *dependence of the solutions of our Hamiltonian system on the initial data* x.

Let us consider the four-dimensional case. In non-resonant systems the Liouville tori and all our theory does not depend on the choice of the concrete Bott integral f, that is, only the fact of its existence is important. See the proof and discussion in the author's papers [**30**], [**33**]. Really we must speak about an *integrable Hamiltonian* H, because the function H "knows everything" about its integrals f. The choice of the second integral f is non-unique. On the other hand, we cannot change the form of the Hamiltonian H, because this energy-function is given by the nature of the physical or mechanical problem. The function H describes, in some sense, the energy of the mechanical system and, consequently, the form of this function is given to us by God. Nevertheless, for calculations it is convenient to fix some concrete integral f. In this case the formulations of some theorems will be simpler.

We will assume in our paper that all considered Hamiltonians (Hamiltonian systems) have Bott integrals and are non-resonant.

2.10. Topological equivalence of integrable Hamiltonian systems.

Let us consider two integrable Hamiltonian systems: v_1 on some isoenergy surface Q_1^{2n-1} (i.e. $Q_1 = \{H_1 = \text{const}\}$) in the phase space M_1^{2n} and v_2 on some other isoenergy surface Q_2^{2n-1} (i.e. $Q_2 = \{H_2 = \text{const}\}$) in the phase space M_2^{2n}. Let us fix the orientation on both manifolds Q_1 and Q_2. The following definition was introduced in [**30**], [**33**].

DEFINITION. *We will call the integrable non-resonant Hamiltonian systems v_1 on Q_1 and v_2 on Q_2 topologically equivalent (or geometrically equivalent) if there exists a diffeomorphism $\tau : Q_1 \to Q_2$ which transforms the Liouville tori of the system v_1 into the Liouville tori of the system v_2 and preserves the orientation of the isoenergy manifolds.*

COMMENT: It is convenient to formulate the stronger notion of topologically equivalent systems by adding the condition of the preservation of the orientations of all critical submanifolds of the Bott integrals. These submanifolds are closely connected with singular fibers of the Liouville foliation on Q.

An integrable non-resonant system defines the foliation of the isoenergy manifold, where non-singular fibers are Liouville tori and singular fibers are the orbits of the Poisson action of the commutative group R^n on the manifold Q. Because we fixed the orientation of the group R^n, we obtain the induced orientation on all of its orbits (regular and singular). The Abelian group R^n is defined by the set of commutative integrals $f_1, \dots, f_n$ (see [**33**]). We return to a more detailed discussion later.

Now let us consider the four-dimensional case M^4. We know here the structure of all critical submanifolds of the Bott integral f on Q^3 (see Lemma 2.1). They are: tori, Klein bottles and circles. We will assume that our systems do not have critical Klein bottles (see above). Then we include critical tori in the set of Liouville tori. Each critical circle of the Bott integral is some periodic solution (integral trajectory) of the Hamiltonian system. Thus, this circle has the uniquely defined orientation (defined by the system v).

DEFINITION. *Two integrable non-resonant systems with Bott integrals will be called topologically equivalent on the orientable isoenergy manifolds Q_1^3 and Q_2^3 if there exists a diffeomorphism $\tau : Q_1^3 \to Q_2^3$ which transforms the Liouville tori of the system v_1 into the Liouville tori of the system v_2, preserves the orientation of the isoenergy surfaces and the orientation of critical circles of the integrals.*

As we know (see [**29**], [**33**]), every non-resonant Liouville torus can be characterized as the closure of any integral trajectory lying in the torus. Consequently, the set of all non-resonant tori does not depend on the choice of the second Bott integral f. Every resonant torus T and every connected component R of an arbitrary critical level surface of the integral f can be approximated by the non-resonant tori. Consequently (see [**29**], [**33**]), the surfaces T and R do not depend on the choice of the integral f.

Thus, the Liouville foliation on Q^3 (and, in particular, the critical isolated circles of the integral f) do not depend on the choice of the integral f.

The following definition is evidently equivalent to the previous one, but emphasizes the fact that *the equivalence or non-equivalence of two integrable systems does not depend on the choice of concrete Bott integrals.*

DEFINITION. *We call two integrable non-resonant Hamiltonian systems v_1 and v_2 on the isoenergy surfaces Q_1 and Q_2 topologically equivalent if there* *exists a diffeomorphism $\tau : Q_1 \to Q_2$, which preserves the orientation of the isoenergy manifolds, transforms the Liouville foliation of the system v_1 into the Liouville foliation of the system v_2 and preserves the orientation of all isolated critical circles of the integrals.*

IMPORTANT REMARK: The level surface of the integral f_1 and Q_1 can be non-connected, i.e. can be the union of several Liouville tori. These tori can be mapped in different level surfaces of the integral f_2 (Figure 7).

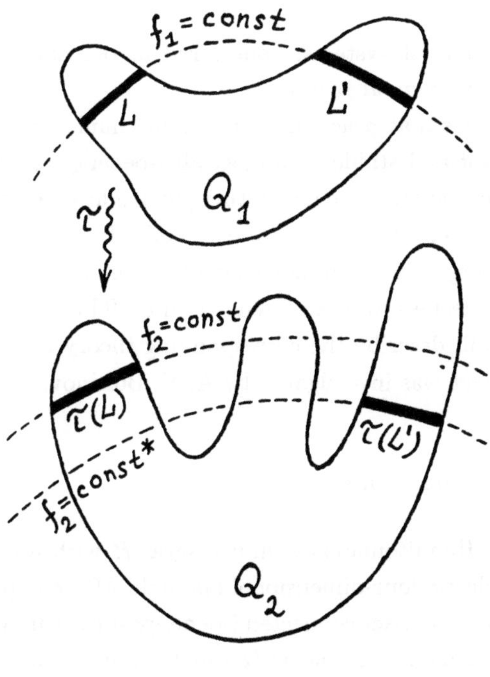

Figure 7

Consequently, the diffeomorphism τ does not preserve, generally speaking, the levels of the integral, but must map a Liouville torus in some Liouville torus.

Let us consider a non-resonant system v on Q^3 in M^4, where $Q^3 = Q^3_h = (H = \text{const} = h)$. Let us consider a small perturbation of h, i.e. $h \to h + \varepsilon$. Then Q_h deforms to $Q_{h+\varepsilon}$ and the system v deforms to the system v_ε on $Q_{h+\varepsilon}$.

DEFINITION. *We will say that the non-resonant system v on Q^3_h is stable (or is in general position) if the system v_ε on $Q^3_{h+\varepsilon}$ is non-resonant and topologically equivalent to the system v on Q^3_h for all sufficiently small values of ε.*

We will assume that all systems, considered in our theory, are *non-resonant and stable (in general position).*

The experience shows that practically all real mechanical integrable systems are non-resonant and stable on almost all isoenergy surfaces Q. On the other hand, we can show some interesting particular cases, when a system v is non-resonant on some Q, but is *non-stable*. In this case the small perturbation of h changes the word-molecule, i.e. the molecules on $Q_{h+\varepsilon}$ are different from the molecules (of the system v) on Q. This situation is very interesting and we will describe the corresponding theory in a subsequent publication. This effect was investigated by A. V. Bolsinov.

3. Formulations of the problems.

We will consider a Hamiltonian system $v = \operatorname{sgrad} H$ with smooth Hamiltonian H on a symplectic four-dimensional manifold M^4 (compact or non-compact). Let Q^3 be a compact connected isoenergy surface in M. Suppose that v is non-resonant and integrable on Q and f is the second independent Bott integral. Let us formulate the main problems of the present paper.

1. ENUMERATION PROBLEM. *Does there exist an algorithm which enumerates all integrable Hamiltonian systems up to topological equivalence?*

2. RECOGNITION PROBLEM. *Does there exist an algorithm which solves the following problem: are two integrable Hamiltonian systems topologically equivalent or not?*

3. PROBLEM OF ALGORITHMICAL CLASSIFICATION OF ALL INTEGRABLE HAMILTONIAN SYSTEMS IN GENERAL POSITION UP TO TOPOLOGICAL EQUI-

VALENCE. *Does there exist such an effective algorithm which can be implemented on the computer?*

To be correct, in the formulation of all these problems, we must comment on how we can define the Liouville foliation a finite method. We suggest the following procedure. Let us call the piecewise linear three-dimensional manifold Q_{PL} with a fixed subdivision in the union of sub-polyhedrons *the piecewise linear Liouville foliation*, if there exists a piecewise smooth homeomorphism of the manifold Q_{PL} on the "smooth original", that is, on the manifold Q with a given integrable Hamiltonian system v. This homeomorphism must transform the elements of the subdivision of the manifold Q_{PL} into the fibers of the Liouville foliation of the system v. It is evident that we can define the piecewise linear Liouville foliations in a finite way (algorithmically).

We prove in the present paper that all these problems have positive answers.

THEOREM 3.1.

1) *There exists an algorithm of enumeration of all classes of topologically equivalent integrable Hamiltonian systems (in general position and with two degrees of freedom).*

2) *There exists an algorithm of recognition of topologically equivalent (and topologically non-equivalent) integrable Hamiltonian systems.*

3) *There exists an algorithmical classification of all integrable Hamiltonian systems up to topological equivalence.*

4. Skeletons and complexity of integrable Hamiltonian systems.

4.1. Skeletons.

The proof of the Theorem 3.1 and the concrete classification of all integrable Hamiltonian systems of low complexity is based on the important notions of the *skeleton* and *complexity* of integrable Hamiltonian systems and on the *classification theorem for isoenergy surfaces* of integrable systems.

The notion of the *skeleton* of the system was discovered by A. T. Fomenko in **[29]**, **[30]**, **[33]** and coincides with the notion of the topological invariant

$I(H,Q)$ of integrable systems, introduced and investigated in these papers. The notion of *marked invariant* $I(H,Q)^*$ was introduced by A. T. Fomenko and H. Zieschang in [**43**]. The notion of *complexity* of an integrable system is similar to the important notion of the *complexity of a compact three-dimensional manifold*, introduced by S. V. Matveev [**118**], [**119**]. In both cases the complexity is measured by the number of singularities and in both cases the important role is played by the graphs of degree 4.

The classification theorem for isoenergy surfaces was proved by the author in [**29**].

Let us consider the Liouville foliation of the three-dimensional manifold Q. Let us cut the manifold Q along some Liouville torus T and then glue the two copies of the torus T (which appear after cutting) using some diffeomorphism (preserving the orientation). See Figure 8. We obtain some new three-dimensional manifold Q_1 with some new Liouville foliation. We will say that the *foliation on* Q_1 *is obtained by the twisting from the foliation on* Q.

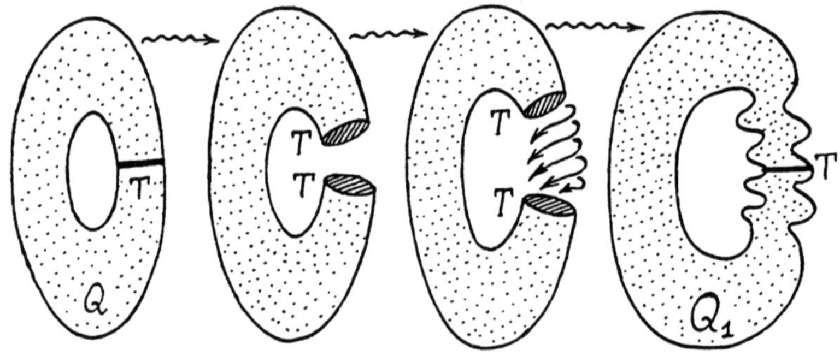

Figure 8

DEFINITION. *Let us call two integrable Hamiltonian systems rougly equivalent if their Liouville foliations are obtained one from another by some twisting along Liouville tori.*

DEFINITION. *We will call the class of all systems rougly equivalent to a given system the (abstract) skeleton of a system.*

Indeed, the introduced notion has very simple and visual geometrical sense. Let us introduce the notion of *geometric skeleton.*

DEFINITION. *The geometric skeleton is the pair (P, K), where P is a compact closed orientable two-dimensional surface and K is a graph in P. The graph K is as follows:*

1) Each vertex of the graph K is either isolated or has degree 2 or 4 (Figure 9).

2) The surface $P \backslash K$ (the complement to K in P) is homeomorphic to the union of several open rings $S^1 \times (0,1)$.

3) The set of boundary circles of the rings (which form the surface $P \backslash K$) can be separated in two parts: positive circles and negative ones, in such a way that exactly one positive circle and exactly one negative circle are glued to each edge of the graph K.

4) The graph K does not contain loops (circles) without vertices, except in the case when P is the torus and K is one circle (loop) without any vertices.

Two skeletons are considered as identical if they are homeomorphic as two topological spaces (with preserved orientation).

REMARK: The graph K can be non-connected. This graph K coincides with the "reduced graph K" which is defined below. The exceptional case (T^2, S^1) (see item 4) is trivial (in some sense) and will be considered separately.

REMARK: If the surface P is non-orientable, then the item 4 in the definition above must be changed as follows: *the graph K does not contain two loops (circles) without vertices which bound the ring (i.e., the circles are the boundary of one ring).* Here K can contain the circles corresponding to the Möbius bands in P.

For an example of a geometric skeleton see Figure 9.

As we will see later, the isolated vertices of K represent the minimal and maximal critical circles of the Bott integral f on Q. Other connected components of K correspond to the connected components of the critical level surfaces of the integral f, which contain the *saddle critical circles.* Thus, the saddle circles correspond to the vertices of the graph. The orientable

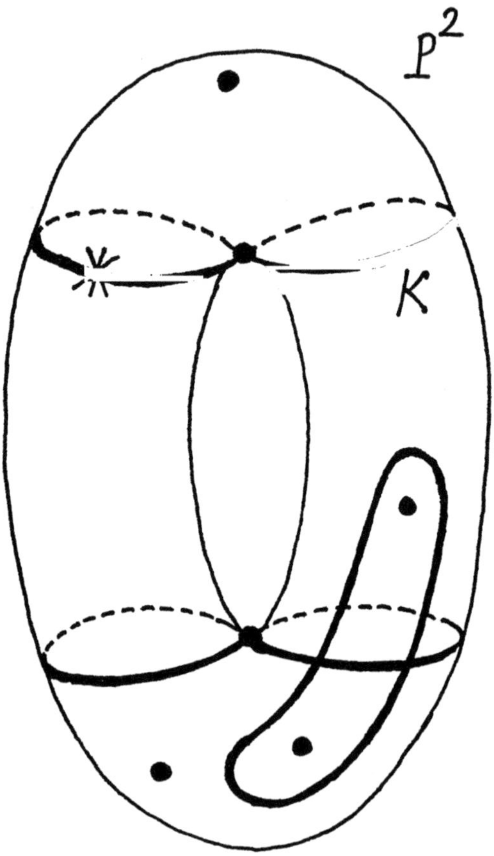

Figure 9

saddles correspond to the vertices of degree 4, the non-orientable saddles to the vertices of degree 2 (we mark these vertices by *stars* = *asterisks*).

The connected components of the manifold $P \backslash K$ (namely, the rings) correspond to the one-parameter families of Liouville tori (including the maximal and minimal critical tori). These tori move from one critical level in direction of another critical level. The following theorem is one of the central points of the whole theory.

THEOREM 4.1. *There exists a natural one-to-one correspondence between the set of all geometric skeletons and the set of all skeletons of integrable Hamiltonian systems.*

It follows from this theorem, that the problems of algorithmical enumeration, of algorithmical recognition and algorithmical classification of the skeletons are solved in the positive sense.

4.2. Complexity.

Let v be the integrable Hamiltonian system on a closed orientable three-dimensional manifold Q with some Bott integral f. Denote by m the total number of all minimal, maximal, and saddle critical circles of the integral. Let us remove from the manifold all isolated critical circles and connected components of all critical level surfaces of f containing the saddle circles. In other words, we remove all singular fibers of the corresponding Liouville foliation (i.e. the fibers different from Liouville tori). As a result, the manifold Q goes to pieces, transforms into the union of a finite number of open manifolds homeomorphic to the direct product $S^1 \times S^1 \times (0,1)$. Let us denote the total number of such manifolds by n.

DEFINITION. *The pair of non-negative integer numbers (m, n) will be called complexity of a given integrable Hamiltonian system v.*

It follows easily from the definition of complexity, that the complexities of two roughly equivalent integrable Hamiltonian systems coincide. Thus, the complexity is an *invariant of the skeleton* of the system. The complexity of the geometric skeleton (P, K) is the pair (m, n), where m is the number of the vertices of the graph K and n is the number of the connected components of the manifold $P \backslash K$.

IMPORTANT REMARK: Thus, the notion of complexity of the integrable system is very natural and reflects the deep properties of the system of Hamiltonian differential equations. In some sense this notion naturally appears in a more or less "unique way" from the point of view of symplectic topology of integrable systems.

THEOREM 4.2.

a) *The number of different skeletons of a given fixed complexity (m, n) is finite.*

b) *The set of all topologically non-equivalent integrable Hamiltonian systems with the same skeleton of complexity* (m, n) *is parametrized by independent parameters*

$$r_i,\ \varepsilon_i,\ n_k,\ 1 \leq i \leq n,\ 1 < k < s,$$

where r_i *are rational numbers,* $0 \leq r_i < 1$ *or if* $r_i = \infty$*, then* $\varepsilon_i = \pm 1$*;* n_k *are integers, and* $s \leq m$*.*

The statement a) of this theorem evidently follows from Theorem 4.1 and from the interpretation of the complexity of the geometric skeleton (see above). Let us comment on item b). We introduce the notion of *framed geometric skeleton* as a geometric skeleton with some numerical marks. For example, the rational numbers and the numbers $\varepsilon_i = \pm 1$ (see Theorem 4.2) are located with the open rings which form the manifold $P \backslash K$. The rule of construction of the integer marks will be discussed below.

Thus, we can say that an integrable Hamiltonian system can be uniquely defined (up to topological equivalence) by the framed skeleton. Conversely, each framed geometric skeleton is the skeleton of some integrable Hamiltonian system.

THEOREM 4.3. *There exists a natural one-to-one correspondence between the set of all framed geometric skeletons and the set of all integrable Hamiltonian systems (considered up to topological equivalence).*

This theorem fixes the answer of all three questions (see above), because Theorem 3.1 is a direct corollary of Theorem 4.3.

5. The list of all integrable Hamiltonian systems of low complexity.

5.1. Letters-atoms.

We need some procedure for the coding of integrable systems. Let us separate the pieces of the geometric skeleton (P, K) in pieces of two types: a) the regular neighborhoods of the connected components of the graph K, b) the remaining rings. Let us call the regular neighborhoods of the connected components of the graph K *letters-atoms*. The complexity of

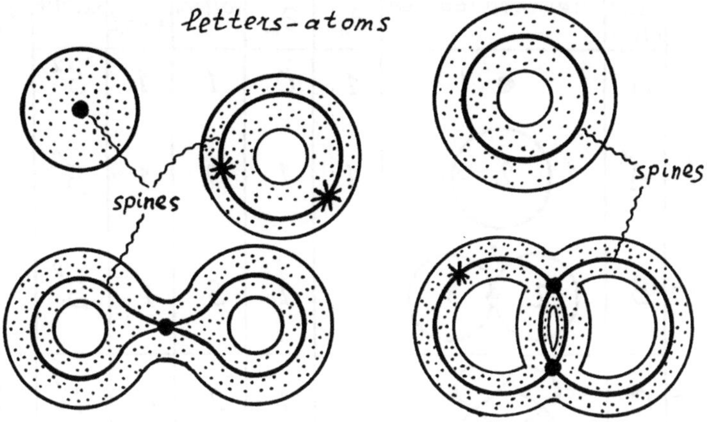

Figure 10

the *letter-atom* is equal to the number of the vertices in its *spine.* The *spine* is the connected component of the graph K, which is the deformation retract of the letter-atom. See some simple examples in Figure 10.

The list of all letters-atoms of complexity no more than 3 can be found in Table 1.

	letter atom	representation	∂_+	∂_-	valency	weight	surface	graph K_c^*
A–	A		1	0	1	1	S^2	
–A^*–	A^*		1	1	2	1	S^2	
>B–	B		2	1	3	1	S^2	
–C_1–	C_1		1	1	2	2	T^2	
>C_2<	C_2		2	2	4	2	S^2	
⩾D_1–	D_1		3	1	4	2	S^2	
>D_2<	D_2		2	2	4	2	S^2	
>B^*–	B^*		2	1	3	2	S^2	
–A^{**}–	A^{**}		1	1	3	2	S^2	

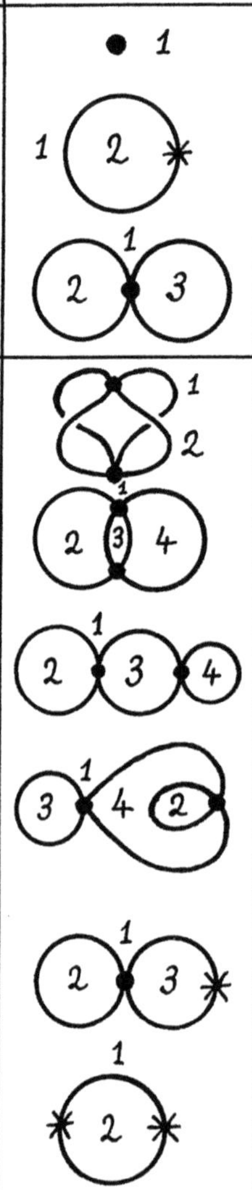

Table 1(a)

	atom	representation	∂_+	∂_-	val.	weight	surf.	graph K_c^*
$>E_1-$	E_1		2	1	3	3	T^2	
$>E_2-$	E_2		2	1	3	3	T^2	
$\geqslant E_3<$	E_3		3	2	5	3	S^2	
$>F_1-$	F_1	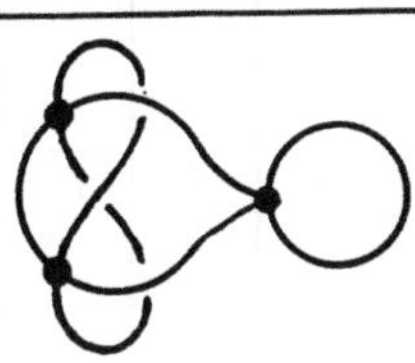	2	1	3	3	T^2	
$\geqslant F_2<$	F_2		3	2	5	3	S^2	
$\geqslant G_1-$	G_1		4	1	5	3	S^2	
$\geqslant G_2<$	G_2		3	2	5	3	S^2	
$\geqslant G_3<$	G_3		3	2	5	3	S^2	
$\geqslant H_1-$	H_1		4	1	5	3	S^2	

Table 1(b)

atom		representation	∂_+	∂_-	valency	weight	surface	graph K_c^*
$\geq H_2 \leq$	H_2	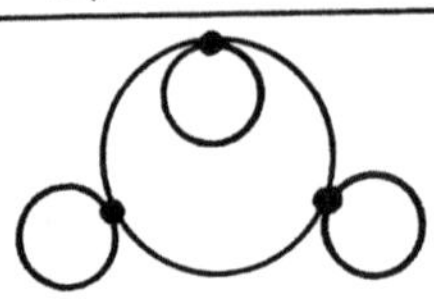	3	2	5	3	S^2	
$> B_1^{**} -$	B_1^{**}	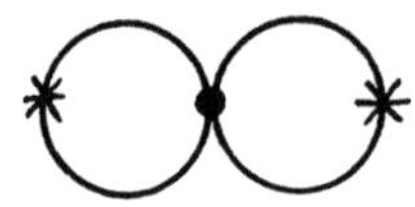	2	1	3	3	S^2	
$> B_2^{**} -$	B_2^{**}		2	1	3	3	S^2	
$- C_1^{*} -$	C_1^*		1	1	2	3	T^2	
$> C_2^{*} <$	C_2^*	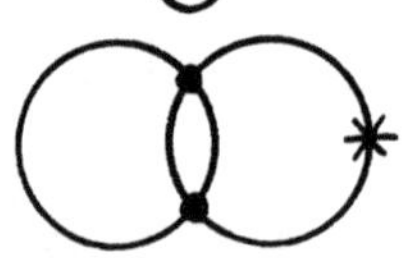	2	2	4	3	S^2	
$\geq D_{11}^{*} -$	D_{11}^*		3	1	4	3	S^2	
$\geq D_{12}^{*} -$	D_{12}^*		3	1	4	3	S^2	
$> D_{21}^{*} <$	D_{21}^*	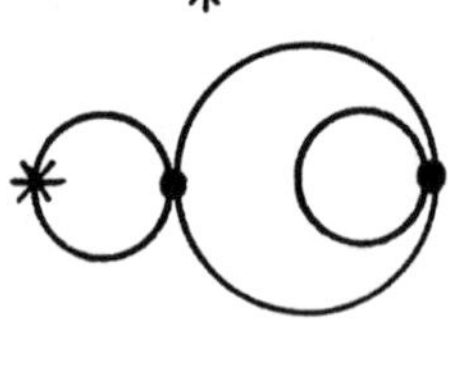	2	2	4	3	S^2	

Table 1(c)

	atom	representation	∂_+	∂_-	valency	weight	surface	graph K_c^*
$>D^*_{22}<$	$\mathcal{D}^*_{22}$		2	2	4	3	S^2	
$>D^*_{23}<$	$\mathcal{D}^*_{23}$		2	2	4	3	S^2	
$-A^{***}-$	A^{***}		1	1	2	3	S^2	

Table 1(d)

Table 2 shows all orientable spines of letters-atoms of complexity no more than 5. Here we marked the number of letters-atoms corresponding to each spine.

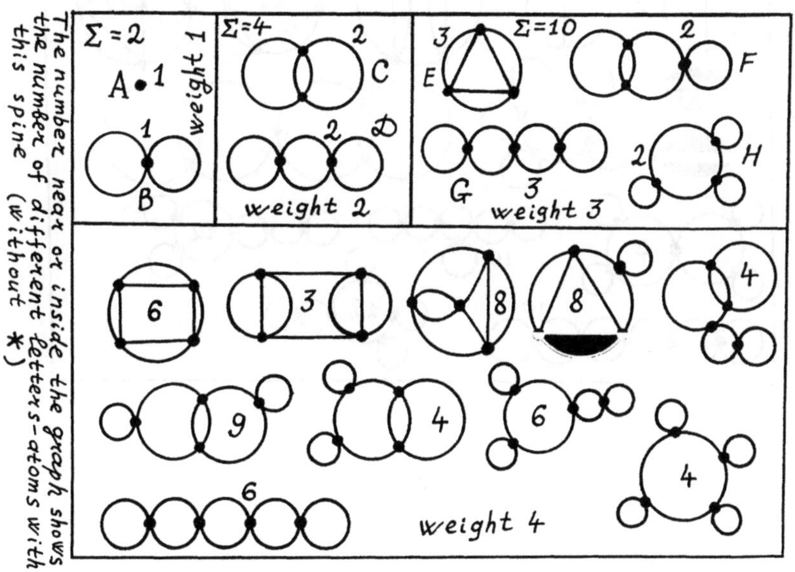

Table 2(a)

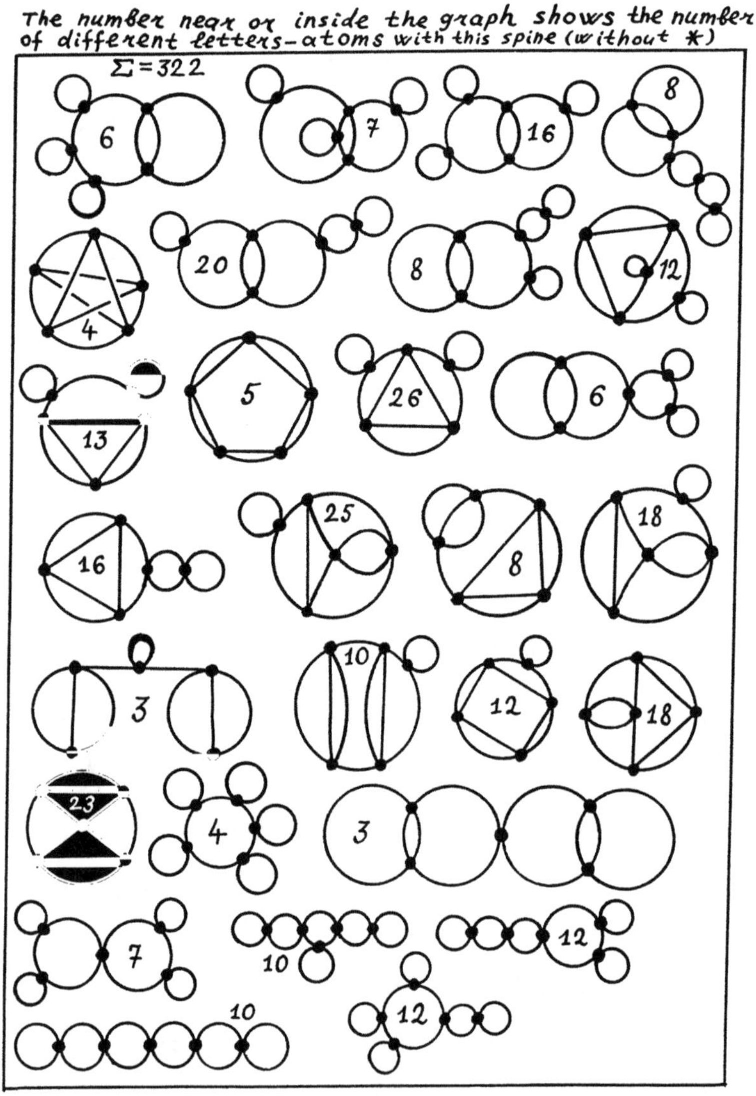
The number near or inside the graph shows the number of different letters-atoms with this spine (without *)
Σ = 322

Table 2(b)

An *orientable spine* is a spine without vertices of the multiplicity 2 (i.e. without non-orientable saddles). *Non-orientable spines* can be easily obtained from orientable ones by introducing one or several *star*-vertices on the edges of the orientable spine (Figure 11).

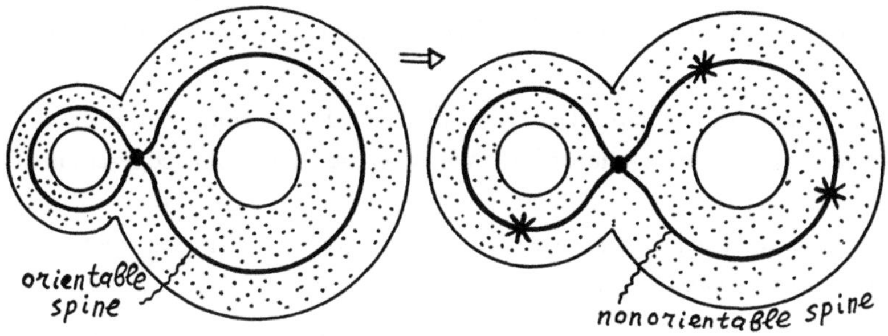

Figure 11

The Tables 1 and 2 were calculated on a computer.

5.2. Words-molecules.

The rings (which form the manifold $P\backslash K$) can be interpreted as segments-edges-connections in some new object which can be called *word-molecule*. A word-molecule is formed by several letter-atoms connected by edges (which correspond to the rings). The word-molecule defines the geometric skeleton of an integrable Hamiltonian system. Figure 12 shows the word-molecule

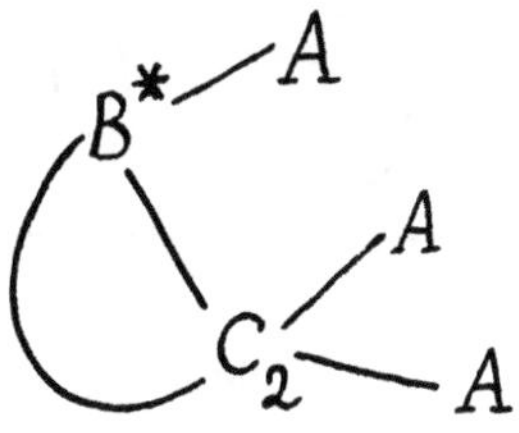

Figure 12

corresponding to the geometric skeleton in Figure 9. The letter A encodes (corresponds to) the isolated vertex of the skeleton. The letter B^* (with the "eight-figure" spine and with the star on it) and the letter C_2 (with the spine: "two intersecting circles") can be seen in Table 1.

IMPORTANT REMARK: Indeed, the segments-edges-connections coming out from the letter-atom, are not equivalent in general. We need some special enumeration of all edges coming out from the letter-atom to reconstruct (in unique way) the word-molecule. Here we suppose that all letters-atoms are enumerated in Tables 1 and 2 by some fixed method.

It is convenient to enumerate the all edges-connections, coming out from one letter-atom, in cyclic order. We will start from the edge-connection, which is vertical (upwards), and we will rotate in the positive direction. Table 3 shows the list of all words-molecules corresponding to the integrable Hamiltonian systems of complexity (m, n), where $m \leq 2$.

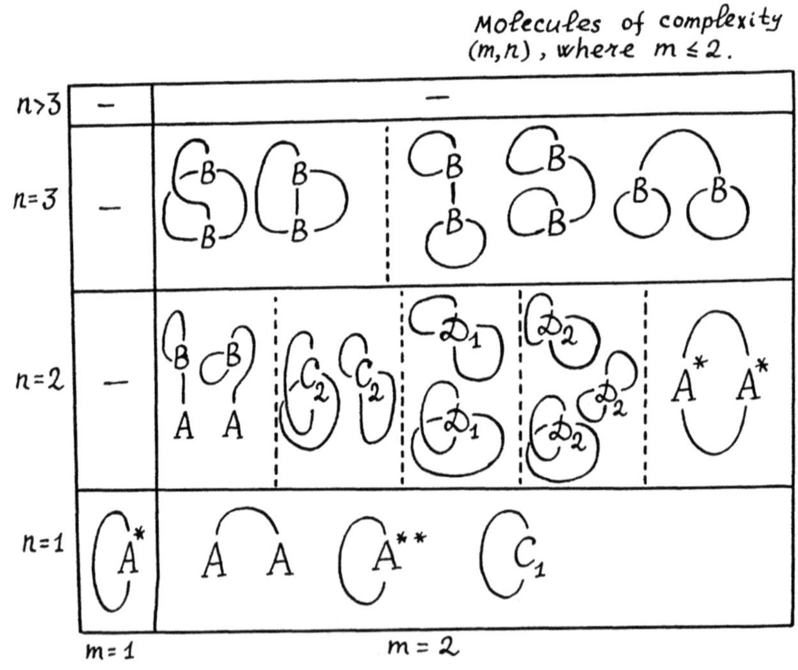

Table 3

Let us denote by $\lambda(m, n)$ the total number of all skeletons of integrable Hamitonian systems of a given complexity (m, n). This number is finite, as follows from Theorem 4.2. The list of the values of the function $\lambda(m, n)$, where $m \leq 4$, is given in Table 4.

n \ m	1	2	3	4
⋮				
7	0	0	0	0
6	0	0	0	54
5	0	0	0	247
4	0	0	11	530
3	0	5	24	561
2	0	10	24	128
1	1	3	2	8

The total number of molecules with complexity (m, n), where $m \leq 4$.

$\lambda(m,n)$

Table 4

5.3. Atomic weight and valency of the word-molecule.

We will call the total number of the vertices in the spine of the letter-atom *atomic weight* (of a given letter-atom). Then we will call the total number of edges-connections in the word-molecule *valency* (of a given word-molecule).

The sum of the valencies of a word-molecule and the atomic weights of all its letters-atoms is always even. Consequently, if m is odd, then all integrable Hamiltonian systems of complexity (m, n) must have critical circles of non-orientable saddle type.

We calculate the number $\Lambda(m)$, where $\Lambda(m)$ is the upper bound of all numbers n such that $\lambda(m,n) \neq 0$ and m is fixed: $\Lambda(m) = \{\sup n : \lambda(m,n) \neq 0\}$. It turns out that $\Lambda(m) = [3m/2]$.

6. Real physical systems and their place in the table of complexity of all integrable Hamiltonian systems.

It is extremely interesting and important to calculate the cells (m, n) in our table which contain the real mechanical and physical integrable systems. A. T. Fomenko, A. A. Oshemkov, L. S. Polyakova and A. V. Bolsinov calculated (based, in particular, on the imporant papers of M. P. Charlamov and his pupils) the words-molecules for the following important physical systems: integrable cases of the equations of the rigid body motion (Kovalevskaya case, Goryachev–Chaplygin case, Sretensky case, Clebsch case, Euler case, Lagrange case); some integrable cases in the Toda lattice, integrable cases for the motion equations of the four-dimensional rigid body (the system on the Lie group $SO(4)$) and so on. In many cases the numerical marks were calculated, i.e. the framed skeletons of integrable systems. Some results are presented in Table 5.

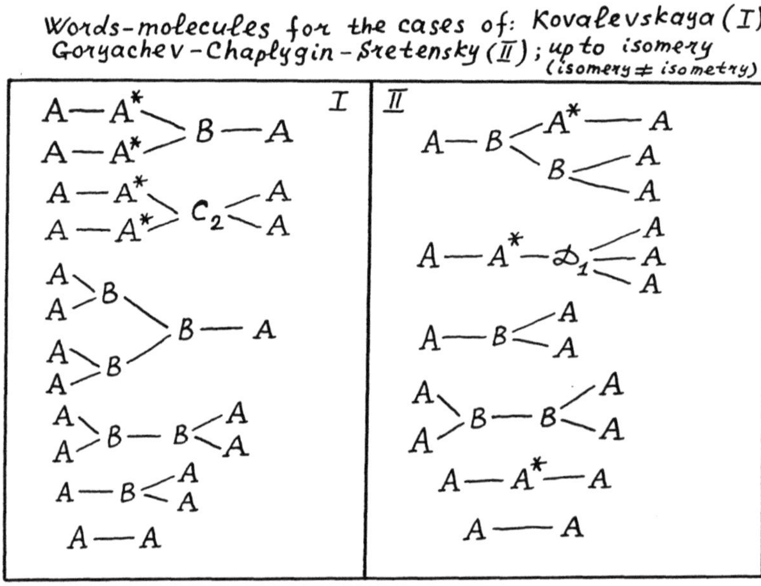

Table 5

In the previous discussion we assumed that the isoenergy surface Q was fixed. But we actually need to consider all isoenergy surfaces of a given integrable system which correspond to different values of the energy function (Hamiltonian) H. This problem arises naturally when we study real physical systems. The energy function H is given to us "by God" (we cannot deform this function). Consequently, it is natural to start from the minimum of the Hamiltonian H and move then in the direction of its maximum.

The change of the Hamiltonian's value changes the isoenergy surface Q. It is natural to collect all types of surfaces Q which appear in this process. This set of integrable surfaces represents the total topology of a given integrable system. In our language we must collect all corresponding words-molecules. The word-molecule evidently does not change when we change the value of the Hamiltonian near a regular value. Some transformation of the word-molecule can occur only in the neighborhood of a singular value of H. The word-molecule can change only when we cross the singular value.

Thus, we obtain a "one-parameter" family of words-molecules. Let us represent these molecules by the points in the corresponding cells of the plane (each such cell corresponds to the concrete complexity (m, n)). Then let us connect each two consecutive points by a segment. As a result, we obtain some curve in the plane (m, n). The curve corresponds to the motion of the value of the Hamiltonian from the minimum to the maximum. Thus, each real physical system is represented by some curve (sequence of segments) in the plane (m, n). Table 6 contains the curves corresponding to some real physical integrable Hamiltonian systems mentioned above. The cells, corresponding to these systems, are shaded. It is clear that the further analysis of another real mechanical system will fill some new cells on the plane. We obtain, as a result, some remarkable "physical zone". Important problem: describe the more or less precise boundary (the form) of this zone. Where are the real physical integrable systems located? The next problem: how to fill "the empty cells" close to the physical zone or enveloped by this zone? (Figure 13). There is the problem of prediction of the properties of

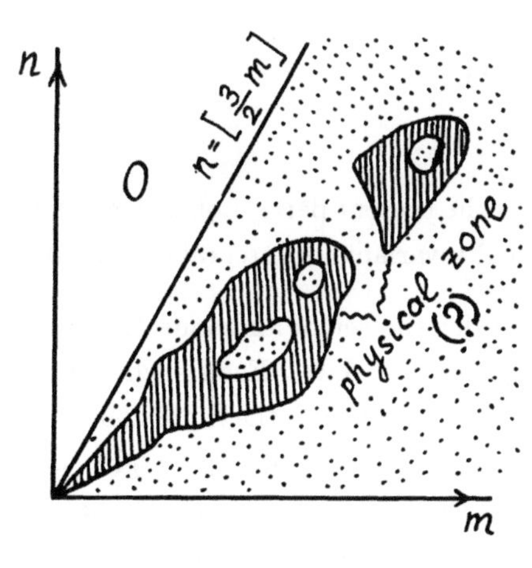

Figure 13

integrable systems, close to the "physical zone", the problem of discovering of some mechanical systems with prescribed complexity etc. It seems to us that this table is similar in spirit to Mendeleev's table of chemical elements.

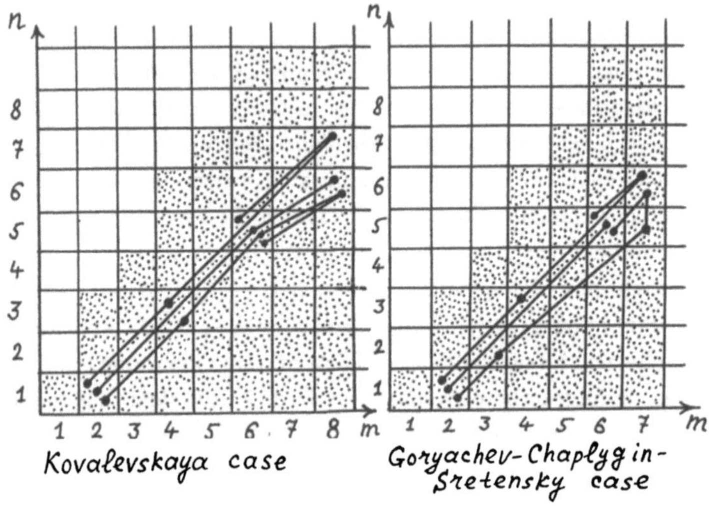

Table 6

7. Formal description of the topological invariants of integrable Hamiltonian system.

We will describe in the present paragraph a slightly different construction of the new topological invariant for integrable systems introduced above.

STEP 1: Let us consider the set of all connected graphs K_c with vertices of multiplicity zero, two and four. Each such graph is obtained by the following procedure. Let us consider the finite collection of circles and form the graph with vertices of multiplicity 4: we glue these circles in several points, which we call the vertices. Important: only two circles intersect in each vertex (Figure 14). Then let us put on some edges an arbitrary number of stars, which are the vertices of multiplicity two. The vertices of multiplicity zero are the isolated points.

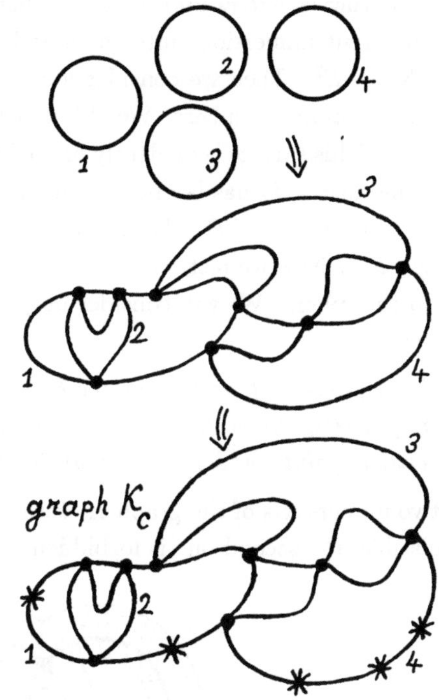

Figure 14

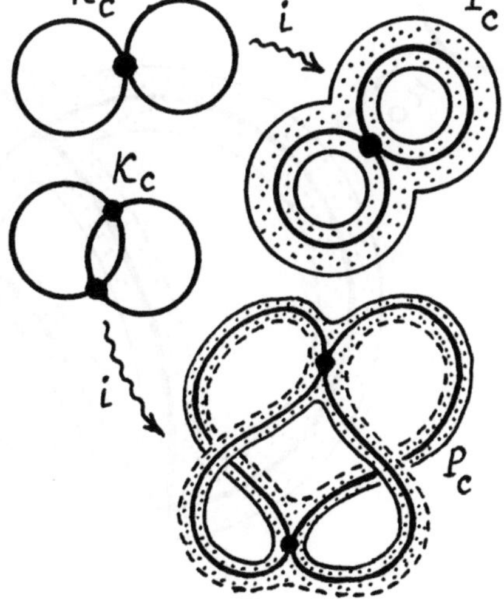

Figure 15

STEP 2: Let us construct a *letters-atoms* (the alphabet). Consider the graph K_c and its different immersions into the standard two-dimensional sphere S^2, i.e. $i : K_c \to S^2$. Then we consider a small tubular neighborhood of such an immersion in the sphere. We obtain some two-dimensional surface P_c (Figure 15). This surface is evidently uniquely defined by a given immersion of K_c. The surface P_c has boundary and contains the graph K_c, which is the deformation retract of P_c (the surface P_c can be retracted, deformed on the graph K_c). Now we restrict the class of possible immersions of the graphs K_c in the sphere. We will consider only the immersions with the following property:

> *there exists a Morse function f on the surface P_c such that its critical saddle points are precisely the vertices of multiplicity 4 of K_c, f does not have other critical points, and $f =$ const on the boundary ∂P_c.*

Figure 16 shows two immersions of the graph K_c into the sphere. The first immersion is admissible, the second one is forbidden.

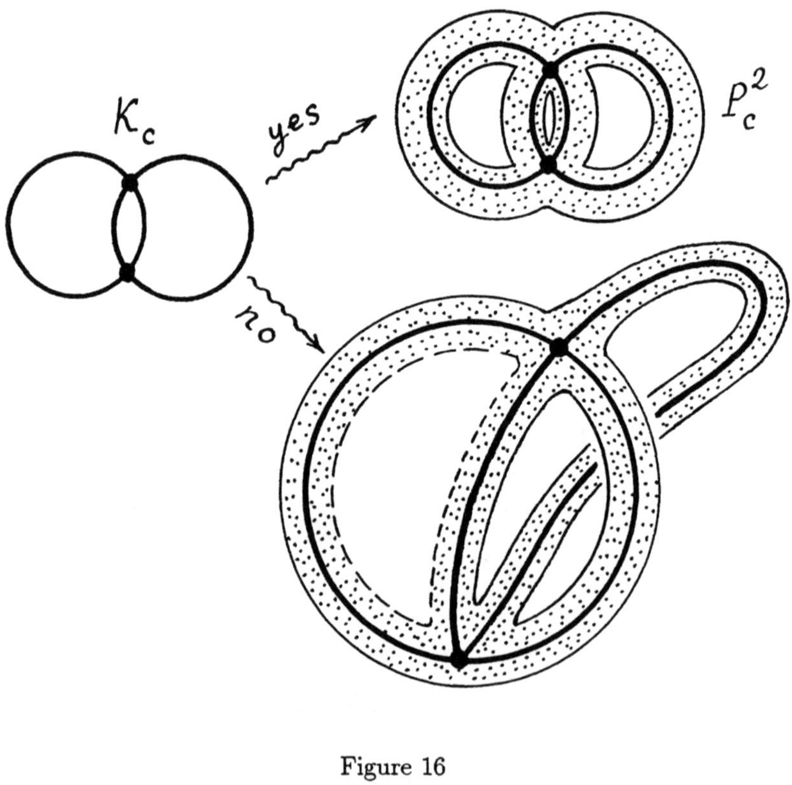

Figure 16

STEP 3: We call two admissible immersions i_1 and i_2 of the graph K_c into the sphere S^2 *equivalent*, if i_1 can be transformed in i_2 by the following operations: a) smooth isotopies in the sphere, b) the operation shown in Figure 17. The equivalence relation is evidently symmetric. The operation *b* allows us to remove and include *loops* on arbitrary edges of the immersed graph.

Figure 17

STEP 4: Let the *letter-atom* be the class of equivalent immersions of the graph K_c into the sphere. We obtain some set of letters-atoms. They form some "alphabet". Our next step: we will "write the words" using this alphabet. We need a "grammatical rule" to write the words.

STEP 5: Let us consider an arbitrary letter-atom and the corresponding surface P_c. The boundary of P_c is the union of circles. Let us call these circles *ends* of the letter-atom and let us represent this object by some letter with several segments–edges coming out of the letter. See some examples in Figure 18. Here the letter C_1 has two ends and we draw the letter C_1 with two segments-edges. The letter C_2 has four ends and we draw the letter C_2 with four segments-edges.

STEP 6: Let us glue the word-molecule from some finite set of the letters-atoms by the following grammatical rule: we connect the pairs of free ends of the letters in such a way that the final graph does not have free ends (Figure 19). Two ends belonging to the same letter can also be connected. Let us denote this word-molecule by W. Now we can construct some closed two-surface P using the word-molecule W. Because each letter-atom is defined by a two-surface P_c (with boundary), we can construct the new surface P by gluing the boundary circles of the surfaces P_c in correspondence with the structure of the word W. We will write $P = \Sigma P_c$ and $K = \Sigma K_c$, $K^{\sim} = \Sigma K_c + \Sigma K_r$, where K_r are the circles obtained by gluing

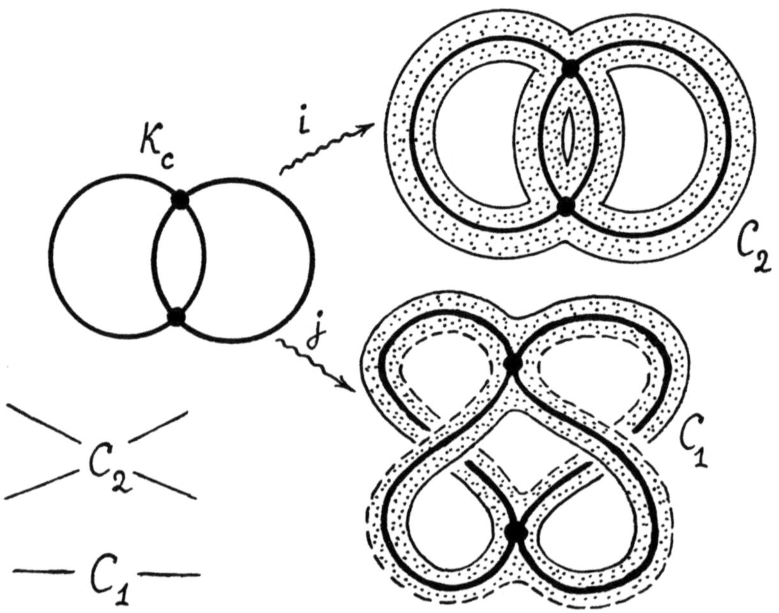

Figure 18

of boundary circles of the surfaces P_c. Here we identify the boundary circles of P_c by some diffeomorphisms. It is evident that we obtain a closed compact two-dimensional surface P with some graph K. For an example see Figure 20. The surface P is the two-sphere and the graph $K^{\sim}$ is the intersection of two equators with four circles K_r (and 4 points) inside each sector. In general, the graph K is not connected; see also the example in Figure 21.

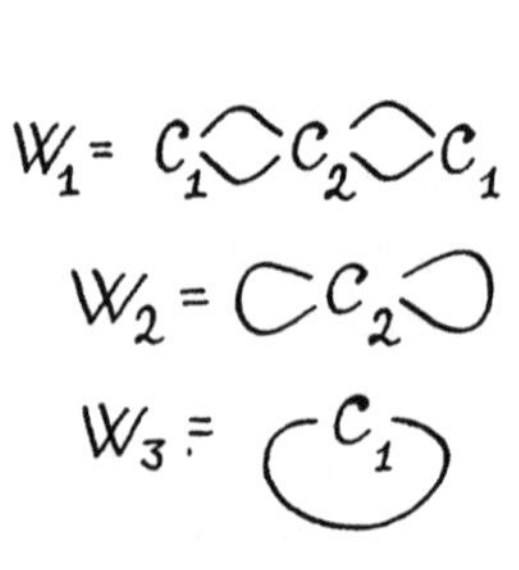

Figure 19

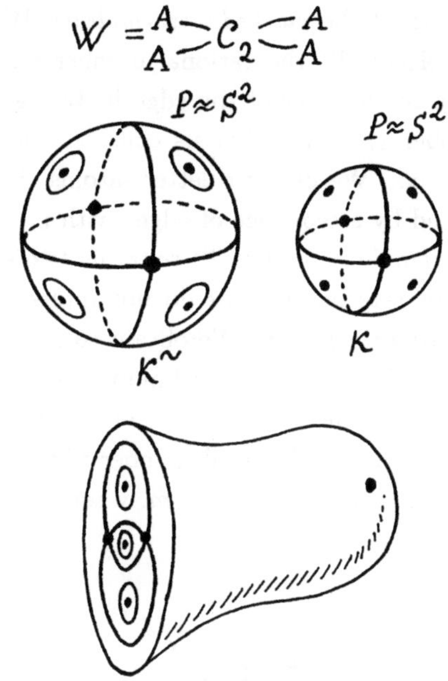

Figure 20

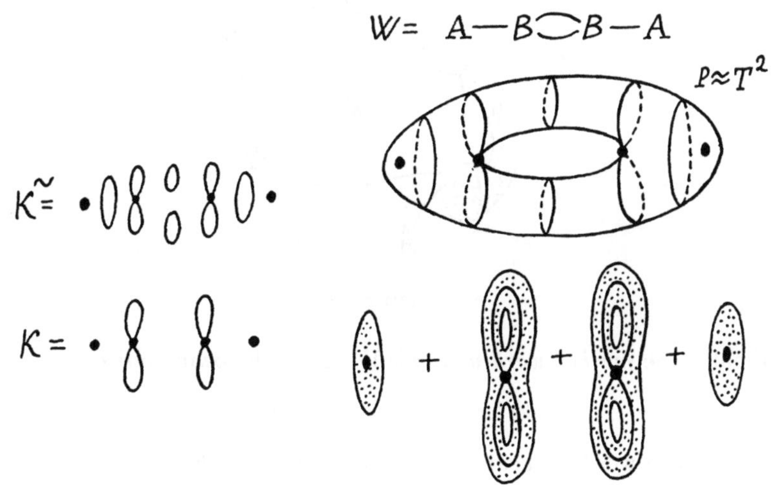

Figure 21

Important Remark: The graph $K^{\sim}$ is uniquely defined by the graph K.

STEP 7: Let us construct the *marked word-molecule* W^*. We put on each edge of the word-molecule W the rational number r_i, where $0 \leq r_i < 1$ or $r_i = \infty$ (here i is the number of the edge in W), and then put on the same edge the number $\varepsilon_i = \pm 1$. Let us consider the set of all edges of W which have $r_i = \infty$. We call two letters-atoms (inside W) *relative* if they can be connected by a sequence of edges with $r_i = \infty$ (on each edge) (Figure 22). The example in Figure 22 contains two sets of relatives. It is clear that different sets of relatives do not intersect. Let us call the connected set of relatives a *family*. We obtain some set of *families* inside the word-molecule W. We put the natural number n_k on each family in W. Here $1 \leq i \leq n$, where n is the total number of edges in the word-molecule W; then $1 \leq k \leq s$, where s is the total number of different *families* in the word-molecule W. It is evident that $s \leq m$, where m is the total number of letters-atoms in W.

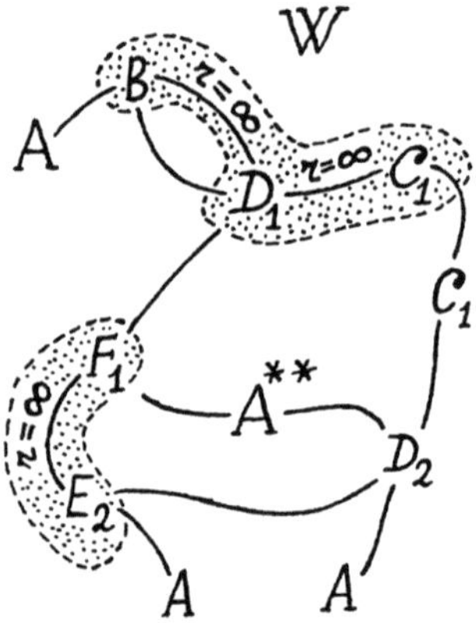

Figure 22

Finally, the *marked word-molecule* W^* is the following object:

$$W^* = (W, \{r_i\}, \{\varepsilon_i\}, \{n_k\}).$$

STEP 8: The word-molecule W is indeed the pair (P, K), where K is the graph embedded in the closed two-surface P. This embedding is fixed (different embeddings can produce different words-molecules). Consequently we can put all the numerical parameters r_i, ε_i, n_k on the surface P. Thus, we can represent the word-molecule in the form:

$$W^* = (P, K, \{r_i\}, \{\varepsilon_i\}, \{n_k\}).$$

We need to define the *equivalent marked word-molecules.*

DEFINITION. *We will call two words-molecules W^* and W'^* equivalent, if there exists a diffeomorphism $\lambda : P \to P'$ such that $\lambda : K \to K'$ (i.e. λ transforms the graph K in the graph K') and the numerical parameters on the corresponding objects coincide, that is: $r_i = r'_i$, $\varepsilon_i = \varepsilon'_i$, $n_k = n'_k$.*

The word-molecule W is the object $I(H,Q)$ discovered by the author in [**30**], and the marked word-molecule W^* coincides with the object $I(H,Q)^*$ discovered by A. T. Fomenko and H. Zieschang in [**43**]. *These objects are topological invariants of integrable Hamiltonian systems.* Now Theorem 4.3 can be formulated as follows.

THEOREM 4.3*. *There exists a natural one-to-one correspondence between the set of all different marked word-molecules W^* and the set of all integrable Hamiltonian systems (considered up to topological equivalence) with two degrees of freedom and in general position.*

Chapter 2. The Classification of Isoenergy Surfaces of Integrable Hamiltonian Systems. The Remarkable Class (H) of Three-Dimensional Manifolds

8. Topological structure of isoenergy surfaces of integrable Hamiltonian system.

8.1. The class (M) of three-dimensional manifolds. The class (H) of isoenergy surfaces of integrable systems.

Let us consider the class (M) of all compact connected closed orientable three-dimensional manifolds M^3. The natural question arises: can an arbitrary three-manifold (from (M)) be the isoenergy surface of some Hamiltonian system (non-integrable, in general)?

THEOREM 8.1 (A. T. Fomenko, S. V. Matveev). *Every compact closed orientable three-manifold can be represented as the isoenergy surface of some Hamiltonian system* $v = \operatorname{sgrad} H$ *for some smooth Hamiltonian* H *on some smooth symplectic manifold* M^4. *The system* v *is, generally speaking, non-integrable. The manifold* M^4 *can be non-compact.*

This theorem immediately follows from the following statement of S. V. Matveev and A. T. Fomenko.

STATEMENT 8.1. *Let* M^3 *be an arbitrary smooth compact closed orientable three-manifold and* I *be the unit interval. Then the direct product* $M^3 \times I$ *is a symplectic manifold (can be endowed with some symplectic structure).*

This theorem shows that "non-integrable and integrable" isoenergy surfaces cover the total class of all three-manifolds. The manifold M^4 in Theorem 8.1 is constructed in a simple way: M^4 is the direct product $M^3 \times I$; the function H is induced on M^4 by the linear function on the interval I (here we use natural projection M^4 on I).

Let us introduce the new class (H) of all connected closed compact three-manifolds which are isoenergy surfaces of *integrable* Hamiltonian systems on M^4 with Bott integral f. Each such manifold is orientable. We have the inclusion $(H) \subset (M)$. Important question: does the class (H) coincide with the class (M)?

THEOREM 8.2 (A. T. Fomenko, [**29**]). *The class (H) does not coincide with the class (M).*

In other words, not every compact closed orientable three-manifold may play the role of a constant-energy surface of a Hamiltonian system, integrated by means of a smooth Bott integral. We obtain the important corollary.

STATEMENT 8.2. *There are some new topological obstacles to the integrability of a Hamiltonian system of general type in the class of Bott integrals.*

Indeed, let us consider some smooth Hamiltonian system. Let us suppose that we can get the information about the topological structure of some isoenergy surface Q^3. If we prove that the surface Q does not belong to the class (H), then automatically we obtain that our system is non-integrable on a given isoenergy surface Q (in the class of Bott integrals). We will demonstrate below concrete applications of this criterion.

Consequently we need more topological information about the structure of the isoenergy surfaces from the class (H).

8.2. Five elementary blocks in the isoenergy surface theory.

We shall describe five types of simple three-dimensional manifolds that appear as "elementary bricks", of which an arbitrary constant-energy surface Q of the integrable system is glued together. Let D^n be the n-dimensional disk.

Type 1. The direct product $S^1 \times D^2$ will be called a *full torus.* Its boundary is one torus T^2 (Figure 23).

Type 2. The direct product $T^2 \times D^1$ will be called a *cylinder (or orientable cylinder).* Its boundary consists of two tori (Figure 23).

Type 3. The direct product $N^2 \times S^1$ will be called an *oriented saddle* or, more descriptively, "trousers", where N^2 is a two-dimensional sphere with three disks removed (or the disk with two holes). This manifold is homotopy equivalent to a figure eight, i.e. to a bouquet of two circles. Its boundary consists of three tori (Figure 23).

Type 4. Consider a full torus which is embedded in R^3 in a standard way. Let us drill in this full torus a thin full torus, which winds twice around the generator of the large full torus (Figure 23). This manifold will be called a *non-orientable saddle.* Its boundary consists of two tori. Let

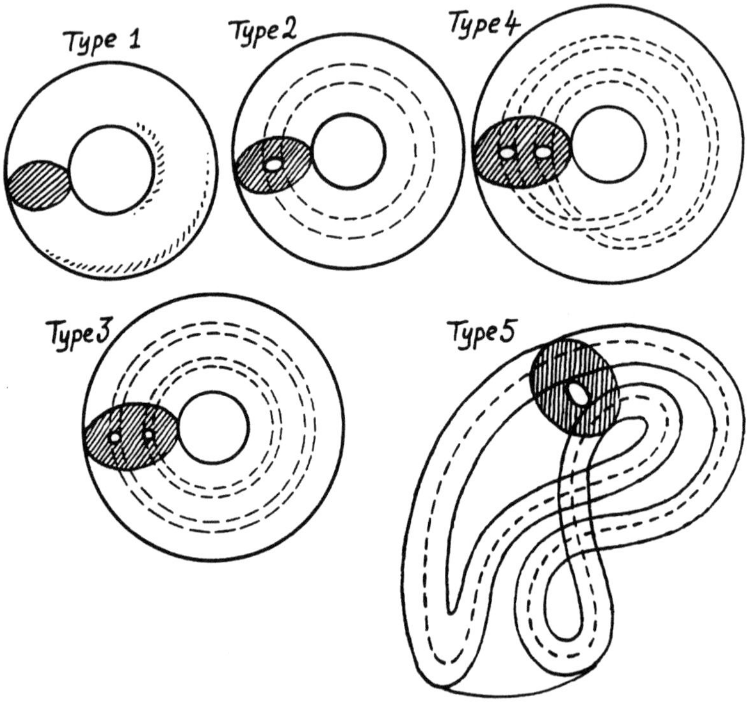

Figure 23

us denote this manifold by A^3. From the topological point of view the manifold A^3 is not new, however. It is obtained by glueing a full torus and an orientable saddle through diffeomorphism of the torus. This may be conditionally written as follows: $A^3 = I + III = (S^1 \times D^2) + (N^2 \times S^1)$. The proof will be given later on.

REMARK: The manifold A^3 has another interpretation. It is the tubular neighborhood of an immersed Klein bottle in R^3. It is clear, that over the circle there exist only two non-equivalent fiber bundles with fiber N^2. These are the direct product $N^2 \times S^1$ (see Type 3 above) and a fiber bundle A^3. In Type 4 the figure eight moves along a circle in such a way that after a revolution the two circles 1 and 2 exchange places (the figure eight reverses). A small neighborhood of the circle (base) S^1 is homeomorphic in this case to two Möbius strips which intersect transversally along their common axis. We will denote this fiber bundle by $N^2 \widetilde{\times} S^1$.

Type 5. We denote by K^2 a Klein bottle and by K^3 the space of an oriented skew product of K^2 by a segment, i.e. $K^3 = K^2 \tilde{\times} D^1$. The boundary of K^3 is a torus. We call the manifold K^3 a *non-orientable cylinder*. The manifold K^3 can be realized as the tubular neighborhood of the Klein bottle K^2 immersed in R^3 as shown on Figure 23. This immersion is different from the immersion of the Klein bottle in Type 4. Thus, two different immersions of the Klein bottle give us two different "elementary bricks" of the Types 4 and 5. From the topological point of view, the manifold K^3 is not new, because (see the proof below) it is represented as the following glueing: $K^3 = I + IV = (S^1 \times D^2) + A^3 = 2 \times I + III = 2(S^1 \times D^2) + (N^2 \times S^1)$.

REMARK: All elementary manifolds I, II, III, IV, V are orientable three-manifolds. The term *non-orientable* in the definition of the non-orientable saddle IV and the non-orientable cylinder V is related with the properties of corresponding separatrix diagrams.

REMARK: The manifold of Type 2 can also be represented as the glueing: $II = I + III$. In all cases the sign "+" denotes the glueing of the boundary tori using some diffeomorphism.

Thus, out of the five types of manifolds listed above, only two are really topologically independent: Type I and Type III. The three others are decomposed into combinations of the manifolds of the Types I and III. But the manifolds of the Types II, IV, V are of great interest for the analysis of trajectories of the system v, because they correspond to peculiar and interesting motions of a mechanical system.

8.3. Stable integral trajectories.

DEFINITION. *Let γ be a closed integral trajectory of a system v on a surface Q (that is, a periodic solution). We say that the trajectory γ is stable if it admits a tubular neighborhood which is fibered into two-dimensional tori which are invariant with respect to the system v, that is, all integral trajectories close to γ "are located" on invariant two-dimensional tori whose common axis is the circle γ (Figure 24).*

Stability of a trajectory implies that a normal two-dimensional disk (of small radius) is fibered fully, without gaps, into *concentric* circles. Since we are primarily concerned with integrable systems, the above definition of stability coincides with the traditional notion of strong stability. The fact is

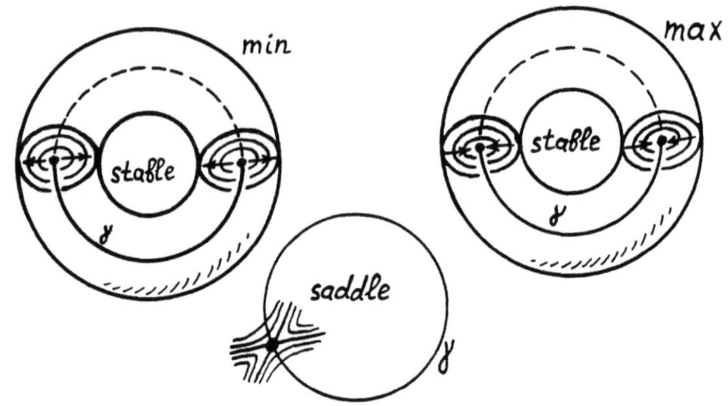

Figure 24

that non-singular level surfaces of a second integral f of such a system are two-dimensional tori, and therefore stable trajectories are those on which the integral f attains its local minima or maxima.

A system may be integrable without having one single closed stable trajectory (although it may have many closed trajectories). A simple example is a geodesic flow on a two-dimensional flat torus, that is, a torus T^2 with the metric $g_{ij} = \delta_{ij}$. It is easy to see that this geodesic flow has an additional linear integral *but that all closed trajectories of the system are unstable* (in our sense).

8.4. The topological classification theorem for three-dimensional isoenergy surfaces of integrable Hamiltonian systems.

Let M^4 be a smooth symplectic manifold (compact or non-compact) and let $v = \operatorname{sgrad} H$ be a Hamiltonian system that is Liouville-integrable on a certain non-singular compact three-dimensional isoenergy surface Q by means of a Bott integral f (simple or complicated, i.e. f is of general type). Let m be the number of periodic solutions of the system v on the surface Q on which the integral f attains a strictly local minimum or maximum (then the solutions are stable). Next, let p be the number of two-dimensional critical tori of the integral f (the minima or maxima of the integral); q the number of critical circles of the integral f (saddles, unstable trajectories of the system) with an orientable separatrix diagram; s the number of critical circles of the integral f (saddles, unstable trajectories of the system) with a non-orientable separatrix diagram; r the number of critical Klein bottles. This is the exhaustive list of all possible connected critical submanifolds of the integral f on Q. The numbers m, p, q, s, r are uniquely defined by a given integral f (simple or complicated).

THEOREM 8.3. 1) (A. T. Fomenko, **[29]**). *The isoenergy manifold Q is represented in the form of glueing (by some diffeomorphisms of the boundary tori) of the following elementary bricks:*

$$\begin{aligned} Q &= m \times I + p \times II + q \times III + s \times IV + r \times V \\ &= m(S^1 \times D^2) + p(T^2 \times D^1) + q(N^2 \times S^1) + s(N^2 \tilde{\times} S^1) + r(K^2 \tilde{\times} D^1). \end{aligned}$$

(Let us call this decomposition ***Hamiltonian****. The Hamiltonian decomposition of Q is uniquely defined if the integral f is* ***simple****, given, and fixed. In the case of a general Bott integral the Hamiltonian decomposition is non-unique).*

2) *(A. V. Brailov and A. T. Fomenko,* **[19]***). Any compact orientable three-manifold of the form*

$$mI + pII + qIII + sIV + rV,$$

i.e. obtained by glueing full tori, orientable and non-orientable cylinders, orientable and non-orientable saddles (trousers), may be realized as the isoenergy surface Q^3 of some Hamiltonian system integrated (by a Bott simple integral f) on an appropriate symplectic manifold M^4. Moreover, the initial decomposition of Q will coincide with the Hamiltonian decomposition of Q corresponding to the integral f.

IMPORTANT REMARK: There are some relations between the numbers m, p, q, s,r: not every collection of these numbers is realized for some isoenergy compact surface Q. We need the surface Q to be closed (without boundary). This gives some restrictions on m, p, q, s, r.

Thus, in the above canonical representation of the manifold Q (when the system v is fixed), all non-negative integers m, p, q, s, r admit a clear interpretation, namely, they let us know how many critical submanifolds of each type a given integral f has on a given manifold Q. If we ignore this interpretation of the numbers m, p, q, s, r and require a simpler topological representation of the isoenergy surface Q, our requirement is met by the following theorem.

THEOREM 8.4. 1) (A. T. Fomenko, [**29**]). *Let Q be a compact non-singular isoenergy surface of a Hamiltonian system $v = \operatorname{sgrad} H$ integrated by some Bott integral f (the concrete realization of this integral is non-important). Then Q admits the following representation:*

$$Q = \alpha I + \beta III = \alpha(S^1 \times D^2) + \beta(N^2 \times S^1),$$

where α and β are some non-negative integers. These numbers are related to the numbers m, p, q, s, r from Theorem 8.3 as follows:

$$\alpha = m + p + s + 2r, \qquad \beta = q + s + r + p.$$

(Let us call this decomposition topological).

2) (A. V. Brailov and A. T. Fomenko, [**19**]). *Any compact orientable three-manifold Q of the form $\alpha I + \beta III$, i.e. obtained by glueing full tori and orientable saddles (orientable trousers) may be realized as an isoenergy surface Q^3 of some Hamiltonian system integrated by some simple Bott integral f on an appropriate symplectic manifold M^4.*

Thus, a compact orientagle three-manifold is the isoenergy surface of some integrable Hamiltonian system of differential equations *if and only if* this manifold has the form $\alpha I + \beta III$ for some α and β.

We obtain the following corollary of Theorem 8.4.

THEOREM 8.5. *Let Q be some isoenergy surface of some Hamiltonian system v. If the manifold Q cannot be represented in the form $\alpha I + \beta III$, then the system v is non-integrable on Q (in the class of Bott integrals).*

Let us demonstrate an application of this theorem. A smooth manifold M is called *hyperbolic*, if M is endowed with a Riemannian metric having constant negative sectional curvature. Let us consider the class of all closed compact three-dimensional hyperbolic manifolds. For the modern theory of these manifolds see W. Thurston's works [**99**], [**100**]. Let us recall that every compact closed oriented three-manifold can be realized as isoenergy surface of some smooth Hamiltonian system [**72**].

THEOREM 8.8 (A. T. Fomenko, S. V. Matveev). *Let us consider an arbitrary closed oriented compact hyperbolic three-manifold Q. Then every Hamiltonian system (having Q as its isoenergy surface) is non-integrable on Q in the class of Bott integrals.*

The topological decomposition is non-unique (even if the integrable system and Bott integral are fixed).

Thus, for an isoenergy surface Q^3, there occur two decompositions. It is clear that the Hamiltonian decomposition is more "detailed". It "remembers" the structure of the critical submanifolds of the given Bott integral f. The topological decomposition is rougher. Its elementary blocks have already partially "forgotten" the original Hamiltonian picture. In the sequel we will use one or the other decompositions subject to the problem of interest.

8.5. Each isoenergy surface determines some graph.

Let us consider some isoenergy surface Q of some integrable system with some Bott integral f (or general type: simple or complicated). Let us consider the corresponding Hamiltonian decomposition of the three-manifold Q (see above). This decomposition is non-uniquely defined in the general case (but is unique if f is *simple*). Then the three-manifold Q can be determined by some graph Z. Let us represent each of the five elementary three-manifolds of the Types I, II, III, IV, V by an elementary graph as shown in Figure 25. The free ends of these graphs correspond to the boundary tori of the three-manifolds I, II, III, IV, V.

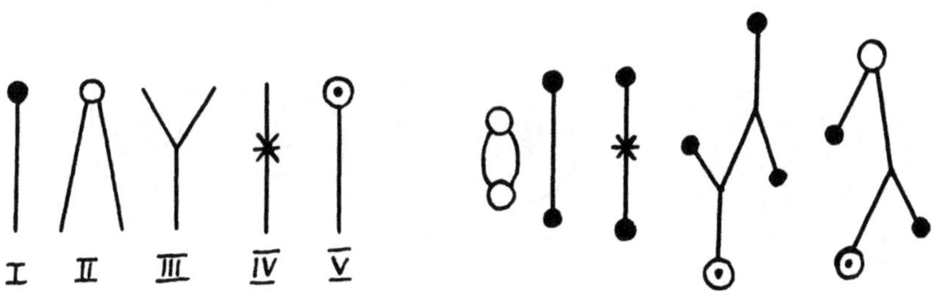

Figure 25

Then each Hamiltonian decomposition $Q = mI + pII + qIII + sIV + rV$ is represented by a graph Z if we glue the free ends of corresponding geometrical figures. The order of these figures is determined by the topology of the manifold Q. We glue the pair of free ends by some diffeomorphism of corresponding boundary two-tori.

For some examples of such graphs see Figure 25.

8.6. Detailed description of the five elementary bifurcations of Liouville tori in integrable systems.

Now we may suggest a complete detailed classification of all types of surgery of Liouville tori for varying values of the *simple* integral f on Q (Theorems 2.2 and 2.2* above).

REMARK: When H and f exchange places, one can speak of bifurcations of Liouville tori as soon as they pass through the critical energy levels when the second integral f is fixed.

We shall examine the following five types of surgery on a torus T^2 which correspond to the manifolds I, II, III, IV, V specified above. We realize the torus T^2 as one of the components of the boundary of a corresponding three-manifold. Then, carried away by the change of the integral f, the torus T^2 transforms into a union of Liouville tori which are the remaining components of the boundary. This surgery acquires the following form.

1) A torus T^2 is contracted to the axial circle of a full torus and then "vanishes" from the level surface of the integral f (Figure 26-I). Denote this surgery as $T^2 \to S^1 \to 0$.

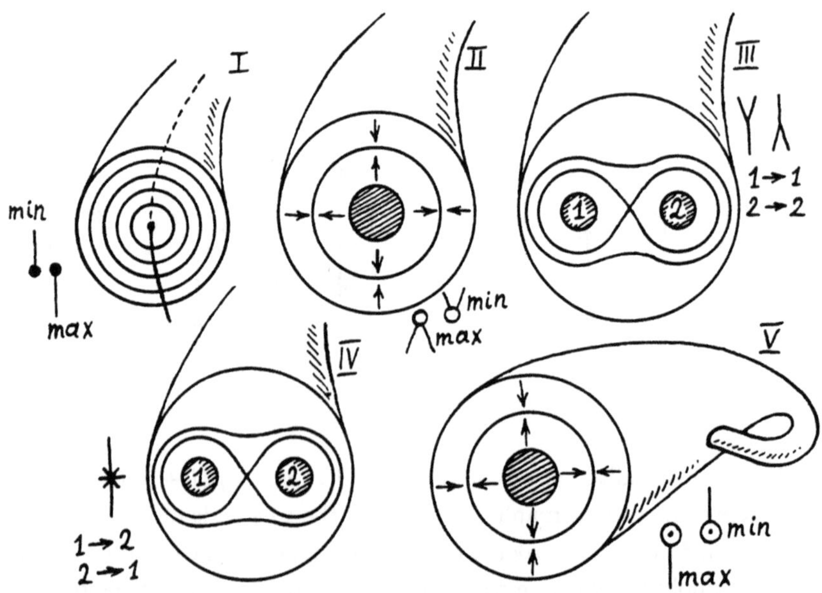

Figure 26

2) Two tori T^2 move toward each other along a cylinder, flow together into one torus, and "vanish" (Figure 26-II). The notation is: $2T^2 \to T^2 \to 0$.

3) A torus T^2 splits into two tori as it passes through the center of the trousers (oriented saddle) when they "stay" on the level surface of the integral f (Figure 26-III). The notation is: $T^2 \to 2T^2$.

4) A torus T^2 spirals twice around a torus T^2 (following the topology of the non-oriented saddle $A^3 = N^2 \widetilde{\times} S^1$) and then stays on the level surface of the integral f (Figure 26-IV). The notation is: $T^2 \xrightarrow{2} T^2$.

5) A torus T^2 transforms into a Klein bottle (covering it twice) and "vanishes" from the level surface of the integral f (Figure 26-V). The notation is: $T^2 \to K^2 \to 0$.

The five types of surgery obtained by reversing the arrows will not be treated as new. *Let f be a Bott integral (simple of complicated) on a regular isoenergy surface Q. Then any surgery of a Liouville torus, which occurs when the torus passes through the critical level of the integral f, is a composition of the elementary Types* 1–5 *of the surgery specified above. The transformations* 2, 4, 5 *split into compositions of the transformations* 1 *and* 3 *(Theorems* 2.2 *and* 2.2**).*

Let us consider the *simple* integral f on the isoenergy three-surface Q. The manifold Q is convenient to be represented in the form of a certain graph $\Gamma = \Gamma(H, Q, f)$. This graph is really the graph Z described above.

Since f is a simple Bott function on Q, the picture of splitting or annihilation on non-singular Liouville tori near critical submanifolds is strictly definite (see above).

1) In Figure 27 a black disk (the vertex of the graph) denotes the connected component of the singular level of the integral f, which is the critical minimax circle (i.e. minimum or maximum of the integral).

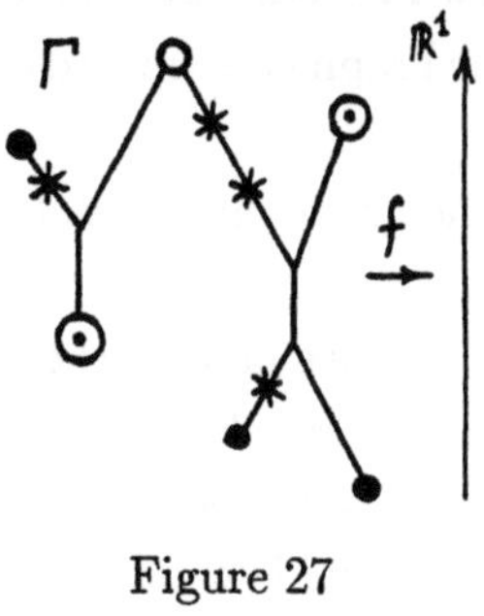

Figure 27

2) A white disk denotes the critical torus (the minimum or maximum of

the integral), Figure 27.

3) The center of the "trefoil" denotes the connected critical level of the integral which contains exactly one critical saddle circle with oriented separatrix diagram (Figure 27).

4) An asterisk indicates such a connected component of a singular level of the integral which contains exactly one critical saddle circle with a non-orientable separatrix diagram (Figure 27).

5) A disk centered at a point (Figure 27) is the critical Klein bottle (the minimum or maximum of the integral).

Every ordinary point on each non-singular edge of the graph (Figure 27) stands for a non-singular Liouville torus. Thus, each non-critical level surface of the function f on Q is shown by a set of ordinary points. Varying the values of f, we let these points shift vertically.

As a result, we obtain a one-dimensional graph Γ. It is clear that this graph may be realized in R^3. *Obviously there exists a continuous mapping h, carrying Q onto the graph* Γ.

1) Exactly one edge of the graph goes into the black vertex of the graph. This edge represents a one-parameter family of Liouville tori connected with a critical min-max circle.

2) Exactly two edges of the graph go up from each white vertex, since the critical torus splits into exactly two non-singular tori and generates two one-dimensional families of regular Liouville tori.

3) Each saddle circle with oriented diagram generates a trefoil, where the edge of the graph, coming into the vertex from below, splits into two outgoing edges, provided that one torus splits into two tori (creation). In case two tori flow into one torus (annihilation), the two edges merging into one vertex of the graph flow into one outgoing edge.

4) Exactly one edge of the graph goes up and one down from each asterisk on the graph Γ.

5) Exactly one edge of the graph emanates from each white vertex with a dot.

It is not accidental that we have labeled these different types of vertices of the graph Γ (in the case of a *simple* integral f) by the numbers 1, 2, 3, 4, 5. The point is that these five types of vertices of the graph Γ and five types of elementary three-manifolds enumerated above are in one-to-one correspondence. It turns out that a sufficiently small neighborhood of the vertex of the graph of type i (from the point of view of Q) is homeomorphic to an elementary manifold of type i.

Indeed, we can construct the graph Γ in the case of an arbitrary Bott integral (simple or complicated). This construction will be discussed below. This "general graph" is closely connected with *letters-atoms* and *words-molecules*.

Let us repeat that (according to Nguen Tyen Zung's theorem) each integrable system $v = \operatorname{sgrad} H$ with Bott integral f on Q can be deformed (by some small perturbation of H and f) in a new Hamiltonian integrable system $v' = \operatorname{sgrad} H'$ with a *simple* Bott integral f' on $Q' = Q$.

9. Different representations of the remarkable class (H).

9.1. Homology properties of integrable isoenergy three-manifolds.

Here we call the three-manifolds Q corresponding to the Bott integrals f *integrable isoenergy three-manifolds*. We saw that the topology of the isoenergy manifold may serve as an obstruction to the integrability of a Hamiltonian system in the class of smooth Bott integrals. The problem is: which topological properties separate integrable isoenergy manifolds Q from the "non-integrable" isoenergy surfaces, that is, how does one describe the class (H) inside the class (M)? It turns out that from the point of view of *homology groups*, the class (H) cannot be recognized in the class (M) (see [**33**]).

STATEMENT 9.1 (Mamedov). *Glueing oriented saddles $N^2 \times S^1$ and full tori $D^2 \times S^1$ one can obtain a three-dimensional oriented compact manifold with any possible (for oriented three-dimensional compact three-manifolds) groups of integer-valued homologies.*

This result follows from the theory of Seifert manifolds, but Mamedov suggested an elementary proof (see [**33**]).

In other words, we cannot effectively compute the obstructions to integrability on the basis of homology groups only: we need some additional information about a given Hamiltonian system. We will demonstrate below the interesting example of new *homological obstacles to integrability* combined with the information about the number of stable periodic solutions of the system v.

9.2. The important classes (H), (Q), (W), (S), (T), (R) and (R_0) of three-manifolds. Morse-type theory for Hamiltonian systems integrated by means of non-Bott integrals.

The author introduced in **[28]**, **[29]** the class (H) of integrable isoenergy three-surfaces.

DEFINITION. *Let us denote by (H) the class of all orientable compact closed three-manifolds which are the isoenergy surfaces of integrable Hamiltonian systems with smooth Bott integrals.*

DEFINITION. *Let us denote by (Q) the class of all orientable compact closed three-manifolds which are represented in the form $\alpha I + \beta III$, where I and III are the elementary manifolds: $I = D^2 \times S^1$, $II = N^2 \times S^1$, α and β are non-negative integer numbers, "+" denotes the glueing of the boundary tori with respect to some diffeomorphisms. (The numbers α and β are chosen in such a way that the glueing produces a closed manifold).*

DEFINITION. *Let us denote by (W) the class of all orientable compact closed three-manifolds W such that the manifold W contains a set of non-intersecting two-dimensional tori whose omission leads to an open three-manifold in which each connected component is foliated as an S^1 bundle over a certain two-dimensional manifold (possibly with boundary). This class was introduced in the important paper of F. Waldhausen* **[107]**. *These manifolds were called Graphenmannigfaltigkeiten and were introduced in the context of three-dimensional topology (without connections to Hamiltonian mechanics). We will call these three-manifolds graph-manifolds.*

Recently, on the advice of the author, S. V. Matveev investigated the class (S) of all three-dimensional compact closed orientable manifolds on which there exist a smooth function g, such that all its critical points are arranged in non-degenerate critical *circles* and all non-singular level surfaces of the function g are unions of non-intersecting two-dimensional tori.

DEFINITION. *The manifold S belongs to the class (S) if and only if there exists a Bott function on it, such that all its critical manifolds are circles and all non-singular level surfaces are composed of tori. It is important to note that here the function g need not be an integral of any Hamiltonian system.*

In Chapters 1 and 2 we have considered the case where the Hamiltonian system has an additional Bott integral f. As shown by concrete studies of real physical systems, this case is actually typical. But situations sometimes occur when a second additional integral f is not Bott. It is natural to ask whether the above results are valid in this case. Of course, the integral should, as before, satisfy some reasonable restrictions because smooth functions may have extremely complicated singularities. Our study of real mechanical systems has shown that it suffices to require that the integral f be *tame*. The term is taken from topology where it denotes objects organized "rather well" as distinct from pathological "wild" spaces. An exact definition is given below. Many of the results that we have obtained prove to be valid also in the case when the smooth integral is only *tame*, and not Bott.

Consider a four-dimensional symplectic smooth manifold and suppose that $v = \operatorname{sgrad} H$ is a Hamiltonian system with a smooth Hamiltonian H, which is Liouville-integrable by means of a certain smooth integral on a non-singular compact isoenergy surface $Q^3 = \{H = \text{const}\}$. Restricting the integral f from the manifold M^4 (or from a neighborhood of the submanifold Q^3 in M^4) to the three-manifold Q^3, we obtain a smooth mapping $f : Q \to R$.

DEFINITION. *A smooth integral f will be called tame if for any critical value $\alpha \in R$ the set $f^{-1}(\alpha)$ is tame, i.e., if there exists a homeomorphism of the whole manifold Q onto itself, which shifts (transforms) this set to a polyhedron. Denote by (T) the class of all three-dimensional compact closed manifolds which are isoenergy surfaces of Hamiltonian systems integrable by means of a tame integral f.*

Since each Bott integral f is clearly tame, a trivial inclusion $(T) \supset (H)$ takes place. In other words, extending the class of considered integrals, we have also extended the class of three-dimensional manifolds which are isoenergy surfaces of integrable systems. From the point of view of three-dimensional topology and its applications to Hamiltonian mechanics, the question is of interest whether or not the classes (T) and (H) coincide. Phrased differently: to what extent is the assumption on the Bott character of the integral essential in the theorems of our theory?

We shall also indicate another class (R) of three-dimensional manifolds which is being actively studied at present. These are manifolds that can be decomposed into a sum of *round handles* (for exact definitions see [8] (D. Asimov)). Now it is more convenient for us to describe these manifolds in another way. The *Morse round (circular) function* is a Bott function of all whose critical submanifolds are circles. It is proved by S. Miyoshi in [**75**] that the manifold M^n (of arbitrary dimension n) can be decomposed as a sum of *round handles* if and only if M^n admits a round (circular) Morse function.

DEFINITION. *Let us denote by (R) the class of all orientable compact closed three-manifolds which admit a round Morse function h.*

The function h does not have a priori any connection with the integrals of Hamiltonian systems.

The manifold M^n is called *topologically reducible* if it can be represented as a connected sum $M = M_1 \# M_2$, where M_1 and M_2 are not homeomorphic to the sphere S^n. In other words, M contains an $(n-1)$-sphere S^{n-1} such that $M = M_1' \cup M_2'$, $M_1' \cap M_2' = S^{n-1}$ and for $i = 1, 2$ we have: $M_i = M_i' \cup D_i$, $M_i' \cap D_i = \partial M_i' = \partial D_i = S^{n-1}$, where D_i is the n-ball (Figure 28). Otherwise the manifold M is called *irreducible*.

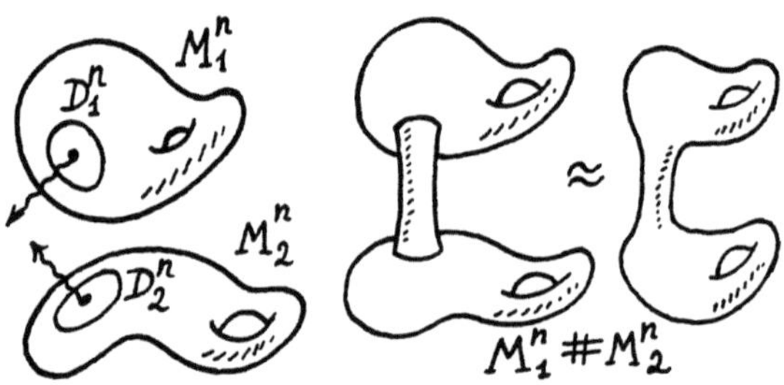

Figure 28

DEFINITION. *Let us denote by (R_0) the class of all orientable compact closed irreducible three-manifolds. It is clear that $(R) \supset (R_0)$ and these classes do not coincide. In other words, the manifold belongs to (R_0) if and only if it is irreducible and admits a round Morse function.*

Let us denote by (H_0), (Q_0), (W_0), (S_0), (T_0) and corresponding subclasses formed by *irreducible* manifolds.

9.3. Five interpretations of the class (H) of isoenergy three-surfaces. Six interpretations of the class (H_0) of irreducible isoenergy surfaces.

The following theorem is the combination of the results of several authors.

THEOREM 9.1. a) *The five classes of three-dimensional orientable closed compact manifolds described above coincide, that is,*

$$(H) = (Q) = (W) = (S) = (T).$$

b) *The six classes of three-dimensional irreducible three-manifolds described above coincide, that is,*

$$(H_0) = (Q_0) = (W_0) = (S_0) = (T_0) = (R_0).$$

This theorem is non-trivial and has important corollaries.

A. T. Fomenko in [**29**] proved that $(Q) \supset (H)$. Then A. V. Brailov and A. T. Fomenko in [**19**] proved that $(Q) \subset (H)$ and consequently, $(H) = (Q)$.

Then it was proved by A. T. Fomenko and H. Zieschang (see [**42**]) that the classes (Q) and (W) coincide, consequently, $(H) = (Q) = (W)$.

Later it was proved by S. V. Matveev (see [**70**]) that the classes (Q) and (S) coincide.

The identity $(H) = (T)$ was proved by A. T. Fomenko and S. V. Matveev in [**71**].

The relation $(H_0) = (R_0)$ was proved by S. V. Matveev and independently by A. T. Fomenko and H. Zieschang (see [**70**], [**42**]).

9.4. Some corollaries.

Theorem 9.1 leads to the following important corollary.

STATEMENT 9.2. *Not every compact closed oriented three-manifold can serve as an isoenergy three-surface of some integrable Hamiltonian system with smooth Bott (or tame) integral.*

We obtain from Theorem 9.1 that Theorem 8.6 is valid in the class of tame integrals. Namely, *if Q is an arbitrary closed oriented compact hyperbolic three-manifold, then every Hamiltonian system (having Q as its isoenergy surface) is non-integrable on the manifold Q in the class of tame integrals.*

In some sense the class (H) has "measure zero" in the class (M) of all three-manifolds. We will discuss this statement later.

Statement 9.3 (A. T. Fomenko, S. V. Matveev, H. Zieschang). *The class* (H) *has the following remarkable property:*

a) *If* Q_1 *and* Q_2 *belong to the class* (H), *then their connected sum* $Q = Q_1 \# Q_2$ *also belongs to* (H).

b) *If* Q *belongs to the class* (H) *and* $Q = Q_1 \# Q_2$, *then the manifolds* Q_1 *and* Q_2 *also belong the class* (H).

Let us comment on Theorem 9.1 and its corollaries. The Statement 9.2 follows from Theorem 9.1 as well as Waldhausen's theorem [**107**], according to which the class (W) does not coincide with the class (M) of all compact closed orientable three-manifolds.

It was found by D. Asimov in [**8**] that a manifold M^n of dimension $n \geq 4$ can be decomposed into a sum of round handles if and only if the Euler characteristic $\chi(M)$ is equal to zero. The situation is more complicated for three-dimensional manifolds, see S. Miyoshi [**75**], J. Morgan [**76**], D. Asimov [**8**], W. Thurston [**98**]. We have: $(R) \supset (H)$, that is, *any isoenergy surface* Q *of an integrable Hamiltonian system admits a round (circular) function.* This follows from Theorem 9.1, since the class $(S) = (H)$ actually consists of manifolds with round functions (moreover, all their non-singular level surfaces are tori). It can be proved that the class (R) is strictly larger than the class (H), that is, there exist manifolds which admit round functions but are not realizable as isoenergy three-surfaces.

The distinction between the classes (R) and (H) is due to the difference in their behavior relative to the operation of taking the connected sum # of two manifolds. According to J. Morgan [**76**], if M is any three-dimensional manifold (we shall not stipulate each time that we are dealing with a class of closed compact oriented manifolds), the connected sum with a sufficiently large number of copies of the manifold $S^1 \times S^2$ eventually leads to the class (R). In other words, we shall have for a sufficiently large number m that the manifold

$$M \# (\#_{i=1}^{m} S^1 \times S^2)$$

belongs to the class (R). On the other hand, we saw that if $M = M_1 \# M_2$ belongs to (H), then the manifolds M_1 and M_2 also belong to (H). This means that the class (R) is strictly larger than the class (H). Indeed, it is sufficient to start with an arbitrary manifold $M_0 \in (M)$ and $M_0 \notin (H)$. We know that such manifolds exist (see Statement 9.2). Taking the connected sum of the manifold M_0 with a large number of manifolds $S^1 \times S^2$, we arrive at the class (R). The manifold thus obtained, however, does not belong to the class (H), since otherwise the original (initial) manifold M_0 must belong to (H), which is contrary to the choice of M_0.

9.5. We can construct new integrable systems: the "sum" of integrable systems is again an integrable Hamiltonian system.

Let us consider two triples (Q_1, H_1, f_1) and (Q_2, H_2, f_2), where $v_1 = \operatorname{sgrad} H_1$ is some Hamiltonian system which is integrable on the isoenergy surface Q_1 with some Bott integral f_1, and $v_2 = \operatorname{sgrad} H_2$ is another Hamiltonian system which is integrable on the isoenergy surface Q_2 with Bott integral f_2.

THEOREM 9.2 (A. T. Fomenko, H. Zieschang, [**42**]). *Let us consider the connected sum $Q = Q_1 \# Q_2$ of the manifolds Q_1 and Q_2. Then there exists a new Hamiltonian system on Q with Hamiltonian H which is constructed from the Hamiltonians H_1 and H_2 and can be called "connected sum" $H = H_1 \# H_2$. This new system is integrable on Q with some Bott integral $f = f_1 \# f_2$ which can be called "connected sum" of the integrals f_1 and f_2. The calculation of the "connected sums" $H_1 \# H_2$ and $f_1 \# f_2$ is algorithmically effective.*

COROLLARY. *If (Q_1, H_1, f_1) and (Q_2, H_2, f_2) are two integrable systems, then there exists the effective algorithm to construct the new integrable Hamiltonian system:*

$$(Q_1 \# Q_2,\ H_1 \# H_2,\ f_1 \# f_2) = (Q_1, H_1, f_1) \# (Q_2, H_2, f_2).$$

This new system will be called "connected sum" of two integrable systems.

One can, for instance, form the connected sum of the integrable Euler case and integrable Lagrange case (in dynamics of rigid body) and obtain the new mixed integrable case "Euler–Lagrange".

10. Lower estimates for the number of stable periodic solutions of integrable Hamiltonian systems.

10.1. Stable periodic solutions, homology groups and the fundamental group of an isoenergy surface.

There exists a simple and important qualitative connection among the following three objects:

a) The Bott integral f on a isoenergy surface Q (of the integrable Hamiltonian system v).

b) Stable periodic solutions of the system on this surface.

c) The group of one-dimensional integer-valued homologies $H_1(Q, \mathbf{Z})$ or a fundamental group $\pi_1(Q)$ of this three-surface.

Let M^4 be a smooth symplectic manifold and $v = \operatorname{sgrad} H$ be a Hamiltonian system on M integrable with some smooth Bott integral f on $Q = \{H = \text{const}\}$. Let $k = k(H, Q)$ be the number of stable periodic solutions of the system v on the isoenergy surface Q. Let us recall that the number k does not depend on the *concrete* choice of the integral f. Let $r = r(H, Q, f)$ be the number of critical Klein bottles of the integral f on Q.

THEOREM 10.1 (A. T. Fomenko, [**29**]). *The number k of stable periodic solutions of the system v on Q is estimated from below through the topological invariants of the surface Q as follows:*

1) *In case the integral f does not have critical Klein bottles on Q, we have:*

a) *$k \geq 2$ if the homology group $H_1(Q, \mathbf{Z})$ is finite;*

b) *$k \geq 2$ if the fundamental group $\pi_1(Q) = \mathbf{Z}$.*

2) *In the general case (when f has critical Klein bottles on Q) we have:*

a) *$k + r \geq 2$ if the homology group $H_1(Q, \mathbf{Z})$ is finite;*

b) *$k \geq 2$ if $H_1(Q, \mathbf{Z}) = 0$ (here the group $\pi_1(Q)$ may be infinite);*

c) *$k \geq 1$ if $H_1(Q, \mathbf{Z})$ is a finite cyclic group;*

d) *$k \geq 1$ if $\pi_1(Q) = \mathbf{Z}$ or $\pi_1(Q)$ is a finite group;*

e) *$k \geq 2$ if $H_1(Q, \mathbf{Z})$ is a finite cyclic group and if the surface Q does not belong to a "small series" of manifolds $Q_0 = I + sIV + rV$ which are easily described in an explicit form (see the description below and in [**29**]).*

3) *If the homology group $H_1(Q, \mathbf{Z})$ is infinite, that is, $\operatorname{rank} H_1 \geq 1$, then the system v has no stable periodic solutions on the surface Q.*

The criterion obtained is rather efficient because it is usually not difficult to check whether an integral is Bott and to calculate the rank of a one-dimensional homology group. In many mechanical integrable systems, the isoenergy surfaces Q are often diffeomorphic either to a sphere S^3 or to the projective space $\mathbf{RP}^3$ or to the direct product $S^1 \times S^2$. For instance, for the equations of motion of a heavy rigid body at high velocities, one may assume after some factorization, that the isoenergy surface Q is homeomorphic to $\mathbf{RP}^3$.

Besides, if a Hamiltonian H has an isolated minimum or maximum (an isolated equilibrium of the system) on M^4, then all sufficiently close level surfaces $Q = \{H = \text{const}\}$ are three-dimensional spheres.

We denote by $L_{p,q}$ the so-called *lens spaces* (i.e. factors of the sphere S^3 under the action of the cyclic group). The cases of interest for Hamiltonian mechanics are singled out as a special assertion.

STATEMENT 10.1 (See [**29**]). *Let a Hamiltonian system $v = \operatorname{sgrad} H$ be integrable by means of a Bott integral f of some isoenergy surface Q homeomorphic to one of the following three-dimensional manifolds: S^3, $\mathbf{RP}^3$, $S^1 \times S^2$, $L_{p,q}$. Then:*

1) *If the integral f does not have critical Klein bottles, we always have $k \geq 2$, that is, the system v necessarily has at least two stable periodic solutions on each of these three-surfaces.*

2) *In the general case (when f can have critical Klein bottles), we have $k \geq 2$ for the sphere S^3, and for the manifolds $\mathbf{RP}^3$, $S^1 \times S^2$, $L_{p,q}$ we have $k \geq 1$. In particular, on the sphere S^3 an integrable system always has at least two stable periodic solutions for any Bott integral.*

Thus, an integrable Hamiltonian system has two stable periodic solutions not only on three-dimensional small spheres close to an isolated equilibrium position (minimum or maximum of the energy H), but also on all extending level surfaces of the function H as long as they are homeomorphic to the sphere S^3. To the best of our knowledge, this qualitative result is also new.

The criterion of Theorem 10.1 is exact in the following sense. Some cases are known where an integrable system v on a surface $Q = \mathbf{RP}^3$ (or $Q = S^3$) has *exactly* one (resp. two) stable periodic solution (V. V. Kozlov, [**52**]).

10.2. Applications and some criteria of non-integrability.

From the results of Anosov, Klingenberg, and Takens, it follows that in the set of all geodesic flows on a smooth Riemannian manifold there exists an open everywhere dense subset of flows without closed stable integrable trajectories [**5**], [**49**]. This means that the *property of a geodesic flow to have no stable periodic solutions (trajectories) is a property of general position.* Recall once again that here we mean "strong stability", defined above.

STATEMENT 10.2 (See [**29**]). *Let the two-dimensional sphere have a Riemannian metric in general position which means that on the sphere there is not a stable closed geodesic line. Then the geodesic flow corresponding to this metric is non-integrable (on each non-singular isoenergy surface Q) in the class of smooth Bott integrals (and tame integrals).*

Statement 10.2 follows from Theorem 10.1 because here we have $Q = \mathbf{RP}^3$ and $H_1(Q, \mathbf{Z}) = \mathbf{Z}_2$.

The rank R of a fundamental group $\pi_1(Q)$ is the least possible number of generators (in the corepresentation of this group).

So, if $\operatorname{rank} \pi_1(Q) = 1$, then a system v integrated on Q by means of a Bott integral f will necessarily have *at least one stable periodic solution on Q.* For instance, in several integrable cases of inertial motion of a four-dimensional rigid body with a fixed point [**33**], some three-dimensional isoenergy surfaces Q are diffeomorphic to $S^1 \times S^2$ (the remarks of Brailov and then the analysis of Oshemkov), that is, $\pi_1(S^1 \times S^2) = \mathbf{Z}$ and $R = 1$. Similarly, as is well known, in the integrable case of Kovalevskaya (for a three-dimensional heavy body), certain isoenergy surfaces are also homeomorphic to $S^1 \times S^2$ (after an appropriate factorization).

STATEMENT 10.3. *Let a Hamiltonian system v be integrable on Q^3 with Bott integral. If the system v has no stable periodic solutions on Q, then:*

1) *The group $H_1(Q, \mathbf{Z})$ is not a finite cyclic group;*

2) Rank $\pi_1(Q) \geq 2$, *with at least one of the generators in the group $\pi_1(Q)$ having an infinite order.*

Consider a geodesic flow of a flat two-dimensional torus, that is, a torus with a locally Euclidean metric. This flow is integrable in the class of Bott integrals and obviously has no periodic stable trajectories. By virtue of Statement 10.2, we must have: $\operatorname{rank} \pi_1(Q) \geq 2$. Indeed, the non-singular surfaces Q are diffeomorphic here to a three-dimensional torus T^3, for which $H_1(T^3, \mathbb{Z}) = \mathbb{Z} \oplus \mathbb{Z} \oplus \mathbb{Z}$.

STATEMENT 10.4 (See [**29**]). *Let* $v = \operatorname{sgrad} H$ *be a smooth Hamiltonian system on* M^4 *and let* Q *be a compact non-singular isoenergy three-surface. Let the following two conditions be fulfilled:*

a) *the system* v *on* Q *has no stable periodic solutions;*

b) *the homology group* $H_1(Q, \mathbb{Z})$ *is a finite cyclic group or* $\operatorname{rank} \pi_1(Q) \geq 1$.

Then the system v *is non-integrable in the class of smooth Bott integrals (or tame integrals) on a given three-surface* Q.

An application of this assertion is seen, for instance, from Statement 10.2 on non-integrability of geodesic flows in general position on a two-dimensional sphere.

In the case of a geodesic flow on a flat torus T^2, we have $Q = T^3$, $H_1(T^3, \mathbb{Z}) = \mathbb{Z}^3$, $R = \operatorname{rank} \pi_1(T^3) = 3$ (that is, the conditions of Statement 10.4 are not fulfilled), and although the flow has no closed stable trajectories on Q, it is nonetheless integrable in the class of Bott integrals.

10.3. Another homological estimate for the number of stable periodic solutions of an integrable system.

Let $\beta = \beta(Q)$ be the Betti number of the isoenergy three-surface Q and $R = R(Q)$ the *rank of* $\pi_1(Q)$, that is, the smallest number of generators of the group $\pi_1(Q)$. Let $m(H, Q)$ be the number of isolated critical minmax circles of the Bott integral f on Q. Let us remind that the number $m(H, Q)$ depends only on H and Q and does not depend on the concrete choice of the integral f. In other words, m is the number of full tori in the Hamiltonian decomposition of $Q = mI + pII + qIII + sIV + rV$. The number m is totally defined by the Hamiltonian foliation of Q determined by H. If $k =$ (number of stable periodic solutions of the system v), then $k \geq m$.

Let $m_0(\beta) = \min(m(H, Q) : \beta(Q) = \beta)$ and $m_1(R) = \min(m(H, Q) : R(Q) = R)$, where the minimum is taken over all three-surfaces Q from the class $(H) = (Q)$ (with prescribed β or R) and over all their decompositions into a sum of elementary "bricks" of type I, II, III, IV, V. The result proved by A. T. Fomenko in [**29**] (the estimation of the number of stable periodic solutions) can be reformulated in the following way.

STATEMENT 10.5. *The following relations are valid:* $m_0(0) = 2, m_0(1) = 0,$ $m_0(2) \geq 0$, $m_0(\beta) = 0$ *for* $\beta \geq 3$; $m_1(0) = m_1(1) = 2$, $m_1(2) \geq 0$, $m_1(R) = 0$ *for* $R \geq 3$.

As A. T. Fomenko and H. Zieschang proved in [**42**], the inequalities displayed above are in fact equalities.

THEOREM 10.2 (See [**42**]). *For the numbers* $m_0(\beta)$ *and* $m_1(R)$ *there are the following equations:*

β and R	0	1	2	3	4	5	...
$m_0(\beta)$	2	0	0	0	0	0	...
$m_1(R)$	2	2	0	0	0	0	...

Theorem 10.1 (see [**29**]) obtains lower bounds for the number k of stable periodic solutions of the integrable system v on Q. This bound has been expressed using topological invariants of Q, namely the first homology group $H_1(Q, \mathbf{Z})$ or the fundamental group $\pi_1(Q)$ when they are "not very large". In the paper [**42**] A. T. Fomenko and H. Zieschang proved a theorem giving the lower bound for the number k in the case when *the group* $H_1(Q, \mathbf{Z})$ *is arbitrary*. We represent the Abelian group $H_1(Q, \mathbf{Z})$ as a direct sum of a free Abelian group B and a finite Abelian group A. Let β be the one-dimensional Betti number of Q, i.e., the *rank* of B. Let ε be the number of elementary divisors of the subgroup A, that is the number of factors of the decomposition of the group A in ordered direct sum of cyclic subgroups where the order of any such subgroup divides the order of the preceeding one. Thus, each group $H_1(Q, \mathbf{Z}) = B + A$ determines two integer numbers β and ε.

Let us consider the Hamiltonian decomposition of the isoenergy three-surface $Q = mI + pII + qIII + sIV + rV$ and the topological decomposition of the same surface $Q = m'I + q'III$.

THEOREM 10.3 (See [**42**]). *Let* Q *be the isoenergy compact closed three-surface of integrable Hamiltonian system with Bott integral. Then the following inequalities are satisfied:*

$$m' \geq \varepsilon - 2\beta + 1, \ q' \geq m' - 2 \text{ for } m + r + s + q > 0.$$

If $m = r = s = q = 0$, *then* $\varepsilon - 2\beta \leq 0$ *with equality actually holding for some pairs* (Q, f). *The case* $m = r = s = q = 0$ *is realized if and only if* Q *is a fiber space with base a circle* S^1 *and fiber a torus* T^2. *If the integral* f

is "totally orientable" (i.e. does not have critical Klein bottles and saddle circles with non-oriented separatrix diagram), i.e. $s = r = 0$, *then we obtain the lower estimate*

$$k \geq m' \geq \varepsilon - 2\beta + 1$$

for the number k *of stable periodic solutions of the Hamiltonian system.*

For $p = 1$ and $m = r = s = q = 0$ we have: $0 \geq \varepsilon - 2\beta$.

Theorem 10.3 supplements Theorem 10.1. Besides, if the integral f is "totally orientable" (i.e. does not have Klein bottles and saddle circles with non-oriented separatrix diagrams) and the homology group $H_1(Q, \mathbf{Z})$ is finite, then we have $\beta = 0$ and $m \geq \varepsilon + 1$, i.e. $m \geq 1$. In this particular case, we "come across" one of the statements of Theorem 10.1. However, the corresponding statement of Theorem 10.1 is stronger, since it contains no assumptions concerning the orientability of the separatrix diagrams. In the general case, Theorems 10.1 and 10.3 are independent. From Theorem 10.3 we have: $m + r + 2s \geq \varepsilon - 2\beta + 1$ and $q \geq m + s - 2$ when $m + r + s + q > 0$.

We see that the integrable system has more periodic stable solutions if the isoenergy surface Q has more one-dimensional homology cycles of finite order (that is, when the number ε increases). Consequently, there is correspondence between the number of stable periodic solutions and the number of generators in the finite part of the group $H_1(Q, \mathbf{Z})$.

Chapter 3. Detailed Description of a New Topological Invariant of Integrable Hamiltonian Systems of Differential Equations

11. Topological construction of the invariant $I(H,Q)$ (the skeleton of an integrable system).

11.1. Fibered full tori and Seifert fibrations.

A fibered full torus (= fibered solid torus) is obtained by a glueing of a cylinder $D^2 \times D^1$, mapping

$$(z,0) \mapsto (e^{2\pi ib/a}, 1),$$

where z is the complex coordinate on $\mathbb{C} = \mathbb{R}^2$, $|z| \leq 1$, a and b - some integer numbers, $a > 0$ and $\gcd(a,b) = 1$.

The fiber is:

$$\{(z_0,t), (e^{2\pi ib/a} z_0, t), (e^{2\pi 2ib/a} z_0, t), \dots, (e^{2\pi i(a-1)b/a} z_0, t)\},$$

where $0 \leq t \leq 1$.

Hence, on the full torus $D^2 \times S^1$ this defines a fibration with fiber S^1 over the disk D^2 which is locally trivial for all points different from the point 0, but the fiber

$$\{(0,t), \text{ where } 0 \leq t \leq 1\}$$

is *exceptional* if $a > 1$ (Figure 29).

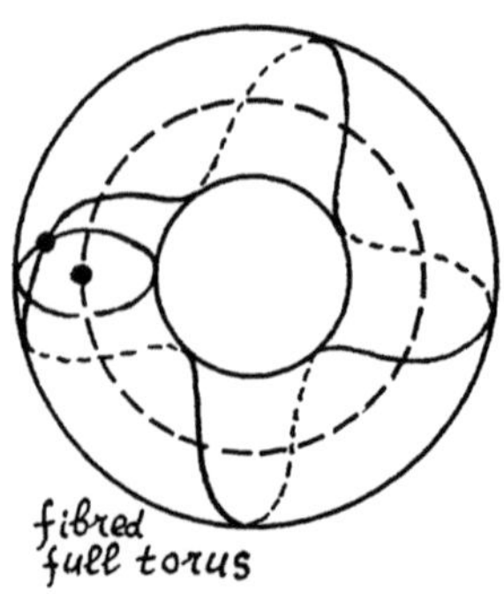

Figure 29

A meridian of the full torus is not a fiber. Fibrations of a full torus with parameters (a, b) and (a', b') are equivalent with respect to the homeomorphisms of the full torus preserving the orientation and the fiber of the full torus if and only if

$$a = a' \text{ and } b \equiv b' \pmod{a}.$$

An important example can be seen in Figure 30, where the fibered full torus of type $(2, 1)$ is represented.

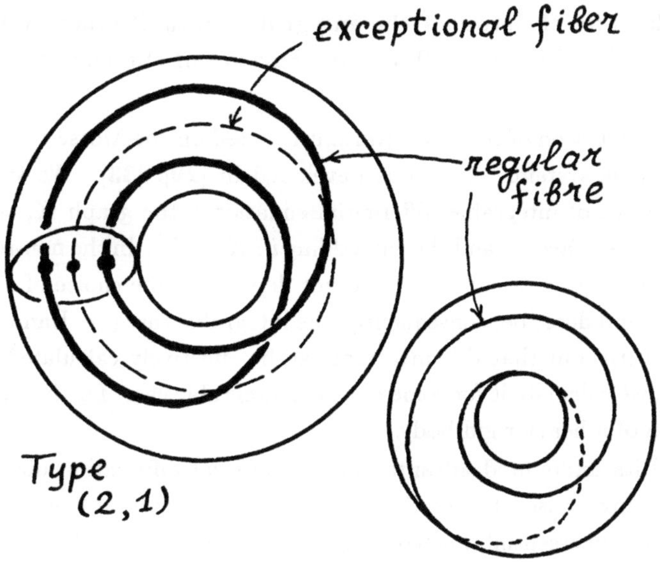

Figure 30

DEFINITION OF SEIFERT SPACES (FIBRATIONS). *Let* $p : M^3 \to B^2$ *be a bundle (fibration) with fiber* S^1 *and base* B. *Let the torus* T^2 *be a component of the boundary* ∂M^3. *Glue to* T^2 *the boundary of a full torus* V *in such a way that the fiber of the bundle in* T^2 *does not become a meridian of* V. *Then* V *admits a fibration as described above such that the fibers on* ∂V *coincide with the fibers of the bundle* $M^3 \to B^2$ *on* T^2. *This can be done with several components of the boundary* ∂M^3. *The manifold obtained is called a Seifert space or Seifert fibration.*

The example of Seifert fibration is given in Figure 30 representing the fibered full torus of type $(2, 1)$. The Seifert fibrations including the fibered full tori of type $(2, 1)$ are very important for Hamiltonian mechanics, because only such Seifert fibrations appear in the isoenergy surfaces Q of integrable systems with Bott integrals [**29**].

In this sense "integrable Hamiltonian mechanics with Bott integrals" corresponds to the Seifert fibrations of the type $(2,1)$ only. Other Seifert fibrations are more general types (a,b) appear in the integration theory with degenerate integrals (of non-Bott type).

11.2. The skeleton of the integrable Hamiltonian system as its topological invariant. The construction of the surface P.

The construction of the new invariant is based on the Morse-type theory for integrable systems which was developed in [**29**], [**33**]. We suggest a new invariant of integrable differential equations: the graph K, the two-dimensional surface P^2 and the embedding $i: K \to P^2$. In the non-resonant case all these geometrical objects do not depend on the choice of a second integral f and describe, consequently, the integrable case (i.e. Hamiltonian) itself. It turns out that this invariant can be effectively calculated. As an example, we calculate it for some classical integrable cases for the equations of motion of a heavy rigid body.

For all its undoubted advantages, the classical bifurcation diagram Σ, which usually is used to describe the topology of integrable cases in mechanics, has an essential shortcoming—it depends on the choice of a second integral f, i.e., it is not a topological invariant of a system of differential equations. The newly discovered object (P,K,i) is, first, a topological invariant and, second, using this object one can reconstruct the bifurcation diagram (if one fixes a concrete form of a second integral f). Thus, the invariant (P,K,i) is a more primary concept than the bifurcation diagram Σ. The object (P,K,i) coincides with the *skeleton* of the integrable system (see 4.1 above). Finally, $(P,K,i) = \text{skeleton} = I(H,Q)$.

Suppose $f: Q \to \mathbf{R}$ is a Bott integral, $\alpha \in \mathbf{R}$ and $f_\alpha \subset f^{-1}(\alpha)$ is a connected component of the level surface of the integral. The surface f_α can be regular or singular. If $\alpha = a$ is a regular (non-critical) value for f, then f_α is a single regular Liouville torus and $f^{-1}(a)$ is a union of a finite number of Liouville tori (Figure 31).

Denote the critical values of f by c, the *connected component* of the critical level surface $f^{-1}(c)$ of the integral by f_s, and the set of critical

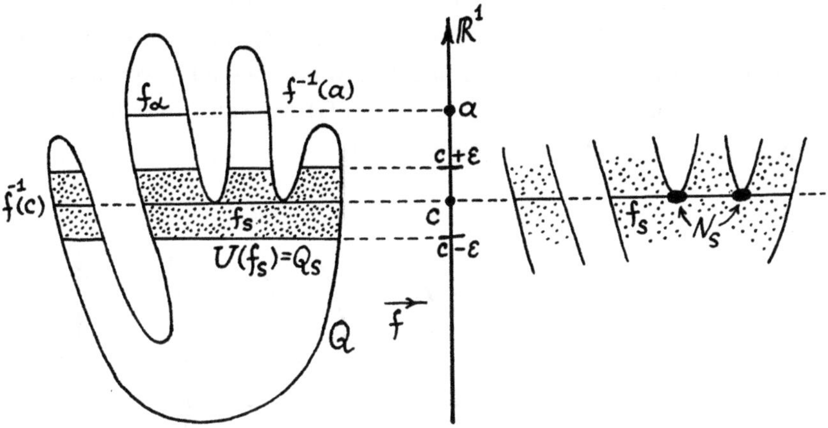

Figure 31

points of the integral f on f_s by N_s (Figure 31). The set N_s can consist of several connected components (inside the connected set f_s).

It should be mentioned that in the general case we have on the same critical level c several connected components f_s. We assume that the index s enumerates all connected components *of all critical levels*.

As proved in [**29**], the connected components of the sets N_s can be only of the following types.

Type I – a minimax circle S^1 (a local minimum or maximum for f); then the connected component of $N_s = f_s = S^1$.

Type II – a minimax torus T^2; then the connected component of $N_s = f_s = T^2$.

Type III – a saddle critical circle S^1 with an orientable separatrix diagram; then the connected component of $N_s \doteq S^1 \neq f_s$.

Type IV – a saddle critical circle S^1 with a non-orientable separatrix diagram; then the connected component of $N_s = S^1 \neq f_s$.

Type V – a minimax Klein bottle K^2, then the connected component of $N_s = f_s = K^2$.

Denote a regular connected small tubular neighborhood of the component

f_s in a manifold Q^3 by $U(f_s)$, where s corresponds to some critical value c of the integral f (Figure 32). The manifold $U(f_s)$ may be assumed to be a connected three-dimensional manifold whose boundary consists of a disconnected union of Liouville tori. For $U(f_s)$ one can take a connected component of the manifold $f^{-1}[c-\varepsilon,\, c+\varepsilon]$.

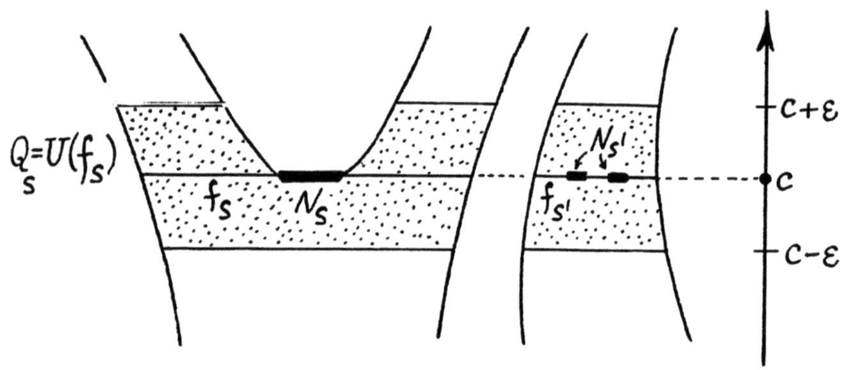

Figure 32

We may assume that $Q = \Sigma_s U(f_s)$, that is, Q is obtained from all the three-manifolds $U(f_s)$ by glueing together their boundaries through certain diffeomorphisms of boundary tori.

The three-manifolds $U(f_s)$ are separated into five types.

Type I: $U(f_s)$ has type I if the corresponding critical set N_s is the minimax circle.

Type II: $U(f_s)$ has type II if the corresponding critical set N_s is the minimax torus.

Type III: $U(f_s)$ has type III if the corresponding critical set N_s consists of the saddle circles with oriented separatrix diagrams.

Type IV: $U(f_s)$ has type IV if the corresponding critical set N_s consists of the critical saddle circles and among them there exists at least one with non-oriented separatrix diagram.

Type V: $U(f_s)$ has type V if the corresponding critical set N_s is the minimax Klein bottle.

THEOREM 11.1 (A. T. Fomenko, [**29**]). *Let Q be a compact non-singular isoenergy surface of a system v, with a Hamiltonian H (not necessarily* *non-resonant), integrable by means of a certain Bott integral f. Then the three-manifolds $U(f_s)$ entering in the decomposition $Q = \Sigma_s U(f_s)$ admit the following representations depending on the type of the critical set N_s.*

Type I: $U(f_s) = P_s^2 \times S^1$, where $P_s^2 = D^2$ is the disk;

Type II: $U(f_s) = P_s^2 \times S^1$, where $P_s^2 = S^1 \times D^1$ is the cylinder (ring);

Type III: $U(f_s) = P_s^2 \times S^1$, where P_s^2 is a certain two-dimensional surface with boundary;

Type IV: $U(f_s) = P_s^2 \hat{\times} S^1$, where P_s^2 is a certain two-dimensional surface with boundary and $P_s^2 \hat{\times} S^1$ is the total space of a Seifert bundle (fibration) with a base P_s^2 and a fiber S^1; all fibered full tori in $U(f_s)$ have type $(2,1)$ (see definition above);

Type V: $U(f_s) = P_s^2 \tilde{\times} S^1$, where $P_s^2 = \mu =$ Möbius band (strip), and $\mu \tilde{\times} S^1$ denotes a skew product (with a torus T^2 as a boundary).

Let us denote $U(f_s)$ by Q_s for simplicity. Then $Q = \Sigma_s Q_s$.

Comment to the Type IV. In the case $P_s^2 = N^2$ (disk with two holes, see above) we have two different representations of this three-manifold Q_s:

$$Q_s \xrightarrow{N^2} S^1, \text{ that is } N^2 \tilde{\times} S^1 \text{ (skew product)},$$

$$Q_s \xrightarrow{S^1} N^2, \text{ that is } N^2 \hat{\times} S^1 \text{ (Seifert product)}.$$

Each three-manifold Q_s from Theorem 11.1 can be represented as a Seifert fibration:

$$\pi_s : Q_s^3 \to P_s^2$$

with the fiber S^1. It is evident that direct and skew products are special types of Seifert fibrations.

COROLLARY. *The integrable system defines the decomposition $Q = \Sigma_s Q_s$, where each three-manifold Q_s is a Seifert fibration with the base a two-dimensional surface P_s^2, and the fiber a circle S^1. The type of Seifert fibration is "no more" than $(2,1)$. This decomposition is uniquely defined by a given Bott integral f. In the cases III and IV the fibers of the Seifert fibrations are uniquely defined (up to homeomorphism). In the cases I, II, and V the fibers are defined non-uniquely.*

It is important that in our theory the following fact is valid: *all Seifert fibrations $P_s^2 \hat{\times} S^1$, which appear in Hamiltonian mechanics with Bott integrals, can be obtained in the following way.*

A full torus obtained by glueing together the two bases of a full cylinder $D^2 \times D^1$ by the mapping $(z,0) \mapsto (-z,1)$ where $z \in \mathbf{C}$, $|z| \leq 1$, is called a *fibered full torus of type* $(2,1)$. The fiber is $\{(z,t),(-z,t)\}$, where $0 \leq t \leq 1$. Then the full torus $D^2 \times S^1$ becomes a fiber bundle with fiber S^1 above D^2, which is locally trivial for all non-zero points of the disk D^2. The circle fiber $\{(0,t),\ 0 \leq t \leq 1\}$ is singular (= exceptional). Now consider the manifold P_s^2 with points $x_1, \ldots, x_m$ marked on it and encircle them with small disks $D_1^2, \ldots, D_m^2$. Then take the direct product $P_s^2 \times S^1$ and discard from it m full tori $D_i^2 \times S^1$, where $1 \leq i \leq m$. In place of these full tori glue in fibered full tori of type $(2,1)$. Denote the manifold obtained by $P_s^2 \hat{\times} S^1$. It is a Seifert bundle (fibration) and P_s^2 is its base.

The next step is very important. The three-manifold Q is obtained from the three-manifolds Q_s by their glueing with respect to some diffeomorphisms of boundary tori. Each boundary torus is fibered over a boundary circle of the surface P_s^2. If the boundary tori belong to some three-manifolds Q_s which are identified by some diffeomorphism λ (see Figure 33), then λ induces some diffeomorphism h which identifies two corresponding boundary circles (Figure 33). Consequently the decomposition $Q = \Sigma_s Q_s$ induces the glueing of the pieces P_s^2 and we obtain some two-dimensional surface $P^2 = \Sigma_s P_s^2$.

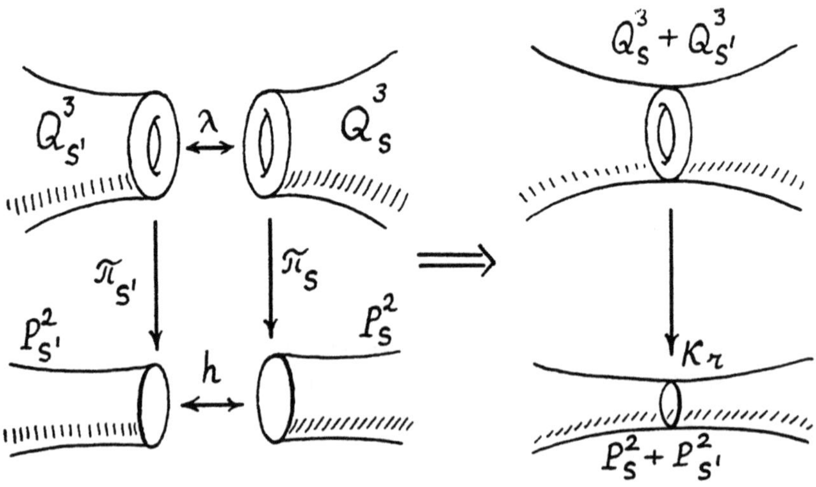

Figure 33

STATEMENT 11.1 (See [**30**]). *Each isoenergy surface Q of an integrable Hamiltonian system with Bott integral f defines (up to homeomorphisms) a certain closed two-dimensional (compact if Q is compact) surface $P^2(H, Q, f) = \Sigma_s P_s^2$ obtained by such glueing together of the surfaces P_s^2 which is induced by the glueing $Q = \Sigma_s Q_s$.*

Let us consider the following diagram:

$$\begin{array}{c} Q = \Sigma_s\, Q_s \\ \downarrow \\ P = \Sigma_s\, P_s \end{array}$$

IMPORTANT REMARK: Each Q_s is fibered over P_s, but nevertheless, in general, Q cannot be fibered over P (as a Seifert fibration). In other words, the projections $\pi_s : Q_s \to P_s$ cannot be glued, in general, to form a total projection $\pi : Q \to P$.

11.3. The construction of the graph K and its embedding in the surface P.

Let us consider the connected component f_s of the critical level $f^{-1}(c)$ which is contained in Q_s (see above). The set f_s can be one-dimensional or two-dimensional.

THEOREM 11.2 (See [**29**]). *Each set f_s is a "Seifert fibration" with the fiber a circle S^1 and the base a one-dimensional graph K_s. In other words, the fibration*

$$\pi_s : f_s \to K_s$$

can be considered as subfibration in the fibration $\pi_s : Q_s \to P_s$. In particular, $K_s \subset P_s$. This embedding $i_s : K_s \to P_s$ is uniquely defined by the Hamiltonian system and by the Bott integral f.

Let us define the graph K as the union of all K_s; i.e., $K = \{K_s\}$ or $K = \Sigma_s K_s$.

COROLLARY. *The integrable system $v = \operatorname{sgrad} H$ on Q with Bott integral f uniquely defines some graph K embedded in the closed surface P. For the construction of the graph, see below.*

Let us consider in the surface P all circles K_r which are the result of glueing of the boundary circles of the surfaces P_s when we construct the surface P (Figure 33). Now let us consider the graph $K^{\sim}$ which is the union of all graphs K_s and all graphs K_r, that is,

$$K^{\sim} = \{K_s\} + \{K_r\}.$$

Then we can construct the following diagram:

$$\begin{array}{ccccl} Q = \Sigma_s Q_s & \supset & \Sigma_s f_s & = & \{f^{-1}(c),\ c - \text{singular}\} \\ \downarrow & & \downarrow & & \\ P = \Sigma_s P_s & \supset & \Sigma_s K_s & = & K \end{array}$$

REMARK: The surface P need not necessarily be embedded into the three-manifold Q. But, as we will see below, each surface P_s can be embedded (non-uniquely) in a corresponding three-manifold Q_s.

The embedding $i : K \to P$ is totally defined by the Hamiltonian H and integral f. The graph K can contain isolated points (isolated vertices).

The graph $K^{\sim}$ is uniquely determined by the graph K.

STATEMENT 11.2. *The connected components K_s of the graph K can be only of the following types.*

Type 1. Isolated point. Here we have: $P_s = D^2$ (disk) (Figure 34). The vertex $= K_s$ is in the center of the disk. These vertices correspond to the minimax critical circles of the integral f.

Type 2. Isolated circle. Here we have: $P_s = S^1 \times D^1$ (ring = cylinder) or $P_s = \mu$ (Möbius band). This component K_s does not have any vertices (Figure 34). The isolated circle is the "axis" of the ring or Möbius band. This circle corresponds to the critical minimax torus or Klein bottle in Q_s.

Type 3. Several circles which intersect in some points in such a way that only two circles meet in one point. These points are the vertices of multiplicity 4 in the graph K_s (4 edges in each such vertex). The surface P_s is oriented and retracts on K_s (Figure 34). The surface P_s can be non-flat. These vertices correspond to the saddle critical circles with oriented separatrix diagram.

Type 4. Several circles joined as in type 3, but there are the additional vertices of multiplicity 2 on some edges of the graph K_s. These vertices will be marked by stars and they correspond to the critical circles of the integral with non-oriented separatrix diagram. The surface P_s is non-oriented and retracts on K_s. Consequently, here we have vertices of two kinds and they correspond to oriented critical circles and non-oriented ones.

Let us describe *the reduction operation* of the graph K and the graph $K^{\sim}$. This operation makes sense only in the case when the integral f has

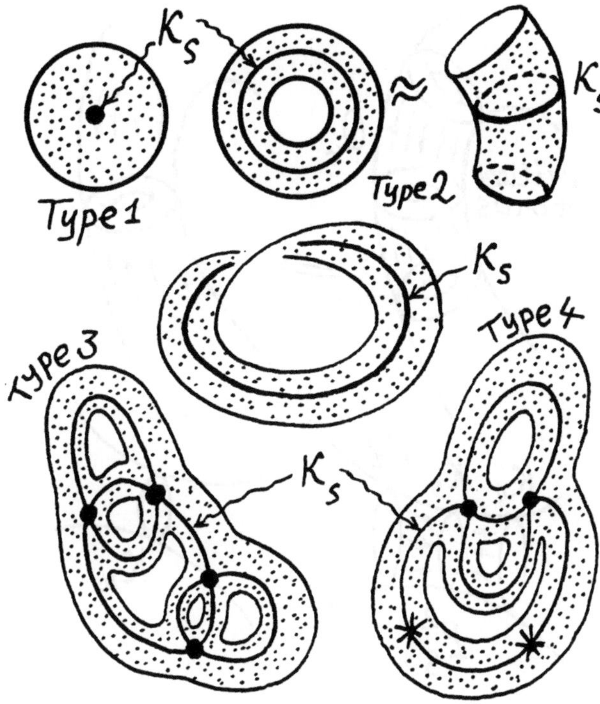

Figure 34

the minimax critical tori on Q. Each critical torus determines the three-manifold Q_s which is evidently homeomorphic to the direct product $T^2 \times D^1$ (Figure 35).

After the corresponding projection $\pi_s : Q_s \to P_s$ we obtain $P_s = S^1 \times D^1$ = cylinder, ring (Figure 35). Let us retract the three-manifold $T^2 \times D^1$ on the torus T^2 and hence retract the cylinder $S^1 \times D^1$ on the circle S^1 (Figure 35). Then we remove this Q_s from the initial decomposition of Q and replace Q_s by one torus. Thus, we replace the cylinder in P by one circle of type K_r. This operation is called *reduction* and after all reductions we obtain the *reduced graph K in the surface P*. The reduction does not change Q and P but changes the graph K. Some triples $(K_r, K_s, K_{r'})$ are replaced by one circle K_r (Figure 35).

We will assume that all graphs K in our forthcoming theorems are *reduced.* We will call the corresponding triplet (P, K, i) *reduced.*

THEOREM 11.3 (A. T. Fomenko, [**30**]). *The surface P and the graph K*

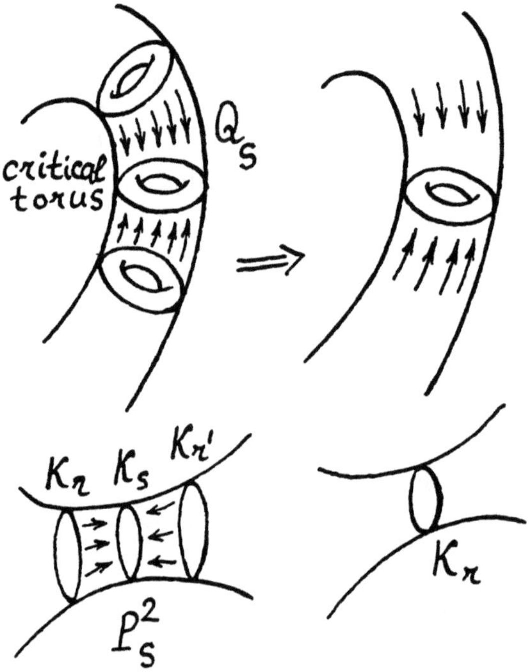

Figure 35

coincide with the surface P and the graph K introduced above in Chapter 1, item 4. If a Hamiltonian H is non-resonant on the isoenergy surface Q, then the reduced triplet (P, K, i) does not depend on the choice of a second Bott integral f. Namely, if f and f' are any Bott integrals of the system v, then the corresponding surfaces P and P' and the graphs K and K' are homeomorphic, and the following diagram is commutative:

$$\begin{array}{ccc} i: K & \rightarrow & P \\ \downarrow & & \downarrow \\ i': K' & \rightarrow & P' \end{array}$$

COROLLARY. *In the non-resonant case the reduced triplet (P, K, i) is a topological invariant of the integrable case (integrable Hamiltonian) itself and makes it possible to classify integrable Hamiltonians by their topological type and complexity. The object $I(H, Q) = (P, K, i) = W$ is the skeleton of an integrable system (on Q) and is the word-molecule introduced in Chapter 1. In particular, each integrable system v on some isoenergy* *surface Q uniquely defines some remarkable integer number, which we will call the genus of the system. This number is the genus of the invariant surface P.*

REMARK: The surface P can be orientable or non-orientable. More precisely, P is orientable if and only if the integral f does not have critical Klein bottles.

It is very important that in the non-resonant case the whole picture does not depend on the concrete choice of the second integral f. It turns out that the integrable Hamiltonian H *knows everything* about the topology of the corresponding Liouville foliation on the isoenergy surface Q. We do not need the concrete formula for f to investigate the topology of a given integrable system. We need only the abstract information that *some Bott integral exists.*

We will say that the skeletons (P, K, i) and (P', K', i') of two integrable systems *coincide* if and only if there exists a homeomorphism $\lambda : P \to P'$ which transforms K onto K' and embedding i into i'.

The division of the surface $P(H, Q) = P$ into regions (this division being determined by the graph K in P) is also *a topological invariant.*

COROLLARY. *If the reduced skeletons (P, K, i) and (P', K', i') of two integrable Hamiltonian systems v and v' do not coincide, then these systems are not topologically equivalent.*

11.4. Topological structure of the critical level surfaces of Bott integarls on the isoenergy three-manifold.

Let us consider the arbitrary critical value c of the integral f and connected component f_s of the critical level surface $f^{-1}(c)$.

STATEMENT 11.3 (See **[29]**, **[30]**). *A cell complex f_s is one of the following types.*

Type 1. *Single circle. (Minimax of the integral f).*

Type 2. *Torus. (Minimax of the integral f.)*

Type 3. *The two-dimensional set (cell complex) f_s is obtained by glueing together several two-dimensional tori along circles that realize non-zero cycles on the tori. If there are several such circles on one torus, they do not intersect. The circles along which the tori entering the f_s "are tangent",* *are saddle critical for f. They are homologous and cut f_s into a union of several flat rings. Two cases are possible:*

Type 3-a. If all saddle critical circles have oriented separatrix diagrams, then the complex f_s is the direct product $K_s \times S^1$ of some graph K_s on the circle (Figure 36).

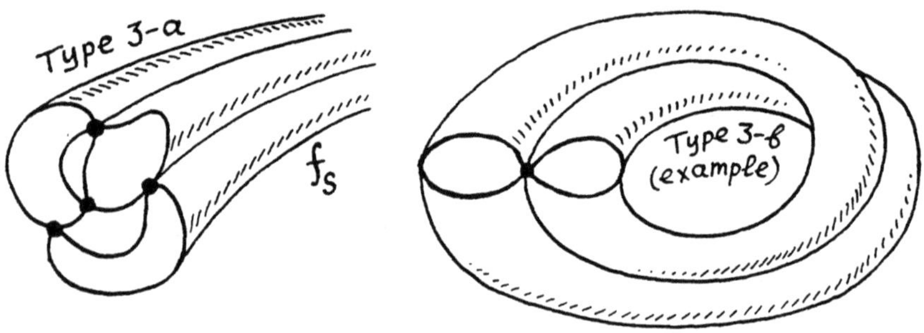

Figure 36

Type 3-b. If among saddle critical circles there exists at least one with non-oriented separatrix diagram, then f_s is the "Seifert fibration" over some graph with a fiber S^1, where exceptional fibers have type $(2,1)$ only. The example is seen in Figure 36.

Type 5. Klein bottle. (Minimax of the integral f.)

The example on Figure 36 shows the case of one saddle critical circle with non-oriented separatrix diagram. The complex f_s is the immersion of the Klein bottle in $\mathbf{R}^3$.

IMPORTANT REMARK: The graph described in item 3-b of the Statement 11.3 does not coincide with the graph K_s, but can be obtained from the corresponding graph K_s by its duplication (see the details below).

11.5. The graph $K^{\sim^*}$ dual to the graph $K^\sim$ in the surface P.

Let us consider the *reduced* graph $K^\sim$ in the closed surface P. Then the graph $K^\sim$ divides the surface P into the union of some open regions. The boundary of each such region is the union of circles; some of them can intersect (Figure 37). It is evident that the boundary of a region consists

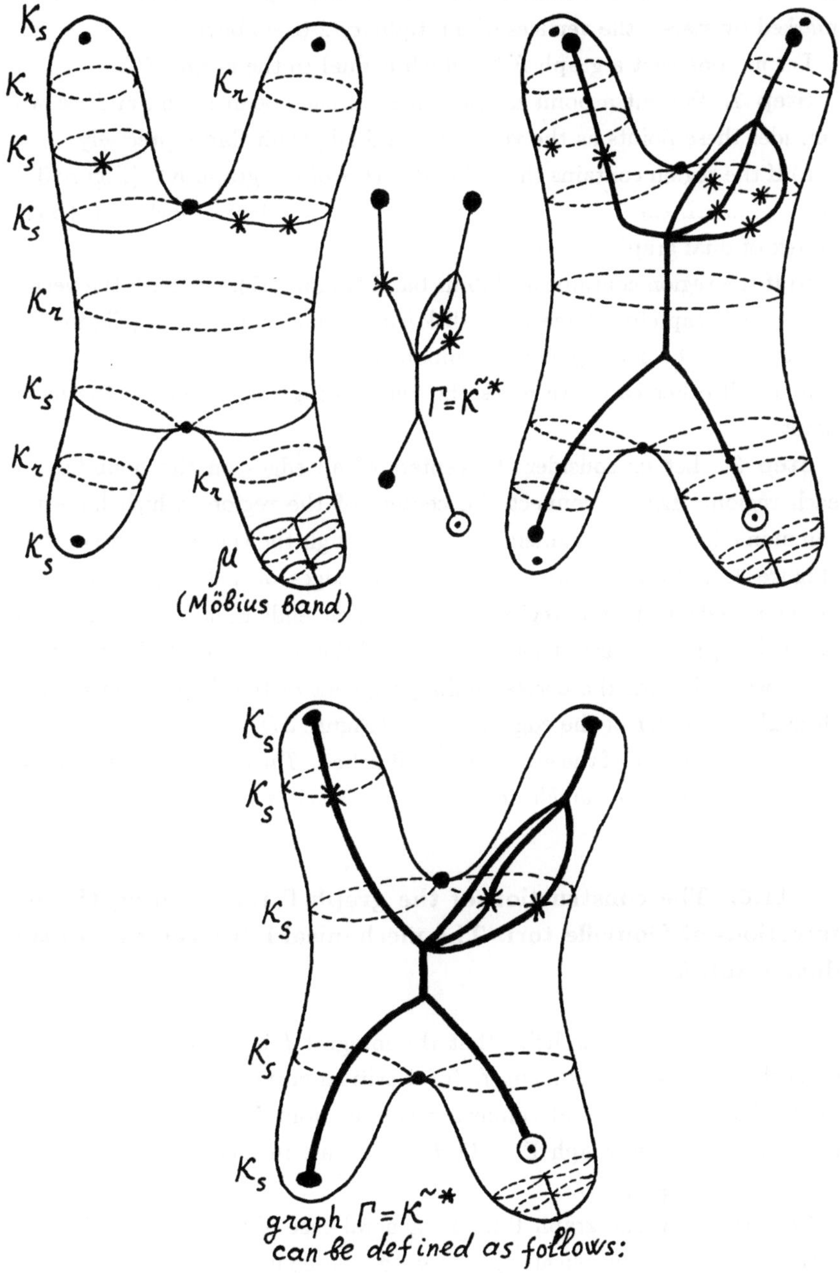

Figure 37

of some circles K_r and some graphs K_s. Some edges of the boundary are marked by stars—the vertices of multiplicity 2 (see above).

Let us construct a graph $K^{\sim^*}$ which is dual to the graph $K^{\sim}$.

Step 1. We put a point in the center of each open region and we will consider these points as the vertices of a dual graph. More precisely:

a) If the region contains the isolated vertex of the graph $K^{\sim}$ (Figure 37), then we mark the vertex of the dual graph by a large black disk (black vertex of dual graph).

b) If the region contains a Möbius band (Figure 37), we mark the vertex of the dual graph by a large white disk with a point in its center (Figure 37). (In this case the corresponding component K_s is a circle).

c) In all other cases we mark the vertex of a dual graph by the usual point.

Step 2. Let us consider the centers of all edges on the boundary of each region. Let us connect the centers of the regions which have the common edges by the segments passing through the centers of these edges (Figure 37). Consequently, we define the edges of a dual graph. Some boundary edges of some region can have their ends in star-vertices of the graph K. (In this case at least one end of the edge is marked by a star). Then we will mark the corresponding segment of the dual graph passing through the center of the edge *by a star* (Figure 37).

The construction of the graph $K^{\sim^*}$ is finished. *The graph $K^{\sim^*}$ is uniquely defined by a given integrable Hamiltonian system (up to homeomorphism).*

11.6. The construction of the graph Γ representing the bifurcations of Liouville tori. The mechanical interpretation of the dual graph $K^{\sim^*}$.

First suppose for simplicity that the integral f is *simple* on Q, that is on each connected component f_s (of a critical surface level $f^{-1}(c)$) there exists exactly one critical connected submanifold N_s. For this case the construction of the graph $\Gamma = \Gamma(H, Q, f)$ was, in fact, described in the item 8.5 as the graph Z.

Construct now the graph Γ in the general case of an arbitrary Bott integral (simple or complicated). Now on one connected component of a critical level may lie several critical manifolds. As distinguished from ordinary Morse functions, critical manifolds of a Bott integral which lie on one level cannot, generally speaking, be taken onto different levels by way of small perturbations of the integral only (and remain in the class of integral) without perturbation of H or ω (see the theorem of Nguen Tyen Zung above). The perturbation of H is forbidden because we do not want to change the initial Hamiltonian system.

Let $f^{-1}(a)$ be the level surface of an integral, where a is a regular value. Then $f^{-1}(a)$ is a union of a finite number of tori. Let us plot them as points in $\mathbf{R}^3$ on the level a, where the axis $\mathbf{R}$ has an upward direction (Figure 38).

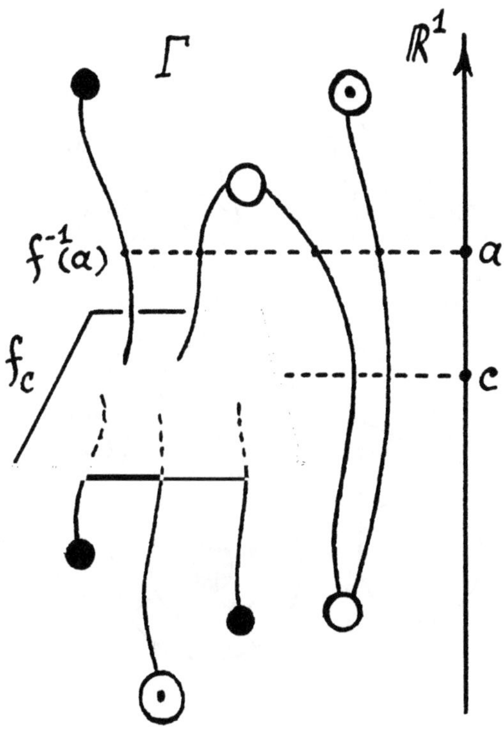

Figure 38

Varying a within the range of regular values, we make these points sweep some arcs—part of an edge of a future graph Γ.

Let N_s be a set of critical points f of f_s, where s corresponds to some singular value c. We will mention two cases:

$$\text{a) } N_s = f_s; \text{ b) } N_s \subset f_s \text{ and } N_s \neq f_s.$$

We know all possibilities for N_s (see [**29**] and above).

Let us consider the case a). Only three types of critical sets are possible here.

a-1. **A minimax circle.** Here $N_s = f_s = S^1$ and its tubular neighborhood $S^1 \times D^2$ is a full torus in Q. As $a \to c$, non-singular tori contract to the axis of the full torus and for $a = c$ they degenerate into S^1. This situation is shown schematically in Figure 39, where the black disk is the vertex of the graph into which (or from which) there goes one edge of the graph.

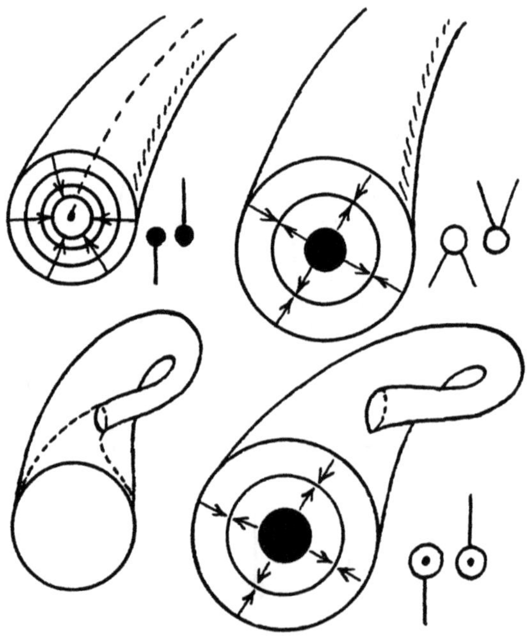

Figure 39

a-2. **A torus.** Here $N_s = f_s = T^2$ and f reaches its minimax on T^2. The tubular neighborhood of T^2 is homeomorphic to a cylinder $T^2 \times D^1$. The boundary of the cylinder is represented by two tori. As $a \to c$, they move toward each other and for $a = c$ they become one torus. We will depict this situation with a light circle (the vertex of the graph) into which (or from which) there go two edges of the graph (Figure 39).

a-3. **A Klein bottle.** Here $N_s = f_s =$ Klein bottle K^2 on which f reaches its local minimax. Its tubular neighborhood $K^2 \tilde{\times} D^1$ is homeomorphic to a skew product of K^2 by a segment. The boundary of $K^2 \tilde{\times} D^1$ is one torus. As $a \to c$, it tends to K^2 and is a two-fold covering when $a = c$. We will show this situation with a light circle with a point inside (the vertex of the graph) into which (or from which) there goes one edge of the graph (Figure 39).

As one can see, these symbols coincide with the same ones introduced above.

Consider now the case b) (more complicated).

Here f_s is a two-dimensional cell complex and N_s is a disconnected union of non-intersecting circles. Each of them is a saddle circle for f. The component f_s will also be called a *saddle*. It may be schematically shown as a flat horizontal square lying on the level c in the space $\mathbf{R}^3$ (Figure 38). Some edges of the graph stick into it from below (when $a \to c$ and $a < c$), the other edges go upward (when $a > c$).

Thus, we have defined a certain graph A containing regular arcs (edges) some of which stick into squares and others end with vertices of the types described above.

Fix the saddle value c and construct a graph T_c showing the exact manner in which the edges of the graph A interacting with f_c join with one another. Consider a vector field $w = \operatorname{grad} f$ on Q. (Here we need some arbitrary Riemannian metric; its choice is up to you). The trajectories of w entering into the critical points of the integral or going out of these points are called *separatrices*. Their union is a *separatrix diagram* of the critical submanifold. Consider on the level f_c its separatrix diagram emanating from each saddle critical circle S^1. If it is orientable, it is obtained by glueing together two flat rings (cylinders) along the axial circle (Figure 40). If it is non-orientable, it is obtained by glueing together two Möbius bands along their axial circle (Figure 40).

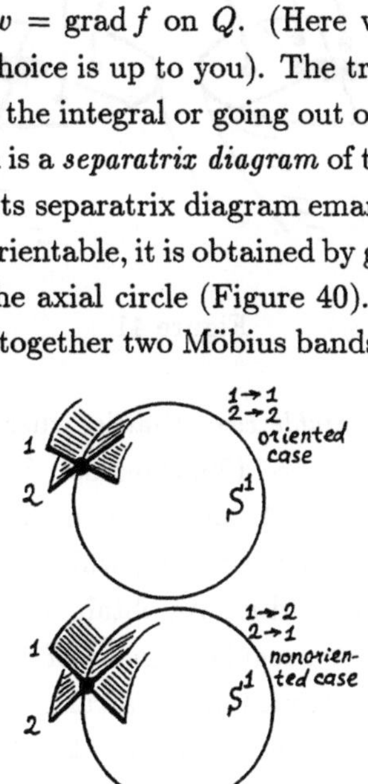

Figure 40

Consider non-critical values $c - \varepsilon$ and $c + \varepsilon$ close to c. The surfaces $f_{c+\varepsilon}$ and $f_{c-\varepsilon}$ consist of tori (Figure 41). The separatrix diagrams of the critical circles lying in f_c intersect these tori transversally along certain circles and divide the tori into unions of regions which we will call *regular*. In each region on a level $f_{c-\varepsilon}$ choose a point and consider the integral trajectories of the field w with these initial conditions. They will pass by the critical circles on the level f_c and get into some other regular regions of tori which form $f_{c+\varepsilon}$. It is obvious that in this way we obtain a homeomorphism between the open regular regions from $f_{c-\varepsilon}$ and the open regular regions from $f_{c+\varepsilon}$.

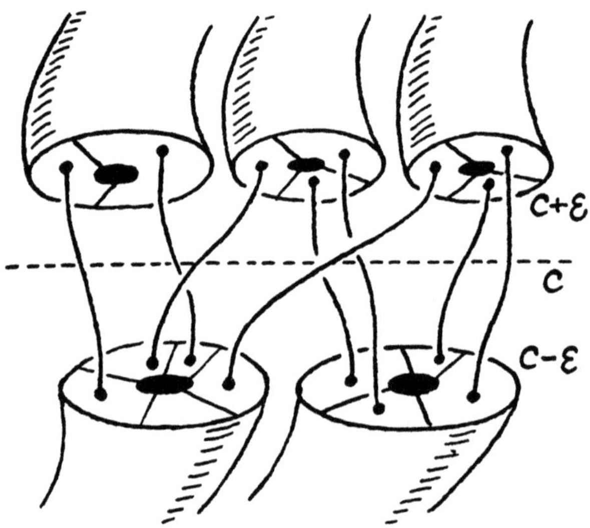

Figure 41

First, consider *the orientable case*, that is, where all separatrix diagrams are orientable (i.e., there are no Möbius bands). Since each regular Liouville torus is a point in the graph A, one can join the points on the level $f_{c-\varepsilon}$ and on the level $f_{c+\varepsilon}$ by arcs (segments) which stand for the bunch of the integral trajectories of the field w. We obtain a certain graph T_c. Its edges show us the motion of open regular regions of the tori. The tori fall into pieces which then rise (descend) and rearrange themselves into new tori. Each upper torus is made of pieces of lower tori (and vice versa).

We now consider *the non-orientable case*, that is, when on f_c there is at least one critical circle with non-orientable separatrix diagram. On each torus that comes up to f_c we mark with asterisks (stars) the regular regions incident with non-orientable separatrix diagrams (that is, with Möbius bands). Consequently, we may also mark with asterisks the corresponding edges of the graph under construction. Thus, we construct a graph by the scheme of the orientable case, after which we mark with asterisks those of its edges which show the motion of asterisked regular regions. The graph obtained will be denoted by T_c (Figure 42).

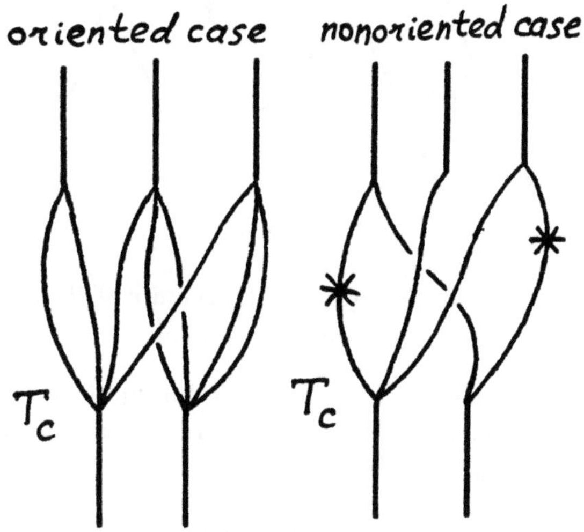

Figure 42

It is clear that the ends of the edges of the graph T_c are identified with the ends of some edges of the graph A. We finally define the graph Γ as a union (glueing)

$$\Gamma = A + \Sigma_c T_c,$$

where $\{c\}$ are critical saddle values of the integral f.

Now let us define the *reduced graph* Γ. Let us consider all white vertices of the graph Γ representing the critical minimax tori of the integral. We replace all these vertices by ordinary points of the corresponding edges (Figure 43), i.e., we consider all these tori as regular Liouville tori. The graph Γ obtained will be called *reduced.*

THEOREM 11.4 (See [**30**]). *The reduced graph Γ is canonically isomorphic (homeomorphic) to the graph $K^{\sim^*}$. In particular, there exists the canonical embedding of the graph Γ into the surface P. From a mechanical point of*

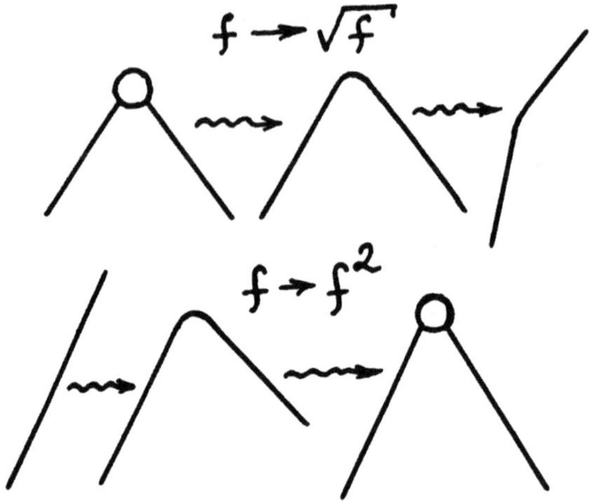

Figure 43

view the graph Γ shows the topological structure of the isoenergy surface Q and the whole picture of bifurcations of Liouville tori inside Q.

In other words, there exists the homeomorphism which canonically maps the edges and the vertices of the graph Γ in the edges and the vertices (of the same type) of the graph $K^{\sim^*}$.

COROLLARY. *In the non-resonant case the reduced graph Γ is a topological invariant of integrable Hamiltonian systems: the topologically equivalent systems have the isomorphic graphs Γ. In particular, the reduced graph Γ does not depend on the choice of the second integral f. Thus each integrable non-resonant system uniquely defines the invariant object (P, K, i, Γ).*

11.7. Some properties of the non-reduced graph $\Gamma = K^{\sim^*}$.

Let us describe what happens with the non-reduced graph Γ if we change the integral f. Let f and f' be any two Bott integrals. Then there exists the homeomorphism $\tau : (P, K, i) \to (P', K', i')$ which maps K onto K'. Then the homeomorphism τ induces the homeomorphism $\Gamma(H, Q, f) \to \Gamma(H, Q, f')$, where the saddle subgraphs T_c for the integral f homeomorphically pass over into saddle subgraphs T'_c for the integral f'. The asterisks of the graph Γ become asterisks of the graph Γ'. Vertices of the type *minimax circles* and *Klein bottles* of the graph Γ passes over into vertices of the same type (respectively) on the graph Γ'. The *torus* type vertices of the graph Γ can be mapped into ordinary interior points of edges of the graph Γ'. On the contrary, ordinary interior points of the edges of the graph Γ can be mapped into *torus* type vertices of the graph Γ'.

From the analytical point of view, the change of the type of vertex (in non-reduced case) corresponds locally to the operation $f \to f^2$ or, vice versa, to the inverse operation $f \to \sqrt{f}$.

We now briefly dwell on the scheme of the proof of the above theorems. Suppose f and f' are two Bott integrals on Q. Since the Hamiltonian is non-resonant, each non-singular torus with an irrational winding is simultaneously a component of the level surface for each of the integrals because integrals are constant on the trajectories of the system v, and therefore constant on the closure of each trajectory. Each non-singular torus with a rational winding is approximated arbitrary close by non-singular tori with irrational windings. Therefore, it is also a component of the level surface for each of the integrals. Consider fields $w = \operatorname{grad} f$ and $w' = \operatorname{grad} f'$. They define one and the same one-dimensional foliation of the surface Q into integral trajectories orthogonal (in a fixed Riemannian metric) to all non-singular level surfaces, i.e., to tori. The velocity of motion along trajectories (determined by the fields w and w') can be different and can become zero simultaneously on the whole torus for one of the fields (whereas the other field is non-zero). Since singular components of the levels are approximated by non-singular tori (when they move along w and w'), then singular connected components of the level surfaces for f', which are other than a torus, coincide with analogous components for f'. If a level surface was a critical torus for f, it becomes non-critical for f' (and vice versa). However, each singular component for f is a singular component for f' as well (and vice versa). Tori form a one-dimensional family in the neighborhood of each non-singular torus. Locally, considering operations $f \to f^2 = f'$ or $f \to \sqrt{f} = f'$, we can transform a non-critical torus into a critical one and vice versa. All other functional operations locally preserving the Bott character of the integral reduce to these two because a smooth Morse function on a one-dimensional manifold may have only quadratic singularities.

From the geometrical point of view, the operation $f \to f^2$ corresponds to folding an edge of the graph Γ and rising on it a minimum or maximum.

The operation $f \to \sqrt{f}$ corresponds to straightening the edge of the graph Γ and eliminating the local minimum.

Construct a homeomorphism of a graph Γ onto the graph Γ'. To define it on a subgraph A, it suffices to consider each non-singular torus for the integral f as a torus for the integral f' (singular or non-singular). This mapping also continues to subgraphs of the form T_c because one-dimensional foliations of the surface Q which are generated by w and w', coincide everywhere except, maybe, some critical or non-critical tori (where the velocity of motion along w and w' becomes zero). As a consequence we note that the total level surfaces of the integrals f and f' (containing one and the same connected component) may comprise a different number of connected components. The values of f and f' may be distinct on one and the same connected component.

11.8. Another geometrical definition of the surface P and the graph K. The embeddings of the surfaces P_s into the manifolds Q_s as cross-sections of Seifert fibrations.

Construct now a surface P in a slightly different way. We will define it as a union (glueing) of the form $P(A) + \Sigma_c P(T_c)$, where $P(A)$ and $P(T_c)$ are two-dimensional surfaces with boundary. Now define

$$P(A) = (S^1 \times \operatorname{Int} A)^+ + \Sigma D^2 + \Sigma\mu + \Sigma S^1 \times D^1.$$

Here $\operatorname{Int} A$ is the union of all open edges of the graph A. Consequently, $(S^1 \times \operatorname{Int} A)$ is a union of open cylinders. The manifold $(S^1 \times \operatorname{Int} A)^+$ is obtained from it by adding boundary circles. Let an edge of the graph A end with a black vertex. The corresponding boundary circle on the boundary of the manifold $(S^1 \times \operatorname{Int} A)^+$ will be glued with a disk D^2. The glueing of such disks will be denoted by ΣD^2.

Let two edges of the graph A meet at the light (white) vertex. It determines two boundary circles on $(S^1 \times \operatorname{Int} A)^+$ which we will glue (join) with a cylinder $S^1 \times D^1$. This operation is denoted by $\Sigma S^1 \times D^1$.

Let an edge of the graph A end with a white vertex with a point within it. Now glue the corresponding boundary circle on $(S^1 \times \operatorname{Int} A)^+$ with a Möbius band μ. This operation is denoted by $\Sigma\mu$.

Thus, ΣD^2, $\Sigma S^1 \times D^1$, $\Sigma\mu$ correspond to minimax circles, tori and Klein bottles.

Now construct $P(T_s) = P_s$, where T_s corresponds to the connected component f_s. Let us recall that each critical level f_c consists of several connected components of the type f_s (Figure 44).

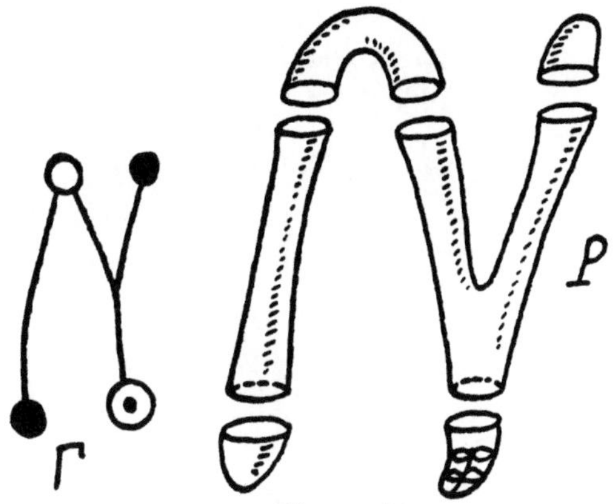

Figure 44

Orientable Case.

First consider the orientable case, where all critical saddle circles on f_s have orientable separatrix diagrams. As shown in [**29**], f_s is homeomorphic to a direct product $K_s \times S^1$, where K_s is a certain graph obtained from several circles through identification of some pairs of points on them. As follows from Statement 11.3, a cell complex f_s is obtained by glueing together several tori along circles that realize non-trivial cycles on the tori. They are homologous and cut f_s into a union of several flat rings.

Thus, the cycle γ is uniquely defined on f_s. Choose on each torus in f_s a certain circle (generator) α complementary to γ (i.e., intersecting with γ exactly at one point). We call it an *oval.* We may assume that ovals intersect one another at points lying on critical circles of the integral. In the orientable case, the union of ovals gives a graph K_s. It need not necessarily be flat.

Let x be a point of intersection of two ovals, that is, a critical point for f. Then segments of the integral trajectories of the field w and the level lines of the function f determine (on a two-dimensional disk centered at the point x, lying in Q and orthogonal to the critical circle on which the point x lies) near the point x a "coordinate cross" on each end of which there is an arrow indicating the direction of w (Figure 45).

Let us construct such normal two-dimensional crosses in each vertex of the graph K_s. Different crosses are joined by segments which are parts of ovals. Join now the ends of the crosses by narrow strips going along arcs of ovals. These strips consist of segments of integral trajectories of the field

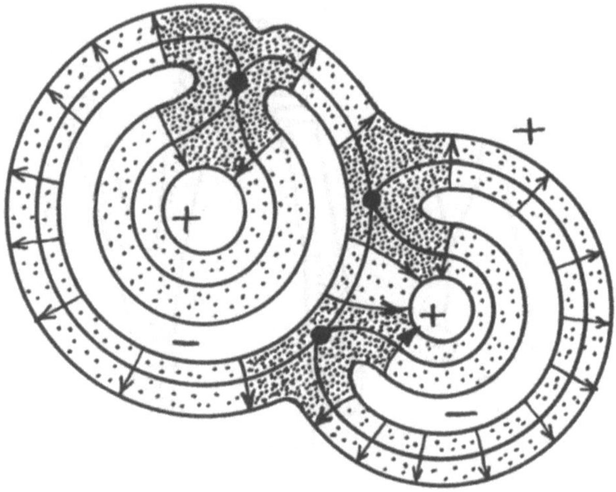

Figure 45

w which orthogonally intersect the ovals (outside the critical points). Arcs of ovals go along the axis of these strips. As a result we obtain a smooth two-dimensional surface with boundary (Figure 45).

The sign "−" denotes the boundary circles corresponding to the tori which came up to f_s from below. The sign "+" stands for the boundary circles corresponding to the tori approaching f_s from above.

As an example, Figure 45 demonstrates three positive (upper) and two negative (lower) tori. The number of negative (resp. positive) tori equals the number of the edges of the graph Γ approaching f_s from below (resp. from above). The obtained surface will be denoted by $P(T_s) = P_s$. Its boundary circles are divided into two classes: lower (negative) and upper (positive). The graph K_s is unique (up to a surface homeomorphism) embedded into P_s^2.

Non-orientable case.

We now construct the surface P_s^2 in the non-orientable case. The scheme of arguments is basically the same. At first suppose for simplicity that exactly one critical circle with a non-orientable separatrix diagram lies on f_s. Then f_s contains the cell complex shown in Figure 46. Consequently, each oval α lying on such f_s must be *doubled*. Any other critical circle, with an oriented diagram, lying on this f_s encounters two copies of the cycle α at two points. This makes us double the cycles α also on those tori which are "tangent" to each other along circles with orientable diagrams but enter in the composition of a connected f_s that contains a circle with a non-orientable diagram.

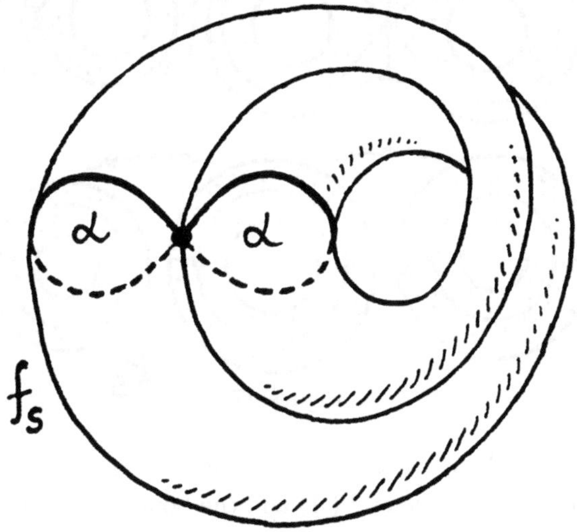

Figure 46

The number of circles with non-orientable diagrams on a connected f_s does not affect the doubling process, i.e., cycles should be doubled only once. The further constructions repeat the orientable case. As a result we obtain the surface P_s^+. The only distinction lies in that now each boundary circle of the surface P_s^+ (i.e., corresponding to one torus) is encountered exactly in two copies (it is doubled). Now construct a surface $P(T_s)$.

From the construction of P_s^+ it is seen that on it a smooth action of the group $\mathbf{Z}_2$ (involution σ) is well defined. In particular $\sigma(x) = x$ if and only if the point x belongs to a critical circle with non-orientable separatrix diagram. Denote such points by $x_1, \ldots, x_m$ and the corresponding circles by $S_1, \ldots, S_m$. Consider the factor-space $P_s = P_s^+/\mathbf{Z}_2$. It is clear that $P_s = P(T_s)$ is a two-dimensional manifold with boundary. Now each of its boundary circles corresponds exactly to one Liouville torus (upper or lower). It is readily seen that the points $x_1, \ldots, x_m$ are interior points on the surface P_s. Mark them with asterisks. In the particular case where f_s has the form shown in Figure 46 (that is, $m = 1$), the surface P_s is a two-dimensional cylinder with one asterisk (Figure 47). Now we can construct the whole surface $P(H, Q, f)$.

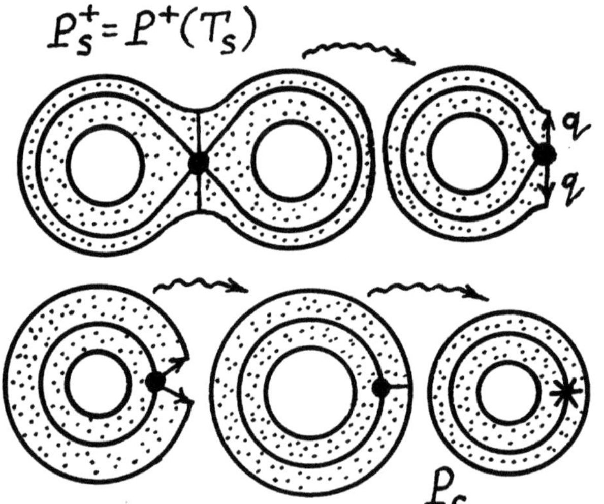

Figure 47

It is clear that there exists a one-to-one correspondence between the boundary circles of the surface $P(A)$ and the boundary circles of the union of surfaces P_s. This correspondence is implemented by edges of the graph A. Identify the corresponding circles by means of certain homeomorphisms and uniquely obtain a closed two-dimensional surface $P(H, Q, f)$ (orientable or non-orientable). Clearly, the replacement of glueing homeomorphisms by some others induces only a homeomorphism of the surface onto itself.

On $P(H, Q, f)$ there is a unique (up to a surface homeomorphism) generally speaking, disconnected graph, which we denote by $K(H, Q, f)$. Consider circles that cut in half the cylinders entering in the surface $S^1 \times \text{Int}\, A$. One may assume that each of them has the form $S^1 \times p$, where p is the middle of the corresponding edge of the graph A. Now define the graph K as a disconnected union of all graphs K_s and the circles of the form $S^1 \times p$. The graph K has only vertices of the following types: a) isolated points, b) the vertices of multiplicity 2 (stars = asterisks), c) the vertices of multiplicity 4.

The construction of the surface $P(H, Q, f)$ is completed. *If f and f' are any Bott integrals on Q in the non-resonant case, then the surfaces $P(H, Q, f)$ and $P(H, Q, f')$ are homeomorphic* [**30**]. Consequently, we will omit the notation f and will write $P(H, Q)$.

STATEMENT 11.4 (See [**30**]). *The above construction determines the embedding of each 2-surface P_s into a three-manifold Q_s as a cross-section of* *the Seifert fibration $\pi_s : Q_s \to P_s$. This embedding is defined non-uniquely, in general. After this embedding the submanifold P_s is realized as the normal section of the Liouville foliation (in the tubular neighborhood of the critical level f_s).*

11.9. Skeleton of an integrable system and the space of connected components of the pre-image of the momentum mapping.

The graph Γ in the case of a *simple* integral f is connected with the "function's tree" introduced by A. S. Kronrod in [**54**]. In the case of a *complicated* integral the construction of the graph Γ is done differently (see above). Kronrod investigated the space of components of the set of level surfaces of the function. In the case of complicated integrals f we are forced to represent the singular level by some special (and very natural) graph which is more complicated than a point.

As has already been pointed out, the classical bifurcation diagram $\Sigma = \Sigma_{(H,f)}$ of the momentum mapping $F : M^4 \to \mathbf{R}^2$ is not a topological invariant of an integrable system. The *skeleton* $I(H,Q)$ is a more natural object from this point of view because it does not depend on the choice of a second integral f and is completely determined only by the non-resonant Hamiltonian H itself. In the case of general position, the bifurcation diagram Σ is shown by a certain curve (graph) with singularities which lies in a two-dimensional plane. Our invariant $W = I(H,Q)$ is also depicted by a certain one-dimensional graph K (or Γ) but embedded into a certain two-dimensional surface. The plane is replaced by a two-surface P, and the "curve" Σ is replaced by a graph Γ which has a "finer organization".

In other words, the invariant $I(H,Q)$ "knows" about the system of differential equations more than the non-invariant bifurcation diagram Σ.

Fixing the value h of the energy H, we obtain one isoenergy three-manifold $Q = Q_h$. Varying h, we vary the three-manifold Q_h, and therefore vary the value of the invariant $I(H,Q_h) = W_h$. The graph $\Gamma_{(H,Q_h)}$ becomes deformed and sweeps a certain two-dimensional cell (simplicial) complex. Denote it by $X(H)$.

Consider the momentum mapping $F : M^4 \to \mathbf{R}^2(H,f)$ for fixed H and f. If $a \in \mathbf{R}^2(H,f)$, then the complete pre-image $F^{-1}(a)$ consists of a certain number of closed connected components. If a is a regualr value for the mapping F, then compact connected components of the pre-image $F^{-1}(a)$

are tori. Replacing each connected component by a point, we consider the *space of connected components* of pre-images of a momentum mapping. Clearly, it coincides with $X(H)$. This object depends only on the symplectic representation of the Abelian group $\mathbf{R}^2$, which is generated by the integrals H and f. Thus, each symplectic action (representation) of the group $\mathbf{R}^2$ on M^4 defines in a unique manner the topological invariant $X(H)$. Here we assume that among the functions defining the representation of $\mathbf{R}^2$ there exists at least one non-resonant integral.

Cutting the space $X(H)$ by "planes" $H = \text{const}$ (Figure 48), in the "cross-section" we obtain graphs $\Gamma_{(H,Q_h)}$. The surface $X(H)$ is two-dimensional, and its singularities are of great interest. It is therefore relevant to formulate here the following problem:

to describe singularities of the space of connected components of pre-images of the momentum mapping of an integrable system of differential equations.

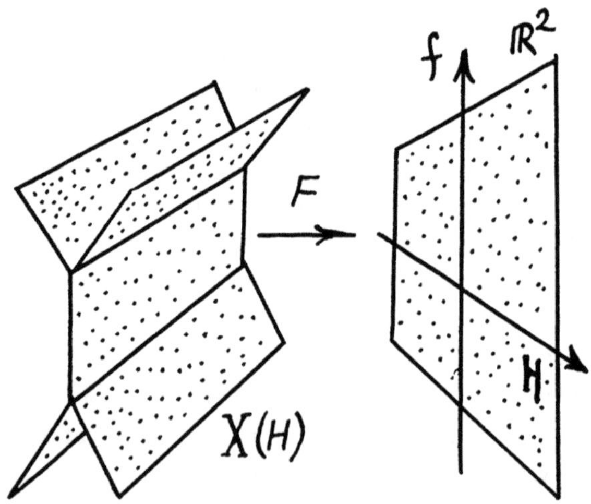

Figure 48

The classification of singularities of codimension 1 has, in fact, been obtained above (in the non-resonant case). It remains to classify singularities of codimension 2.

12. Construction of the marked invariant $I(H,Q)^*$ for integrable systems (framed skeleton W^*).

12.1. The summary of some previous constructions.

Let Q be a three-dimensional isoenergy compact closed manifold and f be a Bott integral. We consider a regular connected ε-neighborhood of a connected component f_s of a critical level. Denote this three-manifold by Q_s. Suppose that Q_s lies between the levels $f^{-1}(c-\varepsilon)$ and $f^{-1}(c+\varepsilon)$. Remember that the subscript c enumerates the critical values of the integral f while the index s enumerates the connected components of all critical levels $f^{-1}(c)$. We have $Q = \Sigma_s Q_s$. Here the sum (union) is taken, in fact, over all critical values c (the index s runs through all connected components for all critical values).

We defined the fibration $\pi_s : Q_s \to P_s$, where the base space is a two-surface with boundary. The boundary consists of a finite number of smooth circles denoted by $S_{s,i}$. The fiber of the fibration is a circle. Over each of the boundary circles $S_{s,i}$ there is a uniquely defined Liouville torus $T_{s,i} = \pi_s^{-1}(S_{s,i})$ which is one of the boundary components of the three-manifold Q_s. Since Q is glued from the pieces Q_s, for every boundary torus $T_{s,i}$, a diffeomorphism to another boundary torus determined by the decomposition $Q = \Sigma_s Q_s$ is defined. Hence, there is a uniquely determined one-to-one corresondence between the boundary circles of the pieces P_s. Glueing together corresponding circles one obtains a closed surface P. On it there is unique (up to isotopy) finite system of circles K_r which result from the glueing of boundary circles $S_{s,i}$ and $S_{s',i'}$ of pieces P_s and $P_{s'}$. The circles of type K_r do not intersect. They form a graph which we denote by $\{K_r\}$. We will use the expression "cut a surface". This has the following meaning. The surface P is glued together from the pieces P_s. The boundary circles of the pieces are identified by diffeomorphisms which we consider as fixed. The operation "cutting" is inverse to "glueing".

Let us simplify the notation. By glueing the circles $S_{s,i}$ and $S_{s',i'}$ (when glueing the pieces P_s and $P_{s'}$) we obtain a circle of type K_r in the graph $K^\sim$. Change the pair of subscripts (s,i) in the expression $S_{s,i}$ to one index. For this we will assume that, by cutting P along the circle K_r, we obtain two circles (the two blanks of the cut): a circle K_r on the piece P_s and a circle $K_{r'}$ on the piece $P_{s'}$. Instead of $S_{s,i}$ and $S_{s',i'}$ we will write K_r and $K_{r'}$ respectively.

The graph $K^{\sim}$ introduced above, is built as the union $K^{\sim} = \{K_r\}+\{K_s\}$ or $K^{\sim} = \Sigma_r K_r + \Sigma_s K_s$. Each component of the type K_r or K_s is connected. In the cases I, II, III we have: $Q_s = P_s \times S^1$. Here $Q_s \cap f^{-1}(c) = K_s \times S^1$. In the case IV we have: $Q_s = P_s \hat{\times} S^1$. Here $Q_s \cap f^{-1}(c) = K_s \hat{\times} S^1$. In case V we have: $Q_s = P_s \tilde{\times} S^1$. Here $Q_s \cap f^{-1}(c) = S^1 \tilde{\times} S^1$ (= Klein bottle).

The presentation of three-manifolds of type I, II and V as a fibration (with fiber a circle) is *non-unique*. This non-uniqueness cannot be avoided even using properties of the integral trajectories of the system. In contrast to this, the three-manifolds Q_s of types III and IV are *uniquely* (up to isotopy) represented as a direct product $P_s \times S^1$ and a Seifert fibration $P_s \hat{\times} S^1$, respectively. The critical submanifolds of the integral f are the following types in the cases I, II, V: circle torus, Klein bottles respectively. Therefore the fiber-circle can be wound around the "axis" several times and this affects the presentation of Q_s as a product. Namely, for replacing two such representations one by another one has to apply some diffeomorphism (Figure 49). It is also responsible for the non-uniqueness of the choice of the fiber. In cases III and IV *the system of critical circles of the integral is uniquely determined* on the surface $f^{-1}(c)$ (being periodic saddlelike solutions of the system). They divide the level $f^{-1}(c)$ in the cases III and IV into a collection of annuli. Thus each annulus uniquely (up to isotopy) decomposes into circles parallel to its boundary. In particular, in all cases I, II, III, IV, V the manifold $Q_s \backslash f^{-1}(c)$ uniquely decomposes into Liouville tori. This fibration intersected with the normal surface P_s uniquely determines a foliation into circles there.

Consider all Q_s of the type II, i.e., the cylinders $T^2 \times D^1$. Two-dimensional cylinders $P_s = S^1 \times D^1$ correspond to them on P^2. A critical torus T^2 (corresponding to a white vertex of the non-reduced graph Γ) lies in the interior of Q_s. "Downstairs", on P^2, a circle K_s (the soul of the cylinder P_s) corresponds to this torus. We contract in Q every manifold Q_s of type II onto its critical torus. Then we contract "downstairs", on P^2, the corresponding annulus onto its soul K_s = circle. Hence, we delete from the composition of Q all cylinders $T^2 \times D^1$ and delete from the decomposition of P^2 all annuli of type I. In the language of the graph Γ the following is going on: a white vertex disappears and is replaced by an ordinary point of an edge.

This *reduction* may alter the integral, but does not change the foliation of

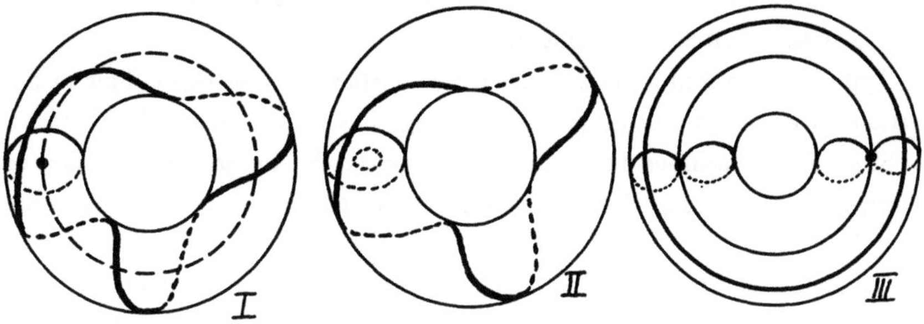

Figure 49

the manifold Q into tori (in the non-resonant case). Since we are interested in the topological equivlence of systems, it is important, that *the reduction brings us to a system topologically equivalent to the original one.* After reduction we obtain a new decomposition of the surface and a new graph. We denote the graph, as before, by K (and $K^{\sim}$) in order not to introduce new notations, but we will speak of it as a *reduced graph* K (and $K^{\sim}$). In the same way we also preserve the old notations for the manifolds Q_s, but we will speak of *reduced three-manifolds. In the following we assume that all our objects are reduced.*

12.2. The choice of coordinate systems on the boundary of Liouville tori.

We start with the study of the *orientable* case. Assume that the reduced surface P is orientable (fix an orientation). In other words, the integral f does not have critical Klein bottles. The general case (i.e., when P^2 may be non-orientable = integral f has Klein bottles) is only technically more complicated. The general scheme of the considerations remains the same as for the orientable case. Moreover, considering isoenergy three-surfaces up to two-fold coverings one may assume that the integral f does not have critical Klein bottles.

After the reduction of an orientable surface P it is only glued of pieces of the following type.

Type I: a disk D^2 with boundary K_r and a central point K_s.

Type II: a more complicated two-surface P_s (corresponding to the critical circles with oriented separatrix diagrams).

Type IV: also a more complicated two-surface P_s (corresponding to the critical circles such that at least one of them has a non-orientable diagram).

Since P^2 is assumed to be orientable we fix the induced orientation of every piece of type P_s. Fix the orientation on each boundary circle $S_{s,i}$ such that it is induced by the orientation of the surface P_s. Let two pieces P_s and $P_{s'}$ be glued together. We consider some pair of circles $K_r = S_{s,i}$ and $K_{r'} = S_{s',i'}$ glued together. Consider Q_s fibered over the base P_s with fiber S^1. Let these fibrations be given and fixed in some way. For type III and IV this procedure is unique, for type I not. Since there is a fixed orientation on Q_s and on the base P_s, there is uniquely determined orientation of the fiber-circle. Fix this orientation of the fiber on each Q_s.

Consider over each circle of type K_r the Liouville torus $T^2 = \pi_s^{-1}(K_r)$ fibered over K_r with fiber S^1. From above follows that we can uniquely define an orientation on the torus. Moreover, we may assume that we fix on the torus a basis of two cycles—meridian and parallel. The parallel corresponds to the fiber of the projection π_s and the meridian corresponds, for instance, to the circle K_r.

REMARK: Consider the manifolds Q_s of type I, III, IV. They contain critical circles of the integral f (periodic solutions of the system v). Since the system is given on all of Q, it uniquely determines orientations on all its solutions. In the cases I, III, IV these solutions (critical circles) are fibers of the fibration π_s. We will discuss these orientations below.

Consider the fibrations $\pi_s : Q_s \to P_s$ and $\pi_{s'} : Q_{s'} \to P_{s'}$. Let P_s and $P_{s'}$ be surfaces of type III or IV (independent one of the other). In particular, P_s and $P_{s'}$ differ from the disk. The pre-images of the circles K_r and $K_{r'}$, i.e., the tori $\pi_s^{-1}(K_r)$ and $\pi_{s'}^{-1}(K_{r'})$, are fibers of the fibration, i.e., the circles $S_r^1 = \pi_s^{-1}(x_r)$ and $S_{r'}^1 = \pi_{s'}^{-1}(x_{r'})$ where x_r and $x_{r'}$ are points on the circles K_r and $K_{r'}$, respectively (Figure 50).

We introduce suitable coordinate systems on the tori T_r^2 and $T_{r'}^2$. The circles S_r^1 and $S_{r'}^1$ are generators of the tori T_r^2 and $T_{r'}^2$. Complete the generator S_r^1 on the torus by a transversal circle D_r^1 and the generator $S_{r'}^1$ by a transversal circle $D_{r'}^1$. The pairs of circles S_r^1, D_r^1 and $S_{r'}^1$, $D_{r'}^1$ obviously form coordinate systems on T_r^2 and $T_{r'}^2$.

The mappings π_s and $\pi_{s'}$ define fibrations of the tori T_r^2 and $T_{r'}^2$ into circles (uniquely determined for types III and IV). It is clear that D_r^1 and

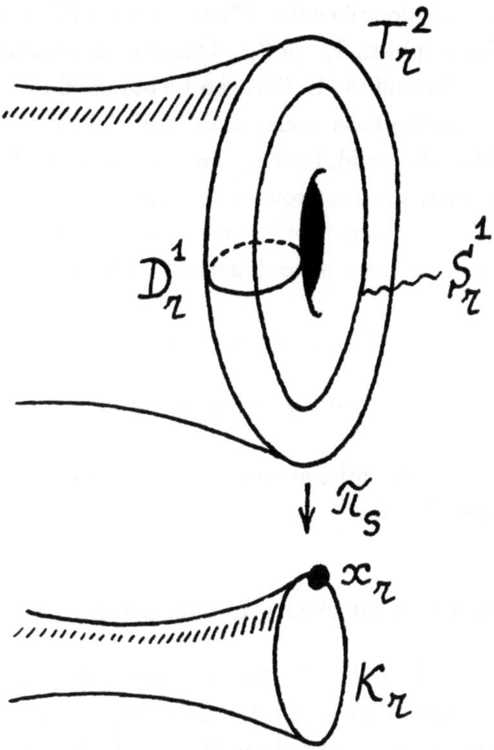

Figure 50

$D^1_{r'}$ are sections of these fibrations of the tori T^2_r and $T^2_{r'}$. The choice of the sections D^1_r and $D^1_{r'}$ is *not unique*; however, this arbitrariness does not affect our final construction.

12.3. The rational parameters on the edges of the word-moledule $W = I(H,Q)$. The case of the regions of type III and IV.

Since P is glued from the pieces P_s, it follows that "upstairs" (i.e. in Q) the tori T^2_r and $T^2_{r'}$ must be glued together by some diffeomorphism λ. It is given by the structure of Q and the system v. Taking into account the introduced coordinate systems on the tori we obtain two uniquely determined matrices with integer entries

$$M_{rr'} = \begin{pmatrix} \alpha(r,r') & \beta(r,r') \\ \gamma(r,r') & \delta(r,r') \end{pmatrix}, \quad M_{r'r} = \begin{pmatrix} \alpha(r',r) & \beta(r',r) \\ \gamma(r',r) & \delta(r',r) \end{pmatrix}.$$

The matrix $M_{rr'}$ describes the diffeomorphism $T^2_r \to T^2_{r'}$ and the matrix $M_{r'r}$ the diffeomorphism $T^2_{r'} \to T^2_r$. One may assume that both matrices have determinant equal to -1. This can be obtained by a suitable choice of the orientation on the base circles of the tori. The matrices are related by the relation $M_{rr'} M_{r'r} = \text{id}$, i.e. they are inverses of each other.

In the following, the first row of the matrix $M_{rr'}$ (or $M_{r'r}$) plays an exceptional role, i.e. the pair of integers $\alpha(r, r')$, $\beta(r, r')$. Below we denote this pair by (α, β) if there is no danger of confusion between the matrices $M_{rr'}$ and $M_{r'r}$.

The numbers α and β may be assumed to be coprime and to satisfy

$$0 \leq \alpha < |\beta|, \text{ if } \beta \neq 0.$$

We can consider the rational number $r = \alpha/\beta$ (when $\beta \neq 0$); if $\beta = 0$ we will assume that $r = \infty$.

12.4. Thin and fat circles in the graph $K^\sim$.

Exhibit among all circles K_r those for which the pair of numbers (α, β) has the form $(\pm 1, 0)$, that is, $\beta = 0$ (or $r = \infty$).

Clearly, these circles emerge from the set of all circles of type K_r. For intuition we will say (and express it on the figures) that these circles K_r are drawn by *thin lines*. The other circles of type K_r (i.e., those with $(\alpha, \beta) \neq (\pm 1, 0)$) will be expressed by *fat lines*. Thus the graph $\{K_r\}$ decomposes into the union of thin and fat circles.

Let us label *each fat circle* by the rational number $r = \alpha/\beta$ (here we assume that $0 \leq \alpha < |\beta|$), if $\beta \neq 0$, and $r = \infty$ if $\beta = 0$.

12.5. The rational parameters in the case of regions of type I.

Assume now that at least one of the regions P_s and $P_{s'}$ having K_r in its boundary is homeomorphic to a disk. Two cases are possible:

1) both regions P_s and $P_{s'}$ are homeomorphic to a disk;

2) only one of the regions is homeomorphic to a disk and the other is of type III or IV. In case 2 let the region P_s be homeomorphic to a disk.

The component K_s of the graph K consists of one vertex—the center of the disk (the boundary of the disk is the circle K_r). Our aim now is to construct for K_r an analogue to the rational number $r = \alpha/\beta$. The construction becomes more complicated because of the fact that over the disk $P_s = D^2$ the fibration $\pi_s : Q_s \to P_s$ into circles is not uniquely determined. We assume first that the region $P_s = D^2$ meets (along K_r) a region $P_{s'}$ of type III or IV. In the language of the graph Γ this means that the only edge going out from the black vertex leads to a saddle vertex (component) of the graph Γ. As we will show below this situation is typical. In this case the choice of a coordinate system on the torus $T_r^2 = \pi_s^{-1}(K_r)$ is done as follows. Since the torus T_r^2 is at the same time a boundary torus $T_{r'}^2$ of the manifold $Q_{s'} = \pi_{s'}^{-1}(P_{s'})$, on the torus $T_{r'}^2$ which is identified with the torus T_r^2 by some fixed diffeomorphism, there is already constructed a fibration into circle-fiber and a chosen section. This fibration and section can be moved to the torus T_r^2. Hence, on the torus T_r^2 we introduce coordinates by taking them "from the neighbor", i.e. from the three-manifold $Q_{s'}$. This operation is uniquely determined. Consider on the solid torus Q_s the meridian which is uniquely given by the following condition. It is defined (up to isotopy and orientation) as a simple curve lying on the boundary torus of the solid (full) torus contractible inside the solid torus but not on the torus. Express this meridian in terms of the basis we just chose on the torus T_r^2 (fiber and section from the neighbor). As a result we obtain two integers p and q, the coefficients of the meridian with respect to the basis. We label the circle K_r by the rational number p/q. This fraction is the analogue of the fraction α/β.

When can this construction be applied, i.e. when can we assume that a region P_s of type I has common border with a region $P_{s'}$ of type III or IV?

LEMMA 12.1. *Consider the orientable reduced case. Then, on the connected surface P, each region P_s of type I has a common border (boundary) with a region of type III or IV except in the single case when the graph Γ has the form shown in Figure* 51-a, *that is when the three-manifold Q is obtained by glueing together two solid tori by a diffeomorphism of their boundary tori.*

Indeed, under the assumptions of the lemma there are no regions of type II and V in the collection of P^2. Consider two cases: 1) the graph Γ contains exactly two vertices, 2) the graph Γ contains more than two vertices. In case 1 the form of the graph is shown in Figure 51-a (two black vertices connected by an edge). In case 2 the graph Γ is drawn in Figure 51-b, that is, all its vertices are black and each edge going off from

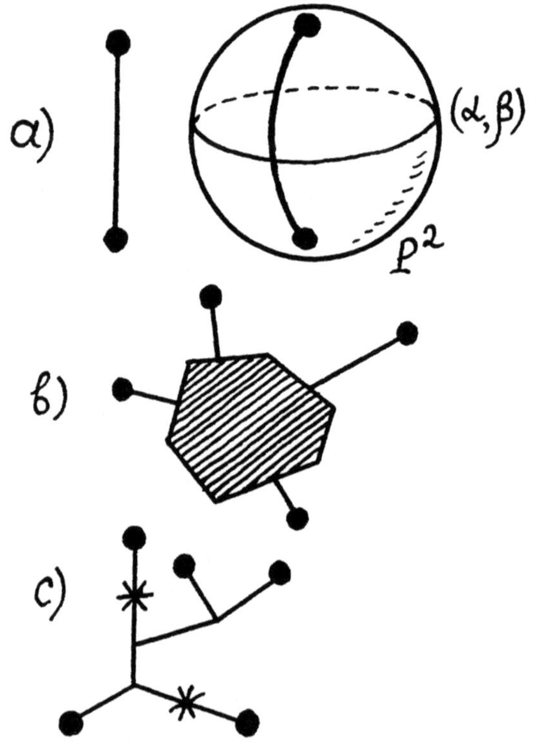

Figure 51

the black vertex arises from some saddle (orientable or non-orientable). For an explicit example of such a graph Γ see Figure 51-c. The lemma is proved.

Hence, the construction of the previous section can be applied in all cases except one, namely when Q is glued from two full tori (Figure 51-a). It is easy to see that the invariant $I(H, Q)$ for case 1 looks as shown in Figure 51-a, that is, it is a sphere divided into two halves by the equator. Here $Q = Q_1 + Q_2$, $P = P_1 + P_2$, and there is exactly one circle K_r (equator), exactly two components of the graph K_s, namely the two centers of the disks.

This case is simple and may be considered separately. For the classification of all systems with such a simple value of the invariant $I(H, Q)$ one can introduce coordinate systems on every full torus, repeat the construction described above and obtain (α, β) etc. Now we turn out main interest to case 2.

12.6. The integer parameters introduced in the word-molecule.

Since we assumed orientability of the reduced surface P, every region P_s is homeomorphic either to a disk (type I) or to an orientable surface of type III or IV. Look at all *fat* circles K_r, and cut the surface P along them. We obtain a collection of regions some of which coincide with the old P_s, but others are unions of some old regions P_s. Call these *new* regions "big" and denote them by R_b, where the subscript b runs through some finite set of integers. The boundary of each domain R_b consists of circles of type K_r (fat ones).

All old thin circles of type K_r are contained in the interior of the regions R_b. On every region R_b we obtain a fibration with the circle as a fiber. Indeed, when $\beta = 0$ (thin circle) we can glue the fibrations along the thin circle. The total space Q_b of the fibration either coincides with some manifold Q_s or is a union of some Q_s. The boundary of Q_b consists of Liouville tori. On each of them there is defined a section of the fibration, given by a circle D_r^1 (Figure 50).

Thus, on the boundary of Q_b there is given a section of the fibration

$$Q_b \xrightarrow{S^1} R_b.$$

Therefore there is a uniquely determined integer invariant—the obstruction to extending this section to the interior of Q_b.

Denote this numerical invariant by n_b.

Furthermore, to each "big" region R_b we adjoin a new rational number defined by $\mu_b = n_b - \Sigma_r \alpha/\beta$. Here the sum is taken over all thick (fat) circles K_r, forming the boundary of the region R_b. We drop the index r in writing the numbers α/β, but we have in mind that the numbers α/β depend on the circles K_r.

Finally, to match the orientations on different blocks of Q, we need to consider the element σ from the Abelian cohomology group $H^1(\Gamma, \mathbb{Z}_2)$.

The same constructions can be realized in the general case of a non-orientable surface P. We omit the details.

SUMMARY: We obtain the following object:

1) A closed compact reduced surface P^2 divided into the union of "big" regions R_b the boundaries of which are the fat circles K_r.

2) A reduced graph $K^{\sim} = \{K_r\} + \{K_s\}$ on the surface P^2 where some of the circles K_r are drawn by thin lines and others by fat ones. All thin circles K_r lie in the interior of the regions R_b.

3) Each fat circle K_r admits a rational number $r = \alpha/\beta$ with $0 \leq \alpha < |\beta|$ where $\beta \neq 0$. Instead of this one may simply write $\alpha/\beta \mod 1$. Here we assume that the regions R_b bounded by K_r are not disks. But if at least one of the regions R_b bounded by K_r is a disk then we assign a number p/q to the circle K_r (see above).

4) To each region R_b, different from the disk, are assigned a rational number $\mu_b = n_b - \Sigma_r \alpha/\beta$.

5) The element $\sigma \in H^1(\Gamma, \mathbf{Z}_2)$ is defined. The object $I(H,Q)^* = W^* = (P^2 = \Sigma_b R_b; K; \{\alpha/\beta, p/q\}; \{\mu_b\}; \sigma\}$ is called a *marked word-molecule* or a *marked topological invariant* (on an isoenergy suface Q). We will say that the values of the invariants $I(H,Q)^*$ and $I(H',Q')^*$ of two systems v and v' are *equal* if there exists a homeomorphism $\varphi : P \to P'$ mapping the graph K onto the graph K' (Γ onto Γ'), the regions R_b into regions $R_{b'}$, the numerical labels $\alpha/\beta \mod 1$ into $\alpha'/\beta' \mod 1$, the numbers μ_b into $\mu_{b'}$ and the cohomology element σ into σ'. In the contrary case we say that the values of the invariants $I(H,Q)^*$ and $I(H',Q')^*$ of the systems v and v' are different.

13. Equivalent interpretations of the marked invariant $I(H,Q)^*$. The realization theorem.

13.1. Letters-atoms and their Seifert realizations.

DEFINITION. *The pair (P_s, K_s), where K_s is a non-empty finite connected graph which is contained in a compact orientable two-surface P_s, will be called a letter-atom, if the following conditions are fulfilled:*

1) The degree of each vertex of the graph K_s is equal to 0, 2, or 4.

2) Each connected component of the two-manifold $P_s \backslash K_s$ is homeomorphic to the ring $S^1 \times (0,1]$.

3) The set of the rings, which form the two-surface $P_s \backslash K_s$, can be separated in two parts (positive and negative rings) in such a way that exactly one positive ring and exactly one negative ring correspond to each edge of the graph K_s.

The number of the vertices in the graph K_s is called *atomic weight* of the letter-atom (P_s, K_s).

It follows from condition 2) that the surface P_s is a regular neighborhood of the graph K_s or, in other words, K_s is the *spine* of the surface P_s. See the example of the letter-atom in Figure 52. This letter-atom has the notation "E_1" in the Table of all letters with atomic weight no more than 3 (see the Table above). The unique letter-atom (P_s, K_s) with the graph K_s having exactly one vertex of degree 0 is the disk with its center. We denote this letter-atom by "A".

Figure 52

Two letters (P_s, K_s) and $(P_{s'}, K_{s'})$ are considered as *identical* if there exists the orientation preserving homeomorphism of the surface P_s onto $P_{s'}$ which maps the graph K_s onto the graph $K_{s'}$.

DEFINITION. *The three-dimensional oriented Seifert manifold Q_s (with oriented fibers) is called a Seifert realization of the letter-atom (P_s, K_s), if:*

1) *the surface P_s is the base of the three-manifold Q_s;*

2) *the orientations of Q_s, its fibers and the base all match;*

3) *all exceptional fibers in Q_s have type $(2,1)$ and the fiber is exceptional if and only if its projection is a vertex of degree 2 in the graph K_s.*

The Seifert realization Q_s of a given letter-atom (P_s, K_s) always exists and is defined uniquely up to diffeomorphisms (which preserve orientation).

The structure of the Seifert fibration on Q_s is defined uniquely (up to isotopy) in all cases except the letter-atom A. Indeed, in this case its Seifert realization has the countable number of Seifert structures (which differ by the twisting along the meridional disk). For general facts about Seifert manifolds see, for instance, in [**82**] (Orlik).

DEFINITION. *The representation of the Seifert manifold as the union of a finite number of compact three-submanifolds Q_s, $1 \leq s \leq n$, is called Seifert subdivision iff:*

1) Each two manifolds Q_s and Q_j, $s \neq j$, either do not intersect or intersect along several common components of their boundaries.

2) Each manifold Q_s (with induced orientation) is a Seifert realization of some letter-atom (P_s, K_s).

Two Seifert subdivisions are called equivalent if they are diffeomorphic (where the diffeomorphism preserves the orientation).

Let us recall that we consider an integrable Hamiltonian non-resonant system v on some orientable three-manifold Q. The system is integrable with a Bott integral f which does not have critical Klein bottles.

Let us consider the connected component R of the level $f^{-1}(c)$ which is a saddle, i.e., does not coincide with the critical torus. Then all critical circles in R are oriented because they are the periodic trajectories of v. Let us recall that $R = f_s$.

LEMMA 13.1 (See [**15**]). *We can introduce the structure of Seifert fibration (with oriented fibers) on the regular neighborhood $U(R)$ of the singular fiber R of the Liouville foliation in such a way that the following conditions are fulfilled:*

1) $U(R)$ is a Seifert realization of some letter-atom (P_s, K_s).

2) $R = \pi_s^{-1}(K_s)$, where $\pi_s : U(R) \to P_s$ is a projection.

3) The orientations of all critical circles induced by v, coincide with their orientations as the fibers (of the fibration).

LEMMA 13.2 (See [**30**]). *Each integrable Hamiltonian system on Q induces a Seifert fibration on the manifold Q. The topologically equivalent integrable systems induce equivalent Seifert fibrations.*

Here we really assume that f has at least one isolated critical circle. In the opposite case, the Liouville foliation on Q is a locally trivial fibration over circle with the fiber a torus. We exclude this case from our analysis because of its triviality. The corresponding skeleton is the pair $(S^1 \times S^1, \emptyset)$ and the word-molecule is the circle without letters.

13.2. The realization theorem.

THEOREM 13.1 (See [**15**]). *Every Seifert subdivision of any closed compact oriented three-dimensional manifold Q is induced by some integrable non-resonant Hamiltonian system on Q (with Bott integral).*

13.3. The identification of the abstract and geometric skeletons. (Proof of Theorem 4.1).

Let us prove Theorem 4.1 about the existence of the natural one-to-one correspondence between the set of all geometric skeletons and the set of all abstract skeletons.

Let $Q = \Sigma_s Q_s$ be the Seifert subdivision of the oriented three-dimensional manifold Q; (P_s, K_s)– the letters-atoms corresponding to the manifolds Q_s, $1 \leq s \leq n$. Let us glue the letter-atoms according to the scheme which duplicates the scheme of the glueing of the three-manifold Q from the manifolds Q_s. The corresponding homeomorphisms of boundary circles must reverse the orientations of the circles. As a result we obtain the pair $(P, K^\sim)$, where $P = \Sigma_s P_s$ is an oriented closed surface and $K^\sim = \Sigma_s K_s + \Sigma_r K_r$– the graph (where K_r are the glued boundary circles), i.e. we obtain the *geometric skeleton*.

Now let us consider an arbitrary integrable Hamiltonian system v on Q. We see from Lemma 13.2 that v generates the Seifert subdivision of the manifold Q which is in correspondence with the geometric skeleton (P, K). It is clear that the twisting along the Liouville tori does not change the skeleton (P, K). Thus, the same geometrical skeleton corresponds to all roughly equivalent integrable systems. We obtain some mapping φ of the set of all abstract skeletons of integrable Hamiltonian systems into the set of all geometric skeletons. The inverse mapping ψ is constructed as follows.

Let (P, K) be any geometric skeleton. Let K_s, $1 \leq s \leq n$, be the connected components of the graph $K^\sim (\neq K_r)$. We divide P into the union of the regular neighborhoods P_s of the graphs K_s. The pairs (P_s, K_s) are the letters-atoms. Consequently, we can construct their Seifert realizations Q_s. Let us glue the manifolds Q_s using any diffeomorphisms of the boundary tori (inversing the orientation). The scheme of glueing must duplicate the scheme of glueing of the skeleton (P, K) from the letters-atoms (P_s, K_s). As a result, we obtain the oriented three-manifold Q with a Seifert subdivision $Q = \Sigma_s Q_s$. This subdivision is induced (see Theorem 13.1) by some integrable Hamiltonian system v on Q. We assign its abstract skeleton to the initial geometric skeleton. The identities $\varphi\psi = 1$ and $\psi\varphi = 1$ are evident.

13.4. Framed skeletons. Admissible coordinate systems.

As we knew, it is necessary to label the geometric skeleton of an integrable Hamiltonian system by some numerical marks. It is convenient to mark the skeleton by some *matrix marks*. Some of this information will be superfluous. Then we will delete the superfluous part of this information.

Let Q_s be a Seifert realization of the letter-atom (P_s, K_s) (different from the letter A) and $\pi : Q_s \to P_s$ be the projection. Let us denote by P_s^0 the complement to the union of regular neighborhoods of the vertices of multiplicity 2 of the graph K_s. Then let us consider $Q_s^0 = \pi^{-1}(P_s^0)-$ the component to the union of regular neighborhoods of singular (exceptional) fibers. Let $z : P_s^0 \to Q_s^0$ be an arbitrary section of the fibration. We choose the coordinate system μ_i, λ_i on each torus $T_i \subset \partial Q_s^0$. Here the meridian μ_i is the image $z\pi(T_i)$ of the oriented circle $\pi(T_i) \subset \partial Q_s^0$. The parallel λ_i is the oriented fiber of the fibration. Let us say that the coordinate systems μ_i, λ_i are *induced* by the section z.

DEFINITION. *The section $z : P_s^0 \to Q_s^0$ is called admissible if the meridian of each regular neighborhood of the exceptional fiber has type $\pm(2,1)$ in the coordinate system which is induced by this fiber on the boundary of this neighborhood.*

In particular, if the manifold Q_s does not have exceptional fibers, then all sections are *admissible*. The existence of admissible sections in the general case (i.e., when Q_s has the exceptional fibers) follows from the fact that all exceptional fibers must have type $(2,1)$.

DEFINITION. *Coordinate systems on the boundary of Seifert realization Q_s are called admissible if they are induced by some admissible section.*

LEMMA 13.3 (See [**15**]). *Let (μ_i, λ_i) be the admissible coordinate systems on the boundary components T_i of a Seifert realization Q_s of the letter-atom (P_s, K_s), $1 \le i \le n$. Another coordinate system (μ_i', λ_i') is admissible* *if and only if the following equalities are valid for every i:*

$$\mu_i' = \mu_i \lambda_i^{k_i}, \ \lambda_i' = \lambda_i, \ \textit{where } \Sigma_{i=1}^n k_i = 0.$$

PROOF: Let z, z' be two admissible sections defining the coordinate systems (μ_i, λ_i) and (μ_i', λ_i'). The equalities $\lambda_i' = \lambda_i$ are evident because the parallels coincide with the oriented fibers of the fibration. The numbers k_i are interpreted as indices of intersection of the meridians μ_i and μ_i'. The sum of the k_i is the total index of intersection of the curves $z(\partial P_s)$ and $z'(\partial P_s)$, because the admissible sections have the same non-intersected images on the boundary of the neighborhoods of exceptional fibers. The curves $z(\partial P_s)$ and $z'(\partial P_s)$ bound the two-dimensional films $z(P_s)$ and $z'(P_s)$ and consequently their intersection index is equal to 0. Thus, $\Sigma_{i=1}^n k_i = 0$.

Let us prove the inverse statement. Let the coordinate systems (μ_i, λ_i) be induced by the admissible section z. We choose the simple arc $a \subset P_s$ which connects the i-th component of the boundary of the surface P_s with the j-th component, where $i \neq j$. Let us cut the manifold Q_s along the ring $\pi^{-1}(s)$, then twist by 360° one of the boundaries of the cut and glue again. We obtain the fibered homeomorphism of the manifold Q_s onto itself, which transforms the admissible section z into another admissible section z'. The corresponding numbers k_l are as follows:

$$k_l = \pm 1 \text{ if } l \neq i, j, \ k_i = \pm 1, \ k_j = \mp 1.$$

It is clear that we can realize by such operations any collection of the numbers k_l with a zero sum. The lemma is proved.

We assumed above that the letter-atom (P_s, K_s) does not coincide with the letter A, where $A =$ (disk with a point). The admissible coordinate system on the boundary of the full torus (which is the Seifert realization of the letter A) must be constructed in another way.

DEFINITION. *The coordinate system (μ, λ) on the boundary of a Seifert realization of the letter-atom A (= full torus) is called **admissible** if λ is the oriented meridian of the full torus and μ—its parallel.*

REMARK: Of course, it is uncomfortable to denote the meridian by λ and the parallel by μ (usual practice is inverse). It is convenient to represent the full torus as the complement to the trivial knot in S^3. In this case μ transforms in the meridian of the knot and λ—in its parallel.

The following lemma is evident.

LEMMA 13.4. *Let (μ, λ) be the admissible coordinate system on the boundary of the Seifert realization of the letter-atom A. This coordinate system is admissible if and only if for some integer number k the following equalities are fulfilled:* $\mu' = \mu\lambda^k$, $\lambda' = \lambda$.

13.5. Superfluous framed skeletons.

Let (P, K) be a geometric skeleton. For simplicity let us replace the skeleton (P, K) by the graph $\Gamma = \Gamma_{(P,K)}$. Its vertices are the connected components of the graph K and its edges correspond (represent) the rings from the $P \backslash K$ which connect the pieces of the graph K. Let us fix the orientation of the graph Γ by choosing the direction on each edge e_i, $1 \leq i \leq n$.

DEFINITION. *The superfluous framed skeleton* (P, K) *is defined if we assign to each edge* e_i *of the graph* Γ *the integer-valued matrix* $A_i = \begin{pmatrix} \alpha_i & \beta_i \\ \gamma_i & \delta_i \end{pmatrix}$ *with* $\det A_i = -1$.

We will say that the correspondence $e_i \mapsto A_i$ defines the *frame (the superfluous frame)* for a given skeleton. Each skeleton can be endowed with different *superfluous frames.*

REMARK: If we change the orientation of the edge e_i, then the matrix A_i is changed into its inverse.

Let us introduce some *equivalence relations* on the set of all *superfluous frames* of the skeleton (P, K). We will say that two frames *are equivalent* if they can be obtained from each other by a finite number of operations of the following two types:

(A) Let v be a vertex of the graph Γ with multiplicity more than 1. Let us assign to each edge e_i, which is incident with this vertex, the integer number $k(e_i)$ in such a way that the sum of all these numbers is equal to zero. Then we will change the matrix A_i (which corresponds to e_i) in the new matrix A'_i. The matrix A'_i is obtained from the matrix A_i by addition of the second column to the first one with the coefficient $k(e_i)$ in the case when the edge e_i goes out of the vertex v. This this case, e_i goes into the vertex v, we will subtract the first row from the second one with the coefficient $k(e_i)$. If the edge e_i is the loop with the vertex v, then we will produce both transformations (their order is unessential).

(B) Let v be the vertex of multiplicity 1 in the graph Γ and k—any integer number. Then the new matrix A'_i corresponding to the edge e_i (which is incident with this vertex) is obtained from the matrix A_i by addition of the second column to the first one with the coefficient k in the case when the edge e_i goes out of the vertex v. In the case when e_i goes into v, we will subtract the first row from the second one with the coefficient k.

STATEMENT 13.1. *There is a natural one-to-one correspondence between the classes of equivalent superfluous frames of a given skeleton (P, K) and the classes of topologically equivalent integrable Hamiltonian systems (of differential equations) with the skeleton (P, K).*

The proof can be found in [**15**].

13.6. Framed skeletons. Exact frames of the skeleton.

Our goal is to replace the superfluous frames (of a given skeleton) by the *exact frames*. We do not need in the superfluous information contained in the superfluous frame of geometric skeleton.

Let us describe *the exact frame* of the geometric skeleton (P, K). Let us assign to each edge e_i of the oriented graph $\Gamma = \Gamma_{(P,K)}$ the pair (r_i, ε_i), where r_i is the rational number such that $0 \leq r_i < 1$ or $r_i = \infty$; and $\varepsilon_i = \pm 1$.

The edge will be called *thin* if the corresponding number r_i is equal to ∞. Let us consider the *thin subgraph* Γ^0 inside the graph Γ, which consists of all vertices of Γ and all its *thin* edges. Let us denote by $\Gamma_1^0, \Gamma_2^0, \ldots, \Gamma_s^0$ those connected components of the graph Γ^0 which do not contain the vertices of multiplicity 1 of the graph Γ (i.e., the letter-atom A). Let us denote by $\Gamma_{s+1}^0, \ldots, \Gamma_l^0$ those connected components which contain such vertices.

Let us assign to the components $\Gamma_1^0, \Gamma_2^0, \ldots, \Gamma_s^0$ the integer numbers $n_1, n_2, \ldots, n_s$ respectively.

DEFINITION. *The collection of numbers r_i, ε_i, n_k, where r_i are rational numbers, $0 \leq r_i < 1$, $\varepsilon_i = \pm 1$ $(1 \leq i \leq n)$; n_k are integer numbers $(1 \leq k \leq s)$, will be called an exact frame of the graph Γ (or an exact frame of the skeleton (P, K)).*

REMARK: If all $r_i \neq \infty$ (i.e.,—no *thin edges*), then the number s coincides with the number of the components of the graph K which are different from the isolated vertex. In this case the number 1 is equal to the number of its components.

STATEMENT 13.2. *There exists the natural one-to-one correspondence between the classes of equivalent superfluous frames of the geometric skeleton (P, K) and its exact frames.*

The proof is non-trivial, see [**15**] (and below). Now we will only describe the correspondence between the numerical parameters.

Let us denote by e_i, $1 \le i \le n$, the edges of the graph $\Gamma = \Gamma_{(P,K)}$. Let us recall that the orientations of the edges e_i are fixed. The elements of the matrix A_i are denoted, as before, by α_i, β_i, γ_i and δ_i. Let us introduce the numbers α'_i such that $0 \le \alpha'_i < |\beta|$, $\alpha'_i \equiv \alpha_i \mod \beta_i$. The numbers δ'_i are defined in a similar way:

$$0 \le \delta'_i < |\beta_i|, \ \delta'_i \equiv \delta_i \mod \beta_i.$$

Let us consider the relations:

$$r_i = \begin{cases} \alpha'_i/|\beta_i|, & \text{if } \beta_i \neq 0; \\ \infty, & \text{if } \beta_i = 0; \end{cases}$$

$$\varepsilon_i = \begin{cases} \operatorname{sign} \beta_i, & \text{if } \beta_i \neq 0; \\ \operatorname{sign} \alpha_i, & \text{if } \beta_i = 0. \end{cases}$$

Let us denote the beginning of the edge e_i by $v_0(e_i)$, and the end of e_i by $v_1(e_i)$. Let us assign the integer number f_{ik} to each oriented edge e_i of the graph Γ, where:

$$f_{ik} = \begin{cases} -\gamma_i/\alpha_i, & \text{if } e_i \in \Gamma^0_k; \\ (\alpha_i - \alpha'_i)/\beta_i, & \text{if } e_i \notin \Gamma^0_k,\ v_0(e_i) \in \Gamma^0_k,\ v_1(e_i) \notin \Gamma^0_k; \\ -(\delta_i - \delta'_i)/\beta_i, & \text{if } e_i \notin \Gamma^0_k,\ v_0(e_i) \notin \Gamma^0_k,\ v_1(e_i) \in \Gamma^0_k; \\ (\alpha_i - \alpha'_i - \delta_i + \delta'_i)/\beta_i, & \text{if } e_i \notin \Gamma^0_k,\ v_0(e_i) \in \Gamma^0_k,\ v_1(e_i) \in \Gamma^0_k; \\ 0, & \text{if } e_i \notin \Gamma^0_k,\ v_0(e_i) \notin \Gamma^0_k,\ v_1(e_i) \notin \Gamma^0_k. \end{cases}$$

Let us define $n_k = \Sigma^n_{i=1} f_{ik}$.

Thus, we assign to each superfluous frame of the graph Γ the exact frame of $\Gamma : (r_i, \varepsilon_i, n_k)$, $1 \le i \le n$, $1 \le k \le s$ (the numbers n_k with $k > s$ do not play any role in the exact frame).

REMARK: The number $-\gamma_i/\alpha_i$ is integer because in the case $e_i \in \Gamma^0_k$ the element β_i is equal to zero, and then $\alpha_i = -\delta_i = \pm 1$. Because $\alpha'_i \equiv \alpha_i \mod \beta_i$ and $\delta'_i \equiv \delta_i \mod \beta_i$, then all numbers in the definition of f_{ik} are also integers.

13.7. Influence of the orientations.

Let us consider the orientations of the graph Γ, surface P, manifold Q, and critical circles of an integrable Hamiltonian system v on Q.

1°. The change of the orientation of the edge e_i in the graph Γ replaces the corresponding number $r_i = p/q$ into the number $r_i^* = p^*/q$, which is uniquely defined by the condition:

$$r_i r_i^* \equiv -1/q^2 \mod 1/q.$$

The numbers e_j, n_k, and the other numbers r_j do not change.

2°. The change of the orientation of the manifold Q changes the orientation of the surface P. The same event happens when we change the Hamiltonian H into the Hamiltonian $-H$, i.e., when we change the orientation of trajectories. The simultaneous change of the orientation of Q and of the orientation of trajectories does not change the orientation of P.

3°. Sometimes in the definition of the equivalent integrable Hamiltonian system we can assume that Liouville tori transform into Liouville tori without special agreement concerning the orientation of critical trajectories. In this case the fixed orientation of the surface P is lost, but the surface P remains oriented. The role of the numbers ε_i reduces to the specification of the element σ in the group $H^1(\Gamma, \mathbb{Z}_2)$. The corresponding cocycle σ is equal to zero on the edge e_i if $\beta_i > 0$ or $\beta_i = 0$, $\alpha_i > 0$, and is equal to 1 if $\beta_i < 0$ or $\beta_i = 0$, $\alpha_i < 0$. The change of orientation of critical trajectories corresponding to the vertex v of the graph Γ (i.e., the change of orientation of Seifert realization of the corresponding letter-atom) induces the transformation of the cocycle σ (namely, the difference is measured by the coboundary of the chain which is equal to 1 on the vertex v and to 0 on the other vertices).

14. The coding of integrable Hamiltonian systems.

14.1. The geometrical representation of the letters-atoms.

It is well known in immersion theory that every compact oriented two-surface with boundary can be immersed in the two-plane with the orientation being preserved. Consequently, we can represent any letter-atom

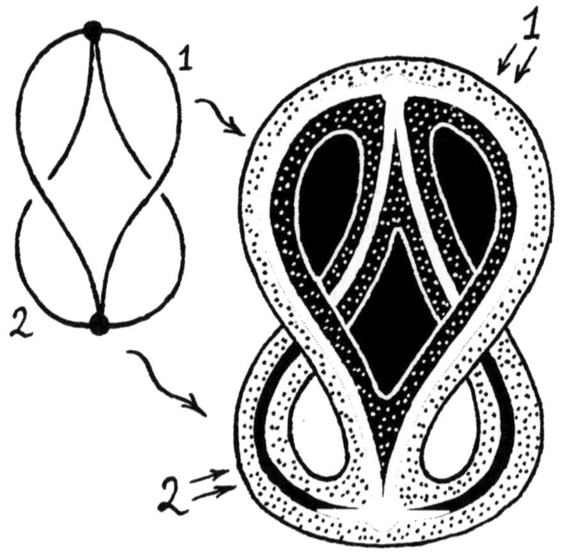

Figure 53

(P_s, K_s) as some surface P_s immersed in a two-plane $\mathbb{R}^2$. Such an immersion is uniquely defined by the immersion of the graph K_s. Thus, it is sufficient to draw only the image of the graph (Figure 53).

Of course, not every immersion of the graph defines the letter-atom: the condition on the item 7 (see step 2) can be broken.

Every two immersions of the surface P_s (with preserved orientations) in the plane are regularly homotopic, if we are allowed to create and remove the loops on the edges of the graph K_s (Figure 54). Let us call such a homotopy *rough*. Thus, we can define the letter-atom as the class of roughly equivalent immersions of the graph K_s into the plane.

The change of the orientation of the surface P_s can produce either old or new letter-atoms. In the first case we will call the letter-atom *mirror symmetric*. Table 1 shows all letters-atoms whose atomic weights are no more than 3. The atomic weight is the number of the vertices of the graph K_s. For each letter-atom (P_s, K_s) the following are shown: a) the number of the positive and negative circles in the boundary ∂P_s, b) their total number (i.e., the *valency* of the letter-atom), c) the type of the surface $\bar{P}_s$, d) the graph $\Gamma_s = K_s^{\sim *}$ (in another notation—the graph T_s or T_c). Here $\bar{P}_s$ denotes the closed surface which is obtained from the surface P_s by the glueing of all its boundary components by disks.

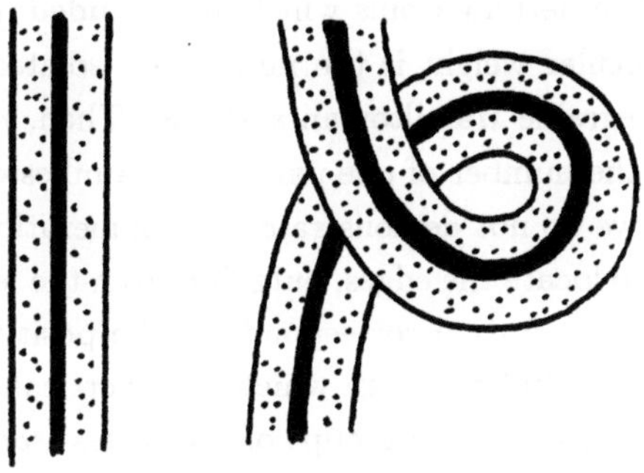

Figure 54

We show also (for $m \leq 2$) the enumeration of the circles from ∂P_s which are necessary for the construction of the words-molecules.

The remarkable fact is as follows: *all letters-atoms with atomic weight no more than 3 are mirror symmetric except for two letters-atoms $D^*_{2,2}$ and $D^*_{2,3}$ which are mirror symmetric to each other.*

Table 1 contains all possible spines (i.e., the graphs K_s) of all letters-atoms with weight no more than 5. The number of all letters-atoms corresponding to this spine can be seen near the graph or inside it (we consider mirror symmetric letters as one letter). By the way, we do not take into account the *isotopes*, that is, the letter-atoms with the spines containing the vertices of multiplicity 2 (stars-asterisks). It is clear, that in this case the operations of annihilation and creation of such vertices do not produce any problems. The partial intersection of the Tables 1 and 2, containing the letters-atoms of weight no more than 3, makes the situation more clear.

14.2. Words-molecules.

Let v be an integrable Hamiltonian system, (P, K)—its geometric skeleton and $\Gamma = \Gamma_{(P,K)}$—the corresponding graph, $\Gamma = K^{\sim^*}$.

The *word-molecule*, corresponding to the system v, is the graph $\bar{\Gamma}$, where each vertex (of $\bar{\Gamma}$) corresponds to a letter-atom.

The skeleton (P, K) can be reconstructed (in a unique way) by its word-molecule, if we indicate the enumeration of all outgoing edges (in each vertex), which correspond to the enumeration of the boundary circles in the Table 1. Table 3 shows all words-molecules of complexity (m, n) with $m \leq 2$. Let us recall that m is the molecular weight, i.e., the sum of all atomic weights of all letters-atoms which are included in the molecule. In other words, molecular weight is the number of isolated critical circles of corresponding integrable Hamiltonian systems. Then, n is the number of connections, i.e., the number of one-parameter families of Liouville tori.

For simplicity, we do not demonstrate the enumeration of the edges because we assume to locate the edges, outgoing from the letter-atom, in such a way that their enumeration corresponds to the positive direction of the motion. We suppose that the edge which goes vertically up is considered as the *first* one. In Table 3 the groups of *molecules-isomers* are separated one from each other. We consdier two molecules as *isomers* iff they differ only by the enumeration of the edges.

Let us denote by $\lambda(m,n)$ the total number of all geometric skeletons (i.e., of words-molecules) of integrable Hamiltonian systems with the complexity (m,n). As follows from Theorem 4.2, this number is always finite. Table 4 shows the values of $\lambda(m,n)$ for $m \leq 4$. It follows from this Table that for every m the function $\lambda(m,n)$ is equal to zero for sufficiently large n. Let us denote by $\Lambda(m)$ the upper bound of non-zero region, i.e., $\lambda(m,\Lambda(m)) \neq 0$ and $\lambda(m,n) = 0$ for $n > \Lambda(m)$. Let us denote by $\lambda^0(m,n)$ the number of all geometric skeletons of integrable Hamiltonian systems such that they do not contain the critical circles of the type of a non-oriented saddle. The corresponding letters-atoms (which form the corresponding word-molecule) do not contain the stars. The following Statement is useful for the understanding of the structure of the functions $\lambda(m,n)$ and $\lambda^0(m,n)$.

STATEMENT 14.1. a) *The following equality is valid for every* $m : \Lambda(m) = [3m/2]$ *(here* [ ] *denotes the integer part).* b) *For all odd* m *we have:* $\lambda^0(m,n) = 0$.

PROOF: Let the geometric skeleton of complexity (m,n) consist of k letters-atoms (P_i, K_i) with atomic weights m_i and with valencies n_i, $1 \leq i \leq k$. Let us denote by w_i and v_i the number of the vertices of the graph K with degree (multiplicity) 4 and 2 (respect.). Let us denote by χ_i the Euler characteristic of the closed oriented surface $\bar{P}_i$, which is obtained from the surface P_i by the glueing of all components of its boundary by disks. Because the graph K_i is the spine of the surface P_i and because the number of its vertices is equal to $2w_i + v_i$, then it follows from the Euler formula that $n_i = \chi_i + w_i$. Let us sum these equalities with respect to i and take into account that $\chi_i \leq 2$, $m_i = v_i + w_i$, $\Sigma_{i=1}^k m_i = m$, $m_i \geq 1$. Then we obtain: $n = (1/2)\Sigma_{i=1}^k n_i \leq (1/2)\Sigma_{i=1}^k (2 + m_i) = k + m/2 \leq (3/2)m$.

Thus, $\Lambda(m) \leq [3m/2]$. The existence of the word-molecule of complexity $(m, \Lambda(m))$ is proved by its demonstration (by construction). See Figure 55 for even m. If m is odd, we need to introduce the letter-atom A^* in an arbitrary edge of this molecule.

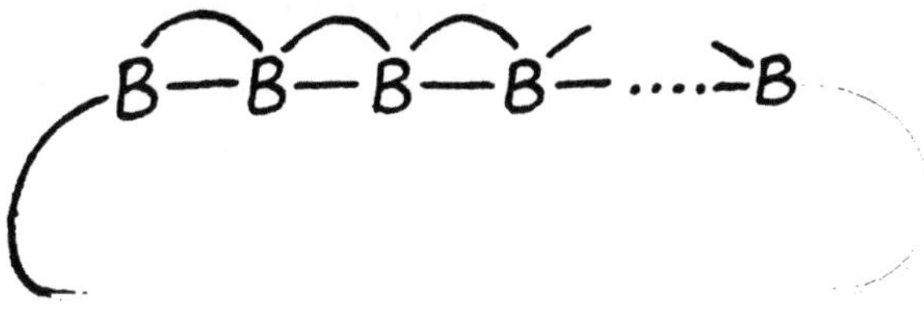

Figure 55

Let us prove b). It follows from the formulas: $n_i = \chi_i + w_i$ and $m_i = v_i + w_i$ that $n_i \equiv m_i \mod 2$ in the case of the absence of the stars (i.e., when all v_i are equal to 0). Because the number $\Sigma_{i=1}^{k} n_i = 2n$ is always even, the word-molecule without stars can exist only for even $m = \Sigma_{i=1}^{k} m_i$.

REMARK: The adding of new stars on the letters-atoms of the molecule increases the number m and does not change n. The inclusion of the letter-atom A^* in some edge changes the pair (m, n) into the pair $(m+1, n+1)$. Consequently, the functions $\lambda(m+t, n)$ and $\lambda(m+t, n+t)$ increase monotonously when the integer parameter t increases. The unique exception: $\lambda(3,1) < \lambda(2,1)$ is induced by the fact that the letter-atom A does not have any edges and consequently we do not have any place to locate the star. In particular, we obtain that $\lambda(m, n) \neq 0$ for $n \leq [3m/2]$.

14.3. Marked words-molecules.

Let us consider the word-molecule with a fixed enumeration of all edges incident to every vertex and with fixed orientations of all edges-connections. Let us assign to each edge e_i the rational number r_i and the sign $\varepsilon_i = \pm 1$, where $0 \leq r_i < 1$ and $1 \leq i \leq n$.

Let us call two letters-atoms *relative* if they are connected (inside the word-molecule) by a sequence (chain) of edges with $r_i = \infty$. The union of all letters-relatives will be called a *family*. Let us assign to each family (without the letter A) the integer number n_k. We will call the resulting object *marked word-molecule.*

The following theorem is the direct corollary of the Statements 13.1 and 13.2.

THEOREM 14.1. *There exists the natural one-to-one correspondence between the set of all marked word-molecules and the set of all integrable Hamiltonian systems up to topological equivalence.*

Let us recall that the word-molecule W coincides with the invariant $I(H,Q) = (P^2, K)$ and the marked word-molecule W^* coincides with the marked invariant $I(H,Q)^* = (P^2, K)^*$.

14.4. Physical integrable systems.

We assumed above that the isoenergy surface Q was fixed. But any analysis of concrete physical systems involves the investigation of different level surfaces of constant energy. Table 5 contains the words-molecules calculated by A. T. Fomenko and A. A. Oshemkov (based on the works of M. P. Charlamov, who described the topological structure of integrable cases for the equations of rigid body motion) for the cases of Kovalevskaya, Goryachev–Chaplygin and Sretensky.

The change of the value of the energy near the regular value preserves the word-molecule in the case, when the system v is *stable*, i.e. *in general position* (see above).

In this case the transformation can be realized only in the moment of crossing of the critical value. Thus we obtain the collection (the sequence) of words-molecules. Let us represent these molecules by points in the corresponding cells of the plane, where each cell (small square) corresponds to the concrete complexity (m,n). We obtain some trajectory (curve) in the plane. Table 6 represents the trajectories corresponding to the mechanical systems mentioned above. In the Kovalevskaya case, the different trajectories correspond to the different values of the constant of area; in the Goryachev–Chaplygin–Sretensky cases—to the different values of gyrostatic momentum. The numerical marks can be more or less easily calculated for the simplest physical integrable Hamiltonian systems. For example, the numerical marks are shown in the Figures 56, 57 for the cases of Euler and Lagrange (the important cases in the classical dynamics of rigid body). The marks ε_i are not shown in these figures because by changing of orientation we can reduce all these marks to the value $+1$.

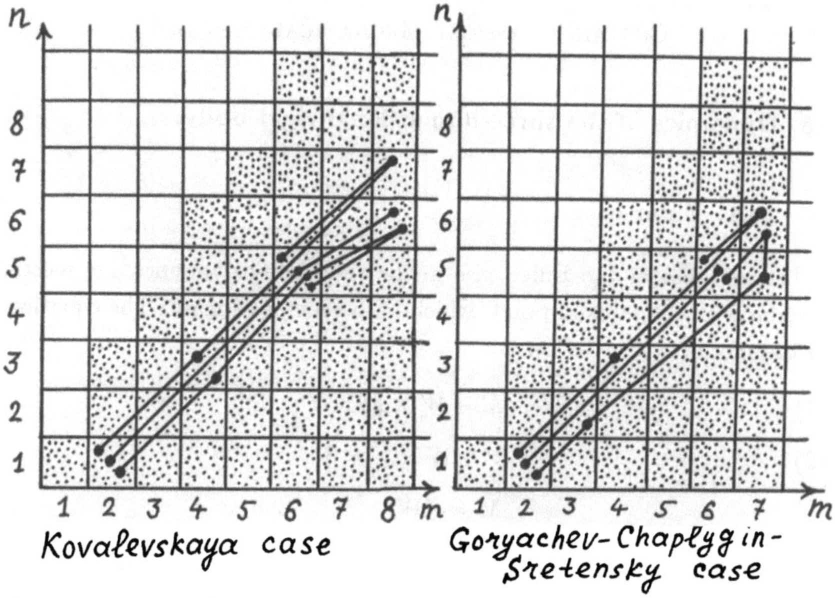

Table 6

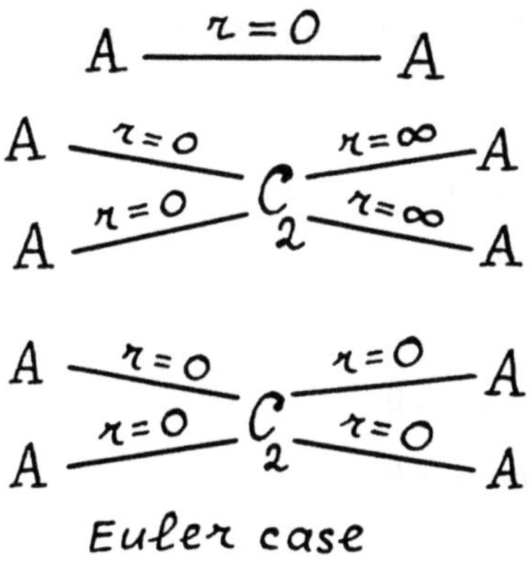

Figure 56

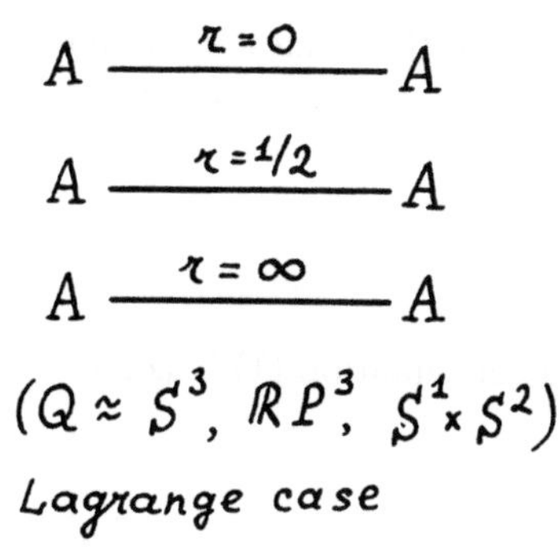

Figure 57

Chapter 4. Calculation of Topological Invariants of Certain Classical Mechanical Systems

15. Dynamics of the three-dimensional rigid body.

15.1. Euler case.

Let us consider the Euler case in the rigid body dynamics, i.e. inertial motion with a one fixed point, which is the center of mass. The equations of the motion are as follows:

$$\dot{M} + W \times M = 0 \tag{1}$$

$$\dot{V} + W \times V = 0 \tag{2}$$

$$M = AW.$$

Here $W = (w_1, w_2, w_3)$ is the vector of angular velocity of the body, $V = (v_1, v_2, v_3)$ is the unit vertical vector, A is the inertia operator, and $M = (A_1 w_1, A_2 w_2, A_3 w_3)$ is the kinetic momentum vector.

Let us consider the motion in the body coordinate system which is defined by the three axes of the inertia ellipsoid. The inertia operator has in this coordinate system the form:

$$A = \begin{pmatrix} A_1 & 0 & 0 \\ 0 & A_2 & 0 \\ 0 & 0 & A_3 \end{pmatrix}.$$

Then equation (1) becomes:

$$\begin{aligned} A_1 \dot{w}_1 + (A_2 - A_3) w_2 w_3 &= 0 \\ A_2 \dot{w}_2 + (A_1 - A_3) w_1 w_3 &= 0 \\ A_3 \dot{w}_3 + (A_2 - A_1) w_1 w_2 &= 0. \end{aligned} \tag{3}$$

Equation (2) has a so-called geometric integral $\langle V, V \rangle = \text{const}$. The constant is really equal to 1 because V is a unit vector. Since the vectors V and W are fixed in space, $\langle V, M \rangle = g = \text{const}$.

Thus, the differential equation (1) can be considered on the four-dimensional manifold:

$$M^4 = TS^2 = \begin{pmatrix} S = v_1^2 + v_2^2 + v_3^2 = 1 \\ G = A_1 w_1 v_1 + A_2 w_2 v_2 + A_3 w_3 v_3 = g \end{pmatrix}$$

The equation (1) has two integrals:

$$H = A_1 w_1^2 + A_2 w_2^2 + A_3 w_3^2 = h - \text{ energy integral, and}$$

$$f = A_1^2 w_1^2 + A_2^2 w_2^2 + A_3^2 w_3^2.$$

In the Euler case all eigenvalues of the inertia operator are different. In the Lagrange case two of them coincide.

The topology of the momentum mapping in the Euler, Lagrange and Kovalevskaya cases were described by M. P. Charlamov in [**22**]; the calculation of the words-molecules was done by A. A. Oshemkov and L. S. Polyakova [**84**]–[**87**]. The numerical parameters r, n were calculated by A. V. Bolsinov.

There are several methods of calculation of the marked words-molecules. We demonstrate one of them based on the investigation of the bifurcation diagram of the momentum mapping.

The general idea is as follows.

Step 1. We calculate the bifurcation diagram of the momentum mapping $F : M^4 \to \mathbb{R}^2$. As a result we obtain (in general case) some curve with singularities in the two-plane $\mathbb{R}^2$ (Figure 58).

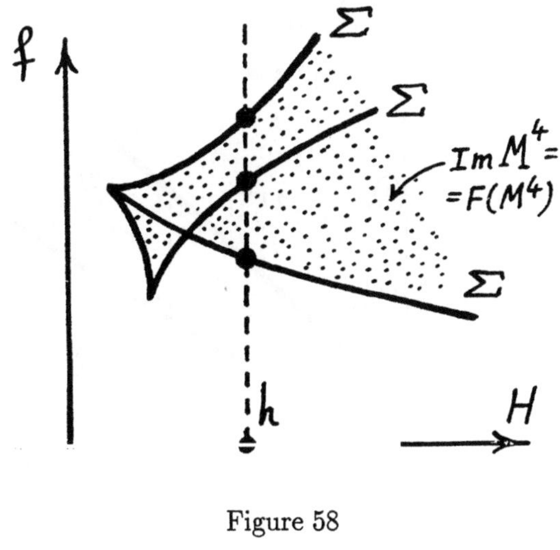

Figure 58

Step 2. We consider the line $H = h = \text{const}$ on the plane (H, f) (Figure 58) and the points where the line intersects Σ. The preimage of the line F^{-1} (line) is a three-manifold Q in M.

Step 3. We investigate all points of intersection $\Sigma\cap$ (line) and calculate the *Hessian* of f for the corresponding critical submanifold in Q.

Step 4. We can describe the type of surgery of Liouville tori on these critical submanifolds based on the main theorem of classification of all possible transformations of Liouville tori (see above).

Step 5. As a result we calculate the graph Γ and reconstruct all other topological characteristics of our invariant.

Step 6. The calculation of the numerical parameters r_i, ε_i and n_k. This work is delicate in many cases, where we need additional geometric information on the topology of the Liouville foliation.

Let us return to concrete examples.

At first we calculate the bifurcation diagram Σ. In the Euler case this diagram is shown in Figure 59. Here h_1, h_2, h_3 are the critical values of the Hamiltonian H on $M^4 = TS^2$. Let us suppose that $0 < A_1 < A_2 < A_3$. The variation of g does not change the topological type of bifurcation diagram. The critical values of H are (except the case $g = 0$):

$$h_1 = g^2/A_3, \quad h_2 = g^2/A_2, \quad h_3 = g^2/A_1.$$

Consequently, three different types of Q_h are possible depending on which of the three intervals (h_1, h_2), (h_2, h_3), (h_3, ∞) contains h. Let us denote these types of Q_h by (1), (2), (3) (these types are shown in Figure 60).

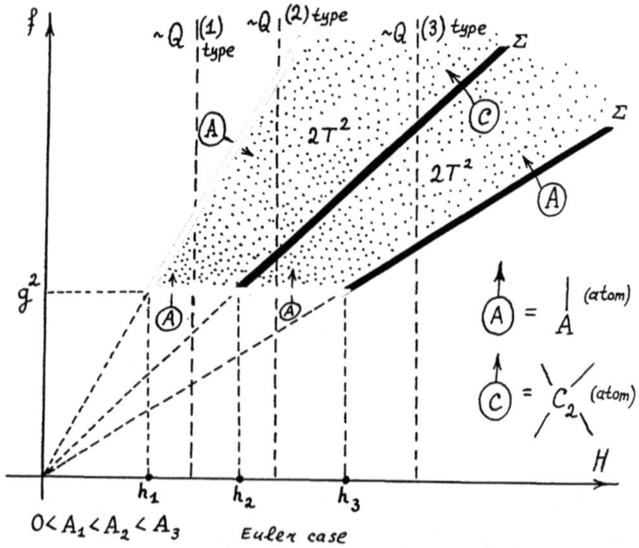

Figure 59

Type	graph atom	Q_h^3	Geometric skeleton (Euler case)
(1)		$2S^3$	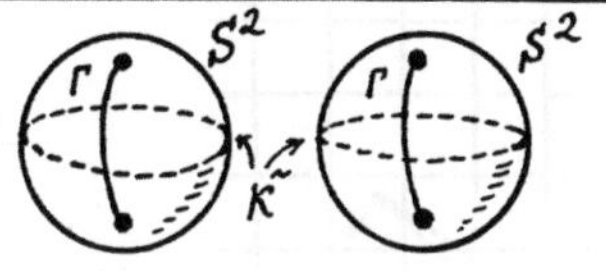 Fig.60-a
(2)	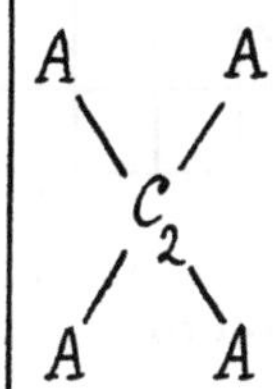	$S^1 \times S^2$	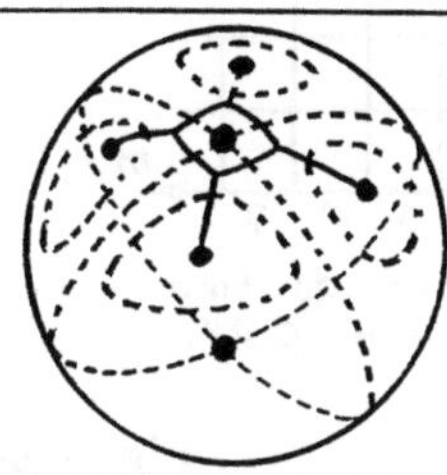 Fig.60-b
(3)	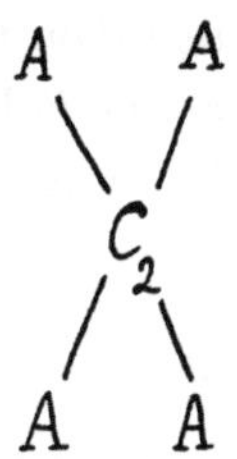	$\mathbb{R}P^3$	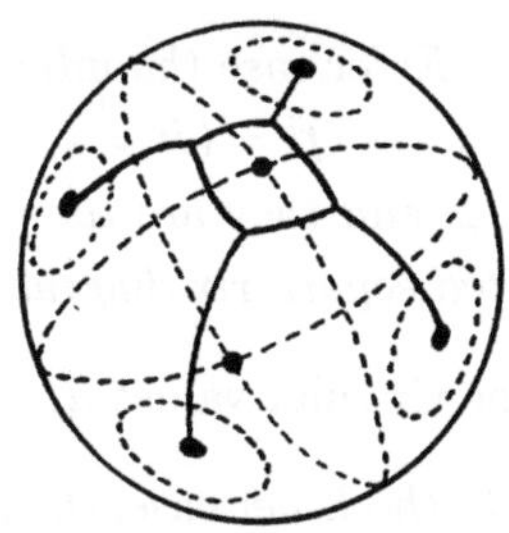 Fig.60-b

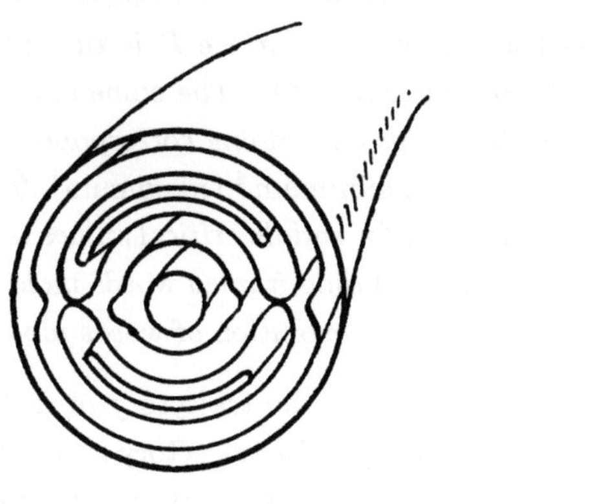

Fig.60-c

Figure 60a-c

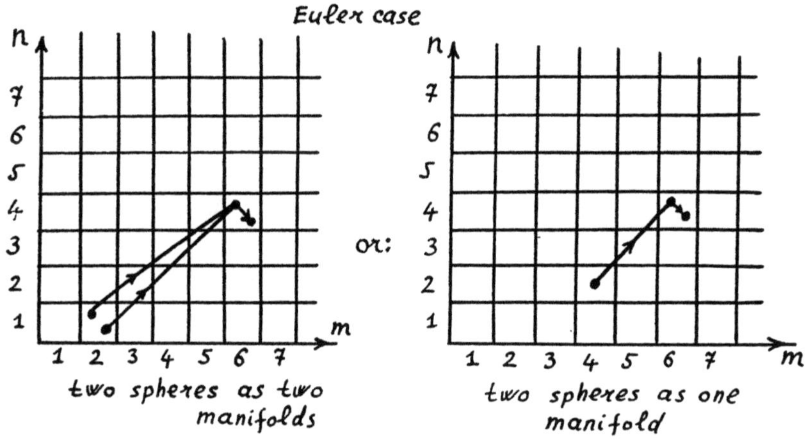

Figure 60d

LEMMA 15.1. *In the Euler case the integral f is a Bott integral on the all regular isoenergy surfaces $Q^3 = (H = \text{const})$. All critical submanifolds are non-degenerate circles and they are the minimax and saddle circles of the integral (with oriented separatrix diagrams).*

Let us consider the singular values h_1 and h_2 in Figure 59.

STATEMENT 15.1. *In the Euler case, the geometric skeleton (P, K) has the form shown in Figures 56, 60.*

a) *For $h < h_2$ the surface P is the union of two spheres; the graph $K^\sim$ is shown by a dotted line; the graph Γ is represented by two segments with black ends (Figure 60-a). The rational parameter r is equal to zero.*

b) *For $h > h_2$ the surface P is the sphere S^2, the graphs $K^\sim$ and Γ are shown in Figure 60-b. The embedding of the graph Γ in $\mathbb{R}^3$ is seen in Figure 60-c. This embedding corresponds to the evolution of the values of integral f (from minimum to maximum). The picture of the transformation of Liouville tori for saddle critical values is seen in Figure 60-c.*

c) *All corresponding framed words-molecules are in Figure 56.*

d) *For $g \neq 0$ the evolution of words-moledules is seen in Figure 60-d.*

Figure 60-d shows the trajectory of molecules of the Euler case in the table of complexity. The trajectory starts from two three-spheres and consequently we can draw two versions of trajectories. In the first version we consider two points in the cell $(2, 1)$, in the second version—one point in the cell $(4,2)$.

15.2. Lagrange case.

In the Lagrange case we have:

$$\dot{M} + W \times M + r \times V = 0,$$
$$\dot{V} + W \times V = 0,$$
$$M = AW,$$

where $A_1 = A_2 = A$, $A_3 = B$, $r = (0,0,r_3)$ (r gives the coordinates of the center of mass). Figure 61-a shows the subdivision of the plane with coordinates $(B/A, 2Ar_3/g^2)$ into the union of domains. For all points of the domain which are marked by the index i we have that the corresponding bifurcation diagram (of the corresponding system) has the form shown in Figure 61-b. Here the four-dimensional manifold M is diffeomorphic to TS^2, where TS^2 is the tangent bundle of the two-sphere. Here $f = -w_3$.

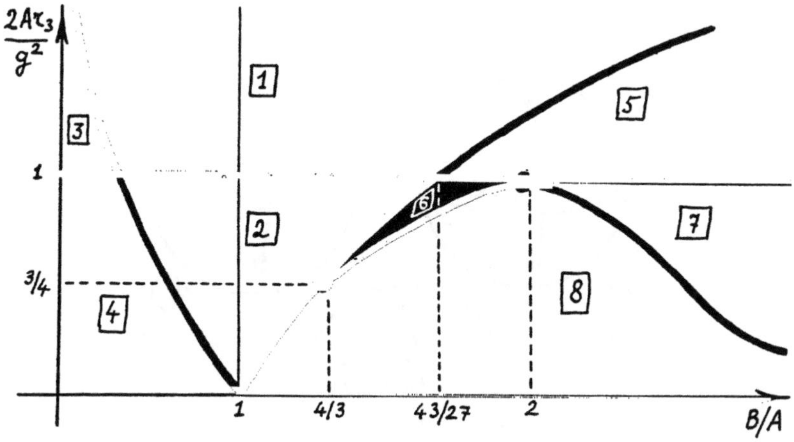

Figure 61-a

LEMMA 15.2. *In the Lagrange case the classical integral f is a Bott integral on almost all regular isoenergy surfaces Q in M. The critical submanifolds on a connected manifold Q are: two minimax circles.*

STATEMENT 15.2. *In the Lagrange case the geometric skeletons are shown in Figure 62. Figure 62-a shows the case when $r = 0$, Figure 62-b shows*

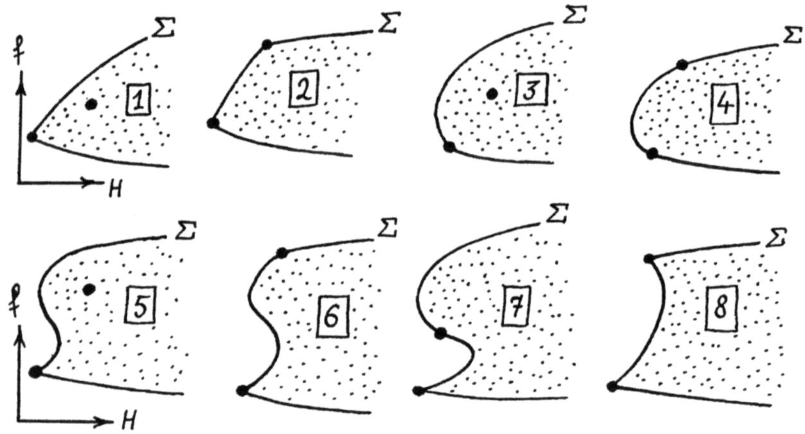

Figure 61-b

the case where $r = \infty$ and Figure 62-c corresponds to $r = 1/2$. The surface P is the two-sphere, the graph Γ is the segment with black ends. The corresponding framed words-molecules are seen in Figure 57 (above).

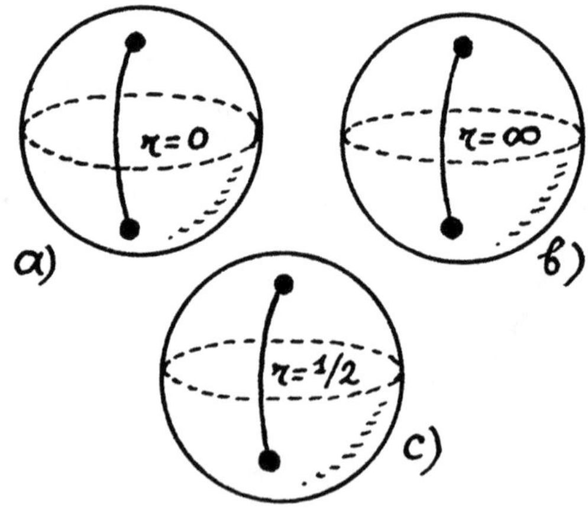

Figure 62

15.3. Kovalevskaya case.

Let us consider the general form of the classical equations of motion of a heavy rigid body:

$$A\dot{\omega} + \omega \times A\omega = \nu \times \operatorname{grad}\Pi,$$
$$\dot{\nu} = \nu \times \omega,$$

where $a \times b$ is the vector product, Π the potential. In this example, the integrals of the Hamiltonian system have the following form:

$$F = \nu_1^2 + \nu_2^2 + \nu_3^2 = 1,$$
$$G = A_1\omega_1\nu_1 + A_1\omega_2\nu_2 + A_3\omega_3\nu_3 = g,$$
$$H = \frac{1}{2}(A_1\omega_1^2 + A_2\omega_2^2 + A_3\omega_3^2) + \Pi(\nu) = h.$$

Here 1, g, h are the constants.

The case of Kovalevskaya is characterized by the fact that

$$A_1 = A_2 = 2A_3 \text{ and } \Pi = -\nu_1.$$

As is well known, there is another additional independent integral

$$f = (\omega_1^2 - \omega_2^2 + \nu_1)^2 + (2\omega_1\omega_2 + \nu_2)^2.$$

Consider the four-dimensional manifold $M^4 = \{F(\omega,\nu) = 1;\ G(\omega,\nu) = g\}$. It is diffeomorphic to the tangent bundle of the sphere: TS^2. On the manifold M^4 a Hamiltonian H and an additional integral f are given. We can therefore apply the construction presented above and calculate the topological invariant = geometric skeleton of the integrable case of Kovalevskaya. This investigation was based on the work of M. P. Charlamov and was carried out by the author in collaboration with A. A. Oshemkov. Take the momentum mapping $H \times f : M^4 \to R^2$ corresponding to the two integrals H and f. For different values of g one obtains different bifurcation diagrams $\Sigma \subset \mathbf{R}^2$ calculated by M. P. Charlamov (Figures 63–66). Here we show (near the dotted lines) the corresponding type of three-manifolds Q_h.

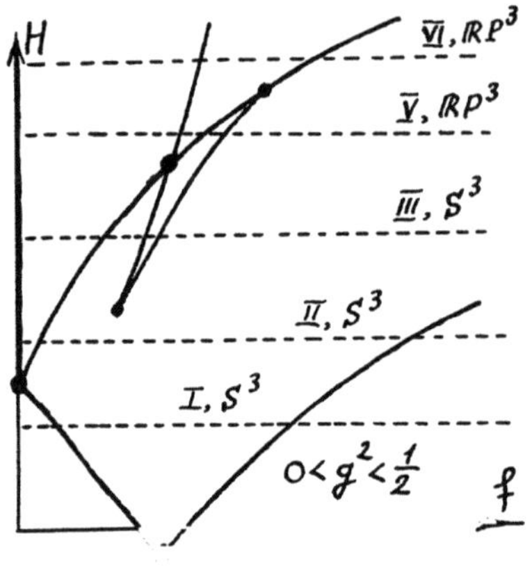

Figure 63

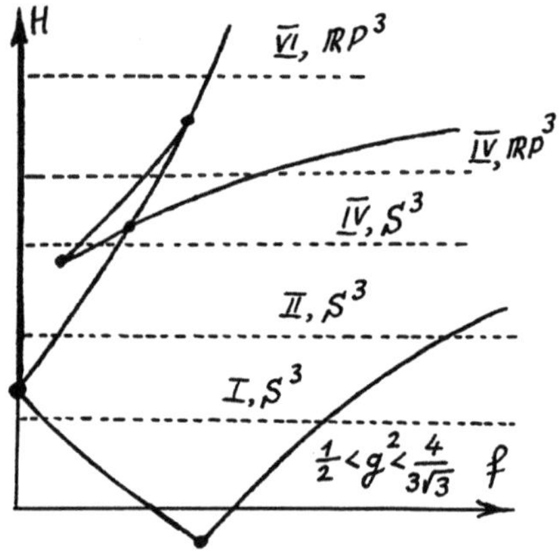

Figure 64

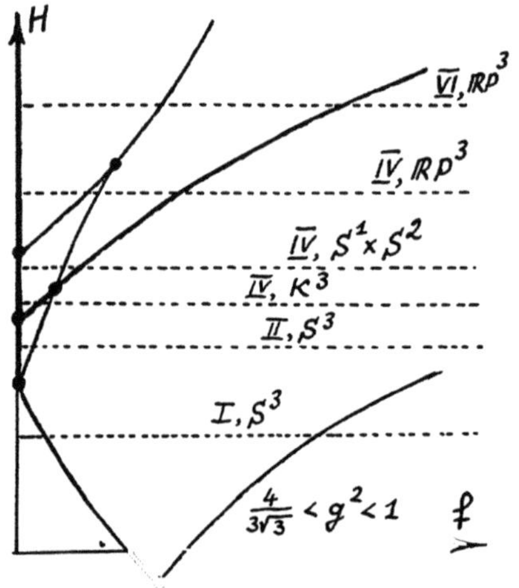

Figure 65

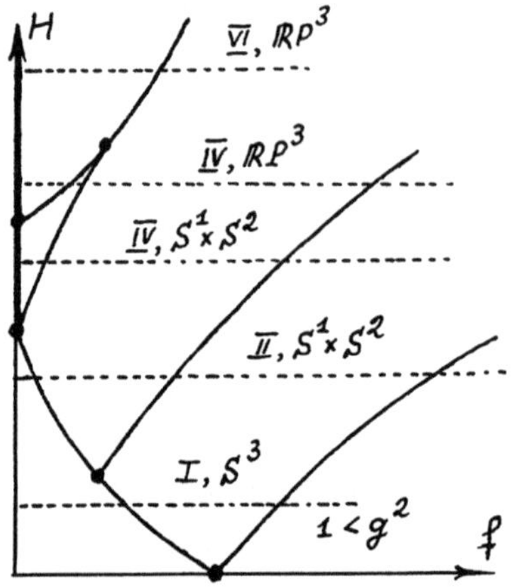

Figure 66

In the example of interest, the three-dimensional connected components of non-singular isoenergy surfaces are defined by the equations $Q^3 = \{F = 1, G = g, H = h\}$. It turns out that the complete list of all graphs Γ encountered in the case of Kovalevskaya consists of six graphs (Figure 67). *The integral f is a Bott integral on almost all isoenergy non-singular three-manifolds Q.*

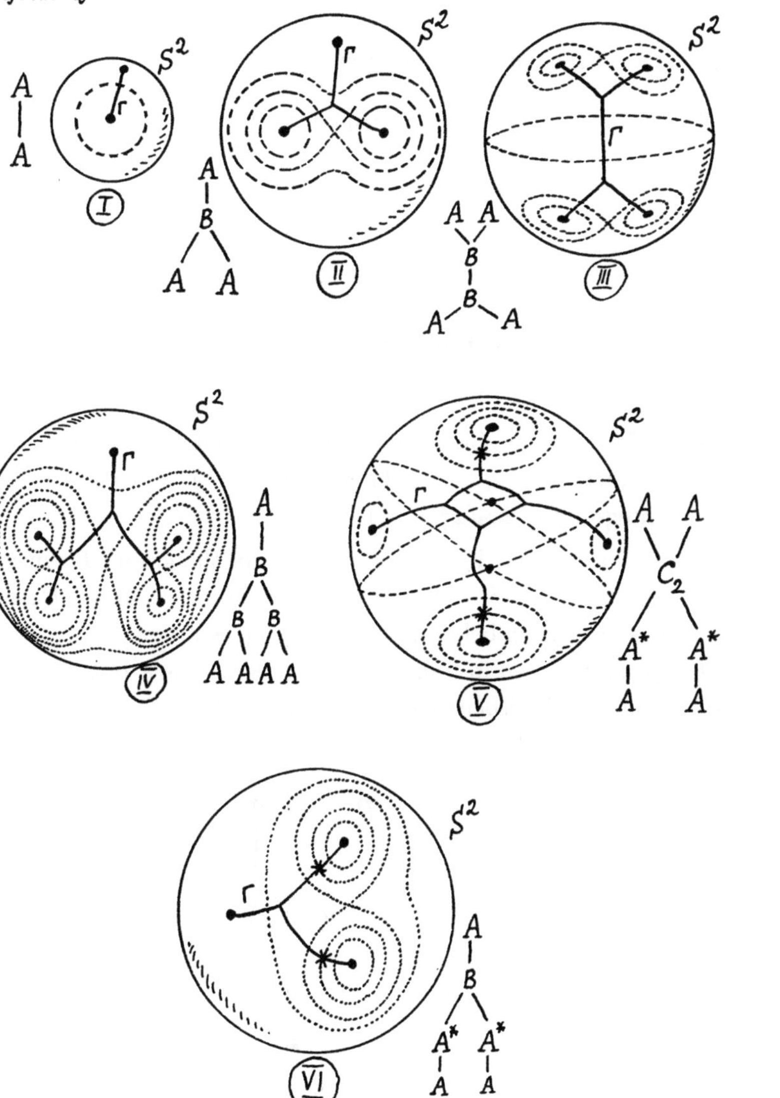

Figure 67

We now present embeddings of the graphs Γ and $K^{\sim}$ into the surfaces P. A detailed analysis of the analytic expressions for integrals makes it possible to construct all these embeddings in an explicit form (Figure 67). The surface P is seen here to be homeomorphic to a sphere in all cases. In Figure 67, the graphs Γ are shown by solid arcs and the graphs K by dashed arcs. It is obvious that $\Gamma = K^{\sim^*}$.

The corresponding words-molecules are seen in the Table 5 (see above). The dynamics of the complexity is seen in Table 6.

REMARK: Systems of algebraic ovals appear in the plane, the further analysis of which can be carried out within the classification theory of algebraic curves.

15.4. Goryachev–Chaplygin case.

The case of Goryachev–Chaplygin is characterized by the fact that $A_1 = A_2 = 4A_3$ and $\Pi = -\nu_1$. It is known that for $g = 0$ there exists an additional integral $f = 2\omega_3(\omega_1^2 + \omega_2^2) + 2\omega_1\nu_3$.

Calculations show that here a complete list of the graphs Γ consists of only two graphs (Figure 68). The bifurcation diagram is shown in Figure 69. The skeletons are shown below and in Table 5 (see above). The dynamics of the complexity is seen in Table 6. The integral f is a Bott integral on almost all non-singular isoenergy three-manifolds Q.

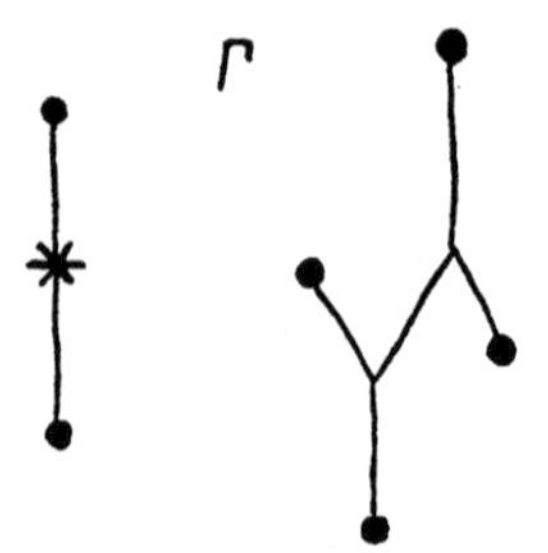

Figure 68

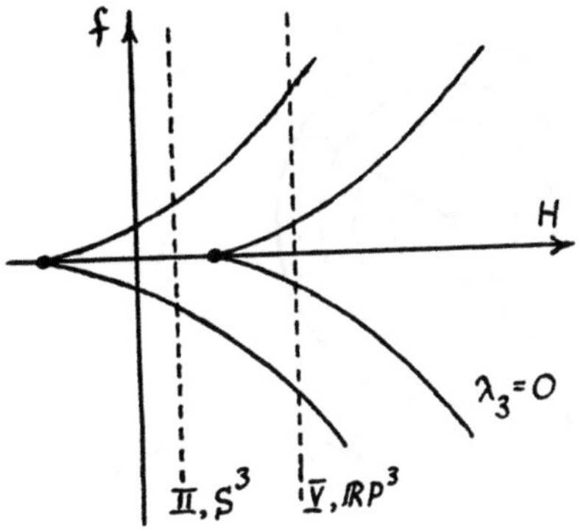

Figure 69

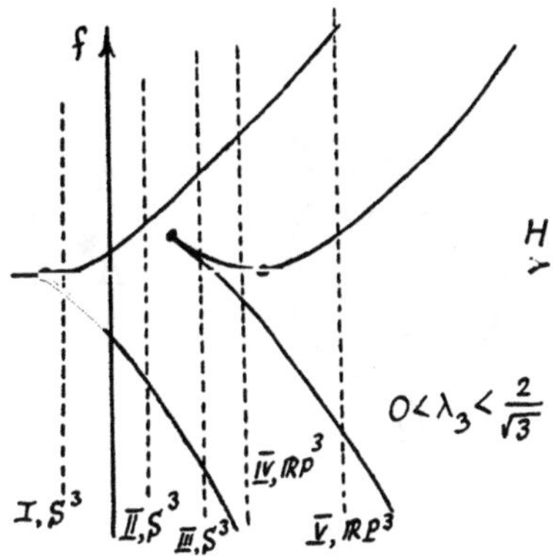

Figure 70

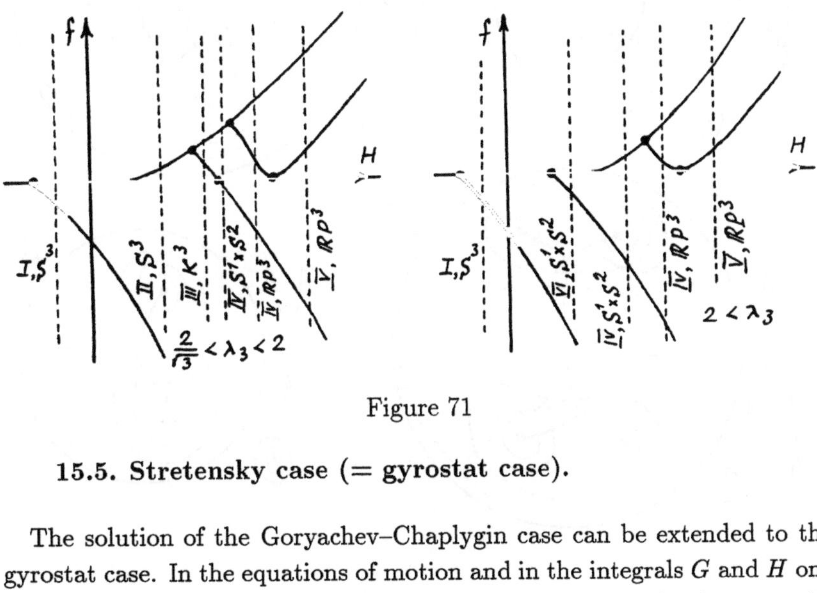

Figure 71

15.5. Stretensky case (= gyrostat case).

The solution of the Goryachev–Chaplygin case can be extended to the gyrostat case. In the equations of motion and in the integrals G and H one should replace $A\omega$ by $A\omega + \lambda$. Then the additional integral is known to be of the form $f = 2(\omega_3 - \lambda)(\omega_1^2 + \omega_2^2) + 2\omega_1\nu_3$.

For $\lambda = 0$ we deal with the usual Goryachev–Chaplygin case. For $\lambda \neq 0$ it turns out that a complete list of the graphs Γ encountered in this integrable case looks like the one shown in Figures 70, 71. Here we again deal with six types of graphs. The skeletons are seen in Figure 72 and in Table 5 (see above). The integral f is a Bott integral on almost all non-singular isoenergy three-manifolds Q.

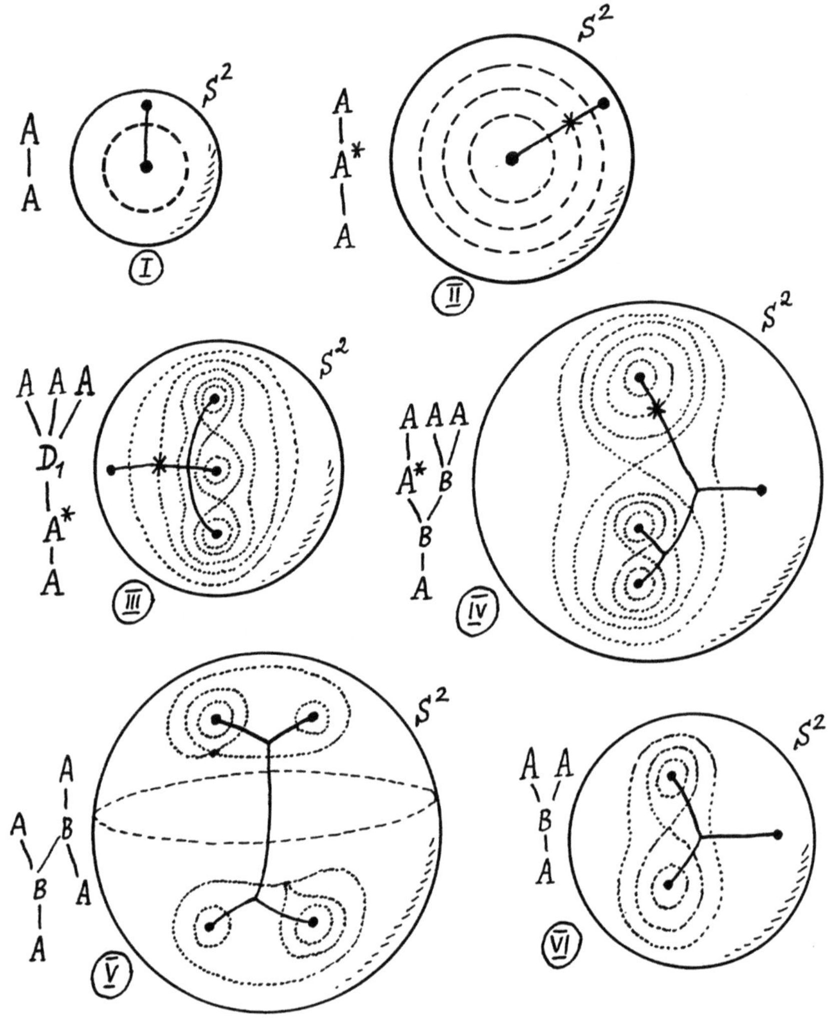
S^2
A
A
I
S^2
A
A*
A
II
S^2
A A A
D_1
A*
A
III
S^2
A A A
A* B
B
A
IV
S^2
A
B
A
A
B
A
V
S^2
A A
B
A
VI

Figure 72

15.6. Clebsch case.

In the Clebsch case we have: $\Pi = A_1\nu_1^2 + A_2\nu_2^2 + A_3\nu_3^2$ and the additional integral has the form:

$$f = A_1^2\omega_1^2 + A_2^2\omega_2^2 + A_3^2\omega_3^2 - (A_2A_3\nu_1^2 + A_1A_3\nu_2^2 + A_1A_2\nu_3^2).$$

The bifurcation diagram of the momentum mapping $F = H \times f$ has the types shown in Figure 73 (it depends on the value of g). The classical integral f is a Bott integral on almost all isoenergy surfaces Q_h, and the system v is non-resonant on almost all Q_h.

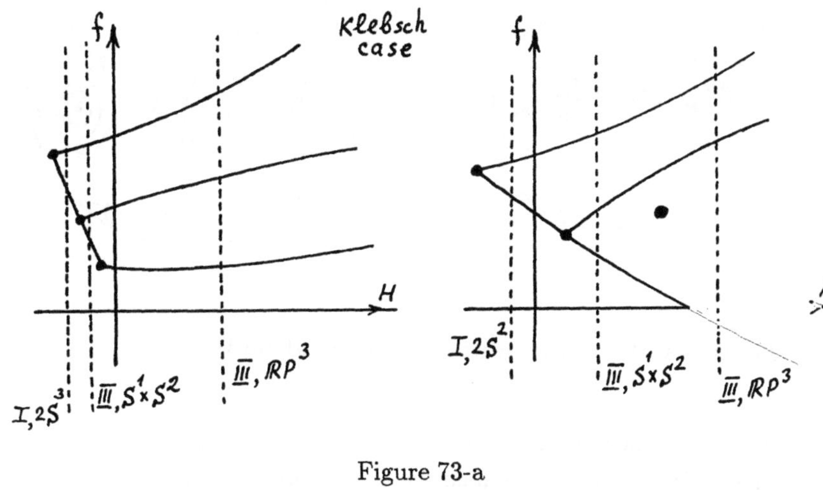

Figure 73-a

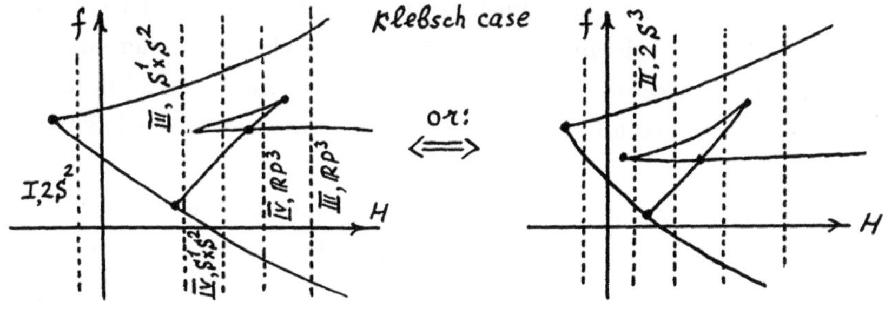

Figure 73-b

The words-molecules are shown in Figure 74.

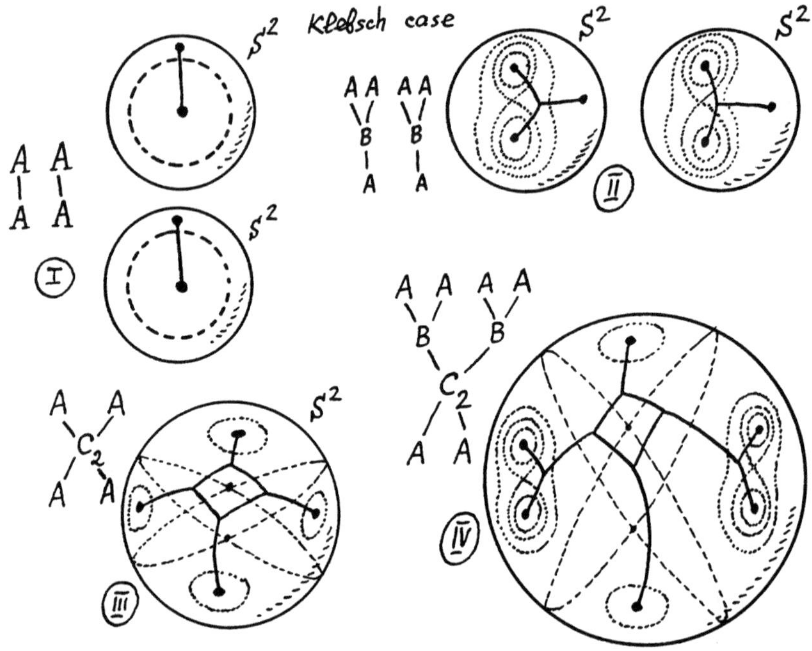

Figure 74

16. Dynamics of the four-dimensional rigid body.

Examine the Euler equations of motion of a multi-dimensional rigid body realized as a Hamiltonian system on the Lie algebra $so(3)$ and $so(4)$ of small dimension. We have, in fact, considered the case of the algebra $so(3)$ when we demonstrated that the equations are integrable on $so(n)$ (see details in [**32**], [**33**]) and coincide for $n = 3$ with the classical Euler equations of motion of a three-dimensional rigid body. See the books [**32**], [**33**] for a detailed review of all results of different authors connected with this problem.

Represent the Lie algebra $so(4)$ as an algebra of skew-symmetric real matrices $X = (x_{ij})$, where $i, j = 1, 2, 3, 4$. Two smooth functions (two polynomials) $f_1 = \Sigma_{i<j} x_{ij}^2$ and $f_2 = x_{12}x_{34} - x_{13}x_{24} + x_{14}x_{23}$ distinguish orbits of the algebra (as common level surfaces of two polynomials on the six-dimensional Lie algebra $so(4)$). Let $f_1 = p_1^2$ and $f_2 = p_2$.

Then under the condition $|2p_2| < p_1^2$ the two indicated polynomials have non-singular common level surfaces of the form

$$S^2 \times S^2 = \{f_1 = p_1^2,\ f_2 = p_2\}.$$

These four-dimensional symplectic manifolds (= orbits of the adjoint action of the group $SO(4)$ on its Lie algebra $so(4)$) are homeomorphic to a direct product of two spheres, $S^2 \times S^2$. On these orbits of the algebra $so(4)$ there appears an important Hamiltonian system

$$\dot{X} = [X, \varphi_{ab}(X)]$$

which is the equation of motion of the four-dimensional rigid body [**32**], [**33**]. Here [,] denotes the standard commutator of matrices. Let us describe these equations in more detail.

Let us consider the basis E_{ij} in the Lie algebra $so(4)$, where E_{ij} are the standard skew-symmetric matrices with the element 1 in the place (i, j) and -1 in the place (j, i); other elements are equal to zero. Then we can write X from $so(4)$ in the form:

$$X = \alpha E_{12} + \beta E_{13} + \gamma E_{14} + \delta E_{23} + \rho E_{24} + \varepsilon E_{34},$$

where all the coefficients are real. Then

$$\varphi_{ab}(X) = \alpha\frac{b_1 - b_2}{a_1 - a_2}E_{12} + \beta\frac{b_1 - b_3}{a_1 - a_3}E_{13} + \gamma\frac{b_1 - b_4}{a_1 - a_4}E_{14} + $$
$$+ \delta\frac{b_2 - b_3}{a_2 - a_3}E_{23} + \rho\frac{b_2 - b_4}{a_2 - a_4}E_{24} + \varepsilon\frac{b_3 - b_4}{a_3 - a_4}E_{34}.$$

For each pair $a = (a_1, a_2, a_3, a_4)$ and $b = (b_1, b_2, b_3, b_4)$ we obtain a flow $\dot{X}$ on the four-manifold $S^2 \times S^2$. The integrals f_1 and f_2 are constant on the orbits (they are invariants of the adjoint action of the group $SO(4)$) and are of the form

$$f_1 = \alpha^2 + \beta^2 + \gamma^2 + \delta^2 + \rho^2 + \varepsilon^2, \quad f_2 = \alpha\varepsilon - \beta\rho + \gamma\delta.$$

The integrals f_3 and f_4 are already not constant on the orbits. They have the form:

$$f_3 = \alpha^2(a_1 + a_2) + \beta^2(a_1 + a_3) + \gamma^2(a_1 + a_4) + \delta^2(a_2 + a_3) +$$
$$+ \rho^2(a_2 + a_4) + \varepsilon^2(a_3 + a_4).$$

$$f_4 = \alpha^2(a_1^2 + a_1a_2 + a_2^2) + \beta^2(a_1^2 + a_1a_3 + a_3^2) + \gamma^2(a_1^2 + a_1a_4 + a_4^2) +$$
$$+ \delta^2(a_2^2 + a_2a_3 + a_3^2) + \rho^2(a_2^2 + a_2a_4 + a_4^2) +$$
$$+ \varepsilon^2(a_3^2 + a_3a_4 + a_4^2).$$

It can easily be checked that the integrals f_1, f_2, f_3, f_4 are functionally independent and the integrals f_3 and f_4 are in involution on the orbits. Thus, the Euler equation $\dot{X} = [X, \varphi_{ab}(X)]$ is completely integrable (in the Liouville sense) for all operators φ_{ab} introduced above.

The general form of the integrable Hamiltonians of these series is as follows (for an arbitrary Lie algebra $so(n)$): $H = \Sigma_{i<j}\lambda_{ij}x_{ij}^2$; $\lambda_{ij} = (b_i - b_j)/(a_i - a_j)$, where $\Sigma_i a_i = 0$ and $\Sigma_i b_i = 0$. A corresponding system of differential equations describes the motion of a four-dimensional rigid body. The integrals of this system of equations (see above) are of the form: $f_3 = \Sigma_{i<j}(a_i + a_j)x_{ij}^2$ and $f_4 = \Sigma_{i<j}(a_i^2 + a_i a_j + a_j^2)x_{ij}^2$.

They are in involution on orbits and are functionally independent almost everywhere.

The geometric skeleton (= word-molecule) of this integrable case of the equations of motion of a four-dimensional rigid body with a fixed center of mass appears to contain *only black vertices* on the graphs Γ. This means that all critical manifolds of the second integral are circles. These circles can be minimax; then the Liouville tori close to them contract onto these circles. Next, critical circles can be saddles. In this case, on the critical level either two tori are transformed into one (or vice versa) or (as shown by calculations) sometimes two tori are transformed into two tori. In the latter case, the critical level surface of a Bott integral has the form $T \times S^1$, where the set T is homeomorphic to the cross-section of the sphere S^2 (embedded into $\mathbb{R}^3$ in a standard manner) with a pair of planes passing through its center.

STATEMENT 16.1 (A. A. Oshemkov, [**84**]). *Given is the integrable Hamiltonian system of a four-dimensional rigid body (of the so-called normal series* [**32**], [**33**]*) on the Lie algebra $so(4)$ with the Hamiltonian f_3 and the non-singular orbits $S^2 \times S^2$ in the algebra $so(4)$ fibered into three-dimensional isoenergy surfaces $Q^3 = \{f_3 = \text{const}\}$. For all these three-surfaces Q, except for a finite set, the function $f = f_4$ (on Q) is a Bott integral. Depending on the values of p_1 and p_2, which set the orbit $S^2 \times S^2 = \{f_1 = p_1^2, f_2 = p_2\}$, the bifurcation diagram for the momentum mapping*

$$F = f_3 \times f_4 : S^2 \times S^2 \to \mathbb{R}^2$$

can be of three types (Figure 75). The digits in Figure 75 indicate the number of Liouville tori forming the complete preimage $F^{-1}(y)$ for points y from a given region.

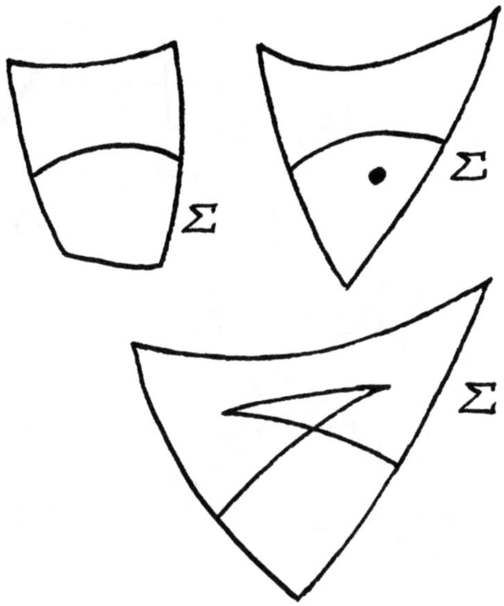

Figure 75

STATEMENT 16.2 (A. A. Oshemkov, [**84**]). *Let a Hamiltonian system of a four-dimensional rigid body (of the normal series) on the Lie algebra* $so(4)$ *with a general integrable Hamiltonian* H *mentioned above be given. For the integral* f *independent of* H *almost everywhere on orbits, choose* f_3 *or* f_4. *Then the non-singular orbits* $S^2 \times S^2$ *in* $so(4)$ *are fibered into isoenergy surfaces* Q, *which possess the property that* f *is a Bott integral for all these three-surfaces except for a finite set. A complete list of all connected graphs* Γ *is presented in Figure 76. All these graphs are realized for a certain Hamiltonian of the form* h *indicated above. All words-molecules are seen in Figure 77.*

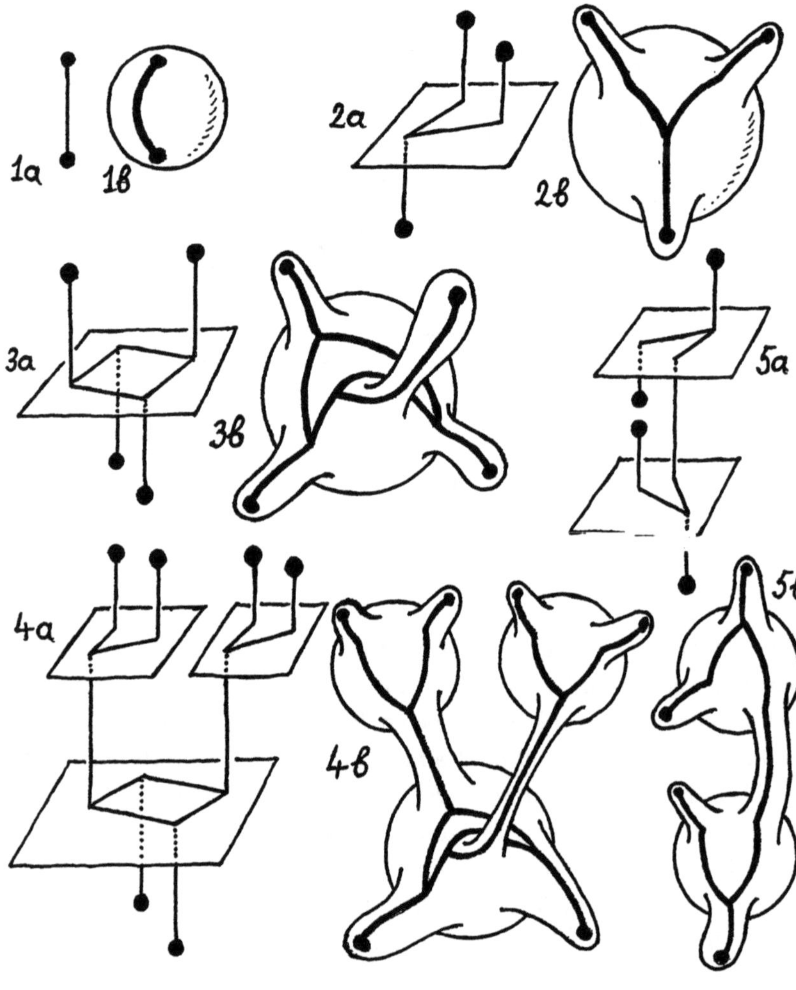

Figure 76a

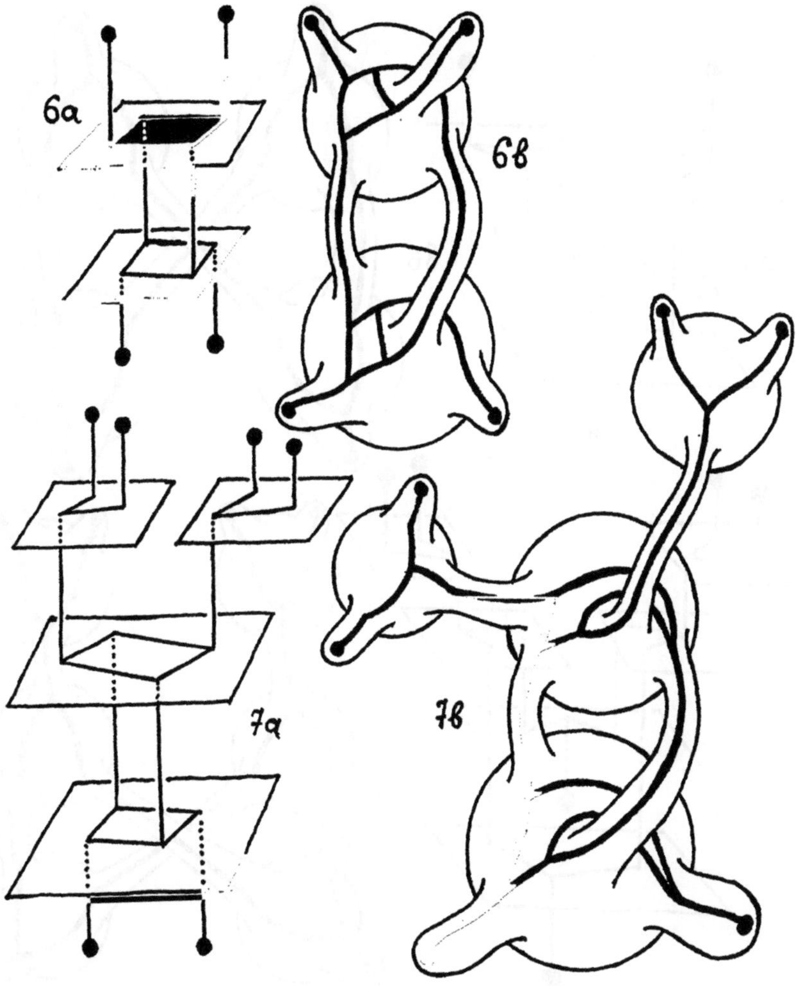

Figure 76b

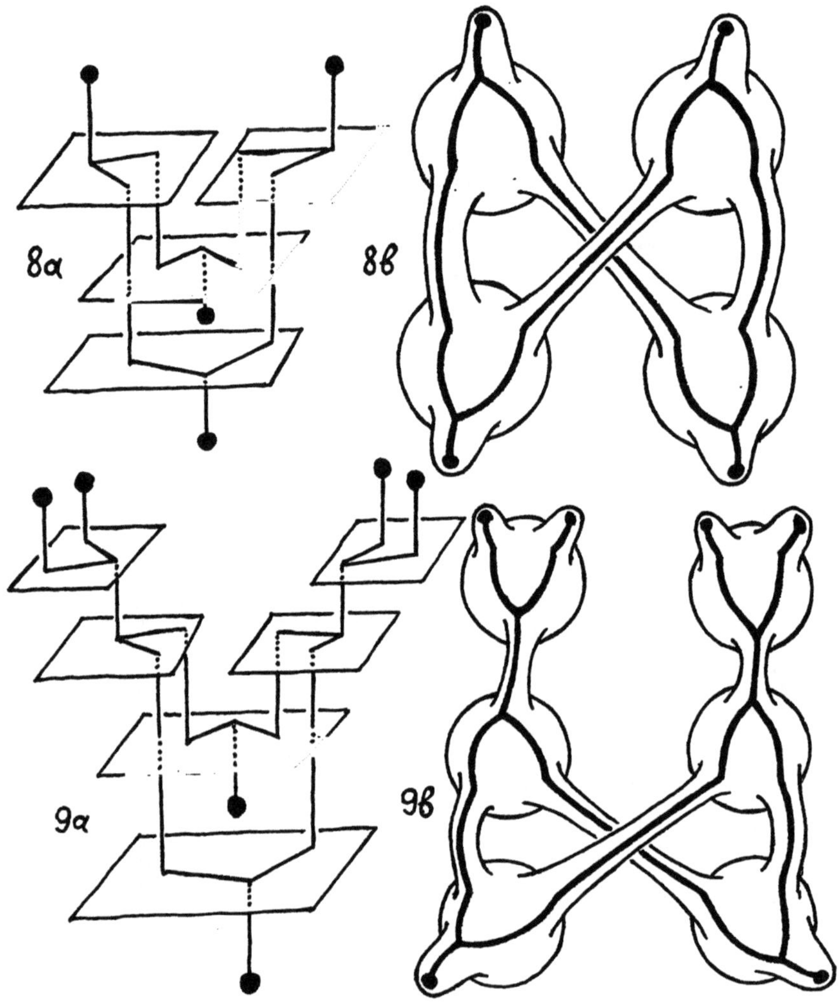

Figure 77

REMARK: An arbitrary integrable Hamiltonian (of the normal series) is represented in the form $H = k_1 f_1 + k_2 f_2 + k_3 f_3$, where k_1, k_2, k_3 do not depend on x_{ij}. Therefore, the bifurcation diagrams for a system with an arbitrary integrable Hamiltonian of this form are obtained from the bifurcation diagrams for a system with the Hamiltonian H (see Figure 75) by means of some non-degenerate linear transformation.

These skeletons (and corresponding words-molecules) are more complicated than those enumerated above. It should be noted that non-orientable edges (i.e., with stars) of the graphs Γ are absent (that is, there are no asterisks on the graph K). The two-dimensional surfaces P are homeomorphic either to a sphere or to a torus.

The integrable Clebsch case for the Euler equations on the Lie algebra $e(3)$ (of the Lie group $E(3)$ of Euclidean motions in $\mathbb{R}^3$) is obtained from the integrable case for the four-dimensional rigid body (see above) (S. P. Novikov). This happens after the contraction of the Lie algebra $so(4)$ on the Lie algebra $e(3)$. The bifurcation diagram of the case $so(3)$ will transform as follows: the right boundary of Σ moves to infinity. In the limit (when we obtain the Clebsch case) the diagram of the case $so(4)$ transforms in the bifurcation diagram for the case $e(3)$.

17. Word-molecules describing some integrable cases of the classical Toda lattice and its integrable perturbations.

The classical periodic Toda lattice has the Hamiltonian

$$(1)\qquad H = \frac{1}{2}\Sigma_{i=1}^{n} p_i^2 + \Sigma_{i=1}^{n-1} \exp(x_i - x_{i+1}) + \exp(x_n - x_1).$$

V. V. Kozlov and V. V. Treschev investigated in [**122**] the Hamiltonian systems with exponential interaction of the form:

$$(2)\qquad \dot{x} = p, \quad \dot{p} = -\Sigma_{k=1}^{m}[c_k \exp(a_k, x)]a_k,$$

where x and p are the n-dimensional vectors of the coordinates and moments (respect.) and $\{a_1, \ldots, a_m\}$ is the "spectrum" of the sum of the exponents.

The following results were obtained by L. S. Polyakova.

STATEMENT 17.1. *Let us consider the Hamiltonian system (2) on* $\mathbb{R}^4$ *with the Hamiltonian*

$$H = \frac{1}{2}(p_1^2 + p_2^2) + c_1 \exp(x_1 - x_2) + c_2 \exp(x_2 - x_1),$$

(where $c_1, c_2 > 0$*) and with an additional integral* $f = p_1 + p_2$*, which is in involution with* H *and functionally independent. Then:*

a) *The manifold* $\mathbb{R}^4$ *is fibered into the sum of three-surfaces* $Q = \{H = \text{const}\} = S^2 \times \mathbb{R}^1$ *such that the integral* f *is a Bott integral on all* Q *except one (which corresponds to the minimum of* H *on* $\mathbb{R}^4$*).*

b) *For all parameters* c_1, c_2 *the bifurcation diagram (of the momentum mapping* $H \times f : \mathbb{R}^4 \to \mathbb{R}^2$*) is shown in Figure 78.*

c) *The geometric skeleton (= word-molecule) does not depend on the values of the parameters* c_1, c_2 *and on the three-surface* Q*, and is shown in Figure 79.*

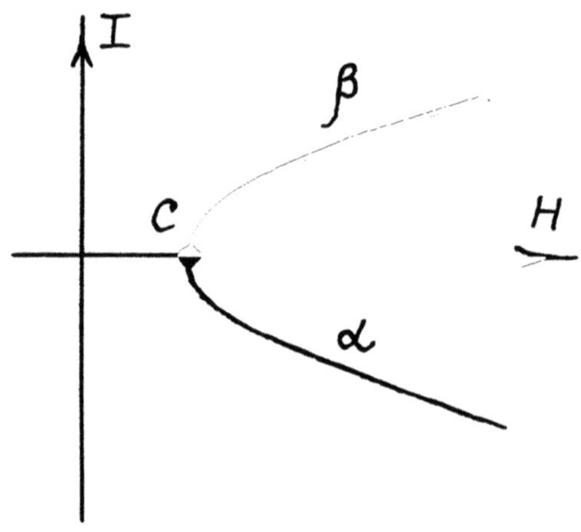

Figure 78

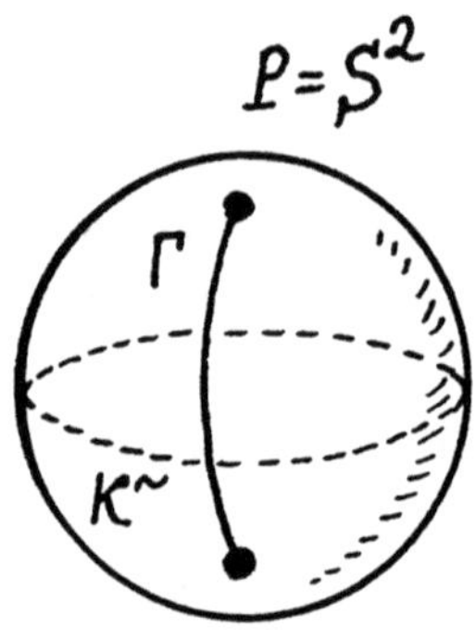

Figure 79

STATEMENT 17.2. *Let us consider the Hamiltonian system (2) on* $\mathbb{R}^4$ *with the Hamiltonian* $H = \frac{1}{2}(p_1^2 + p_2^2) + \Sigma_{i=1}^m a_i \exp(\alpha_i x_2)$ *and with the additional integral* $f = p_1$*. Let us suppose that all* $a_i > 0$ *and there exist* $\alpha_i > 0$ *and* $\alpha_j < 0$*. Then:*

a) *The additional integral* f *is a Bott integral on all isoenergy surfaces* $Q^3 = \{H = \text{const}\} = S^2 \times \mathbb{R}^1$ *except of a finite subset of* Q*.*

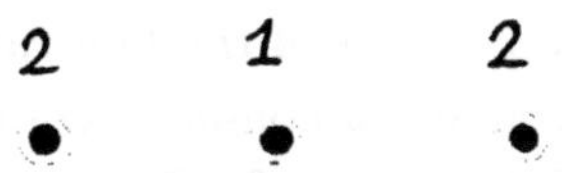

Figure 80

b) *The bifurcation diagram Σ for the momentum mapping is homeomorphic to the bifurcation diagram of the classical Toda lattice (independently of the choice of the parameters m, α_i and a_i).*

c) *The geometric skeleton is the same as in the classical case (see Figure 79).*

18. Word-molecule describing one algebraic analogous of the Toda lattice.

We calculate here the framed geometric skeleton of the Hamiltonian system on $\mathbb{R}^4$ with the Hamiltonian

$$(1)\qquad H = \frac{1}{2}(p_1^2 + p_2^2 + p_3^2) + e^{x_1 - x_2} + e^{x_2} + e^{-x_1 - x_2}.$$

This system is an algebraic analogues of the Toda lattice connected with the Lie algebra b_2 [**123**]. According to the classification of Kozlov and Treschev [**122**], this Hamiltonian system corresponds to the Dynkin diagram shown in Figure 78. The additional integral f can be found from the condition $\{H, f\} = 0$ and from the assumption that f is the polynomial of degree four with respect to the moments, with the coefficients depending on the exponents. Thus,

$$(2)\qquad f = (p_1 p_2 - e^{x_1 - x_2} + e^{-x_1 - x_2})^2 + 2e^{x_2}(p_1^2 + e^{x_1 - x_2} + e^{-x_1 - x_2}).$$

It is easy to verify that H and f are in involution and functionally independent (almost everywhere).

STATEMENT 18.1 (L. S. Polyakova). *Let us consider the Hamiltonian system on $\mathbb{R}^4$ with the Hamiltonian* (1) *which has the additional integral f in the form* (2). *Then:*

a) *The phase space $\mathbb{R}^4$ is fibered into the union of isoenergy three-surfaces $Q = \{H = \text{const}\} = S^3$ (i.e., three-spheres).*

b) *For all Q^3, except one three-manifold (corresponding to the minimum of the function H on $\mathbb{R}^4$), the integral f is a Bott integral on Q^3.*

c) *The bifurcation diagram Σ for the momentum mapping $F = H \times f : \mathbb{R}^4 \to \mathbb{R}^2$ is shown in Figure* 80.

d) *The preimage of every point of the bifurcation diagram (except the point A, see Figure 81) contains a non-degenerate critical circle: the minimum in the preimage of the points of AB (Figure 81), and the maximum (of the integral f) in the preimage of the points of AC.*

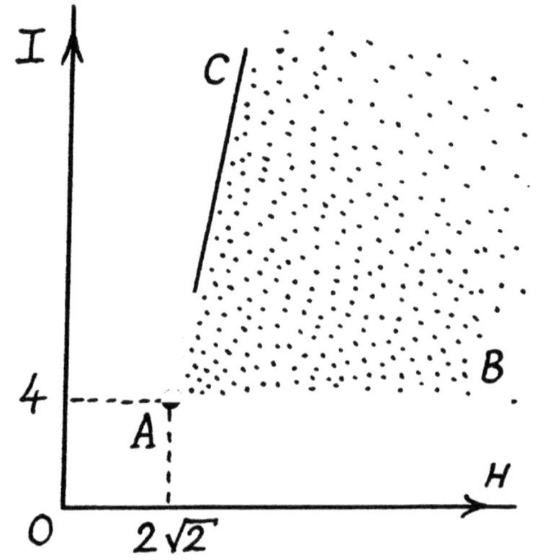

Figure 81

e) *The framed skeleton (P, K) of this integrable case is shown in Figure 82. The surface P is S^2, the graph Γ is the segment, and $(\alpha, \beta) = (0, 1)$, i.e., $r = 0$.*

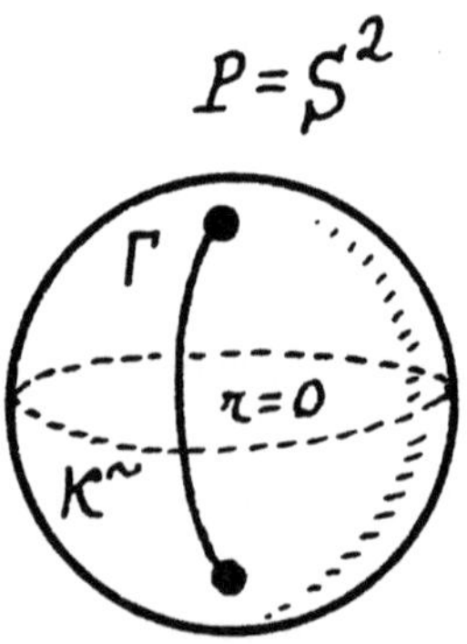

Figure 82

CHAPTER 5. THE CLASS (H) OF ISOENERGY SURFACES OF INTEGRABLE HAMILTONIAN EQUATIONS AND HYPERBOLIC GEOMETRY

19. No hyperbolic three-manifolds in the class (H). The volumes of closed compact hyperbolic three-manifolds.

Let us recall that a compact orientable three-manifold lies in the class (H) iff it can be obtained by glueing copies of $D^2 \times S^1$ and of $N^2 \times S^1$ together via homeomorphisms between some of their boundary components.

Following Thurston [**99**], we call an orientable three-manifold *geometric* if it admits a complete locally homogeneous Riemannian metric. There exist eight types of three-dimensional geometries: *spherical, Euclidean (or flat), the geometries of* $S^2 \times R^1$, $H^2 \times R^1$, $SL_2R^{\sim}$, *Nil, Sol, and the hyperbolic geometry.*

The well-known geometrization conjecture of Thurston asserts that the interior of every compact three-manifold admits a canonical decomposition into pieces having geometric structures. It turns out that the class (H) is closely connected with the first seven types of geometries.

THEOREM 19.1. *A compact orientable three-manifold with (possibly empty) torus boundary belongs to the class* (H) *if and only if its interior admits a canonical decomposition into pieces having geometries of the first seven types. In particular, the class* (H) *of isoenergy surfaces of integrable Hamiltonian systems contains no hyperbolic manifolds.*

The proof of the theorem in one direction can be found in [**70**] (Fomenko, Matveev, Sharko), where the topological structure of manifolds in (H) is described, and in the other direction in Scott's article [**124**], where geometric manifolds of the types 1–7 are classified.

Now we will enumerate all three-manifolds with respect to their complexity. It turns out that the "first two three-manifolds" which do not belong to the class (H) are remarkable hyperbolic manifolds. The first of them was discovered by Thurston, the second—by Weeks (and independently by Fomenko and Matveev).

An exposition of the theory of hyperbolic manifolds is given in Thurston [**99**], [**100**], where in particular a procedure is given for effectively constructing a large series of hyperbolic three-manifolds. Particular examples of hyperbolic manifolds had already been constructed by Seifert and

Weber in 1933, but until recently there have been no very large known classes of *closed* hyperbolic manifolds. The construction of closed hyperbolic three-manifolds is far more complicated than that of non-compact three-manifolds. See details in [**72**], where we give the review of the main ideas of construction of hyperbolic manifolds and of the methods for their volume calculations.

We have proceeded to solve the problem from a fundamentally different point of view. Our procedure is based on a combination of Fomenko's topological theory of classification of the integrable Hamiltonian systems with Matveev's complexity theory [**69**] for three-dimensional manifolds. All manifolds in (H) are not hyperbolic. Matveev and Fomenko have shown that all three-manifolds of small complexity lie in (H), but that the *first manifolds not in (H) are hyperbolic.*

Further, it turns out that, among all irreducible atoroidal manifolds of complexity less than or equal to three with torus boundary, there are exactly two hyperbolic manifolds Q_1^d and Q_2^d of complexity 2, and nine hyperbolic manifolds Q_i^d, $3 \leq i \leq 11$, of complexity 3. Each of these manifolds generate a series of closed compact hyperbolic manifolds $(Q_i^d)_{p,q}$, where p and q are coprime integers. Geometric methods were then used to compute the volumes of over 100 manifolds in each series, that is, *around a thousand in all.* The minimum volume among the manifolds of type $(Q_i^d)_{p,q}$ is that of the remarkable manifolds $Q_1 = (Q_1^d)_{5,-2}$, and the next is that of $Q_2 = (Q_2^d)_{5,1}$.

The manifold Q_2 turns out to be the one studied by Thurston, who proposed it as a candidate for the hyperbolic manifold of minimum volume. Since the volume of Q_1 ($\cong 0.94$) is less than that of Q_2 ($\cong 0.98$), this does not hold. Then we discovered the paper of J. Weeks [**108**], who investigated this remarkable three-manifold and calculated its volume. We now give a more precise statement of the results of our paper [**72**].

1) *All closed orientable three-manifolds of complexity not greater than eight belong to the class (H) of isoenergy surfaces of integrable Hamiltonian systems (Theorem* 10 *in* [**72**]*) and are not hyperbolic.*

2) *All compact irreducible atoroidal orientable three-manifolds of complexity not greater than three with torus boundary belong to the class (H), except for precisely two three-manifolds Q_1^d and Q_2^d of complexity* 2 *and nine three-manifolds Q_i^d, $3 \leq i \leq 11$, of complexity* 2 *(Theorem* 9 *in* [**72**]*). These* 11 *manifolds admit hyperbolic structures, and hence cannot belong to the class (H).*

REMARK: Here we denote by (H) the class of all isoenergy surfaces with torus boundary (corresponding to integrable systems).

3) *A table of approximately* 1000 *volumes of hyperbolic manifolds of the form* $(Q_i^d)_{p,q}$, *computed to five decimal places, is obtained by numerical methods on a computer (Theorem* 14 *in* [**72**]*).*

4) *The remarkable manifold* Q_1 *has the least volume among the* $(Q_i^d)_{p,q}$.

5) *Many non-homeomorphic closed hyperbolic three-manifolds with equal volumes have been found.*

We now state some conjectures.

CONJECTURE 1. *The manifold* Q_1 *has the least volume of all closed hyperbolic manifolds.*

CONJECTURE 2. *The distribution of volumes of the manifolds* $(Q_i^d)_{p,q}$ *given in Theorem* 14 *in* [**72**] *forms an initial segment of the sequence of volumes of all closed hyperbolic three-manifolds.*

CONJECTURE 3. *Within any sequence of manifolds of the form* $M_{p,q}$ *(where* M *is a manifold with torus boundary, see* [**72**]*) the volumes increase with increasing complexity.*

20. The complexity of three-manifolds.

In this section we give a short review of Matveev's theory of complexity. We associate with each compact three-manifold M a non-negative integer $d(M)$ called its *complexity*. The complexity $d(M)$ has a number of useful properties and, as we shall see below, conforms closely to the complexity of the three-manifold in an intuitive sense. To describe the complexity $d(M)$ we shall need two notions: *spines* and *almost special polyhedra*. We recall that a polyhedron Q collapses to a subpolyhedron P if, for any triangulation (K, L) of (Q, P), we can transform K to L by a finite sequence of elementary simplicial collapses, where each elementary simplicial collapse consists in removing a principal open simplex together with a free open face of that simplex (Figure 83).

DEFINITION. *A polyhedron* $P \subset \operatorname{Int} M$ *is called a spine of the compact manifold* M *if* M *(or* M *with an open ball removed if* $\partial M = \emptyset$*) collapses onto* P.

We should mention that a spine carries a lot of information about the structure of the manifold, although in general a manifold cannot be uniquely

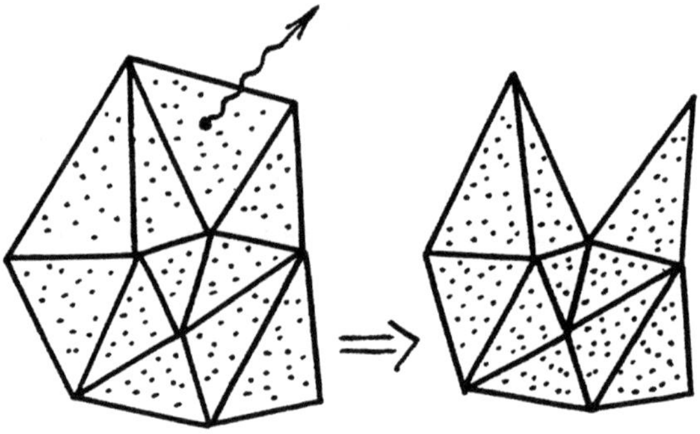

Figure 83

recovered (reconstructed) from its spine. It is useful to bear in mind the following equivalent definition of spine: P is a *spine* of M if M is homeomorphic to the mapping cylinder of a map $p : \partial M \to P$. Let Δ denote the one-skeleton of the standard three-dimensional simplex, that is, a circle with three radii.

DEFINITION. *A compact polyhedron P is called almost special if each of its points has a neighborhood homeomorphic to the cone on a subpolyhedron of Δ. Points whose links are homeomorphic to Δ are called vertices of P.*

The singularities that can be seen in $\operatorname{Con} \Delta$ are stable in nature: they do not disappear with small changes in the attaching maps of cells, for any representation of the polyhedron as a cell complex. It is interesting to note that precisely such singularities arise in the theory of minimal surfaces (soap films). It follows easily from the definition of an almost special polyhedron that this property is hereditary, that is, it is preserved in passing to subpolyhedra.

DEFINITION. *The complexity $d(M)$ of a compact three-manifold M is the number of vertices in a minimal (in the sense of fewest vertices) almost special spine.*

The following theorems were proved by S. V. Matveev.

THEOREM 20.1 (Finiteness property). *For any integer $k \geq 0$ there exists only finitely many distinct closed irreducible three-dimensional manifolds of complexity k.*

THEOREM 20.2 (Additive property). *For any compact manifolds M_1 and*

M_2 *we have:*

$$d(M_1 \# M_2) = d(M_1) + d(M_2).$$

A proof of Theorem 20.1 is given in [**89**]. The proof of Theorem 20.2 in the difficult direction (that is, the inequality $d(M_1 \# M_2) \geq d(M_1) + d(M_2)$ is based on Haken's theory of normal surfaces. Experience with spines shows that an almost special spine can easily be constructed for any given manifold (see for example: Matveev, Savvateev, Three-dimensional manifolds having simple special spines, Colloq. Math. 32 (1974), 83–97). Thus there is no difficulty in finding an upper bound for the complexity $d(M)$. Finding a non-trivial lower bound is a difficult problem. There is as yet no known algorithm for computing the complexity.

DEFINITION. *A compact polyhedron P is called **special** if the link of each of its points is homeomorphic to a circle or to a circle with two or three radii, and if each connected component of its set of non-singular points is homeomorphic to an open two-cell.*

The set of singular points (that is, points of types 2 and 3) of a special polyhedron P is a regular graph of degree 4. The components of its complement (that is, the open two-cells) are called the two-components of P, and the graph itself is called the *singular graph* of P. We shall require the following important fact. In almost all interesting cases, a minimal almost special spine of a closed three-manifold is special. More precisely, the following holds.

THEOREM 20.3 (See [**125**]). *If a closed orientable irreducible three-manifold M is distinct from S^3 and $\mathbb{RP}^3$, then a minimal almost special spine of M is special.*

A three-manifold can be uniquely recovered from a special spine. There is a close connection between the complexity of an orientable three-manifold and the number of three-simplexes which have to be glued together to construct it (not to be confused with the number of three-simplexes in a triangulation!). We take a collection of three-simplexes, partition the faces into pairs, and identify the faces in each pair via one of the three possible orientation-reversing linear homeomorphisms. If we remove regular neighborhoods of the vertices (the images of the vertices of the simplexes) from the resulting polyhedron, then we obtain a compact orientable manifold M. In other words, M is obtained by glueing together three-simplexes with truncated vertices.

STATEMENT 20.1 (S. V. Matveev). *A compact orientable three-manifold M with boundary can be obtained by glueing n simplexes with truncated vertices if and only if it has a special spine with n vertices.*

The proof of the theorem is based on the following observation: if we consider the standard three-simplex as a simplicial complex, and take the two-skeleton of its dual cell decomposition, then we obtain a polyhedron homeomorphic to Con Δ. Under glueing of simplexes, these polyhedra combine to form a special spine of M. Conversely, if the vertices of a special spine are replaced by simplexes with truncated vertices, and these are glued together as dictated by the spine, then we obtain the manifold M.

21. The classification of manifolds of small complexity. The role of the class (H). The exceptional manifolds Q_1 and Q_2.

Here we give a short review of some results from [**72**].

The finiteness property of the complexity stated above (or more precisely a constructive proof of this property) enables one to organize a search for all closed irreducible three-manifolds of complexity not exceeding a given number d. We present the results of such a search, carried out by computer.

THEOREM 21.1 (See [**72**]). *The number $N(d)$ of closed orientable irreducible three-manifolds of complexity d, for $d \leq 6$, is given in the following table:*

d	0	1	2	3	4	5	6
$N(d)$	3	2	4	7	14	13	74

An analysis of the manifolds produced by the computer enabled us to establish the following facts.

1) *All the manifolds of complexity* ≤ 5 *and most of those of complexity* 6 *are elliptic.*

2) *All six flat three-manifolds have complexity* 6.

3) *The remaining six manifolds of complexity* 6 *have a geometric structure modelled on Nil. Hence by Theorem* 19.1 *all closed orientable three-manifolds of complexity not greater than* 6 *belong to the class* (H) *of integrable isoenergy surfaces.*

A further search (for $d > 6$) for manifolds of complexity d is difficult, since it requires a fairly large amount of machine time. On the other hand, the class (H) is well understood and completely classified (see above). It thus makes sense to carry out a search "modulo (H)". For this we have to teach the computer, in constructing in turn all special spines with a given number of vertices, how to recognize at an early stage of the construction whether or not the corresponding manifold belongs to the class (H). We present a theoretical basis for such a (partial) recognition.

We shall say that a two-component α of a special spine P of a closed orientable three-manifold M *has defect* k if its closure contains all the vertices of P with precisely k exceptions. Let M^α denote the three-manifold with torus boundary whose spine is obtained from P by deleting an open two-disk interior to α. An important fact is that the complexity of M^α does not exceed k. The proof of this is obvious, since on puncturing α and collapsing, all the vertices in its closure also disappear. We note also that if M^α belongs to (H), then so does M (see [**70**]).

A computer search of manifolds with torus boundary and a subsequent manual analysis has enabled us to establish the following theorem.

THEOREM 21.2. *Every compact orientable irreducible atoroidal three-manifold of complexity $d \leq 3$ with torus boundary belongs to the class (H) with the exception of precisely two remarkable manifolds Q_1^d and Q_2^d of complexity 2 and nine manifolds Q_i^d, $3 \leq i \leq 11$, of complexity 3, which do not belong to the class (H).*

A description of the two exceptional three-manifolds Q_1^d and Q_2^d is given below. We note that the boundary of each manifold Q_i^d, $1 \leq i \leq 11$, consists of a single torus. If a closed three-manifold M is obtained from Q_i^d, $1 \leq i \leq 11$, by attaching a solid torus to its boundary, then M may or may not belong to the class (H). A special spine for M is easily constructed from a special spine P_i or Q_i^d: essentially, a new two-component (a meridional disk of the attached solid torus) is added to P_i. It turns out that every manifold of complexity $d \leq 8$ obtained from the Q_i^d in this way belongs to (H). This is no longer true for complexity 9: there exist at least two manifolds of complexity 9 which can be obtained from Q_1^d and Q_2^d by attaching solid tori, and which do not belong to (H) (see [**72**]).

Collecting together the above facts, we arrive at the following criterion for a closed three-manifold M to belong to the class (H):

if a special spine P of M has no more than eight vertices and contains a two-component of defect $k \leq 3$, then $M \in (H)$.

The use of this criterion allowed us to complete the search (modulo (H)) for all closed orientable three-manifolds of complexity not greater than eight. The following result turns out to be true [**72**].

THEOREM 21.2. *Every closed orientable three-manifold of complexity no greater than eight belongs to the class* (H) *of integrable isoenergy surfaces.*

Thus the two exceptional manifolds Q_1 and Q_2 mentioned above are the simplest manifolds (in the sense of complexity) that do not belong to the class (H).

We describe in more detail the two manifolds Q_1^d and Q_2^d that we found above. *They are uniquely determined by their special spines*, neighborhoods of whose singular graphs are shown in Figure 84. As a topological manifold, Q_1^d can be represented as the complement of a certain knot in the lense space $L_{5,1}$, and Q_2^d as the complement to the figure eight knot in S^3. The interior of Q_1^d and Q_2^d admit complete hyperbolic structure; in particular, the manifolds do not belong to the class (H). The details are in [**72**].

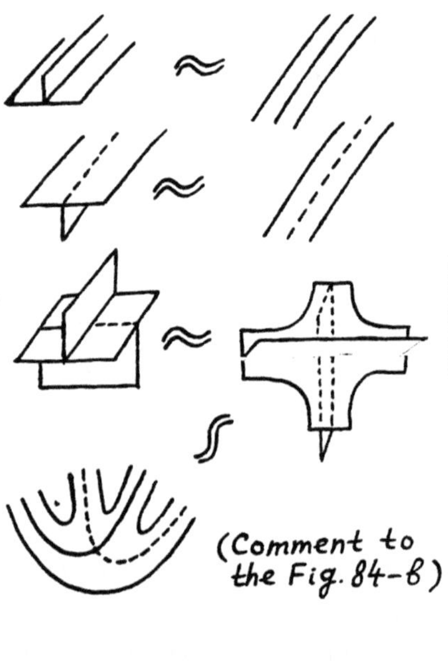

Figure 84-a

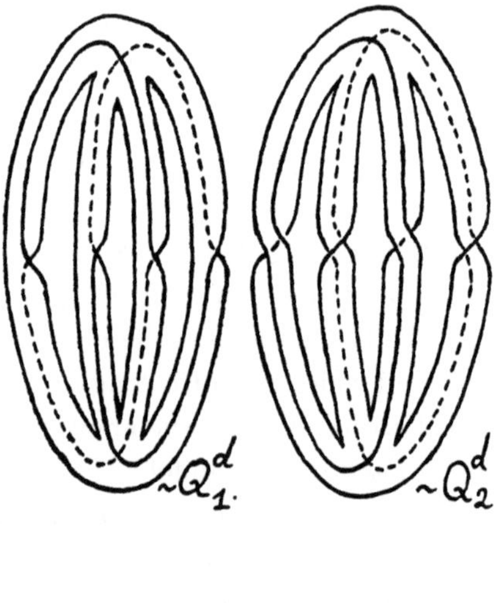

Figure 84-b

It is extremely useful to produce a clear concrete description of the *closed compact exceptional three-manifolds* Q_1 *and* Q_2. Figure 85 shows neighborhoods of the singular graphs of the minimal special spines of Q_1 and Q_2 respectively. As can be seen from Figure 85, the number of singular vertices of each of these special spines is equal to nine. Hence these manifolds have complexity not greater than nine. But since the theorems mentioned above guarantee that there are no hyperbolic manifolds of complexity 8 or less, the complexity must equal 9.

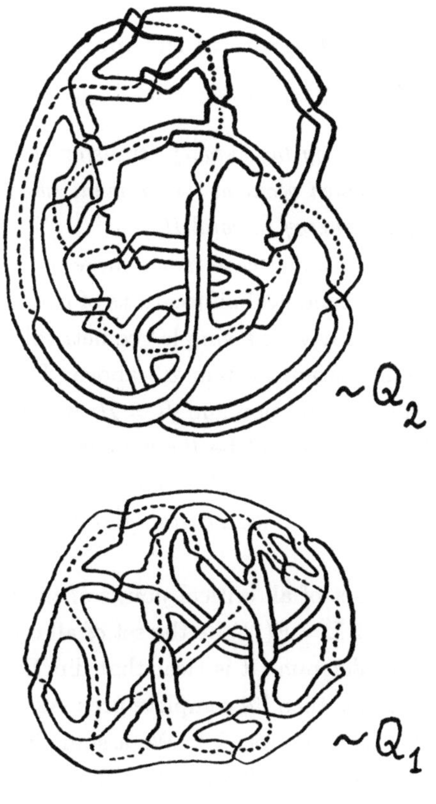

Figure 85

Chapter 6. Multidimensional Integrable Hamiltonian Equations. Classification of the Surgery of Liouville Tori in the Neighborhood of Bifurcation Diagrams

22. Five elementary bifurcations of multidimensional Liouville tori. Five types of elementary manifolds.

22.1. Bifurcation diagram of the momentum mapping in the multidimensional case.

In this section we *classify the surgery (transformations) in general position of Liouville tori, which occur at the moment when a torus "intersects" the critical level of the energy integral H.*

As in the three-dimensional case, such surgery turns out to split into compositions of certain canonical bifurcations of five types; the latter being described explicitly and having a simple geometrical nature. See the definition of bifurcation diagram in item 2.3 above. Suppose that $f_1, \ldots, f_n$ are smooth integrals in involution and $f_1 = H$, where H is a Hamiltonian, $v = \operatorname{sgrad} H$. Let $F : M^{2n} \to \mathbf{R}^n$ be *the momentum mapping*, i.e., $F(x) = (f_1(x), \ldots, f_n(x))$. The point $x \in M$ is called *regular* for F if $\operatorname{rank} dF(x) = n$. Otherwise the point x is called *critical*, and its image $F(x)$ is called *the critical value*.

Let $N \subset M$ be the set of all critical points of the mapping F. Clearly, the set N is closed. Let $\Sigma = F(N)$ be the set of all critical values. This set is called *a bifurcation diagram*. It is clear that $\dim \Sigma \leq n-1$. If the point $x \in \mathbf{R}^n$ is a *regular value*, then its complete preimage $B_a = F^{-1}(a) \subset M$ is the union of regular Liouville tori T^n. We assume here for simplicity that the whole fiber B_a is compact (Figure 86). (The corresponding assertions for non-compact fibers B_a, that is, for "cylinders" will easily follow from the results obtained below).

If $c \in \Sigma$, then the compact level surface (i.e. the fiber) B_c is *singular* (critical) and evidently $\dim B_c \leq n$.

When the regular point a in $\mathbf{R}^n$ becomes deformed, its preimage becomes somewhat deformed too. The fiber B_a is transformed through diffeomorphisms, that is, does not undergo qualitative topological surgery until the

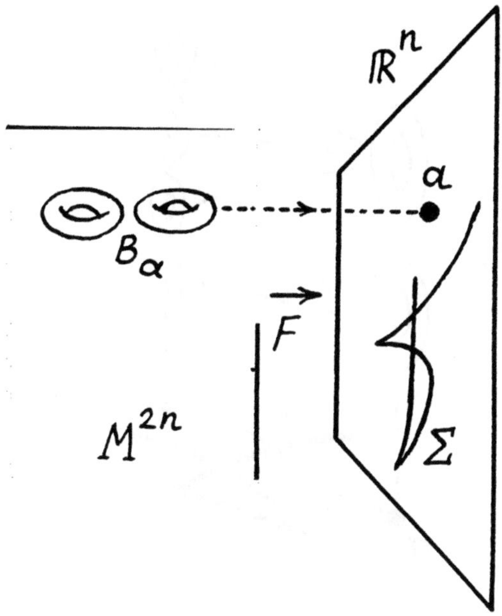

Figure 86

point a, moving along $\mathbf{R}^n$, meets the bifurcation diagram Σ. In particular, any two fibers B_a and B_b, such that the regular points a and b may be joined with a smooth curve $\gamma \subset \mathbf{R}^n \backslash \Sigma$ (that is, a curve that contains none of the critical values of the momentum mapping), are diffeomorphic; they consist of one and the same number of Liouville tori.

If at some point c a smooth curve γ meets a bifurcation diagram Σ, then the fiber B_a and B_b may be distinct. If, when moving, the point a punctures Σ (at the point c), then the fiber B_a undergoes, generally, a qualitative topological transformation, i.e., a surgery (Figure 87).

The general problem is *to describe the topological surgery of Liouville tori occurring at the moment when the point a intersects the bifurcation diagram* Σ.

The study of the indicated surgery makes it possible to describe the mechanism of modifications of the motion of an integrable Hamiltonian equation depending on fixed values of the integrals. Recall that integral trajectories of an integrable system determine rectilinear windings (almost periodic motion).

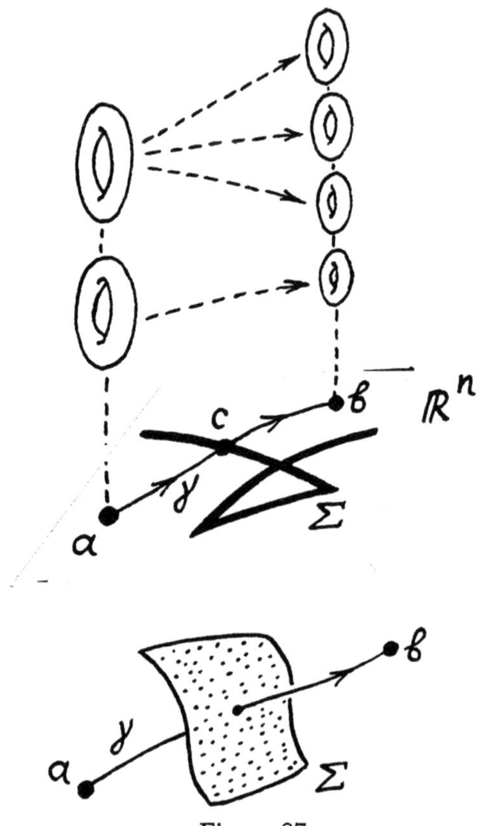

Figure 87

It is clear that there exist the following two cases:

$$1)\quad \dim\Sigma < n-1, \qquad 2)\quad \dim\Sigma = n-1.$$

In case 1 the diagram Σ does not separate $\mathbf{R}^n$, that is, any two regular points a and b (from $\mathbf{R}^n\backslash\Sigma$) are joined by a smooth curve $\gamma\subset\mathbf{R}^n\backslash\Sigma$. Compact non-singular fibers are therefore mutually diffeomorphic and they consist of one and the same number of Liouville tori.

Substantially more complicated is case 2. Here the diagram Σ, generally, separates $\mathbf{R}^n$ into several open non-intersecting domains. Each of them has, generally speaking, its own topology of the regular fibers B_a. This topology changes from domain to domain.

Thus, let $\dim\Sigma = n-1$. Examine a point c on Σ and investigate the surgery on Liouville tori when a smooth curve γ punctures the diagram Σ

at the point c. It suffices to examine only a small neighborhood $U = U(c)$ of the point c in $\mathbf{R}^n$.

We will investigate the case of "general position", that is, when the path γ punctures Σ transversally with non-zero velocity at the point c which lies on an $(n-1)$-dimensional smooth stratum of the diagram Σ. In other words, we will assume $U \cap \Sigma$ to be a smooth $(n-1)$-dimensional submanifold in $\mathbf{R}^n$. In the case of general position, one may assume the set $N \cap F^{-1}(U)$ of critical points to be a union of a finite number of smooth submanifolds in M stratified by the *rank* of dF. This means that N may be represented as a union of non-intersecting smooth submanifolds N_i on each of which the *rank* of dF is exactly equal to i (some of these submanifolds may, of course, be empty).

The notion of *general position* may also be specified as follows. Since in the neighborhood of the point c the set Σ is considered to be an $(n-1)$-dimensional submanifold, one may assume that in the neighborhood of a certain connected component B_c^0 of a singular fiber B_c *the last* $n-1$ *integrals* $f_2, \ldots, f_n$ *are independent*, and the first integral $f_1 = H$ (energy) becomes dependent on these integrals on the submanifold of critical points $L = N \cap B_c^0$. Indeed, let us restrict the mapping F to the submanifold $N \cap F^{-1}(U)$, which, according to the requirement of general position, is a union of a finite number of smooth submanifolds. Since the restriction of F to each stratum, including the maximal one, $N' \cap F^{-1}(U)$, is a smooth mapping of a smooth submanifold, it follows that $dF(x) : T_xN' \dot{\to} T_{F(x)}\Sigma$ is an epimorphism and $\operatorname{rank} dF(x) \geq n-1$ because $\dim U \cap \Sigma = n-1$. At the same time, since $x \in N$ is a critical point, it follows that $\operatorname{rank} dF(x) \leq n-1$. Consequently, $\operatorname{rank} dF(x) = n-1$. One may therefore assume (changing the basis in the set of integrals in case of necessity) that $f_2, \ldots, f_n$ are independent on the set B_c^0. Hence, the integral f_1 becomes dependent on these integrals on the set $L = N \cap B_c^0$.

22.2. Integral surfaces X^{n+1} in phase space M^{2n} as common level surfaces of the integrals $f_2, \ldots, f_n$. The behavior of the energy H and its Bott character on these surfaces. The Bott transformations of Liouville tori.

Now let us formulate the final definition of the surgery in general position of the Liouville torus. Fix the values of the last $n-1$ integrals $f_2, \ldots, f_n$

and examine the obtained $(n+1)$-dimensional surface X^{n+1}. Restricting the first integral = energy $f_1 = H$ to this surface, we obtain a smooth function f on the manifold X^{n+1}.

DEFINITION. *We will say that the surgery on Liouville tori T^n which form a non-singular fiber B_a is the* **surgery in general position** *if in the neighborhood of a modified torus T^n the integral surface X^{n+1} is a compact and non-singular submanifold, and the restriction of the energy function $f_1 = H$ to X^{n+1} is (in the same neighborhood) a Bott function (in the sense of Chapter 1).*

We will call the corresponding toral transformation *Bottian* (Bott transformations). It is clear that the integral surface X^{n+1} is the preimage of the one-dimensional interval (segment) in $\mathbf{R}^n$, namely: $X^{n+1} = F^{-1}(\gamma)$, where γ is a path in $\mathbf{R}^n$ (Figure 88). If we change the integrals $f_2, \ldots, f_n$, the surface X^{n+1} also changes. The energy H is fixed.

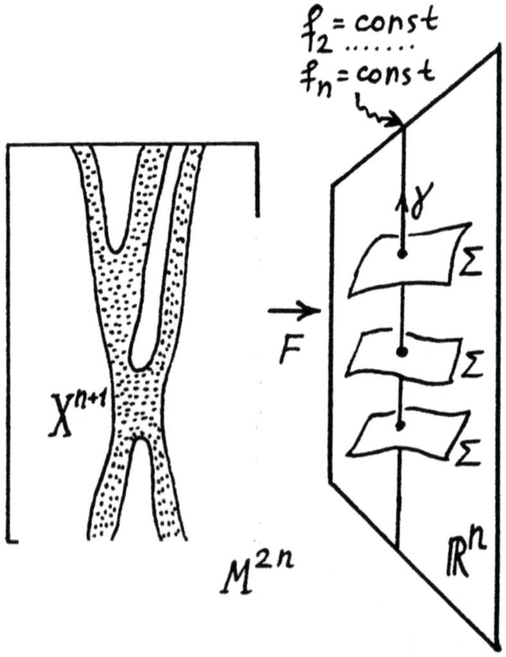

Figure 88

Let v be an integrable system of differential equations on M^{2n}. Fix the values of all last integrals:

$$f_2 = a_2, f_3 = a_3, \ldots, f_n = a_n,$$

(where $a = (a_1, \ldots, a_n)$ is a non-critical value in $\mathbf{R}^n$ for the momentum mapping F) and suppose that the obtained $(n+1)$-dimensional surface X^{n+1} is compact and non-singular, that is, the integrals $f_2, \ldots, f_n$ are independent on X^{n+1}. The integral surface X^{n+1} is an invariant submanifold of the system v. Varying the value of the energy H, we displace the Liouville torus T^n along the surface X^{n+1}. Sometimes there arise limit degeneracies, that is, a torus T^n transforms by some surgery (Figure 89).

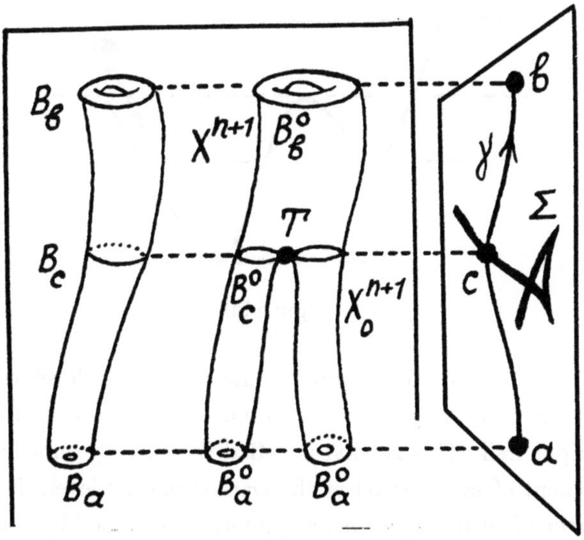

Figure 89

22.3. Mechanical interpretation.

Such limit degeneracies actually arise in concrete mechanical systems with dissipation.

For $n = 1$, the transformations $T^1 \to 2T^1$ and $T^1 \to 0$ can be observed in the problem of the motion of a heavy point in a "two-hump" well. Due to the small energy dissipation, the motion of a point in phase space proceeds along one-dimensional tori (that is, circles) which slightly evolve and finally, meet with a critical energy level, undergo surgery. If *small friction* is introduced into an integrable system, then to a first approximation one may think that the energy dissipation is modelled by a decrease in the energy value $f_1 = H$ and induces therefore, a slow evolution (drift) of the Liouville tori along the level surface X^{n+1}. Examine the motion of a ball in a well under the action of gravity [**53**]. See Figure 90. This motion (in the absence of frictional forces) may be completely described. If we introduce a small friction, we may assume the motion to be integrable, as before, on each sufficiently small time period. But as the time interval increases, the friction becomes progressively pronounced. As a result, the ball will rise at increasingly small height (Figure 90).

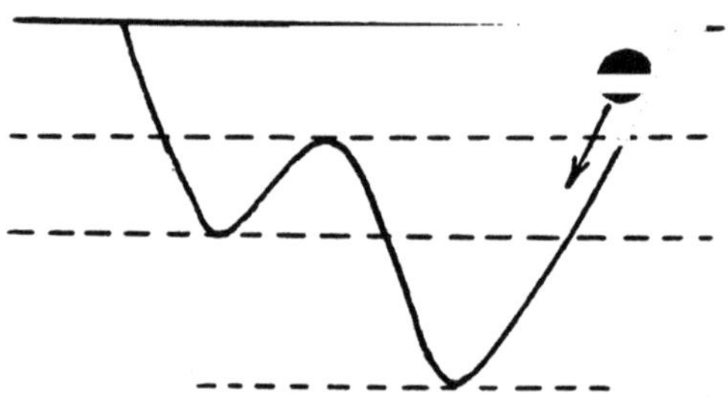

Figure 90

Finally, at a certain time moment the decreasing level will touch the saddle and the motion will change: the ball will find itself either in the left or in the right well. This is what is called the surgery on a Liouville torus at the moment of intersection of the critical energy level. It is clear that the character of such surgery is completely determined by the topology of the level surface X^{n+1}.

22.4. Classification of critical submanifolds of Bott Hamiltonians H on the integral surfaces X^{n+1}.

THEOREM 22.1 (See [**29**], [**19**]). *Let H be a smooth Bott Hamiltonian on the common level surface (= integral surface) X^{n+1}. Then the connected components of its critical submanifolds are as follows:*

a) *Torus T^n or torus T^{n-1} (= the local minimax of the energy H on X^{n+1}).*

b) *Torus T^{n-1} (the saddle critical submanifold of H).*

c) *The n-dimensional non-orientable manifold K_p^n which will be described below (= the local minimax of the energy H on X^{n+1}).*

Now we will describe the non-orientable critical submanifolds of the energy H on X.

22.5. Classification of non-orientable critical submanifolds of Bott Hamiltonians.

We have discovered non-orientable manifolds K_p^n, which are minima or maxima of a Bott Hamiltonian H on an integral surface X^{n+1}. In the present subsection we give a complete classification of such manifolds.

For each p we denote by G_p (for $p = 0, 1$) a group of transformations of a torus $T^n = \mathbf{R}^n/\mathbf{Z}^n$ generated by the involution

$$R_p(x) = \begin{cases} (-x_1, x_2 + \frac{1}{2}, x_3, x_4, \dots, x_n), & p = 0, \\ (x_2, x_1, x_3 + \frac{1}{2}, x_4, \dots, x_n), & p = 1, \end{cases}$$

where

$$x = (x_1, \dots, x_n) \in \mathbf{R}^n/\mathbf{Z}^n.$$

It is assumed here that $n \geq 2$ for $p = 0$ and $n \geq 3$ for $p = 1$. The group G_p acts on the torus T^n without fixed points, and consequently the factor set

$$K_p^n = T^n/G_p$$

is a smooth n-dimensional manifold. The transformation R_p changes orientation, and therefore the manifold K_p^n is non-orientable. The manifolds K_0^n and K_1^n are not homeomorphic because they have different homology groups, namely:

$$H_1(K_0^n, \mathbf{Z}) = \mathbf{Z}^{n-1} \oplus \mathbf{Z}_2 \text{ and } H_1(K_1^n, \mathbf{Z}) = \mathbf{Z}^{n-1}.$$

From the definition of the manifold K_p^n, it immediately follows that

$$K_0^n = K_0^2 \times T^{n-2}, \quad K_1^n = K_1^3 \times T^{n-3}.$$

Here K_0^2 is a usual Klein bottle and K_1^3 is its natural three-dimensional generalization [**19**].

THEOREM 22.2 (A. V. Brailov, A. T. Fomenko). *Let $f_1, \dots, f_n$ be an involutive set of smooth independent functions on a symplectic manifold M^{2n} and let $F : M^{2n} \to \mathbf{R}^n$ be the corresponding momentum mapping. Let $X^{n+1} = \{f_2 = a_2, \dots, f_n = a_n;\ a_i = \text{const}\}$ be an integral submanifold of the momentum mapping. Suppose that the restriction $f_1 \mid X^{n+1}$ is a Bott function and that L is a connected non-orientable critical manifold = the local minimax of the function f_1. Then the manifold L is diffeomorphic either to K_0^n (when $n \geq 2$) or to K_1^n (when $n \geq 3$).*

22.6. Five types of elementary "bifurcation manifolds".

Now we will consider five types of $(n+1)$-dimensional manifolds whose boundaries are tori.

1) Let us consider a "full torus" $D^2 \times T^{n-1}$, whose "axis" is a torus T^{n-1}, embedded in $\mathbf{R}^{n+1}$ in a standard way. Its boundary is a torus T^n. We will call $D^2 \times T^{n-1}$ a *dissipative full (solid) torus.* The reason for this terminology, which comes from mechanics, is given below.

2) The direct product $T^n \times D^1$ will be called *an oriented cylinder.* Its boundary consists of two tori T^n.

3) Let N^2 be a disk with two holes. The direct product $N^2 \times T^{n-1}$ will be called *an oriented saddle.* Its boundary consists of three tori T^n.

4) *Non-orientable saddle.* Over the torus T^{n-1} we consider all non-equivalent fiber bundles with the segment $D^1 = [-1, +1]$ as fiber. They are classified by the elements α of the homology group $H_1(T^{n-1}, \mathbf{Z}_2) = \mathbf{Z}_2 \oplus \mathbf{Z}_2 \oplus \cdots \oplus \mathbf{Z}_2$ ($n-1$ times). The space of the fiber bundle corresponding to the element α will be denoted by Y_α^n. It is clear that Y_α^n is an n-dimensional smooth manifold whose boundary consists of two tori T^{n-1} if $\alpha = 0$ and of one torus if $\alpha \neq 0$. We single out *a zero cross-section* and denote it by T^{n-1}. This cross-section is homeomorphic to the base, that is, to the torus. We have the fiber bundle

$$(*) \qquad Y_\alpha^n \xrightarrow{D^1} T^{n-1}.$$

Now consider a new fiber bundle, associated with the first one, of the form

$$(**) \qquad A_\alpha^{n+1} \xrightarrow{N^2} T^{n-1}$$

with a torus T^{n-1} as the base, and with fiber N^2. This fiber bundle is defined as follows. A disk with two holes is homotopy equivalent to a figure eight. On N^2 consider a line segment $D^1 = [-1, +1]$ which passes through the disk center and joins (after its continuation) the centers of the two removed disks (holes) (Figure 91). With each fiber bundle (*) one may associate the fiber bundle (**) by changing the fiber D^1 with the new fiber N^2.

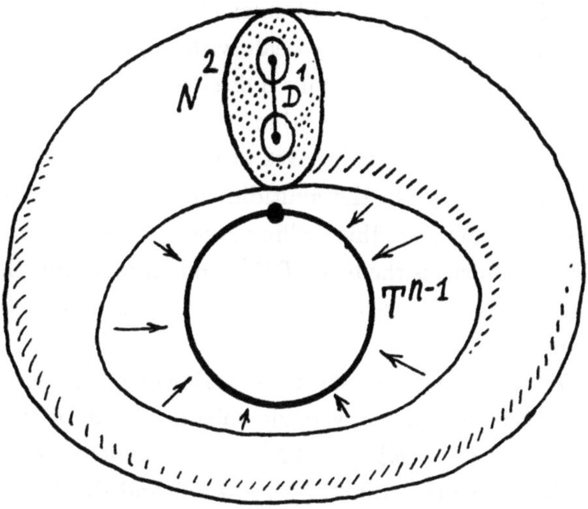

Figure 91

A particular case is the direct product $N^2 \times T^{n-1}$, that is, a manifold of type 3 (see above). It is obtained only if $\alpha = 0$. If $\alpha \neq 0$, then the corresponding fiber bundle A_α^{n+1} is non-trivial. Recall that in case $\alpha \neq 0$ the boundary of an n-dimensional manifold Y_α^n is one torus T^{n-1}. If $\alpha \neq 0$, then the $(n+1)$-dimensional manifold A_α^{n+1} will be called *non-oriented saddle*. The boundary of the manifold A_α^{n+1} consists of two tori T^n if $\alpha \neq 0$.

It is readily seen that for $\alpha \neq 0$ *all manifolds* A_α^{n+1} *are diffeomorphic to one another*, and therefore we will write them as follows:

$$A_\alpha^{n+1} = N^2 \tilde{\times} T^{n-1} \quad \text{(i.e., a skew product)}.$$

This follows from the fact that any non-intersecting closed curve on a torus may be taken as one of the basis generators on the torus. Thus, as in the four-dimensional case, we obtain *only two topologically different manifolds of the type* A_α^{n+1}, *namely:*

$$N^2 \times T^{n-1} \text{ and } N^2 \tilde{\times} T^{n-1}.$$

The first manifold has type 3 in our classification, the second has type 4. In other words, the manifold $A_{\alpha=0}^{n+1}$ is the orientable saddle: $A_0^{n+1} = N^2 \times T^{n-1}$.

Manifolds of type 4 admit a vivid description. For $n = 2$ we obtain the manifold A^3 which we have already dealt with in previous chapters. In the multidimensional case we should take a dissipative full torus (with a torus as the boundary) and drill there a "thin" full torus $D^2 \times T^{n-1}$ which goes (winds) twice around some of the axes of the large full torus. We obtain one of the manifolds A_α. In the multidimensional (as distinguished from the three-dimensional) case, this drilling is not unique because the axis of the dissipative full torus is the torus T^{n-1}, and on this torus there are $n-1$ independent cycles-circles (generators) along which the thin full torus can be wound twice.

5) Let $\pi : T^n \to K^n$ be a two-sheeted covering over a non-orientable manifold K^n. All such coverings π were classified: see [**19**] or the section 22.5 above. We have $\pi : T^n \to K_p^n$. By W_π^{n+1} we will denote the *cylinder of the mapping* π. It is clear that $\dim W_\pi^{n+1} = n+1$ and $\partial W_\pi^{n+1} = T^n$ (torus). We know from section 22.5 that only two different manifolds K_p^n really exist, namely: K_0^n and K_1^n. Let us denote the corresponding manifolds W_π^{n+1} by W_0^{n+1} and W_1^{n+1}.

STATEMENT 22.1. *The elementary manifolds of types 2, 4, and 5 are represented as the glueing of elementary manifolds of the types 1 and 3, i.e., of* $D^2 \times T^{n-1}$ *and* $N^2 \times T^{n-1}$. *In other words, from the topological point of view only the manifolds* $D^2 \times T^{n-1}$ *and* $N^2 \times T^{n-1}$ *are independent "elementary bricks".*

The details of this decomposition are in [**19**]. In the particular four-dimensional case we obtain the decompositions which are described above in the Chapter 1.

22.7. Five types of elementary bifurcations of Liouville tori.

Now we will describe five types of elementary surgeries on the torus T^n (Figure 92).

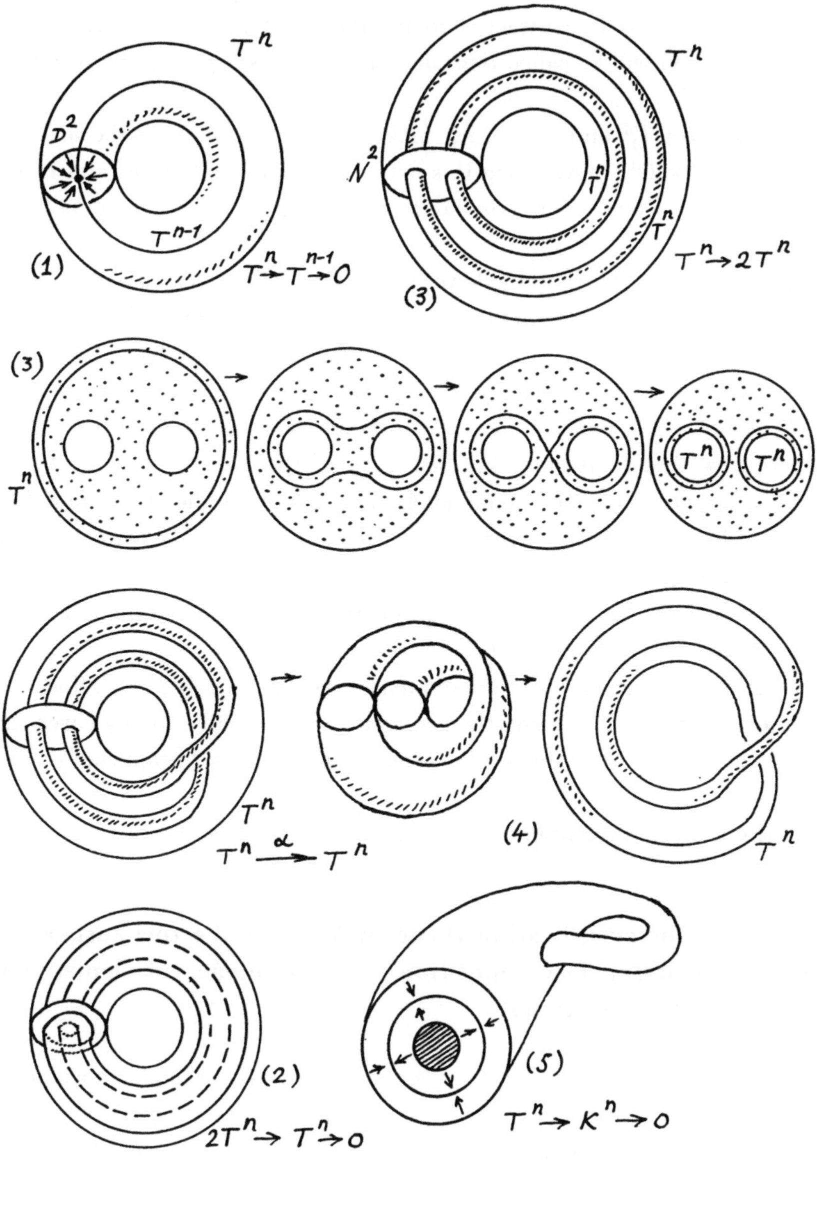

T^n
D^2
T^{n-1}
(1)
$T^n \to T^{n-1} \to 0$
T^n
N^2
T^n
T^n
$T^n \to 2T^n$
(3)
(3)
T^n
T^n
T^n
T^n
$T^n \xrightarrow{\alpha} T^n$
(4)
T^n
(2)
$2T^n \to T^n \to 0$
(5)
$T^n \to K^n \to 0$

Figure 92

1) A torus T^n is given as the boundary of a dissipative full torus $D^2 \times T^{n-1}$ and then it is contracted to its axis, that is, to the torus T^{n-1}. This operation will be called *limit degeneracy.* This type of surgery will be denoted by $T^n \to T^{n-1} \to 0$.

2) Two tori T_1^n and T_2^n, which make up the boundary of the *oriented cylinder* $T^n \times D^1$, move along this cylinder toward each other and *in the middle of the cylinder flow into one torus* T^n. The corresponding notation: $2T^n \to T^n \to 0$.

3) A torus T^n, which is the lower boundary of *the oriented saddle* $N^2 \times T^{n-1}$ ascends, and according to the topology of the manifold $N^2 \times T^{n-1}$ (see above) *splits into two tori* T_1^n and T_2^n. The corresponding notation: $T^n \to 2T^n$.

4) A torus T^n, which is one of the boundaries of the manifold A_α^{n+1}, where $\alpha \neq 0$ (for instance, the boundary of the interior thin full torus, Figure 92), goes "up" the manifold A_α, and in the middle of it is modified into one torus—the upper boundary of the manifold A_α. The corresponding notation: $T^n \xrightarrow{\alpha} T^n$. All such modifications are parametrized by non-zero elements $\alpha \in \mathbb{Z}_2^{n-1} = H_1(T^{n-1}, \mathbb{Z}_2)$, but let us recall that all manifolds A_α are diffeomorphic for all $\alpha \neq 0$.

5) Realize a torus T^n as the boundary of a manifold W_π^{n+1}. Deforming the torus inside W_π^{n+1} along the projection π, we finally doubly cover the manifold K_p^n with the torus T^n. After this the torus "disappears". The corresponding notation: $T^n \to K_p^n \to 0$.

Each of these transformations has an *inverse* transformation.

22.8. The classification theorem for Liouville tori surgery in general position. Decomposition in the sequence of elementary bifurcations of five types.

1) If $\dim \Sigma < n-1$, then all non-singular fibers B_a are diffeomorphic to one another.

2) Let $\dim \Sigma = n-1$. Let the non-degenerate Liouville torus T^n, carried away by variation of the value of the energy function $H = f_1$, move along a common non-singular level surface X^{n+1} of the last integrals $f_2, \ldots, f_n$. This is equivalent to the fact that a point $a = F(T^n)$ moves along a smooth line segment γ toward a bifurcation diagram Σ. Let, at a certain time moment, the torus T^n undergo a topological surgery, i.e., get into the critical

energy level. This happens if and only if the torus T^n meets on its way the critical points N of the momentum mapping $F : M^{2n} \to \mathbf{R}^n$ (that is, at a point c the path γ punctures transversally an $(n-1)$-dimensional stratum of the diagram Σ). Suppose that this is a surgery in general position.

THEOREM 22.3 (See [**29**], [**33**]). *All possible types of Liouville torus surgeries are exhausted by composition of the abovementioned types of canonical (elementary) surgeries 1, 2, 3, 4, 5. In reality, independent from the topological point of view are only the transformations 1 and 3, while types 2, 4, and 5 are compositions.*

COROLLARY. *If the Hamiltonian H is simple on X (that is, only one connected critical submanifold is located on each critical level of the Hamiltonian H), then every surgery of T^n coincides with one of the transformations 1, 2, 3, 4, 5 (or inverse to them).*

In the case of arbitrary complicated Hamiltonians H on X we always can deform (by small perturbations) the symplectic structure ω on M^{2n} and the integrals $f_2, \dots, f_n$ in such a way that the Hamiltonian H becomes simple on the perturbated surface X' close to X.

In the case of *simple* Hamiltonians H on X we have the following picture.

In case 1 (the surgery $T^n \to T^{n-1} \to 0$), as the energy H increases, the torus T^n at first transforms into a degenerate torus T^{n-1}, and then disappears altogether from the constant-energy surface $H = \text{const}$. In this case the torus T^n "does not pierce" the critical energy level, it "is decelerated" and disappears.

In case 2 (the modification $2T^n \to T^n \to 0$), as the energy H increases, the two tori T_1^n and T_2^n at first flow into one torus T^n and then disappear from the surface $H = \text{const}$. Here the torus T^n "does not pierce" the critical energy level either. The torus T^n, may, however, be regarded here as reflected by the critical energy level.

In case 3 (the modification $T^n \to 2T^n$), as the energy H increases, the torus T^n "pierces" the critical energy level and splits into two tori T_1^n and T_2^n on the surface $H = \text{const}$.

In case 4 (the modification $T^n \xrightarrow{\alpha} T^n$), as the energy H increases, the torus T^n also "pierces" the critical energy level and again transforms into the torus T^n (a non-trivial modification of double winding, see above).

In case 5 (the modification $T^n \to K_p^n \to 0$) the torus T^n doubly covers a non-orientable manifold K_p^n, after which it disappears from the surface

$H = \text{const}$.

Changing the direction of motion of the Liouville torus, we arrive at five inverse processes:

1*) Formation of a torus T^n from the torus T^{n-1}.

2*) Trivial formation of two tori T_1^n and T_2^n from one torus T^n.

3*) Non-trivial confluence of two tori T_1^n and T_2^n into one torus T^n.

4*) Non-trivial modification of a torus T^n into a torus T^n (see Figure 92).

5*) Formation of a torus T^n from a "multidimensional Klein bottle" K_p^n.

Some of the above-said types of surgery have already been discovered in concrete important mechanical systems. See for example the papers by Charlamov and Pogosyan [**21**]–[**23**], [**89**].

23. Classification of surgery of Liouville tori in integrable Hamiltonian equations. Classificaion of the integral manifold X^{n+1}.

23.1. Classification of integral manifolds.

Let us consider the phase space M^{2n} (smooth symplectic manifold) and let a system $v = \operatorname{sgrad} H$ be integrated by smooth independent commuting integrals $f_1, \ldots, f_n$ (where $f_1 = H$). Let X^{n+1} be any fixed compact non-singular common level surface (= integral surface) of the last $n-1$ integrals $f_2, \ldots, f_n$. Let the restriction of H on X be a Bott function (= Bott Hamiltonian). Let

m be the number of minimax critical tori T^{n-1} (for H on X),

p—the number of minimax critical tori T^n,

q—the number of critical saddle tori T^{n-1} with oriented separatrix diagram,

s—the number of critical saddle tori T^{n-1} with non-oriented separatrix diagram,

r_0—the number of minimax critical non-oriented n-dimensional "Klein bottles" K_0^n (of the first type),

r_1—the number of minimax non-oriented critical n-dimensional "Klein bottles" K_1^n (of the second type).

THEOREM 23.1 (See [**29**], [**33**]). *The integral surface X^{n+1} has the following form:*

$$X^{n+1} = m(D^2 \times T^{n-1}) + p(T^n \times D^1) + q(N^2 \times T^{n-1}) + s(N^2 \tilde{\times} T^{n-1}) + \\ + r_0 W_0^n + r_1 W_1^n,$$

that is, it results from glueing the boundary tori (by some diffeomorphisms) of the elementary bricks of five types 1, 2, 3, 4, 5. This decomposition is non-unique and is called Hamiltonian decomposition of the integral surface.

COROLLARY. *Each integral surface X^{n+1} has the form:*

$$X^{n+1} = \alpha(D^2 \times T^{n-1}) + \beta(N^2 \times T^{n-1})$$

for some integers α and β (topological decomposition of X).

This decomposition is non-unique.

As in the four-dimensional case, one could distinguish between orientable and non-orientable cases (of critical submanifolds). It turns out that every multidimensional integrable system can be "covered" by some new integrable system, where the new Bott Hamiltonian does not have critical non-oriented "Klein bottles" K_p^n.

STATEMENT 23.1 (See [**29**]). *If $(U(X^{n+1});\ \mathrm{sgrad}\, H;\ f_2, \ldots, f_n)$ is the integrable Hamiltonian system with a Bott Hamiltonian H which has the non-oriented critical minimax submanifolds K_p^n on the integral surface X^{n+1}, then it may always be doubly covered by a new Hamiltonian system*

$$(U'(X'^{n+1});\ \mathrm{sgrad}\, H';\ f_2', \ldots, f_n')$$

with a Hamiltonian H' on a covering surface X'^{n+1} which does not have non-oriented critical submanifolds. Here $U'(X'^{n+1})$ is a two-sheeted covering of the open neighborhood $U(X^{n+1})$ of the manifold X^{n+1}.

23.2. The manifold obtained by any composition of elementary bifurcations of Liouville tori is realized for a certain integrable Hamiltonian system on an appropriate symplectic manifold.

Above we have displayed five (or two) elementary Liouville torus transformations. It remains unclear whether any of their compositions is realized in a certain symplectic manifold on which an integrable system is given.

STATEMENT 23.2 (See [**19**]). *Let X^{n+1} be a smooth compact closed orientable manifold obtained by glueing an arbitrary number of elementary manifolds of type* 1, 2, 3, 4, 5 *(or only* 1 *and* 3*) through any diffeomorphisms of their boundary tori T^n. Then there always exists a smooth compact symplectic manifold M^{2n} with boundary diffeomorphic to a disconnected union of a certain number of manifolds $S^{n-1} \times T^n$ and an integrable Hamiltonian system with Bott Hamiltonian and with independent involutive smooth integrals $f_1, \ldots, f_n$ (where $f_1 = H$) satisfying: $X^{n+1} = \{x \in M^{2n} : f_2(x) = \cdots = f_n(x) = 0\}$.*

REMARK: There are obviously some topological obstacles to the realization of the manifold X^{n+1} in a compact *closed* symplectic manifold M^{2n}.

23.3. Classification of transformations of Liouville tori in the case of arbitrary Bott Hamiltonians (simple or complicated).

We call the Hamiltonian H *complicated* (on some integral surface X) if some critical energy level $\{H = \text{const}\}$ contains *more than one* connected critical submanifold. For the classification of surgery of Liouville tori in the case of *simple* Hamiltonians H, see above.

It is clear that we can define the *geometric skeleton* (P, K) of an integrable system v on each integral surface X^{n+1} in a way similar to the one described above for the four-dimensional case. This fact is based on the theorems of the section 22, where we describe the topological structure of all critical submanifolds of H and the structure of all "elementary bricks" 1, 2, 3, 4, 5. This skeleton is again the two-dimensional surface P with some graph K in it. We have the fibrations (with singular fibers):

$$\begin{array}{rcl} X^{n+1} = \Sigma_s X_s^{n+1} & \supset & \Sigma H^{-1}(c) \\ \downarrow\ T^{n-1}\ (\text{fiber}) & & \downarrow\ T^{n-1}\ (\text{fiber}) \\ (\text{2-surface}):\quad P^2 = \Sigma_s\ P_s^2 & \supset & \Sigma_s K_s = K\ (\text{graph}) \end{array}$$

All construcitons of the previous chapters are valid (with some natural corrections determined by the new multidimensional effects).

Consequently, we can define the letters-atoms, the words-molecules, framed word-molecules, the complexity (m, n) etc. To construct the numerical parameters r_i we need to consider the diffeomorphisms $\lambda : T^n \to T^n$ and corresponding matrices with integer coefficients. Then we must consider these matrices and define the sequence of parameters similar to r_i. We omit here the details.

Finally, *the framed word-molecules (with "multidimensional" rational and integer parameters) classify all integrable Hamiltonian systems v with Bott Hamiltonians on the integral surfaces X^{n+1}. This classification gives us the classification of all transformations of Liouville tori inside the surfaces X^{n+1} (when we change the value of the energy H, where H is an arbitrary Bott Hamiltonian, simple or complicated).*

There arises a natural and important question: how is the topological surgery of Liouville tori organized when the path γ (in $\mathbf{R}^n$) punctures the bifurcation diagram Σ in its *singular* points which do not lie on the interior of its $(n-1)$-dimensional strata? To put it differently, what is the surgery of special types (not in general position)?

A similar topological theory holds for Hamiltonian systems admitting *non-commutative integration* which we investigated in [**33**]. Almost all the assertions formulated in the present paper hold true, with the exception that n-dimensional Liouville tori are everywhere replaced by r-dimensional tori, where $r \leq n$.

24. The saddle critical torus T^{n-1} always realizes a non-trivial homology cycle in the critical level surface of the Hamiltonian H restricted to integral manifold X^{n+1}.

This section can be considered as technical, but really this theorem is very important in the whole theory and we give here its proof.

24.1. Proof of Theorem 22.1.

Let us introduce the following notation. For $c \in \Sigma$ pick a small neighborhood $U(c) \cap \Sigma$ of the point c on Σ to be a smooth $(n-1)$-dimensional submanifold. Let γ be a smooth path transversally puncturing Σ at the point c and joining two non-critical values a and b located on different sides of the hypersurface Σ. Consider a connected component B_c^0 of a singular fiber $B_c = F^{-1}(c)$, where F is the momentum mapping. Let $T = B_c^0 \cap N$ and let X_0^{n+1} be a connected component of the fiber $F^{-1}(U(c))$ lying between two closed non-singular fibers B_a and B_b (Figure 89). Since a and b are non-critical values, it follows that B_a and B_b are unions of Liouville tori. Let $B_a^0 = X_0^{n+1} \cap B_a$ and $B_b^0 = X_0^{n+1} \cap B_b$, that is, $\partial X_0^{n+1} = B_a^0 \cup B_b^0$. Denote by f the restriction of the first integral $f_1 = H$ to the surface $X_0^{n+1} = X_0$ (Figure 89).

LEMMA 24.1. *A point $x \in X_0$ is critical for the function f if and only if the first inegral f_1 depends (on M) on the last $n-1$ integrals $f_2, \ldots, f_n$.*

This fact is evident.

If $T = B_c^0$, then T is a common (singular) level surface of all n integrals $f_1, \ldots, f_n$. Level surfaces R, close to this surface, are non-singular compact Liouville tori. It is clear that R is the boundary of a tubular neighborhood V^{n+1} of the submanifold T in X_0. If $\dim T = k$, then R is homeomorphic to T^n and is fibered over T with fiber S^{n-k}. This may happen only if $n-k$ is equal to zero or to one, that is, if $\dim T = n$ or $\dim T = n-1$. If T_0^n is a connected component of T, then T_0^n is equal to T^n or to K_p^n. If $B_c^0 \neq T$, the consideration becomes more complicated. Clearly, in this case $\dim T < \dim B_c^0 = n$, that is, $\dim T \leq n-1$. From the conditions imposed on the integrals of the system it follows that on T the integrals $f_2, \ldots, f_n$ are independent (they are independent on the entire B_c^0).

Consequently, on T_0 there exists $n-1$ independent commuting vector fields $\operatorname{sgrad} f_i$, $2 \leq i \leq n$. As is known, from this it immediately follows that $T_0 = T^{n-1}$, and the result follows.

REMARK: If a critical torus has the dimension n, then it is either the set of local minima or of the local maxima of the energy H.

24.2. Toric handles in integrable systems.

DEFINITION. *The direct product $T^k \times D^\lambda \times D^{n+1-k-\lambda}$ will be called a toric handle of index λ and of degree of degeneracy k. The part of a handle boundary: $(T^k \times S^{\lambda-1}) \times D^{n+1-k-\lambda}$ will be called the base of a handle. The product $T^k \times S^{\lambda-1}$ will be called the base axis.*

Let us define the *operation of glueing a toric handle* to the boundary V^n of the submanifold $T^k \times S^{\lambda-1}$. Suppose that its tubular neighborhood is homeomorphic to the direct product $(T^k \times S^{\lambda-1}) \times D^{n+1-k-\lambda}$. One may remove this tubular neighborhood whose boundary is homeomorphic to $T^k \times S^{\lambda-1} \times S^{n-k-\lambda}$. On the other hand, the boundary of the base of the toric handle is also homeomorphic to the product $T^k \times S^{\lambda-1} \times S^{n-k-\lambda}$. Identifying this boundary with the boundary of the removed neighborhood will be called *a toric surgery on the boundary V^n*. For further purposes we may assume the smooth path $\gamma = \gamma(t) \subset \mathbf{R}^n$ to be modelled by a segment on the real axis $\mathbf{R}^1$ on which there lie three points: $a < c < b$, where c

is the critical value, a and b are close to c. Let us put $C_a = F^{-1}(t \leq a)$, $C_b = F^{-1}(T \leq b)$; then $C_a \subset C_b$. To say it differently, we may assume that $f : X_0^{n+1} \to \mathbf{R}^1$ and $C_a = (f \leq a)$, $C_b = (f \leq b)$, $B_c^0 = f^{-1}(c)$.

LEMMA 24.3. *Suppose that on a singular fiber B_c^0 there lies exactly one critical (saddle) torus T^{n-1}.*

1) *Let the separatrix diagram of T^{n-1} be orientable. Then C_b is obtained from C_a by glueing to the boundary B_a a toric handle of index 1 and of degree of degeneracy $n-1$, C_b being homotopy equivalent to C_a, to which the manifold $T^{n-1} \times D^1$ is glued along two non-intersecting tori $T_{1,a}^{n-1}$ and $T_{2,a}^{n-1}$.*

2) *Let the separatrix diagram of T^{n-1} be non-orientable. Then the set C_b is homotopy equivalent to the set C_a, to the boundary B_a^0 on which an n-dimensional manifold Y_α is glued along the torus T_a^{n-1}, the manifold Y_α has boundary T^{n-1} and is the fibration $Y_\alpha^n \xrightarrow{D^1} T^{n-1}$ corresponding to the non-zero element $\alpha \in \mathbf{Z}_2^{n-1} = H_1(T^{n-1}, \mathbf{Z}_2)$.*

See the proof in [**29**] or [**33**].

24.3. The theorem on the non-trivial cycle T^{n-1}.

THEOREM 24.1. *Let a torus T_a^{n-1} embed in some non-singular Liouville torus $T_a^n \subset B_a^0$ be either one of the bases of a toric handle (of index 1 and of degree of degeneracy $n-1$) or the boundary of a manifold Y_α^n (in case when the separatrix diagram is non-orientable). Then this torus T_a^{n-1} is always realized as one of the generators in the homology group $H_{n-1}(T_a^n, \mathbf{Z}) = \mathbf{Z}^{n-1}$. If both bases of the toric handle are glued to one and the same Liouville torus T_a^n, then the corresponding axes of these bases, that is, the tori $T_{1,a}^{n-1}$ and $T_{2,a}^{n-1}$, do not intersect and realize one and the same generator of the homology group $H_{n-1}(T_a^n, \mathbf{Z})$ and are, therefore, isotopic within the torus T_a^n.*

PROOF: Examine a critical saddle torus T^{n-1}. Theorem 22.1 implies that this torus is the orbit of an action of an Abelian subgroup $\mathbf{R}^{n-1}$ embedded into a group $\mathbf{R}^n$ generated by the fields $\operatorname{sgrad} f_i$, $1 \leq i \leq n$. The basis in the subgroup $\mathbf{R}^{n-1}$ is formed by the fields $\operatorname{sgrad} f_i$, $2 \leq i \leq n$. Fix this subgroup. Since the action of the group $\mathbf{R}^n$ (and $\mathbf{R}^{n-1}$) is defined on the entire manifold M^{2n}, we may examine the orbits of the group $\mathbf{R}^{n-1}$ close to the orbit T^{n-1}. Consider a non-singular Liouville torus T_a^n which is rather close to the fiber B_c^0 and on which the separatrix diagram cuts out some torus T_a^{n-1}. This torus is not, of course, the orbit of an action of the group $\mathbf{R}^{n-1}$ on the torus T_a^n. But, as we will show, the torus T_a^{n-1}, may be approximated by certain orbit of an action of the group $\mathbf{R}^{n-1}$. To this end, consider the element $\alpha \in H_{n-1}(T^{n-1}, \mathbf{Z}_2)$. From Lemma 24.3 we know that the torus T_a^{n-1} is one of the components of the boundary of the manifold Y_α^n glued to the torus T_a^n. If $\alpha \neq 0$, then $\partial Y_\alpha^n = T_a^{n-1}$; if $\alpha = 0$, then

$$\partial Y_\alpha^n = \partial(T^{n-1} \times D^1) = T_{1,a}^{n-1} \cup T_{2,a}^{n-1} \text{ and } T_{1,a}^{n-1} = T_a^{n-1}.$$

Fixing the element α determines a certain number k of generators in the critical torus T^{n-1}; going around it, a normal segment of the separatrix diagram changes its orientation. Let us single out these generators. In the orientable case, $k = 0$ because $\alpha = 0$.

Since the torus T^{n-1} is the orbit of an action of the group $\mathbf{R}^{n-1}$, then changing the generators in the group $\mathbf{R}^{n-1}$ (if necessary), one may always assume that in the non-orientable case ($k \geq 1$) among the fields $\operatorname{sgrad} f_i$, $2 \leq i \leq n$, there exist exactly k fields $\operatorname{sgrad} f_2, \ldots, \operatorname{sgrad} f_{k+1}$ such that a single detour along the orbits of the point $x \in T^{n-1}$, which are generated by the corresponding one-dimensional subgroups $\mathbf{R}_2^1, \ldots, \mathbf{R}_{k+1}^n$, changes orientation of the normal segment of the separatrix diagram. First, consider the orientable case, where $k = 0$. Then in the subgroup $\mathbf{R}^{n-1}$ one may single out an $(n-1)$-dimensional parallelepiped Π, i.e. a fundamental domain of the action of the group $\mathbf{R}^{n-1}$ on the torus T^{n-1}. This parallelepiped Π covers the whole torus, that is, the torus T^{n-1} is obtained through identification of opposite sides of this parallelepiped. Since the parallelepiped Π consists of transformations on M, we may consider the orbit of this parallelepiped when it is acting on a certain point

$$h \in T_{1,a}^{n-1} \subset T_a^n.$$

This orbit will not, of course, be a closed $(n-1)$-dimensional torus T_a^n. But since the point $h \in T_{1,a}^{n-1}$ is close to the point $x \in T^{n-1}$, one may assume that the orbit $\Pi(h)$ is an "almost-torus", that is, each of the generators of the parallelepiped Π becomes a segment whose ends are close on the torus T_a^n (i.e. we obtain an "almost-circle"). We will choose on the torus T^n some coordinates $\varphi_1, \ldots, \varphi_n$, according to the Liouville theorem. We make use of the fact that points of the parallelepiped Π are represented by symplectic transformations. Then in these coordinates the "almost-torus" $\Pi(h)$ is *a linear, totally geodesic submanifold*, maybe, with a non-empty boundary. Representing the torus T^n_a (in these coordinates) in the form of a standard cube with opposite sides identified, we obtain in it a plane Π', whose intersections with the opposite sides are $(n-2)$-dimensional subspaces which appear to be close after identification of the sides. It is clear that the plane Π' may be slightly turned so that it becomes (after factorization of the cube) a certain $(n-1)$-dimensional linear, totally geodesic torus T^{n-1}_* in the torus T^n_a.

It is clear that the torus T^{n-1}_* is close to the "almost-torus" $\Pi(h)$ and at the same time close to the torus $T^{n-1}_{1,a}$. This implies that these tori are isotopic.

Thus, we have proved that there exists a small isotopy of the torus $T^{n-1}_{1,a}$ in the torus T^n_a, which carries it into a linear torus. But in this case, the torus $T^{n-1}_{1,a}$ realizes a generator in the group $H_{n-1}(T^n_a, \mathbf{Z})$, as required. Thus, this implies the theorem in the orientable case.

Note that moving the plane Π slightly, we have obtained a new plane Π_* whose generators may already include the generator $\operatorname{sgrad} f_1$ which was excluded from the plane Π. It is clear that $T^{n-1}_* = \Pi_*(h)$.

Now consider the non-orientable case.

The reasoning will be more delicate. The point is that in this case the parallelepiped Π itself is insufficient. Indeed, the definition of a non-orientable separatrix diagram implies that the orbits

$$(\Pi \cap \mathbf{R}^1_2)h, \ldots, (\Pi \cap \mathbf{R}^1_{k+1})h$$

of generators $\mathbf{R}^1_2, \ldots, \mathbf{R}^1_{k+1}$ (corresponding to the fields $\operatorname{sgrad} f_i$, $2 \leq i \leq k+1$) are not "almost closed" trajectories on the torus T^n. We will denote the corresponding edges of the parallelepiped Π as Π_i, that is, $\Pi_i \cap \mathbf{R}^1_i$, $2 \leq i \leq k+1$.

When Π_i acts on the point h, the point has time only to make half the revolution on the torus T^n_a. In order that the point make almost the whole revolution, one should once again act on it by an edge of the parallelepiped Π_i. In other words, for the point h to make almost the whole revolution on the torus T^n_a, it needs to be subjected to the action of $2\Pi_i$, that is, the corresponding side of the parallelepiped Π should be doubled. So, we arrive at the following scheme.

One should double all the sides of the parallelepiped $\Pi_2, \dots, \Pi_{k+1}$. As a result, one obtains a new parallelepiped Π extended in k directions. Now act with this parallelepiped Π on the point h. As a result, obtain a certain orbit $\Pi(h)$. It is clear that now this orbit is represented (in the action-angle variables on a Liouville torus) by a linear plane which is "almost-closed" after factorization of the cube to the torus. Further arguments repeat those of the orientable case. This implies the theorem.

In **[33]** one can find also the version of this proof in the four-dimensional case. This version is more visual.

Appendix 1. The Structure of the Fundamental Group of Isoenergy Surfaces of Integrable Hamiltonian Systems

25. The fundamental group of an isoenergy surface given as manifold of class (Q), i.e. $Q = mI + pII + qIII + sIV + rV$.

Here we describe some results of the paper [**42**] (Fomenko, Zieschang).

25.1. Let us consider again the four-dimensional case. We will describe the representation of the fundamental group of the isoenergy three-manifold Q^3, which is given as a manifold in the class $(Q) = (H)$, that is, Q is decomposed in the sum of elementary three-manifolds of five types.

Assume that there are given m full (solid) tori (type I), p oriented cylinders (type II), q oriented saddles (type III), s non-oriented saddles (type IV) and r non-oriented cylinders (type V).

Then (see above) we have the following Hamiltonian decomposition: $Q = mI + pII + qIII + sIV + rV$.

Let us assume *that the oriented cylinders connect orientable or non-orientable saddles and that different boundary components of saddles are not connected without involving a cylinder. We assume that all full tori and non-oriented cylinders are glued to saddles.* Then:

(a) $3q + 2s = 2p + m + r$ if $q + s > 0$;

(b) if $q = s = 0$, then $p \leq 1$.

The following cases belong to (b):

$$p = 0,\ m = 2,\ r = 0; \qquad p = 0,\ m = 1,\ r = 1;$$
$$p = 0,\ m = 0,\ r = 2; \qquad p = 1,\ m = 0,\ r = 0.$$

Assume from now on that $q + s > 0$. Construct a graph Γ_0: the vertices are the saddles and the edges are the cylinders. Next we calculate the fundamental group of the space consisting of all saddles and cylinders. Finally we add the relations corresponding to the non-oriented cylinders and full tori.

25.2. Consider the *orientable saddles*: $S_1, \dots, S_q$. For $1 \leq i \leq q$ we have the following representation of the fundamental group $\pi_1(S_i)$:

(a) *generators*: $a_{i1}, a_{i2}, a_{i3}, h_i$;

(b) *defining relations*: $[a_{i1}, h_i] = [a_{i2}, h_i] = a_{i1}a_{i2}a_{i3} = 1$;

(c) *boundary components*: $\langle a_{ij}, h_i \mid [a_{ij}, h_i] = 1\rangle$, $1 \le j \le 3$.

Here $\langle \ \mid \ \rangle$ denotes the representation of the group by its generators and relations (first part – generators, second one – relations).

25.3. Let us denote the *non-orientable saddles* by $N_1, \dots, N_s$. Then for $1 \le i \le s$ the fundamental group $\pi_1(N_i)$ is given by the generators: $b_{i1}, b_{i2}, b_{i3}, k_i$, and defining relations:

$$b_{i1}b_{i2}b_{i3} = 1,\ \ k_i b_{i2} k_i^{-1} = b_{i1},\ \ k_i b_{i3} k_i^{-1} = b_{i2} b_{i3} b_{i2}^{-1},\ \ k_i b_{i1} k_i^{-1} = b_{i2}.$$

Therefore the group $\pi_1(N_i)$ has:

(a) *generators* b_i, k_i and

(b) *defining relations* $[k_i^2, b_i] = 1$, that is: $\pi_1(N_i) = \langle b_i, k_i \mid [k_i^2, b_i] = 1\rangle$, where $b_i = b_{i1}$.

(c) For the *boundary components* we have:

$$\langle b_{i1}, k_i^2 \mid [b_{i1}, k_i^2] = 1\rangle = \langle b_i, k_i^2 \mid [b_i, k_i^2] = 1\rangle,$$
$$\langle b_{i3}, b_{i2}^{-1} k_i \mid [b_{i3}, b_{i2}^{-1} k_i] = 1\rangle = \langle b_i^{-2}, b_i^{-1} k_i \mid [b_i^{-2}, b_i^{-2} k_i] = 1\rangle.$$

Notice that $b_{i3} = b_{i1}^{-2}$, $b_{i2} = b_{i1}$ in the group $\pi_1(N_i)$.

25.4. To calculate the fundamental group of the union $I \cup II \cup IV$ we define, for every cylinder Z_n:

(a) *a generator* w_n

and we fix some maximal tree $\Delta \subset \Gamma_0$. As a result we obtain the

(b) *relations*: $w_n = 1$ if $Z_n \in \Delta$.

Let p_1 be the number of cylinders which connect two boundary components of orientable saddles, p_2 the number of those that connect boundary components of non-orientable saddles, p_3 the number of cylinders connecting boundary components of orientable saddles with those of non-orientable saddles. The numbers of cylinders which belong to the tree Δ are denoted by p_{1+}, p_{2+}, p_{3+}, respectively. Now we have:

(c) $p = p_1 + p_2 + p_3$;

(d) $q + s - 1 = p_{1+} + p_{2+} + p_{3+}$.

25.5. If the cylinder Z_n connects the boundary components with parameters j and 1 of the orientable saddles with parameters i and k, respectively, then the following relations arise:

(a^I) $w_n^{-1}h_k w_n = h_i^{\alpha(n)} \times a_{ij}^{\beta(n)}$;

(b^I) $w_n^{-1}a_{kl}w_n = h_i^{\gamma(n)} \times a_{ij}^{\delta(n)}$.

If the cylinder Z_n connects two components of non-orientable saddles there arises the following collection of relations:

(a^{II}) $w_n^{-1}k_i^2 w_n = (k_j^2)^{\alpha(n)} \times b_j^{\beta(n)}$,

(b^{II}) $w_n^{-1}b_i w_n = (k_j^2)^{\gamma(n)} \times b_j^{\delta(n)}$; or

(a^{III}) $w_n^{-1}k_i^2 w_n = (b_j^{-2})^{\alpha(n)} \times (b_j^{-1}k_j)^{\beta(n)}$,

(b^{III}) $w_n^{-1}b_i w_n = (b_j^{-2})^{\gamma(n)} \times (b_j^{-1}k_j)^{\delta(n)}$; or

(a^{IV}) $w_n^{-1}b_i^{-2}w_n = (b_j^{-2})^{\alpha(n)} \times (b_j^{-1}k_j)^{\beta(n)}$.

(b^{IV}) $w_n^{-1}(b_i^{-1}k_i)w_n = (b_j^{-2})^{\gamma(n)} \times (b_j^{-1}k_j)^{\delta(n)}$.

If Z_n connects a non-orientable saddle with an orientable one then the following relations are obtained:

(a^V) $w_n^{-1}h_i w_n = (k_j^2)^{\alpha(n)} \times b_j^{\beta(n)}$,

(b^V) $w_n^{-1}a_{il}w_n = (k_j^2)^{\gamma(n)} \times b_j^{\delta(n)}$; or

(a^{VI}) $w_n^{-1}h_i w_n = (b_j^{-2})^{\alpha(n)} \times (b_j^{-1}k_j)^{\beta(n)}$,

(b^{VI}) $w_n^{-1}a_{il}w_n = (b_j^{-2})^{\gamma(n)} \times (b_j^{-1}k_j)^{\delta(n)}$.

Here we assume that $|\alpha(n)\delta(n) - \beta(n)\gamma(n)| = 1$. In the following we will write only (a), (b) when the exact form of the relation is not needed.

25.6. Consider a non-orientable cylinder K_n, $1 \leq n \leq r$. Then one obtains:

(a) *generators* c_n, d_n;

(b) *the defining relation* $c_n d_n c_n^{-1} d_n = 1$.

Hence,

$$\pi_1(K_n) = \langle c_n, d_n \mid c_n d_n c_n^{-1} d_n = 1 \rangle.$$

Furthermore, $\pi_1(\partial K_n) = \langle c_n^2, d_n \mid [c_n^2, d_n] = 1 \rangle$. Glueing K_n to an orientable saddle gives the relations:

(c^I) $h_i = (c_n^2)^{\alpha'(n)} \times d_n^{\beta'(n)}$

(d^I) $a_{il} = (c_n^2)^{\gamma'(n)} \times d_n^{\delta'(n)}$,

but glueing it to a non-orientable saddle gives:

(c^{II}) $k_i^2 = (c_n^2)^{\alpha'(n)} \times d_n^{\beta'(n)}$,

(d^{II}) $b_i = (c_n^2)^{\gamma'(n)} \times d_n^{\delta'(n)}$, or

(c^{III}) $b_i^2 = (c_n^2)^{\alpha'(n)} \times d_n^{\beta'(n)}$,

(d^{III}) $b_i^{-1}k_i = (c_n^2)^{\gamma'(n)} \times d_n^{\delta'(n)}$.

Here $|\alpha'(n)\delta'(n) - \beta'(n)\gamma'(n)| = 1$. We will write only (c) and (d) when the exact form of the relation is not required.

25.8. THEOREM 25.1 (See [**42**]). *If the isoenergy three-surface Q contains saddles, then the fundamental group $\pi_1(Q)$ has the generators:* 25.2(a), 25.3(a), 25.4(a), 25.6(a) *and defining relations:* 25.2(b), 25.3(b), 25.4(b), 25.6(b,c,d), 25.7(a).

In the case $q + s = 0$ the fundamental group $\pi_1(Q)$ is easily determined.

25.9. THEOREM 25.2 (See [**42**]). *Let $q + s = 0$.*

(a) *In the case $p = r = 0$, $m = 2$, we have: $\pi_1(Q) = \langle a, b \mid a^\pi b^\rho = 1,\ b = 1\rangle = \mathbf{Z}_\pi$ for $\pi, \rho \geq 0$, $gcd(\pi, \rho) = 1$ (i.e. are coprime); here $\mathbf{Z}_0 = \mathbf{Z}$. Then $Q = S^3$ for $\pi = 1$, $Q = S^1 \times S^2$ if $\pi = 0$, $Q = L(\pi, \rho)$ if $\pi > 1$ (where $L(\pi, \rho)$ is a lens space). The decomposition into two full tori is a Heegaard splitting of genus 1.*

(b) *For $p = 0$, $m = r = 1$, we have $\pi_1(Q) = \langle c, d \mid cdc^{-1}d = c^{2\pi} d^\rho = 1\rangle$, where $\pi, \rho \in \mathbf{Z}$, $gcd(\pi, \rho) = 1$ (i.e., are coprime).*

(c) *For $p = m = 0$, $r = 2$ we have:*

$$\pi_1(Q) = \langle c_1, d_1, c_2, d_2 \mid c_1 d_1 c_1^{-1} d_1 = c_2 d_2 c_2^{-1} d_2 = 1,$$
$$c_2^2 = c_1^{2\alpha'} d_1^{\beta'},\ d_2 = c_1^{2\gamma'} d_1^{\delta'}\rangle$$

where $\alpha', \beta', \gamma', \delta' \in \mathbf{Z}$ and $|\alpha'\delta' - \gamma'\beta'| = 1$.

(d) *For $m = r = 0$, $p = 1$ we have $\pi_1(Q) = \langle a, b, w \mid w^{-1}aw = a^\alpha b^\beta,\ w^{-1}bw = a^\gamma b^\delta\rangle$, where $\alpha, \beta, \gamma, \delta \in \mathbf{Z}$ and $|\alpha\delta - \beta\gamma| = 1$. Here the isoenergy surface Q is a bundle over a circle S^1 with fiber a torus T^2.*

26. Calculation of the fundamental groups of isoenergy surfaces Q given as manifolds of class (W).

STATEMENT 26.1 (See [**42**], [**48**]). *An arbitrary three-manifold Q from the class $(H) = (W)$ admits a decomposition into a system of maximal Seifert fiber spaces S_i. More precisely: there exists a system T of tori $T_1, \ldots, T_p$ in the manifold Q with the following properties. Let $S_1, \ldots, S_h$ be the connected components of the closure of $Q \backslash U(T)$, where $U(T)$ is a neighborhood of T. Then each S_j is a Seifert space. If the tori $\partial U(T_i)$, $1 \leq i \leq p$, are contained in S_j and S_k (Figure 93) (or both tori are in S_j), then the three-manifold $S_j \cup U(T_i) \cup S_k$ (or $S_j \cup U(T_i)$ respectively) does not admit a Seifert fibration.*

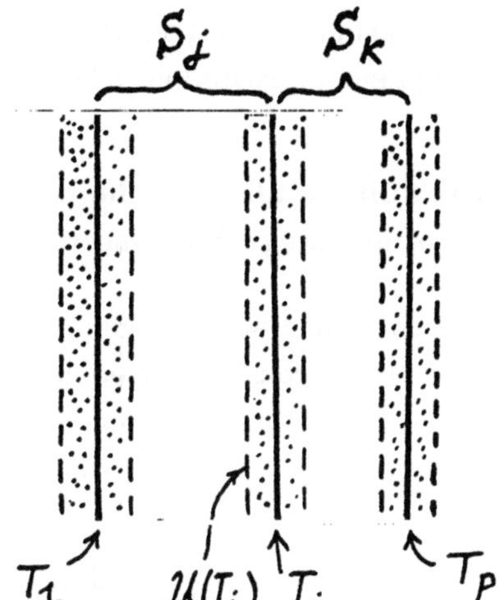

Figure 93

In this sense the pieces S_i are the maximal Seifert subspaces in the total space Q. Here we do not need any integrable system v on Q.

The fundamental group of a Seifert space is easily computed (see [**83**] for instance), and from it a representation of the group $\pi_1(Q)$ is obtained by applying the Seifert–Van Kampen theorem.

The detailed description of this group is in [**42**].

It follows from the theorem of W. Jaco and P. Shalen [**48**], that for irreducible three-manifolds Q its decomposition $Q = \Sigma_i S_i$ is uniquely defined (up to isotopy of Q). In other words, if $Q = \Sigma_i S_i$ and $Q = \Sigma_{i'} S_{i'}$ are two decompositions of Q into the union of maximal Seifert components, then there is some homeomorphism $\rho : Q \to Q$ such that ρ transforms the system T (of tori) into the system T' (of tori) and S_i into $S_{i'}$. Then it follows from our theory that there always exists an integrable Hamiltonian system v on Q such that the tori $T_1, \ldots, T_p$ (of the system T) are the Liouville tori for v. Consequently every two integrable Hamiltonian systems v and v' on Q have a common (up to homeomorphism) system of Liouville tori $T_1, \ldots, T_p$ (other tori can, of course, be different).

Appendix 2. Distribution of Complexity of Integrable Hamiltonian Systems on Different Isoenergy Manifolds

27. The form of the domain filled by the points (m, n), representing the complexity of all integrable systems on a fixed Q.

Let us consider the two-plane with integer lattice (m, n), where $m \geq 0$, $n \geq 0$ (Figure 94). Let us consider all points (m, n) representing the complexity of integrable Hamiltonian systems v on some fixed isoenergy three-manifold Q. We obtain some region on the table (Figure 94).

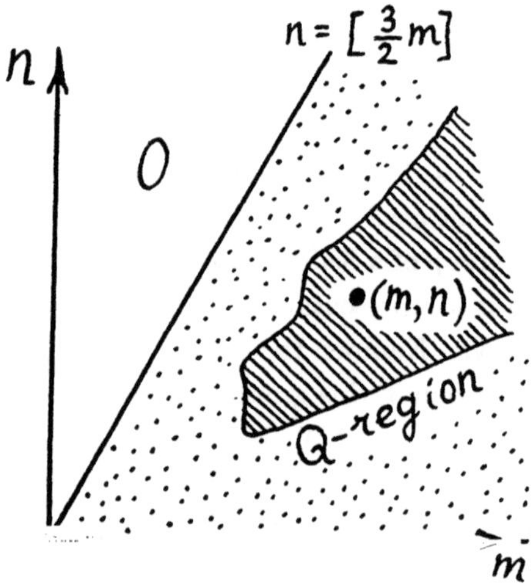

Figure 94

Important Problem: Is it possible to describe the form of this domain? What about the dependence of this domain on the manifold Q?

For example, we know from the analysis of concrete integrable physical systems (see above) that the trajectories of the systems (on the plane (m, n)) corresponding to the three-sphere S^3, move approximately along "the diagonal" $m \cong n$.

Let Q be a fixed isoenergy three-manifold from the class (H).

Definition. *The integer point (m, n) on the two-plane is called Q-point if there exists integrable Hamiltonian system v (with a Bott integral) such* *that its complexity is equal to* (m, n). *This point can be called also Q-admissible. The set of all Q-points for a fixed Q is called Q-region on the plane* $\mathbf{R}^2$.

PROBLEM: Calculate the *Q-region* for the most important three-manifolds Q arising in Hamiltonian mechanics. Among such manifolds are: the sphere S^3, the projective space $\mathbf{RP}^3$, the direct product $S^1 \times S^2$, the torus T^3. Let us describe some other important three-manifold Q_g arising in concrete Hamiltonian mechanics. Let us consider the two-dimensional disk with g holes N_g and its direct product with the circle: $N_g \times S^1$. This three-manifold has the boundary consisting of g tori. If $x \in \partial N_g$, then the circle $x \times S^1$ belongs to the boundary of $N_g \times S^1$. Let us identify this circle with a point (which can be denoted by $*_x$). Then each boundary torus transforms in a circle (consisting of all points $*_x$, where x belongs to one boundary circle in N_g). As a result we obtain some closed oriented three-manifold Q_g. This manifold appears in many important integrable systems (especially for $g = 2$) and is homeomorphic to a connected sum $(S^1 \times S^2) \# \ldots \# (S^1 \times S^2)$ (g times).

Let us consider the fundamental group $\pi_1(Q)$ for the manifold Q. Let us denote by $r(Q)$ the minimal number of generators in the group $\pi_1(Q)$. Then let us denote by $b(Q)$ the maximal number with the following property: there exists the epimorphism

$$\pi_1(Q) \to F_b,$$

where F_b is the free group with b generators. Thus, if Q is fixed, then we can calculate two important numbers: $r(Q)$ and $b(Q)$.

THEOREM 27.1 (Nguen Tyen Zung). *For any integrable system v on a given isoenergy three-manifold Q the following inequalities for its complexity (m, n) are valid:*

$$(*) \qquad m - 1 + b(Q) \geq n \geq \frac{1}{2}(m + 1 - r(Q)).$$

THEOREM 27.2 (Nguen Tyen Zung). *For the isoenergy surface $Q = S^3$ the corresponding S^3-region is precisely the following: $m - 1 \geq n \geq \frac{1}{2}(m + 1)$ plus the point $m = 2$, $n = 1$ (Figure 95). In particular, the boundary of this S^3-region coincides with the boundary of the region $(*)$ (see above).*

In this sense, the inequalities $(*)$ are exact for the sphere S^3. Let us consider the case of $\mathbf{RP}^3$.

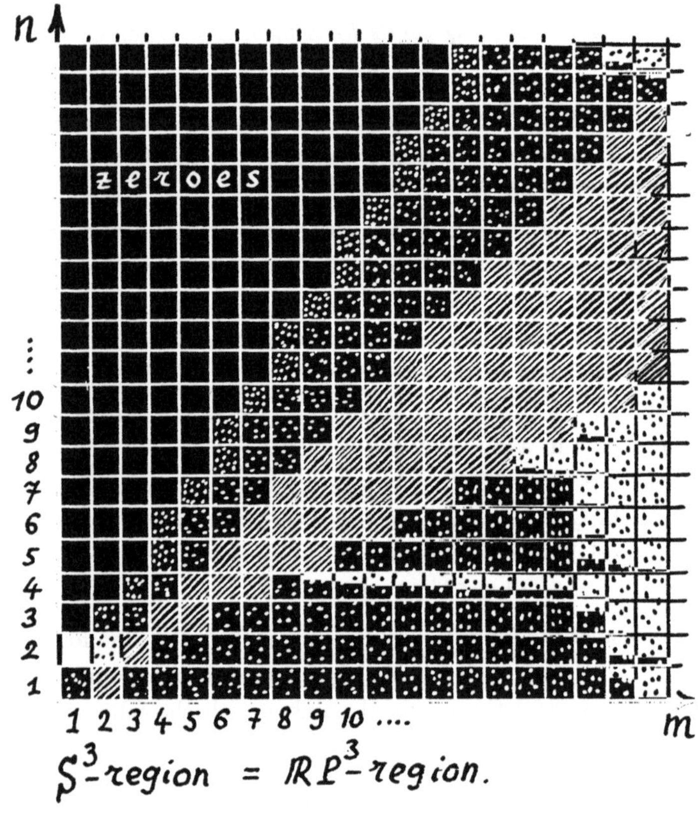

Figure 95

THEOREM 27.3 (Nguen Tyen Zung). *The* $\mathbf{RP}^3$*-region on the plane* (m, n) *has precisely the following form:* $m - 1 \geq n \geq \frac{1}{2}(m+1)$, *plus the point* $m = 2$, $n = 1$, *that is, it coincides with the* S^3*-region.*

In the case of $\mathbf{RP}^3$ the "general region" $(*)$ is larger than the real $\mathbf{RP}^3$-region, because $(*)$ has here the following form: $m - 1 \geq n \geq m/2$.

For $Q = S^1 \times S^2$ we have that the $(S^1 \times S^2)$-region is contained in the region: $m \geq n \geq m/2$. It can be proved that the T^3-region is contained in the region: $m \geq n \geq m/2$. This estimate is stronger than the estimate in the general Theorem 27.1.

The proof of these theorems is based on the careful analysis of the representation of the fundamental group $\pi_1(Q)$ obtained in [**42**].

Our theory permits us to classify all known integrable cases in Hamiltonian mechanics. If you discover some concrete case of integrability in physics or mechanics, you can calculate its topological invariant (= framed skeleton). After this you can discover the place of your system in our Table of all integrable systems. Consequently, you can compare the complexity of your integrable case with the complexity of all other known classical integrable cases. Thus, you can realize the place of your investigations in the general picture of integrable Hamiltonian mechanics. The location of your system in the Table shows to you whether this system is "close" to the collection of known systems, or you discovered quite an original integrable case, totally different from all others (if the corresponding point (m, n) is very "far" from the classical "integrable region").

In my opinion, the exact description of a "physical zone" in the Table of Integrable Systems (TIS) is an important problem.

We will call the table of non-marked molecules *molecular table.*

Appendix 3. Brief Survey of New Results

We include in this Appendix the recent results obtained by participants of the seminar "Modern Geometric Methods", headed by the author at the Department of Mechanics and Mathematics at Moscow State University in 1989–1990. The detailed papers will be published in the volume "Integrable Systems" upcoming in the new series "Advances in Soviet Mathematics", to be published by the American Mathematical Society.

1. In present paper a theory of topological classification of integrable Hamiltonian systems with *two degrees* of freedom was constructed. Although many facts of this theory remain valid for arbitrary dimension, a development of the theory for *multidimensional* systems was restrained by lack of an important object, namely, a topological invariant allowing to classify *multidimensional* systems. Such invariant was discovered by A.T. Fomenko. This invariant classifies the parts of Hamiltonian systems, "which have maximal dimension $2n$" on symplectic manifold M^{2n}. Some sets of zero measure and of dimension $2n-1$ and less can remain and these "small-dimensional strata" are missed by our invariant. Some new additional parameters are necessary for the description of topological behavior of integrable systems near such strata. A search of such parameters is an interesting problem.

2. A.V. Bolsinov discovered the effective algorithm for calculation of numerical marks on marked word-molecules corresponding to the integrable systems with two degrees of freedom. The goal of this work is the investigation of topological of integrable Hamiltonian system in the neighbourhoods of singular points of momentum mapping. A.V. Bolsinov investigated the structure of singular fiber of momentum mapping, containing one or several non-degenerate critical points of the momentum mapping. In particular, A.V. Bolsinov calculated the list of all different singular fibres containing 2 non-degenerate singular points. As the important application of this theory, A.V. Bolsinov calculated the total topological invariants of most interesting integrable systems connected with the dynamics of rigid body motion.

3. G.G. Okuneva proved the following theorem.

(a) There exists the class of smooth Hamiltonian systems of differential equations (with two degrees of freedom) which are integrable in Liouville sense only on one single three-dimensional isoenergy sufrace $Q^3_{h_o}$ in M^4 and are non-integrable on the all others isoenergy 3-surfaces Q^3_h, where $h \neq h_o$.

(b) Moreover, arbitrary integrable smooth Hamiltonian system v on M^4 can be perturbed in such a way that the perturbated system $\tilde{v}$ is close to the initial system v and the system $\tilde{v}$ remains integrable only on some one single isoenergy 3-surface $Q^3_{h_o}$ and became non-integrable on the all others isoenergy 3-surfaces Q^3_h.

4. Nguen Tyen Zung investigated the complexity of integrable Hamiltonian systems on the fixed isoenergy 3-manifold.

(a) He discovered the relation between the fundamental group of isoenergy 3-manifold Q and the complexity of integrable system.

(b) Nguen Tyen Zung described the form of Q-region (in the molecular table) for arbitrary isoenergy 3-manifold.

(c) He discovered the connection between the complexity of integrable systems and the complexity in Matveev's sense of the corresponding isoenergy 3-manifold.

5. A.T. Fomenko and Nguen Tyen Zung obtained the topological classification of all integrable Hamiltonian systems on the isoenergy 3-manifolds Q which are homeomorphic to the standard 3-sphere. In particular, we obtained the total classification of all links formed by stable periodic trajectories of integrable Hamiltonian systems on isoenergy 3-sphere S^3.

6. L.S. Polyakova investigated the topological structure and calculated the topological invariants of three important analogs of famous Toda lattice. These integrable Hamiltonian systems were discovered by V.V. Kozlov, D.V. Treschev and E.K. Sklyanin.

It turns out that the marked word-molecules in all three cases coincide. It follows from this fact that these three Hamiltonian systems are topologically equivalent and are equivalent to the following integrable cases in the dynamics of a rigid body: Euler, Lagrange, Kovalevskaya, Goryachev-Chaplygin (gyrostat) at the corresponding values of the energy H.

7. G.G. Okuneva discovered some applications of the principal constructions of our theory to Hamiltonian systems the integrals of which do not constitute a commutative subalgebra of the space $C^\infty(M^{2n})$. As an example, an integrable case of Lagrange of the motion of a rigid body with a fixed point and quadratic potential where a redundant non-commutative set of integrals and non-trivial (different from tori) invariant surfaces appear, is investigated. It turns out that integrable Lagrange case contains the remarkable 4-manifold which realizes the so called "strong bordism" between two integrable systems on two fixed isoenergy 3-manifolds Q_1^3 and Q_2^3. The theory of the "bordisms of integrable Hamiltonian systems" was developed by A.T. Fomenko in 1990 and then was continued by A.T. Fomenko, A.V. Bolsinov, Nguen Tyen Zung and L.S. Polyakova. This theory will be described in a separate paper.

8. E.N. Selivanova obtained the topological classificaiton of integrable geodesic flows of Liouville's metrics on two-dimensional torus T^2. Recall, that a geodesic flow of Riemannian manifold M with the metric $ds^2 = \sum g_{ij}\, dq_i\, dq_j$ is the Lagrangian system on tangent bundle T_*M with the Lagrange's function $L = \sum g_{ij}\, \dot{q}_i\, \dot{q}_j$. It is well known that any metric on T^2 with integrable geodesic flow (where integral is a quadratic function of the impulses) has the form $ds^2 = (f(x) + g(y))(dx^2 + dy^2)$. Here f and $h-$ any smooth functions. Let us consider the local minima of the graphs of these functions f and g on the circle. The picture of mutual positions of these minima can be described by some vector with integral coordinates. The main theorem of E.N. Selivanova states that two integrable geodesic flows are topologically equivalent if and only if their corresponding vectors are equivalent. This theorem allows us to enumerate all different topological types of integrable geodesic flows (with impuls-quadratic integrals) on 2-torus T^2.

9. The work of Nguen Tyen Zung and L.S. Polyakova develops the method of Selivanova's paper. Nguen Tyen Zung and L.S. Polyakova obtained a topological classification of integrable geodesic flows on a two-dimensional sphere S^2 with linear and quadratic on momenta additional integral. Classical examples of such systems are geodesic flows on standard ellipsoids and the Poisson's sphere. The works of E.N. Selivanova, Nguen Tyen Zung and Polyakova used some results of J.D. Birkhoff and V.N. Kolokoltzov. Let us correspond to each integrable Riemannian metric G_{ij} (of Liouville type) some vector $v(g_{ij})$ with integer coordinates (as in Selivanova's work). The main theorem of Nguen Tyen Zung and L.S. Polyakova states that two geodesic flows on sphere S^2 are topologically equivalent iff the corresponding integer vectors are equivalent. All geodesic flows on a sphere S^2, integrable by the aid of non-trivial quadratic on momenta integrals, can be represented on the molecular table of complexity as the following set:

$$\{(m,n) = (6,4) \text{ or } \frac{m}{2} + 3 \leq n \leq m - 2, \\ \text{where } m > 6, m = 4k+2, n = 2l; k, l - \text{ integer}\}.$$

10. Then A.T. Fomenko formulated the following conjecture, in a certain sense reversing the previous results.

CONJECTURE: Let $\{g_{ij}\}$ be an arbitrary Riemannian metric on sphere S^2 with the integrable geodesic flow. Then the corresponding region (= collection of integer points) on molecular table of complexity is contained in the region described above in the item 9. In other words, from the point of their complexity the geodesic flows with linear and quadratic additional integral exhaust "almost all" integrable geodesic flow on sphere S^2.

It was proved by V.V. Kalashnikov (junior), Nguen Tyen Zung and L.S. Polyakova that this conjecture is "almost correct". In other words, it turns out that the complexity of any integrable geodesic flow is "very close" to the complexity of some known integrable flow corresponding to the Liouville's type metric.

11. A.A. Oshemkov calculated the non-marked word-molecules (= topological invariants) for all known integrable cases in dynamics of heavy rigid body in $\mathbf{R}^3$. On the base of this information Oshemkov obtained the rough topological classification of integrable systems of described type. Several interesting cases of integrability for superfluid ^{3}He were investigated by B. Kruglikov.

References

1. Abraham, R. and Marsden, J. E., "Foundations of Mechanics," Second Ed. Benjamin/Cummings, 1978.
2. Adler, M. and Van Moerbeke, P., *Completely integrable systems, Euclidean Lie algebras and curves*, Adv. Math., Vol. 38, No. 3 (1980), 267–317.
3. ———, *The algebraic integrability of geodesic flow on SO(4)*, Invent. Math. **67** (1982), 297–331.
4. ———, *Linearization of Hamiltonian systems, Jacobi varieties and representation theory*, Adv. Math. **38** (1980), 318–379.
5. Anosov, D., *On typical properties of closed geodesics*, Izvestiya Akad. Nauk SSSR, Vol. 46, No. 4 (1983), 675–709.
6. Arnold, V. I., "Mathematical Methods of Classical Mechanics," Springer-Verlag, New York, 1978.
7. ———, *Hamiltonian properties of the Euler equations of the dynamics of a rigid body in an ideal liquid*, Uspekhi Matem. Nauk., Vol. 26 (1969), 225–226.
8. Asimov, D., *Round handles and non-singular Morse–Smale flows*, Ann. of Math., Vol. 102, No. 1 (1975), 41–54.
9. Atiyah, M. F., *Convexity and commuting Hamiltonians*, Bull. London Math. Soc. **14** (1982), 1–15.
10. Bogoyavlensky, O. I., *On exact function on the manifolds*, Matem. Zametki., Vol. 8, No. 1 (1970), 77–83.
11. ———, *Integrable Euler equations on Lie algebras, which arise in problems of mathematical physics*, Izvestiya Akad. Nauk SSSR, Vol. 48, No. 5 (1984), 883–893.
12. ———, *Some integrable cases of Euler equations*, Dokl. Akad. Nauk SSSR, Vol. 287, No. 5 (1986), 1105–1108.
13. Bolsinov, A. V., *Involutive sets of functions on dual spaces of Lie algebras of the type $G +_\phi V$*, Uspekhi Matem. Nauk, Vol. 42, No. 6 (1987), 183–184.
14. ———, *Criterion of the completeness of involutive sets of functions, which are constructed by argument shift method*, Dokl. Akad. Nauk SSSR, Vol. 301, No. 5 (1988), 1037–1040.
15. Bolsinov, A. V.; Fomenko, A. T. and Matveev, S. V., *Topological classification and arrangement with respect to complexity of integrable Hamiltonian systems of differential equations with two degrees of freedom. The list of all integrable systems of low complexity*, Uspekhi Matem. Nauk, Vol. 45, No. 2 (1990), 49–77.
16. Bott, R., *Non-degenerate critical manifolds*, Ann. of Math. **60** (1954), 248–261.
17. Brailov, A. V., *Several cases of complete integrability of Euler equations and applications*, Dokl. Akad. Nauk SSSR, Vol. 268, No. 5 (1983), 1043–1046.
18. ———, *Complete integrability of some geodesic flows and integrable systems with non-commuting integrals*, Dokl. Akad. Nauk SSSR, Vol. 271, No. 2 (1983), 273–276.
19. Brailov, A. V. and Fomenko, A. T., *The topology of integral submanifolds of completely integrable Hamiltonian systems*, Matematicheskii Sbornik, Vol. 133, No. 3 (1987), 375–385. English translation: Math. USSR Sbornik, Vol. 62, No. 2 (1989), 373–383.
20. Channell, P. J. and Scovel, C., *Symplectic integration of Hamiltonian systems*, Preprint, Los Alamos (1988), LA-UR-88-1828, 1–47, Los Alamos National Laboratory, USA.
21. Charlamov, M. P., *Topological analysis of integrable cases of V. A. Steklov*, Dokl. Akad. Nauk SSSR, Vol. 273, No. 6 (1983), 1322–1325.

22. ————, "Topological Analysis of Integrable Cases in Dynamics of Rigid Body," Leningrad Univ. Press, Leningrad, 1988.
23. ————, *Topological analysis of classical integrable systems in dynamics of rigid body*, Dokl. Akad. Nauk SSSR, Vol. 273, No. 6 (1983), 1322–1325.
24. Dao Chong Thi, *Integrability of Euler equations on homogeneous symplectic manifolds*, Matem. Sbornik, Vol. 106, No. 2 (1978), 154–161.
25. Dubrovin, B. A.; Fomenko, A. T. and Novikov, S. P., "Modern Geometry," Springer-Verlag, Part 1, GTM 93, 1984; Part 2, GTM 104, 1985; Part 3 (in print).
26. Flaschka, H., *The Toda lattice*, I, Phys. Rev. B9 (1974), 1924–1925.
27. ————, *On the Toda lattice*, II, Progr. Theor. Phys. **51** (1974), 703–716.
28. Fomenko, A. T., *Morse theory of integrable Hamiltonian systems*, Dokl. Akad. Nauk SSSR, Vol. 287, No. 5 (1986), 1071–1075. English translation: Soviet Math. Dokl., Vol. 33, No. 2 (1986), 502–506.
29. ————, *The topology of surfaces of constant energy in integrable Hamiltonian systems, and obstructions to integrability*, Izvestiya Akad. Nauk SSSR, Vol. 50, No. 6 (1986), 1276–1307. English translation: Math. USSR Izvestiya, Vol. 29, No. 3 (1987), 629–658.
30. ————, *Topological invariants of Hamiltonian systems integrable in Liouville sense*, Functional Anal. and its Appl., Vol. 22, No. 4 (1988), 38–51. See corresponding English translation.
31. ————, *Symplectic topology of completely integrable Hamiltonian systems*, Uspekhi Matem. Nauk, Vol. 44, No. 1(265) (1989), 145–173. See corresponding English translation.
32. ————, "Symplectic Geometry. Methods and Applications," Moscow Univ. Press, Moscow, 1988. English translation: Gordon and Breach, 1988, Advanced Studies in Contemporary Mathematics, Vol. 5.
33. ————, "Integrability and Non-integrability in Geometry and Mechanics," Kluwer Academic Publishers, 1988.
34. ————, *Qualitative geometrical theory of integrable systems. Classification of isoenergetic surfaces and bifurcation of Liouville tori at the critical energy values*, Lecture Notes in Math., Springer-Verlag **1334** (1988), 221–245.
35. ————, *New topological invariants of integrable Hamiltonians*, Baku International Topological Conference, Abstracts, Part 2, Baku, USSR (1987), p. 316.
36. ————, "Differential Geometry and Topology," Plenum, New York, 1987.
37. Fomenko, A. T. and Fuks, D. B., "Course of Homotopic Topology," Nauka, Moscow, 1989.
38. Fomenko, A. T. and Sharko, V. V., *Round exact Morse functions, Morse type inequalities and integrals of Hamiltonian systems*, in "Global Analysis and Non-linear Equations," Voronez Univ. Press, Voronez, 1988, pp. 92–107. English translation: Lectures Notes in Math., Springer-Verlag.
39. Fomenko, A. T. and Trofimov, V. V., "Integrable Systems on Lie Algebras and Symmetric Spaces," Gordon and Breach, Advanced Studies in Contemporary Mathematics, Vol. 2, 1988.
40. ————, *Non-invariant symplectic group structures and Hamiltonian flows on symmetric spaces*, Selecta Mathematica Sovietica, Vol. 7, No. 4 (1988), 355–414.
41. Fomenko, A. T. and Zieschang, H., *On the topology of three-dimensional manifolds arising in Hamiltonian mechanics*, Dokl. Akad. Nauk SSSR, Vol. 294, No. 2 (1987), 283–287. English translation: Soviet Math. Dokl., Vol. 35, No. 3 (1987), 529–534 (Preprint published in 1985, Preprint No. 55/1985, Ruhr-Universitat Bochum, Germany, 1–10).

42. ________________, *On typical topological properties of integrable Hamiltonian systems*, Izvestiya Akad. Nauk SSSR, Vol. 52, No. 2 (1988), 378–407. English translation: (Preprint published in 1987. Preprint No. 92/1987, Ruhr-Universitat Bochum, Germany, 1–47).
43. ________________, *Criterion of topological equivalence of integrable Hamiltonian systems with two degrees of freedom*, Izvestiya Akad. Nauk SSSR, Vol. 54, No. 3, (1990), 546–575. (Preprint published in 1988. Preprint: Topological classification of integrable Hamiltonian systems, Institut des Hautes Etudes Scientifiques **35**, December, 1988, IHES/M/88/62, 1–45).
44. Frunks, J., *The periodic structure of non-singular Morse–Smale flows*, Comment. Math. Helv., Vol. 53, No. 2 (1978), 279–294.
45. Gel'fand, M. and Dorfman, I. Ya., *Hamiltonian operators and related algebraic structures*, Func. Anal. and its Appl., Vol. 13, No. 4 (1979), 13–30.
46. Guillemin, V. and Sternberg, S., "Symplectic Techniques in Physics," Cambridge University Press, 1985.
47. ________________, *Convexity properties of the moment mapping*, Invent. Math., Vol. 67, No. 3 (1982), 491–513.
48. Jaco, W. and Shalen, P., *Surface homomorphisms and periodicity*, Topology, Vol. 16, No. 4 (1977), 347–367.
49. Klingenberg, W., "Lectures on Closed Geodesics," Springer-Verlag, Berlin, 1978.
50. Kostant, B., *The solution to a generalized Toda lattice and representation theory*, Adv. in Math., Vol. 34, No. 3 (1980), 195–338.
51. __________, *Poisson commutativity and the generalized periodic Toda lattice*, Lect. Notes in Math. **905** (1982), 12–28.
52. Kozlov, V. V., *Integrability and non-integrability in Hamiltonian mechanics*, Uspekhi Matem. Nauk, Vol. 38, No. 1 (1983), 3–67.
53. ___________, "Methods of Qualitative Analysis of Rigid-Body Dynamics," Moscow Univ. Press, Moscow, 1980.
54. Kronrod, A. S., *About functions of two variables*, Uspekhi Matem. Nauk, Vol. 5, No. 1 (1950), 24–134.
55. Kupershmidt, B. A. and Wilson, G., *Modifying Lax equations and the dynamical systems of Calogero type*, Comm. Pure Appl. Math., Vol. 31, No. 4 (1978), 481–507.
56. Kupershmidt, B. A. and Ratiu, T., *Canonical maps between semidirect products with applications to elasticity and superfluids*, Comm. Math. Phys. **90** (1983), 235–250.
57. Lacomba, E., *Mechanical systems with symmetry on homogeneous space*, Trans. Amer. Math. Soc. **185** (1973), 477–491.
58. Landau, L. D. and Lifshitz, E. M., "Mechanics," Pergamon, 1960.
59. Le Hong Van and Fomenko, A. T., *Lagrangian manifolds and Maslov index in minimal surface theory*, Dokl. Akad. Nauk SSSR, Vol. 299, No. 1. See corresponding English translation.
60. Lerman, L. M. and Umansky, Ya. L., *Structure of Poisson actions on four-dimensional symplectic manifolds*, in "Methods of Qualitative Theory of Differential Equations," Gorjky Univ. Press, Gorjky, 1982, pp. 3–19.
61. ________________, *Integrable Hamiltonian systems and Poisson actions*, in "Methods of Qualitative Theory of Differential Equations," Gorjky Univ. Press, Gorjky, 1984, pp. 126–139.
62. Lindon, R. and Schupp, P., "Combinatorial Group Theory," Springer-Verlag, Berlin, Heidelberg, New York, 1977.
63. Manin, Yu. I., *Matrix solutions and fiber bundles over curves with singularities*, Func. Anal. and its Appl., Vol. 12, No. 4 (1978), 53–63.

64. Marsden, J. and Weinstein, A., *Reduction of symplectic manifolds with symmetry*, Reports on Math. Phys., Vol. 5, No. 1 (1974), 121–130.
65. Marsden, J., Ratiu, T. and Weinstein, A., *Semidirect products and reduction in mechanics*, Trans. Amer. Math. Soc., Vol. 281, No. 1 (1984), 147–178.
66. Maslov, V. P., "Perturbation Theory and Asymptotic Methods," Nauka, Moscow, 1965.
67. Matveev, S. V., *Additive complexity and Haken's method in topology of three-dimensional manifolds*, Ukrainsky Matem. J., Vol. 41, No. 9 (1989), 28–36.
68. ——————, *Generalized surgery of three-dimensional manifolds and representations of homology spheres*, Matem. Zametki., Vol. 42, No. 2 (1982), 268–277.
69. ——————, *One procedure of defining three-manifolds*, Vestnik MGU (Moscow Univ.), Vol. 30, No. 2 (1975), 11–20.
70. Matveev, S. V.; Fomenko, A. T. and Sharko, V. V., *Round Morse functions and isoenergy level surfaces of integrable Hamiltonian systems*, Matematicheskii Sbornik, Vol. 135(177), No. 3 (1988), 325–345. See corresponding English translation. (Preprint published in 1986. Preprint 86, 76, Kiev, Institute Matematiki Akad. Nauk Ukr. SSR (1986), 1–32).
71. Matveev, S. V. and Fomenko, A. T., *Morse type theory for integrable Hamiltonian systems with tame integrals*, Matematicheskii Zametki., Vol. 42, No. 5 (1988), 663–671.
72. ——————, *Constant energy surfaces of Hamiltonian systems, enumeration of three-dimensional manifolds in increasing order of complexity, and computation of volumes of closed hyperbolic manifolds*, Uspekhi Matem. Nauk, Vol. 43, No. 1(259) (1988), 5–22. English translation: Russian Math. Surveys, Vol. 43, No. 1, 3–24.
73. McKean, H., *Integrable systems and algebraic curves*, in "Global Analysis," Lecture Notes in Math., **755**, Springer-Verlag, New York, 1976, pp. 83–200.
74. Milnor, J., *Hyperbolic geometry: the first 150 years*, Bull. of the Amer. Math. Soc. **6** (1982), 9–24.
75. Miyoshi, S., *Foliated round surgery of codimension one foliated manifolds*, Topology, Vol. 21, No. 3 (1982), 245–262.
76. Morgan, J., *Non-singular Morse–Smale flows and three-dimensional manifolds*, Topology, Vol. 18, No. 1 (1979), 41–54.
77. Moser, J., *Various aspects of integrable Hamiltonian systems*, in "Dynamical Systems," Prog. in Math. **8**, Birkhauser, Boston, Cambridge, Mass., 1980.
78. ——————, *Three integrable Hamiltonian systems connected with isospectral deformations*, Adv. in Math. **16** (1975), 197–200.
79. Novikov, S. P., *Periodic Korteweg-de Vries problem*, Func. Anal. and its Appl. **8** (1974), 54–66.
80. ——————, *Hamiltonian formalism and a multivalued analogue of the Morse theory*, Uspekhi Matem. Nauk, Vol. 37, No. 5 (1982), 3–49.
81. Novikov, S. P. and Shubin, M. A., *Morse inequalities and Neumann* Π_1*-factors*, Dokl. Akad. Nauk, Vol. 289, No. 2 (1986), 289–292.
82. Orlik, P., *Seifert manifolds*, Lecture Notes in Math., **291** (1972), Springer-Verlag.
83. Orlik, P.; Vogt, E. and Zieschang, H., *Zur Topologie gefaserter dreidimensionalen Mannigfaltigkeiten*, Topology, Vol. 6, No. 1 (1967), 49–65.
84. Oshemkov, A. A., *Topology of isoenergy surfaces and bifurcation diagrams of integrable case of the dynamic of the rigid body on so*(4), Uspekhi Matem. Nauk, Vol. 42, No. 6 (1987), 199–200.
85. ——————, *Bott integrals of some integrable Hamiltonian systems*, in "Geometry, Differential Equations and Mechanics," Moscow Univ. Press, Moscow, 1986, pp. 115–117.

86. ________, *The phase topology of some integrable Hamiltonian systems on* $so(4)$, Baku International Topological Conference, Abstracts, Part 2, Baku (1987), p. 230.
87. ________, *Description of isoenergy surfaces of some integrable Hamiltonian systems with two degrees of freedom*, in "Proc. Semin. Vector. Tens. Analysis," Moscow Univ. Press, Moscow, 1988, pp. 122–132.
88. Perelomov, A. M., *Lax representation for the systems of S. Kovalevskaya type*, Comm. Math. Phys. **81** (1981), 239–241.
89. Pogosyan, T. I. and Charlamov, M. P., *Bifurcation set and integral manifolds in the problem of the motion of a rigid body in linear forces field*, J. of Appl. Math. and Mech., Vol. 43, No. 3 (1979), 419–428.
90. Rais, M., *L'indice des produits semi-directs* $E_\rho \times G$, C. R. Acad. Sci. Paris, Ser. A, Vol. 287, No. 4 (1978), 195–197.
91. Ratiu, T., *The C. Neumann problem as a completely integrable system on an adjoint orbit*, Trans. of the Amer. Math. Soc., Vol. 264, No. 2 (1981), 321–332.
92. ________, *Euler-Poisson equations on Lie algebras and the n-dimensional heavy rigid body*, Amer. J. Math., Vol. 103, No. 3 (1982).
93. Ratiu, T. and Van Moerbeke, P., *The Lagrange rigid body motion*, Annales de l'Institut Fourier, Vol. 32, No. 1 (1982), 211–234.
94. Smale, S., *Topology and Mechanics*, I, Invent. Math. **10** (1970).
95. Sharko, V. V., *Minimal Morse functions*, Trudy Matem. Inst., V. A. Steklova **154** (1983), 306–311.
96. Sternberg, S., "Lectures on Differential Geometry," Prentice Hall, Englewood Cliffs, New Jersey, 1964.
97. Tatarinov, Ya. V., "Lectures on Classical Mechanics," Moscow Univ. Press, Moscow, 1984.
98. Thurston, W. B., *Existence of codimension one foliations*, Ann. of Math., Vol. 104, No. 2 (1976), 249–268.
99. ________, *The geometry and topology of three-manifolds*, Preprint (1981).
100. ________, *Three-dimensional manifolds, Kleinian groups and hyperbolic geometry*, Bull. Amer. Math. Soc., Vol. 6, No. 3 (1982), 357–381.
101. Trofimov, V. V., "Introduction to the Geometry of Manifolds with Symmetry," Moscow Univ. Press, Moscow, 1989.
102. ________, *Euler equations on Borel subalgebras of semisimple Lie algebras*, Izvestiya Akad. Nauk SSSR, Vol. 43, No. 3 (1979), 714–732.
103. Trofimov, V. V. and Fomenko, A. T., *Dynamic systems on orbits of linear representations of Lie groups and complete itnegrability of some hydrodynamic systems*, Func. Anal. and its Appl., Vol. 17, No. 1 (1983), 31–39.
104. ________, *Geometrical and algebraic mechanisms of integrability of Hamiltonian systems on homogeneous spaces and Lie algebras*, Moscow VINITI, Modern Problems in Math. Fundamental Researchs, **16** (1987), 227–299. (Encyclopedia of Mathematical Sciences, Springer-Verlag).
105. ________, *Liouville integrability of Hamiltonian systems on Lie algebras*, Uspekhi Matem. Nauk, Vol. 39, No. 2 (1984), 3–56.
106. Van Moerbeke, P., *The spectrum of Jacobi matrices*, Invent. Math. **37** (1976), 45–81.
107. Waldhausen, F., *Eine Klasse von three-dimensional en Mannigfaltigkeiten*, Invent. Math., Vol. 3, No. 4 (1967), 308–333 (Part 1); and Vol. 4, No. 2, 88–117 (Part 2).
108. Weeks, J., *Hyperbolic structures on three-manifolds*, A dissertation presented to the Faculty of Princeton Univ. Dept. of Math. (1985).
109. Weinstein, A., *Symplectic manifolds and their Lagrangian submanifolds*, Adv. Math. **6** (1971).

110. ———, *The local structure of Poisson manifolds*, J. of Diff. Geom. **18** (1983), 523–557.
111. ———, "Lectures on Symplectic Manifolds," CBMS Reg. Conf. Series, A.M.S., Providence, 1977.
112. Zakharov, V. E. and Faddeev, L. D., *The Korteweg–de Vries equation, a completely integrable Hamiltonian system*, Func. Anal. and its Appl., Vol. 5, No. 4 (1971), 18–27.
113. Zieschang, H.; Vogt, E. and Coldeway, H. D., *Surfaces and planar discontinuous groups*, Lecture Notes in Math., **835** (1980), Springer-Verlag.

Additional Literature

114. Johannson, K., *Topologie und Geometrie von three-Mannigfaltigkeiten*, Jahrsber. Dtsch. Math., Vol. 86, No. 2 (1984), 37–68.
115. Jaco, W., *Lectures on three-dimensional topology*, Amer. Math. Soc., XII (1980), Providence, R.I..
116. Hempel, J., *Three-manifolds*, Ann. of Math. Stud. **86** (1976).
117. Waldhausen, F., *On irreducible three-manifolds which are sufficiently large*, Ann. Math., Vol. 87, No. 1 (1987), 56–88.
118. Matveev, S. V., *Complexity theory of three-dimensional manifolds*, Preprint, Kiev Math. Institute of Ukr. Acad. of Sc., Preprint 88.13 (1988), 1–32.
119. ———, *Complexity of three-dimensional manifolds and their enumeration with respect to increasing complexity*, Dokl. AN SSR (1989). in print.
120. Ziglin, S. L., *Separatrix splitting, solution branching and non-existence of itnegral in rigid-body dynamics*, Trudy Moskovskogo Matematicheskogo Obshchestva, Moscow **41** (1980), 287–303.
121. ———, Functional Analysis and its Appl., Vol. 17, No. 1 (1983).
122. Kozlov, V. V. and Treschev, D. V., *Polynomial integrals of Hamiltonian systems with an exponential interaction*, Izvestiya AN SSR Ser. Matem., Vol. 53, No. 3 (1989).
123. Bogoyavlensky, O. I., *On perturbations of the periodic Toda lattices*, Commun. Math. Phys. **51** (1976), 201–209.
124. Scott, P., *The geometries of three-manifolds*, Bull. London Math. Soc. **15** (1983), 401–487.
125. Matveev, S. V. and Savvateev, V. V., *Three-dimensional manifolds having simple special spines*, Colloq. Math. **32** (1974), 83–97.
126. Seifert, H. and Weber, C., *Die Dodekaederräume*, Math. Z. **37** (1933), 237–253.
127. Adams, C. C., *Hyperbolic structures on link complements*, Preprint, Univ. of Wisconsin, Madison, WI (1983).
128. Iacob, A., *Invariant manifolds in the motion of a rigid body about a fixed point*, Rev. Roum. Math. Pures et Appl., T. 16, No. 10 (1971).
129. Cushman, R. and Knörrer, H., *The energy momentum mapping of the Lagrange top*, Lecture Notes in Math. **1139** (1984).

Monotone Maps of $\mathbb{T}^n \times \mathbb{R}^n$ and Their Periodic Orbits

CHRISTOPHE GOLÉ

Abstract. This article presents a theorem of existence of $n+1$ (2^n if non degenerate) periodic orbits of any given rotation vector for a large class of exact symplectic maps of the n-dimensional annulus $\mathsf{T}^n \times R^n$. This theorem is global and uses the discrete variational approach of Aubry.

§0. Introduction.

In this paper, we present a theorem of existence of periodic orbits with any rational rotation vector for a large class of symplectic diffeomorphisms of the space $\mathsf{A}^n = \mathsf{T}^n \times \mathsf{R}^n$.

A lot of work has been produced recently, after a conjecture by Arnold [**Ar 2**], to relate the study of fixed points of symplectic maps on a manifold to the Morse theory of that manifold. The latter is seen to live in an underlying variational space to which it transmits some of its topology, often yielding far more fixed points than, say, Lefshetz theory would predict.

In our case, the variational setting is the one of Aubry [**A-LeD**] for twist maps of the annulus as generalised by Bernstein and Katok [**B-K**] to higher dimensions. This discrete setting generates symplectic maps that we call monotone here. For these maps, provided some boundary conditions, we find $n+1$ periodic orbits for each rational rotation vector, and 2^n when they are non degenerate. It expands the result of Bernstein and Katok in that it does not require the maps to be close to integrable and that it relaxes their convexity condition.

A source of inspiration in this field is the seminal work of Conley and Zehnder [**C-Z**, 1 and 2] who found homotopically trivial periodic orbits for hamiltonian flows on T^n and A^n. In this setting, Josellis [**J**] proved a theorem to which ours is parallele. We also use some refinement of Conley's Morse theory that Floer [**Fl1**] introduced in his work on the Arnold conjecture.

For $n = 1$, the maps that we consider here (i.e. compositions of monotone maps) are always time-1 maps of time dependent Hamiltonians, by a theorem of Moser (assuming appropriate boundary conditions)[**M**]. It is not clear that his theorem can be generalized for $n > 1$, especially in the non-convex case. The aim of this paper is not, however, to fill that hypothetical gap. Rather, our wish is to promote Aubry's approach as an alternate method for the global study of periodic orbits of exact symplectic maps. This approach, stemming from various physical situations has been used in numerical studies where 2^n orbits were indeed observed ([**K-M**]). Recently, as exposed by Robert MacKay in this conference, it has even yielded the first known nontrivial examples of Aubry-Mather sets in higher dimensions.

In the final section of this paper, we sketch some extensions of the theorem presented here and make a few remarks about the problem of existence of quasiperiodic orbits.

This work is part of the author's thesis at Boston University. It has greatly benefited from discusions with A. Floer, M. Levi, J.C. Sikorav, F. Tangerman, P. Veerman and E. Zehnder, all of whom I thank warmly.

I want to thank R. MacKay, J. Meiss, E. Lacomba, Mme Moulis-Desolneux, C. Simó, and G. Tarantello for the ideas and support they gave me during this conference. Many thanks to Albert Fathi whose help may prove precious in the problem of relating monotone maps and hamiltonian flows.

Finally, I wish to express all my gratitude to my advisor Dick Hall for all the support and inspiration he has given me during the completion of this work.

§1. Monotone maps

Let $\mathbf{T}^n = \mathbf{R}^n/\mathbf{Z}^n$. We consider the space $\mathbf{A}^n = \mathbf{T}^n \times \mathbf{R}^n = T^*\mathbf{T}^n$ with coordinates $(\boldsymbol{\theta}, \mathbf{r}), \boldsymbol{\theta} \in \mathbf{R}^n/\mathbf{Z}^n$, $\mathbf{r} \in \mathbf{R}^n$ endowed with its canonical symplectic form $\boldsymbol{\Omega} = \sum_{i=1}^{n} dr_i \wedge d\theta_i = d(\boldsymbol{\alpha})$, where $\boldsymbol{\alpha} = \sum_{i=1}^{n} r_i d\theta_i$. Corresponding coordinates in the universal covering will be denoted with a tilde.

A diffeomorphism F of $\mathbf{A}^n$ is *symplectic* if

$$F^*\boldsymbol{\Omega} = \boldsymbol{\Omega} \tag{1.1}$$

and *exact symplectic* if

$$F^*\boldsymbol{\alpha} - \boldsymbol{\alpha} \quad \text{is exact} \tag{1.2}$$

We write: $F(\mathbf{z}) = F(\boldsymbol{\theta}, \mathbf{r}) = (\boldsymbol{\Theta}, \mathbf{R}) = \mathbf{Z}$,

$$\text{and: } DF_{\mathbf{z}} = \begin{pmatrix} \boldsymbol{\Theta}_\theta & \boldsymbol{\Theta}_{\mathbf{R}} \\ \mathbf{r}_\theta & \mathbf{r}_{\mathbf{R}} \end{pmatrix} = \begin{pmatrix} \mathbf{a} & \mathbf{b} \\ \mathbf{c} & \mathbf{d} \end{pmatrix} \tag{1.3}$$

1.4 DEFINITION: A diffeomorphism F of $\mathbf{A}^n$ is monotone if it is exact symplectic and if $\det \mathbf{b}(\mathbf{z}) \neq 0$ for all $\mathbf{z} \in \mathbf{A}^n$.

REMARK: This terminology is inspired by, although not quite equivalent to M. Herman's [**H**]. He does not assume *exact* symplectic.

1.5 DEFINITION: A point $\mathbf{z}$ of $\mathbf{A}^n$ is called $(\mathbf{w}, q)$-periodic for F if $(\mathbf{w}, q) \in \mathbf{Z}^n \times \mathbf{Z}$ and:

$$\tilde{F}^q(\tilde{\mathbf{z}}) - \tilde{\mathbf{z}} = \mathbf{w},$$

where $\tilde{F}$ and $\tilde{\mathbf{z}}$ denote some *chosen* lift of F and z to the covering space $\tilde{\mathbf{A}}^n = \mathbf{R}^{2n}$ of $\mathbf{A}^n$. The vector $\mathbf{w}/q$ is called the *rotation vector* of the orbit. When (w_i, q) are relatively prime for some i, we say that $(\mathbf{w}, q)$ is relatively prime.

1.6 REMARK: The rotation vector *is* lift-dependant: We fix a lift once and for all (rotation vectors given by different lifts differ by fixed integer vectors).

(1.7) DEFINITION: A $(\mathbf{w}, q)$-periodic point (or orbit) $\mathbf{z}$ is said to be *non-degenerate* if

$$\det(D(F^q)_{\mathbf{z}} - I) \neq 0$$

§2. Generating Functions: The variational setting.

A Monotone diffeomorphism F is often uniquely determined by a single, real valued function on $\mathbb{A}^n$. Indeed, since F is exact symplectic, we can write (see 1.2):

$$F^*\boldsymbol{\alpha} - \boldsymbol{\alpha} = dS \tag{2.1}$$

or:

$$\mathbf{R}d\boldsymbol{\Theta} - \mathbf{r}d\boldsymbol{\theta} = dS \tag{2.1'}$$

where $S : \mathbb{A}^n \to \mathbb{R}$. If, moreover, S can be expressed as a function of the variables $(\boldsymbol{\theta}, \tilde{\boldsymbol{\Theta}})$, that is, if the map:

$$\psi : (\boldsymbol{\theta}, \mathbf{r}) \to (\boldsymbol{\theta}, \tilde{\boldsymbol{\Theta}}) = (\boldsymbol{\theta}, \partial_1 S(\boldsymbol{\theta}, \mathbf{r}))$$

is a *global* diffeomorphism of $\mathbb{A}^n$, then S is called a *global generating function.*

REMARK: Here and in the sequel, $\partial_1 S$ (resp. $\partial_2 S$) represents the partial derivative with respect to the first (resp. the second) *vector* variable.

If S is C^2, then F is C^1 and we have:

$$\begin{aligned} \mathbf{r} &= -\partial_1 S(\boldsymbol{\theta}, \tilde{\boldsymbol{\Theta}}) \\ \mathbf{R} &= \partial_2 S(\boldsymbol{\theta}, \tilde{\boldsymbol{\Theta}}) \end{aligned} \tag{2.3}$$

If we also denote by S the lift of S to $\mathbb{R}^{2n}$, then (2.3) becomes

$$\begin{aligned} \mathbf{r} &= -\partial_1 S(\tilde{\boldsymbol{\theta}}_1, \tilde{\boldsymbol{\Theta}}) \\ \mathbf{R} &= \partial_2 S(\tilde{\boldsymbol{\theta}}, \tilde{\boldsymbol{\Theta}}) \end{aligned} \tag{2.4}$$

In that setting, S satisfies:

$$S(\tilde{\boldsymbol{\theta}} + m, \tilde{\boldsymbol{\Theta}} + m) = S(\tilde{\boldsymbol{\theta}}, \tilde{\boldsymbol{\Theta}}), \ \forall m \in \mathbb{Z}^n \tag{2.5}$$

Let $\tilde{\mathbf{z}}_i = \tilde{F}^i(\tilde{\mathbf{z}}_0) = (\tilde{\boldsymbol{\theta}}_i, \mathbf{R}_i)$ then the orbit $\{\tilde{\mathbf{z}}_i\}$ is uniquely determined by the sequence $\{\tilde{\boldsymbol{\theta}}_i\}$, since

$$\mathbf{r}_i = -\partial_1 S(\tilde{\boldsymbol{\theta}}_i, \tilde{\boldsymbol{\theta}}_{i+1}) = \partial_2 S(\tilde{\boldsymbol{\theta}}_{i-1}, \tilde{\boldsymbol{\theta}}_i) \tag{2.6}$$

Instead of orbits, we shall therefore look for *critical states*. These are bi-infinite sequences of $\mathbf{R}^n = \tilde{\mathbf{T}}^n : (\mathbf{x}) = \ldots \mathbf{x}_{-1}, \mathbf{x}_0, \mathbf{x}_1, \mathbf{x}_2, \ldots$ satisfying

$$(2.7) \qquad \partial_1 S(\mathbf{x_i}, \mathbf{x_{i+1}}) + \partial_2 S(\mathbf{x_{i-1}}, \mathbf{x_i}) = 0$$

REMARK: This construction is readily generalized to exact symplectomorphisms $F = F_1 \circ F_2 \circ \ldots \circ F_s$ where each F_i is monotone with global generating function S_i. In that setting, (2.7) becomes:

$$(2.8) \qquad \partial_1 S_i(\mathbf{x_i}, \mathbf{x_{i+1}}) + \partial_2 S_{i-1}(\mathbf{x_{i-1}}, \mathbf{x_i}) = 0$$

(which makes sense for $i > m$ by setting $S_{i+km} = S_i$).

For a monotone F, finding $(\mathbf{w}, q)$-periodic points is equivalent to finding critical states within the space

$$(2.9) \qquad \phi^*_{\mathbf{w},q} = \psi_{\mathbf{w},q}/\mathbf{Z}^n$$

of *periodic states*, where:

$$(2.9) \qquad \psi_{\mathbf{w},q} = \{(\mathbf{x}) \in (\mathbf{R}^n)^{\mathbf{Z}} | \mathbf{x}_{i+q} = \mathbf{x}_i + \mathbf{w}\},$$

and $\mathbf{Z}^n$ acts on $\psi_{\mathbf{w},q}$ by $(\mathbf{x} + m)_j = \mathbf{x}_j + \mathbf{m}, \forall \mathbf{m} \in \mathbf{Z}^n$. Alternatively, we shall think of $\psi_{\mathbf{w},q}$ as the codim n plane:

$$(2.10) \qquad \psi_{\mathbf{w},q} = \{(\mathbf{x}_0, \ldots, \mathbf{x_q}) | \mathbf{x_i} \in \mathbf{R}^n, \mathbf{x_q} = \mathbf{x}_0 + \mathbf{w}\} \simeq (\mathbf{R}^n)^q.$$

Finding $(\mathbf{w}, q)$-points is therefore done by solving the equation ("discrete variational problem")

$$(2.11) \quad \nabla L = 0, \text{ for } L(\mathbf{x}) = \sum_{i=1}^{q} S(\mathbf{x_{i-1}}, \mathbf{x_i}),$$
$$\text{and } \nabla L(\mathbf{x}) = \partial_1 S(\mathbf{x_i}, \mathbf{x_{i+1}}) + \partial_2 S(\mathbf{x_{i-1}}, \mathbf{x_i})$$

where $(\mathbf{x}) \in \phi^*_{\mathbf{w},q}$.(When dealing with a composition of F_i's, one uses $\sum S_i$).

Because of the periodicity of S (2.5), the function L defined in (2.11) is invariant under the action of $\mathbf{Z}^n$ on $\psi_{\mathbf{w},q}$, namely:

$$L(\mathbf{x} + m) = L(\mathbf{x}), \quad \mathbf{m} \in \mathbf{Z}^n$$

and hence it is well defined on $\phi^*_{\mathbf{w},q}$.

We will sometimes refer to L as the energy on the set of states. In the case $n = 1$, McKay and Meiss [**McK-M**] derived a formula analogous to the index formula in classical Morse-theory. It gives a relation between the index of $\nabla^2 L(x)$ (the Hessian of L at a critical point $\mathbf{x}$) and the spectrum of $DF^q(\mathbf{z})$ ($\mathbf{z} \in \mathbb{A}$ corresponds to the sequence $\mathbf{x}$). Mather [**Ma**] later completed the study of that relation by interpreting the contribution of $DF^q(\mathbf{z})$ as the rotation number of its action on the tangent circle-bundle. Recently, Kook and Meiss [**K**,M] generalized the above mentioned formula for higher dimensional monotone maps. Even though, when $n > 1$ this formula does not completely characterize the stability type of a periodic orbit, using it one can prove [**G1**]:

LEMMA (2.12). *Let* $\mathbf{x}$ *be a critical state in* $\phi^*_{w,q}$ *corresponding to the orbit of* $\mathbf{z}$. *Then* $\mathbf{z}$ *is non degenerate if and only if* $det\nabla^2 L(\mathbf{x}) \neq 0$ *(i.e. the Hessian of* L *is nondegenerate at* $\mathbf{x}$*).*

2.13 REMARK:

As noted by M.R. Herman [**He**], the condition:

$$\sup_{\mathbf{z} \in \mathbb{A}^n} \|\mathbf{b}^{-1}(\mathbf{z})\| < +\infty \tag{2.14}$$

is enough to ensure the existence of a globally defined generating function for a monotone diffeomorphism F. See [**G1**] for more details.

§3. Examples.

3.0 Monotone Twist Maps of the Annulus (and their compositions)

The theory presented here is an attempt to generalise some of the theory of monotone twist maps to higher dimensions. Monotone twist maps are monotone maps for $n = 1$, that is, on the Annulus $\mathbb{A} = \mathbb{S}^1 \times \mathbb{R}$. In that case the non-degeneracy condition $det\ \mathbf{b} \neq 0$ (see definition 1.4) is, modulo a choice of sign, equivalent to the Twist condition. The main result in our work is an analogue to the Poincaré-Birkhoff theorem in higher dimensions (existence of periodic orbits of all rotation numbers), and one of our aims is to investigate possible generalisations of the Aubry-Mather theorem (existence and topological characterisation of quasi-periodic orbits). The standard example of a twist map is the Standard map of the annulus:

$$f(\theta, r) = (\theta + r + (k/2\pi)sin(2\pi\theta), r + (k/2\pi)sin(2\pi\theta))$$

For references on this vast subject, consult, e.g., [**Ban**], [**Ch**], [**Ha**], [**Ma1**].

3.1 "Completely integrable" diffeomorphisms

We say that F is "completely integrable" if it is of the form:

$$F(\boldsymbol{\theta}, \mathbf{R}) = (\boldsymbol{\theta} + A(\mathbf{r}), \mathbf{r}),$$

Where A is a diffeomorphism with $A(\mathbf{r}) = \nabla V(\mathbf{r})$ for some V. When A is linear, it must be symmetric and the generating function for F is:

$$S(\tilde{\boldsymbol{\theta}}, \tilde{\boldsymbol{\Theta}}) = \frac{1}{2}\langle A^{-1}(\tilde{\boldsymbol{\Theta}} - \tilde{\boldsymbol{\theta}}), \tilde{\boldsymbol{\Theta}} - \tilde{\boldsymbol{\theta}}\rangle$$

indeed $\mathbf{R} = \partial_2 S(\tilde{\boldsymbol{\theta}}, \tilde{\boldsymbol{\Theta}}) = A^{-1}(\tilde{\boldsymbol{\Theta}} - \tilde{\boldsymbol{\theta}}) = -\partial_1 S(\tilde{\boldsymbol{\theta}}, \tilde{\boldsymbol{\Theta}}) = \mathbf{r}$ and hence $\tilde{\boldsymbol{\Theta}} - \tilde{\boldsymbol{\theta}} = A(\mathbf{r})$.
Since A is symmetric, it can be diagonalized with real eigenvalues. Thus, in the right frame for $\mathbf{T}\mathbb{A}^n$, F is a product of shear maps of the annulus $\mathbb{A}$:

$$F = f_1 \times \ldots \times f_n \ , \ f_i : \mathbb{A} \to \mathbb{A}$$

where $f_i(\theta_i, r_i) = (\theta_i + \lambda_i \mathbf{r}_i, \mathbf{r}_i)$, f_i is a positive (resp. negative) twist map if $\lambda_i > 0$ (resp. $\lambda_i < 0$).

For these maps, each torus $\{\mathbf{r} = \mathbf{r}_0\}$ is invariant and a point $(\boldsymbol{\theta}, \mathbf{r}_0)$ has rotation vector $A(\mathbf{r}_0)$. When $A(\mathbf{r}_0)$ is of the form $\frac{\mathbf{w}}{q}$, the whole torus $\{\mathbf{r} = \mathbf{r}_0\}$ is formed by $(\mathbf{w}, q)$-periodic orbits. Irrational $A(\mathbf{r}_0)$ correspond to quasi-periodic orbits. The topological structure of the "remnants" of such orbits for general symplectic maps of $\mathbb{A}^n$ is still a mystery.

The completely integrable diffeomorphisms occur naturally as time-1 maps of the completely integrable Hamiltonian systems with:

$$H(\boldsymbol{\theta}, \mathbf{r}) = \frac{1}{2}\langle A(\mathbf{r}), \mathbf{r}\rangle$$

3.2 The standard map

$$F(\boldsymbol{\theta}, \mathbf{r}) = (\boldsymbol{\theta} + \mathbf{r} + \nabla V(\boldsymbol{\theta}), \mathbf{r} + \nabla V(\boldsymbol{\theta}))$$

with $S(\tilde{\boldsymbol{\theta}}, \tilde{\boldsymbol{\Theta}}) = \frac{1}{2}(\tilde{\boldsymbol{\Theta}} - \tilde{\boldsymbol{\theta}})^2 - V(\tilde{\boldsymbol{\theta}})$.

EXAMPLE (FROESCHLE, [**Froe**]):

$$V(\theta_1, \theta_2) = \frac{1}{(2\pi)^2}\{K_1 \cos(2\pi\theta_1) + K_2 \cos(2\pi\theta_2) + h \cos(2\pi(\theta_1 + \theta_2))\}$$

this gives a 3-parameter family of monotone maps on $\mathbb{A}^2$. This is also studied numerically in Kook-Meiss [**K-M**].

3.3 Hamiltonian systems

If we assume

$$Det\frac{\partial^2 \mathbf{H}}{\partial \mathbf{r}^2} \neq 0,, \text{ and that, for } |\mathbf{r}| > a, \quad \mathbf{H}(\boldsymbol{\theta}, \mathbf{R}) = \frac{1}{2}\langle A\mathbf{r}, \mathbf{r}\rangle + \langle \mathbf{C}, \mathbf{R}\rangle$$

for A symmetric, nondegenerate, and $\mathbf{C} \in \mathbb{R}^n$ (see Conley-Zehnder, Thm. 3 [**C-Z 2**]) then the time-1 map F_1 can always be decomposed into monotone F_{ϵ_i}'s. Moreover, they trivially satisfy the boundary conditions of our theorem. Also note that such Hamiltonian systems fall into the domain of those studied by Josellis [**J**]

§4. Statement of the theorem

In this section, we assume that our map F continues, through a family $\{F_\lambda\}_{\lambda\in\Lambda}$ of monotone maps, to a completely integrable map of the form:

$$F_0(\boldsymbol{\theta}, \mathbf{r}) = (\boldsymbol{\theta} + A^{-1}(\mathbf{r}), \mathbf{r})$$

where A is a symmetric, nondegenerate matrix.

More precisely, we will be working with the following type of family of maps:

(4.1) Consider a family $(F_\lambda)_{\lambda\in\Lambda}$ where each F_λ is a monotone map generated by the function S_λ, and Λ is a *hausdorff, compact* and *connected* topological space (e.g., a compact region of $\mathbb{R}^N$).
Further, assume that:

(4.2) The dependence of F_λ on λ is continuous for the C^1 topology, the dependence of S_λ on λ is continuous for the C^2 topology.

(4.3) there is a point $0 \in \Lambda$ at which:

$$S_0(\mathbf{x}\ X) = \frac{1}{2}\langle A(\mathbf{x} - x), (\mathbf{x} - x)\rangle$$

where A is symmetric, non degenerate (that is, F_0 is as above).

(4.4) $$S_\lambda(\mathbf{x}, X) = S_0(\mathbf{x}, X) + R_\lambda(\mathbf{x}, X)$$

and R_λ is either 0 outside a bounded set, i.e.:

$$R_\lambda(\mathbf{x}, X) = 0 \quad \text{when: } \|\mathbf{x} - x\| \geq K \tag{4.4a}$$

(That amounts to saying that $F_\lambda = F_0$ when $\|A^{-1}(\mathbf{r})\| \geq K$). or assumes sublinear growth in its derivative:

$$\lim_{\|\mathbf{x}-x\|\to\infty} \frac{\|\partial_\alpha R_\lambda(\mathbf{x}, X)\|}{\|\mathbf{x} - x\|} = 0, \; \alpha = 1, 2 \tag{4.4b}$$

4.5 THEOREM. *Let $(F_\lambda)_\lambda \in \Lambda$ be a family as above. then, for any $\lambda \in \Lambda$, F_λ has at least $n+1$ $(\mathbf{w}, q)$-periodic orbits and 2^n when they are non degenerate. If moreover F_λ satisfies 4.4a for some K and $\|\mathbf{w}/q\| \leq K$ then all the $(\mathbf{w}, q)$-periodic orbits are contained in the compact set $\{(\boldsymbol{\theta}, \mathbf{r}) | \; \|A^{-1}(\mathbf{r})\| \leq K\}$.*

4.6 EXAMPLE: Let H_λ be a compact family of time dependant Hamiltonian functions such that:

$$H_0 = \frac{1}{2}\langle A\mathbf{r}, \mathbf{r}\rangle,$$

$$H_\lambda = H_0, \text{ for } \|\mathbf{r}\| > C, \text{ and } Det \frac{\partial^2 \mathbf{H}}{\partial \mathbf{r}^2} \neq 0$$

Then, according to 3.3, the time $\frac{1}{N}$-map at the time $\frac{k}{N}$, say $F_{k,\lambda}$, is monotone, with generating function

$$S_{k,\lambda}(\mathbf{x}, X) = \frac{1}{2}\langle \frac{1}{N} A^{-1}(\mathbf{x} - x), (\mathbf{x} - x)\rangle + R_{k,\lambda}(\mathbf{x}, x)$$

and $R_{k,\lambda} = 0$ when $\langle \frac{1}{N^2} A^{-2}(\mathbf{x} - x), \mathbf{x} - x\rangle \geq C^2$. As noted above, the following theory readily generalizes to the composition of such maps, and hence to the time one map of the Hamiltonian H_λ.

REMARK: To simplify the notation, we only consider a families of monotone maps. The same statements could be made about families of *compositions* of monotone maps provided they satisfy the same boundary conditions (actually, one only needs the "linear parts" A to commute).

We present the proof of theorem 4.5 in the next two sections. The idea is that the gradiant flow ∇L_0 corresponding to F_0, has a normally hyperbolic invariant set (for that *flow*), which is isolated and continues, in the sense of

Conley, to isolated invariant sets for all $\lambda \in \Lambda$. The invariant set for ∇L_0 is in fact diffeomorphic to the F_0-invariant torus of rotation vector $\mathbf{w}/q$. We then use a result by Floer which implies that the *invariant set* itself conserves its cohomology throughout the parameter space. The conclusion follows from Conley-Zehnder's Morse inequalities. In section ,we state a more general theorem where the continuation setting is not needed.

§5. Existence of the isolating block P.

In this section, it is not necessary to assume that our map F continues to a completely integrable one. All we need is that it be monotone with generating function $S(\mathbf{x}, X) = S_0(\mathbf{x}, X) + R(\mathbf{x}, X)$, and R satifies (4.4a). We show that the corresponding gradiant flow ∇L has $P(C)$ (see the definition bellow) as isolating block. In [**G1**], we also find an isolating block $B(C)$ when only the more general condition (4.4b) is assumed. The proof will not be reproduced here. At the end of this section we introduce the setting that will enable us to distinguish periodic *orbits* instead of *points*. Again, the following result holds for compositions of monotone maps (with good boundary conditions): see [**G1**].

Remember (2.9, 2.10) that

$$\begin{aligned} \psi_{\mathbf{w},q} &= \{(\mathbf{x}) \in (\mathsf{R}^n)^{\mathsf{Z}} | \mathbf{x}_{i+q} = \mathbf{x}_i + \mathbf{w}\} \\ &\simeq \{(\mathbf{x}_0, \dots, \mathbf{x_q}) | \mathbf{x_q} = \mathbf{x}_0 + \mathbf{w}\} \end{aligned} \tag{5.1}$$

and

$$\phi^*_{\mathbf{w},q} = \psi_{\mathbf{w},q}/\mathsf{Z}^n$$

One can also see it as:

$$\phi^*_{\mathbf{w},q} = \{(\mathbf{x}_0, \mathbf{a}) | \mathbf{x}_0 \in \mathsf{T}^n, \sum_{i=1}^{q} \mathbf{a_i} = \mathbf{w}\} \simeq \mathsf{T}^n \times (\mathsf{R}^n)^{q-1}. \tag{5.2}$$

Where $\mathbf{a}_i = \mathbf{x}_i - \mathbf{x}_{i-1}$.

We think of q as the period dimension, n the number of degrees of freedom. We want to break the degrees of freedom dimension into the eigenspaces of the matrices A, so as to compare the map F to a product of shear maps (see example 3.1).

Without loss of generality, we can choose coordinates on R^n *for which the matrix* A *(see 4.3) is diagonal. This will be the standing assumption.*

In these coordinates we denote by x_i^k the kth component of $\mathbf{x}_i$. We also denote by λ^k the (real) eigenvalues of A. By permuting the coordinate, we can assume that:

(5.3) $$\lambda_k > 0 \text{ for } 0 \leq k \leq k_0,\ \lambda_k < 0 \text{ for } k_0 < k \leq n$$

note that we do not assume all the λ_k to be of the same sign. When they are, we say that F satisfies a convexity condition (cf. [**B-K**]), or that it is twist ([**K-M**],[**He**]). In that case at least one orbit is trivial (global extremum: see [**B-K**]).

5.4 DEFINITION:

$$P(C) = \{(\mathbf{x}) \in \phi^*_{\mathbf{w},q} \mid \sup_{\substack{i \in \{0,\dots,q\} \\ k \in \{1,\dots,n\}}} |x_i^k - x_{i-1}^k| < C\}$$

It is not hard to see that $P(C) \simeq \mathrm{T}^n \times (\mathrm{D}^n)^{q-1}$.

5.5 LEMMA. *Let F be a monotone map with generating function*

(5.5a) $$S(\mathbf{x}, X) = S_0(\mathbf{x}, X) + R(\mathbf{x}, X),$$

and R satifies

(5.5b) $$R(\mathbf{x}, X) = 0 \text{ whenever } \|\mathbf{x} - x\| \geq K$$

Then there is a constant $C_0 > 0$, independent of q such that, for any $C \geq C_0$, $P(C)$ is an isolating block for the flow ∇L on $\phi^*_{\mathbf{w},q}$, whenever

(5.5c) $$\|\mathbf{w}/q\| < K.$$

5.6 REMARK: condition (5.5c) just restricts our attention to interesting orbits; we know completely the other ones, which lay on invariant tori (large $\|\mathbf{r}\|$).

PROOF:

We will show that the flow ∇L points out of $P(C)$ for points belonging to the following set of boundary components of $\partial P(C)$; for C large enough:

$$(5.7) \qquad \bigcup_{k=1}^{k_0} \bigcup_{i=0}^{q} \{(\mathbf{x}) \in P(C) \cap \phi^*_{\mathbf{w},q} | \; x_i^k - x_{i-1}^k = \pm C\}$$

$$\stackrel{def}{=} \bigcup_{k=1}^{k_0} \bigcup_{i=0}^{q} \partial P^+_{i,k}$$

The same argument shows that ∇L points in at the other components of $\partial P(C)$.

Schematically, the picture is the following:

(5.8) $\mathbf{T}^n \times$

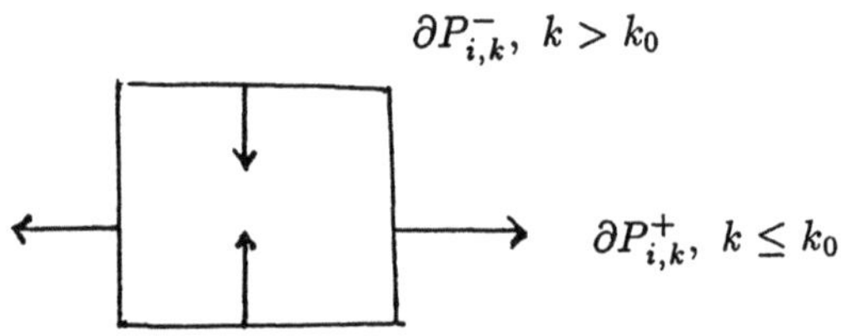

Each $\partial P_{i,k}$ is itself a box, either attracting or repelling.

It is enough to look at one of them, say $\partial P^+_{i,k}$, for $k \leq k_0$ (see def. in 5.7).

Because of the periodicity of S, the set $\phi_{\mathbf{w},q}$ is invariant under the flow $\dot{\mathbf{x}} = \nabla L(\mathbf{x})$. Indeed, if $\mathbf{x} \in \phi_{\mathbf{w},q}$, then

$$\begin{aligned} \dot{\mathbf{x}}_{i+q} - \dot{\mathbf{x}}_i &= \partial_1 S(\mathbf{x}_{i+q}, \mathbf{x}_{i+q+1}) + \partial_2 S(\mathbf{x}_{i+q-1}, \mathbf{x}_{i+q}) \\ &\quad -\partial_1 S(\mathbf{x}_i, \mathbf{x}_{i+1}) - \partial_2 S(\mathbf{x}_{i-1}, \mathbf{x}_i) \\ &= 0, \quad \text{by periodicity of } \partial_\alpha S, \; \alpha = 1,2 \end{aligned} \tag{5.9}$$

Hence ∇L is parallel to $\phi^*_{\mathbf{w},q}$, and it is enough to show that:

$$N^+_{i,k}.\nabla L > 0 \quad (k \leq k_0) \tag{5.10}$$

where $N^+_{i,k}$ is the normal vector to $\partial P^+_{i,k}$, namely:

$$(N^+_{i,k})_{j,l} = \begin{cases} 1 & \text{if } (j,l) = (i,k) \\ -1 & \text{if } (j,l) = (i-1,k) \\ 0 & \text{otherwise} \end{cases} \tag{5.11}$$

(Both ∇L and $N^+_{i,k}$ are thought of as vectors in $(\mathbf{R}^n)^{q+1}$). On the other hand:

(5.12)

$$(\nabla L(\mathbf{x}))_{i,k} = \lambda_k(2x_i^k - x_{i-1}^k - x_{i+1}^k) + \partial_1 R^k(\mathbf{x}_i, \mathbf{x}_{i+1}) + \partial_2 R^k(\mathbf{x}_{i-1}, \mathbf{x}_i)$$

Thus:

$$N^+_{i,k}.\nabla L(\mathbf{x}) =$$
$$\lambda_k(2x_i^k - x_{i-1}^k - x_{i+1}^k) + \partial_1 R^k(\mathbf{x}_i, \mathbf{x}_{i+1}) + \partial_2 R^k(\mathbf{x}_{i-1}, \mathbf{x}_i) \tag{5.13}$$
$$-[\lambda_k(2x_{i-1}^k - x_{i-2}^k - x_i^k) + \partial_1 R^k(\mathbf{x}_{i-1}, \mathbf{x}_i) + \partial_2 R^k(\mathbf{x}_{i-2}, \mathbf{x}_{i-1})]$$

Choose $C \geq K$ (see 4.1a). Then:

$$|x_i^k - x_{i-1}^k| = C \Rightarrow \|\mathbf{x}_i - \mathbf{x}_{i-1}\| \geq K \tag{5.14}$$

and therefore $R^k(\mathbf{x}_i, \mathbf{x}_{i-1}) = 0$. For such a C and for $\mathbf{x}$ in $\partial P^+_{i,k}(C)$, we have (rearranging terms):

$$N^+_{i,k}.\nabla L(\mathbf{x}) =$$
$$(2\lambda_k)(x_i^k - x_{i-1}^k) + \lambda_k(x_i^k - x_{i+1}^k) + \partial_1 R^k(\mathbf{x}_i, \mathbf{x}_{i+1}) \tag{5.15}$$
$$-\lambda_k(x_{i-1}^k - x_{i-2}^k) - \partial_2 R^k(\mathbf{x}_{i-1}, \mathbf{x}_i)$$

Since in $\partial P^+_{i,k}(C)$, $x_i^k - x_{i-1}^k = C$ and that for $k \leq k_0$, $\lambda_j^k > 0, \forall j$, this expression will be positive whenever:

$$\sup_{\mathbf{x} \in P(C)} |\lambda_k(x_i^k - x_{i+1}^k) + \partial_1 R^k(\mathbf{x}_i, \mathbf{x}_{i+1})| \leq |\lambda_k| C \tag{5.16}$$

$$\sup_{\mathbf{x} \in P(C)} |\lambda_k(x_{i-1}^k - x_{i-2}^k) + \partial_2 R^k(\mathbf{x}_{i-2}, \mathbf{x}_{i-1})| \leq |\lambda_k| C$$

for all i.

This in turn will be made possible because $R(\mathbf{x}, X)$ and its derivatives are 0 outside of $\|\mathbf{X} - \mathbf{x}\| \geq K$ (4.1a): their periodicity implies that they are bounded, say

$$\|\partial_\alpha R\| < K_1 \tag{5.17}$$

Hence, (5.16) is satisfied whenever

$$C \geq K + \frac{K_1}{\inf |\lambda_k|} \tag{5.18}$$

The reader will note that this does not guarantee strict positivity of $\nabla L.N^+_{ik}$.

In fact, what we have proven so far is:

$$\nabla L.N^+_{ik} > 0 \text{ on } \partial P^+_{i,k}, \text{ except possibly on}$$

$$E_1 = \partial P^+_{i,k} \cap \{x^k_{i-1} - x^k_{i-2} = C\} \cap \{x^k_{i+1} - x^k_i = C\}$$

where it is null. However, we will show that, because $\mathbf{x} \in \phi^*_{\mathbf{w},q}$, ∇L has to be positively transverse to at least one of the adjacent faces to E_1. We start with the face $\partial P^+_{i-1,k}$:

$$\text{(5.19)} \qquad x^k_{i-1} - x^k_{i-2} = C \text{ on } E_1 \Rightarrow R^k(\mathbf{x}_{i-2}, \mathbf{x}_{i-1}) = 0 = R^k(\mathbf{x}_i, \mathbf{x}_{i+1})$$

Hence:

$$\text{(5.20)} \qquad \begin{aligned} \nabla L\Big|_{E_1} .N^+_{i-1,k} &= (2\lambda_k)(\mathbf{x}^k_{i-1} - \mathbf{x}^k_{i-2}) \\ &\quad + \lambda_k(x^k_{i-1} - x^k_i) - \lambda_k(x^k_{i-2} - x^k_{i-3}) \\ &\quad - \partial_2 R^k(\mathbf{x}_{i-3}, \mathbf{x}_{i-2}) \\ &= \lambda_k C - \lambda_k(x^k_{i-2} - x^k_{i-3}) - \partial_2 R^k(x_{i-3}, x_{i-2}) \\ &> 0 \end{aligned}$$

because of (5.12), except possibly when $x^k_{i-2} - x^k_{i-3} = C$.
Define:

$$E_2 = E_1 \cap \{x^k_{i-2} - x^k_{i-3} = C\}$$

and by induction

$$E_l = E_{l-1} \cap \{x^k_{i-(l)} - x^k_{i-(l-1)} = C\}$$

and:

$$\nabla L\Big|_{E_{l-1}} .N^+_{i,l-1} > 0, \quad \text{except possibly on } E_l.$$

On $\phi^*_{\mathbf{w},q}$, this process comes to an end, since $\mathbf{x}_{i-q} = \mathbf{x}_i - \mathbf{w}$. At the end, we get:

$$\nabla L\Big|_{E_{q-1}} .N^+_{i-(q-1),l} > 0, \quad \text{except possibly on } E_q$$

But $E_q = \{(\mathbf{x}) | x^k_j - x^k_{j-1} = C, \ \forall j \in (0, \dots, q)\}$ cannot be in $\phi^*_{\mathbf{w},q}$, because then:

$$x^k_{j+q} - x^k_j = qC \ ,$$

which contradicts the assumption 5.1c. □

We now turn to the problem of distinguishing geometrically distinct periodic orbits: if $(\mathbf{x}_0 \dots \mathbf{x}_{q-1})$ is critical, so is $(\mathbf{x}_1 \dots \mathbf{x}_q)$. But they represent the same orbit. Hence, to look for distinct periodic orbits, we need to quotient by the action of the following diffeomorphism:

$$(5.21) \qquad T : \phi^*_{\mathbf{w},q} \to \phi^*_{\mathbf{w},q} \text{ defined by}$$

$$T(\mathbf{x})_i = \mathbf{x}_{i+s}$$

then

$$(5.22) \qquad L \circ T = L, \text{ since } S_i = S_{i+s}.$$

Bernstein-Katok ([**B-K**], proposition 1), show that the quotient map:

$$(5.23) \qquad \phi^*_{\mathbf{w},q} \xrightarrow{\pi} \phi^*_{\mathbf{w},q}/T = \phi_{\mathbf{w},q}$$

is an q-fold covering map. In fact, $\phi_{\mathbf{w},q} = \phi^*_{\mathbf{w},q}/T$ is a fiber bundle over T^n with fiber $(\mathsf{R}^n)^{q-1}$ whereas $\phi^*_{\mathbf{w},q}$ is diffeomorphic to $\mathsf{T}^n \times (\mathsf{R}^n)^{q-1}$. For this purpose they define the coordinates $(\mathbf{v}, \mathbf{t}_1 \dots, \mathbf{t}_{q-1})$ of $\psi_{\mathbf{w},q}$ by:

$$(5.24) \qquad \begin{aligned} \mathbf{v} &= \frac{1}{q}(\mathbf{x}_0 + \dots + \mathbf{x}_{q-1}) \\ \mathbf{t}_i &= \mathbf{x}_i - \mathbf{x}_{i-1} - \mathbf{w}/q \end{aligned}$$

In these coordinates, the action of Z^n on $\psi_{\mathbf{w},q}$:

$$(5.25) \qquad (\mathbf{x}_i) \to (\mathbf{x}_i + \mathbf{m}) \text{ becomes } (\mathbf{v}, \mathbf{t}) \to (\mathbf{v} + m, t)$$

and the shift

$$(5.26) \qquad T(\mathbf{v}, \mathbf{t}_1 \dots \mathbf{t}_{q-1}) = (\mathbf{v} + \mathbf{w}/q, \mathbf{T}_2, \dots, \mathbf{t}_{q-1}, -\sum_{i=1}^{q-1} \mathbf{t}_i)$$

L induces a function on $\phi_{\mathbf{w},q}$, that we denote by $\underline{L}$ and the critical points of $\underline{L}$ on $\phi_{\mathbf{w},q}$ are in 1-1 correspondence with the geometrically different *periodic orbits* of rotation vector $\mathbf{w}/q$ of F.

We will have to do some Morse theory on $\phi_{\mathbf{w},q}$. Observe that $\underline{P} = \pi(P)$ is an isolating block for the gradiant flow of $\underline{L}$ on $\phi_{\mathbf{w},q}$, with exit set $\underline{P}^- =$

$\pi(P^-)$. It is important to note that L (resp. $\underline{L}$) has no critical points outside of P (resp. $\underline{P}$).

We end this section with a simple proof of Theorem 4.5 when F satisfies the convexity condition (and either 4.4a or 4.4b). This proof does not assume that F continues to a completely integrable F_0.

5.27 Proof of 4.5 in the convex case

Suppose that the matrix A is positive definite. In this case, the classical Morse-Lyusternick-Schnirelman theory applies (see [**Mi**] and [**Kl**]). Indeed, $\underline{P}^- = \partial \underline{P}$ and $\underline{P}$ is an attractor block for the flow $\dot{x} = -\nabla \underline{L}(x)$ on $\phi_{\mathbf{w},q}$. As remarked above, $\underline{L}$ has no critical points outside of $\underline{P}$. Let $c = \sup \underline{L}(x)$, then $\underline{P} \subset \underline{L}^{-1}(-\infty, c] = M_c$. This set, by a standard argument in Morse theory ([**Mi**,2] Th. 3.1) is a deformation retract of $\phi_{\mathbf{w},q}$, itself homotopically equivalent to $\mathbf{T}^n$ the cup long $l(M_c) = n+1$ and Lyusternick-Schnirelman implies the existence of at least $n+1$ critical points for L.

When the critical points are non degenerate, the classical Morse theory gives at least the sum of Betti numbers $SB(M_c) = 2^n$ critical points inside M_c. (All the above holds with an isolating block B instead of P when the boundary condition 4.4b is used instead of 4.4a: see [**G1**].)

§6. Ghost tori and the proof of Theorem 4.5.

Going back to the general continuation setting, we denote by ∇L_λ the gradient flow associated to S_λ, as in (2.11).

6.1 DEFINITION: We call a ghost-torus and denote by $G^\lambda_{\mathbf{w},q}(C)$ (resp. $\underline{G}^\lambda_{\mathbf{w},q}(C)$) the maximal invariant set for ∇L_λ (resp. $\nabla \underline{L}_\lambda$ in the isolating blocks $P_{\mathbf{w},q}(C)$ or $B_{\mathbf{w},q}$ (resp. $\underline{P}_{\mathbf{w},q}(C)$ or $\underline{B}_{\mathbf{w},q}(C)$) constructed in section 5.

(Note the new rotation vector index). Since we assume Λ to be compact, the constant C as above can be chosen uniformly for all $\lambda \in \Lambda$.

Since the $G^\lambda_{\mathbf{w},q}$'s are the maximal isolated invariant sets for the *same* isolating neighborhood $P_{\mathbf{w},q}$ (or $B_{\mathbf{w},q}$) in the continuous family of flows ∇L_λ, they are *related by continuation* in the sense of Conley. (This notion of continuation is actually much broader and allows for instance, dealing with sequences of such relationships. For the more general definition, we refer to [**C**] or [**Sa**]).

The invariant set, or ghost torus $G^o_{\mathbf{w},q}$ plays a particular role. In fact:

6.2 LEMMA. *(a) $G^o_{\mathbf{w},q}$ is diffeomorphic to $\mathbf{T}^n$ and all its points are critical,*
(b) $G^o_{\mathbf{w},q}$ is a retract of $\phi^*_{\mathbf{w},q}$
(c) It is normally hyperbolic for the flow ∇L_0.

Moreover, $G^o_{\mathbf{w},q}$ corresponds naturally to the invariant torus for F_0, of rotation vector $\mathbf{w}/q$.

PROOF: Unless we state otherwise, we use the intrinsic coordinate system $(\mathbf{x_o}, \dots, \mathbf{x_{q-1}})$ for $\phi^*_{\mathbf{w},q}$. The same coordinates will be used for

$$\mathbf{T}(\phi^*_{\mathbf{w},q}) \cong \mathbf{T}((\mathbb{R}^n)^q/\mathbb{Z}^n) \cong (\mathbb{R}^n)^q$$

In these coordinates, $\nabla L_\lambda(\mathbf{x})$ is:

$$\nabla L_\lambda(\mathbf{x})_i = \partial_1 S_\lambda(\mathbf{x_i}, \mathbf{x_{i+1}}) + \partial_2 S_\lambda(\mathbf{x_{i-1}}, \mathbf{x_i})$$

with $\mathbf{x_q} = \mathbf{x_o} + \mathbf{w}, \mathbf{x}_{-1} = \mathbf{x_{q-1}} - \mathbf{w}$.

(a) To find critical points for ∇L_0, we solve the equation:

$$(6.3) \qquad \begin{aligned} 0 &= \partial_1 S_0(\mathbf{x_i}, \mathbf{x_{i+1}}) + \partial_2 S_0(\mathbf{x_{i-1}}, \mathbf{x_i}) \\ &= A(2\mathbf{x_i} - \mathbf{x_{i-1}} - \mathbf{x_{i+1}}), \quad i \in \{0, \dots, q-1\} \end{aligned}$$

with the constraint $\mathbf{x}_{i+q} = \mathbf{x_i} + \mathbf{w}$.

Since 6.3 is equivalent to $\mathbf{x_i} - \mathbf{x_{i-1}} = \mathbf{x_{i+1}} - \mathbf{x_i}$, the solutions are given by one of the $\mathbf{x_i}$'s, say $\mathbf{x_o}$ and are of the form $\{(\mathbf{x_o}, \mathbf{x_o} + \frac{\mathbf{w}}{\mathbf{q}}, \dots, \mathbf{x_o} + \frac{\mathbf{q-1}}{\mathbf{q}}\mathbf{w})\}$ under the identification $(\mathbf{x_i}) \sim (\mathbf{x_i} + \mathbf{m})$, $\mathbf{m} \in \mathbb{Z}^n$ this gives us a torus. One can see that this torus constitutes *all* of $G^o_{\mathbf{w},q}$: the proof of (c) will show us that, moding out by this eigendirection, ∇L_0 is a linear hyperbolic flow. Hence it can not contain more than a fixed point as isolated invariant set in the space $\phi^*_{\mathbf{w},q}/G^o_{\mathbf{w},q}$. Thus:

$$(6.4) \qquad G^o_{\mathbf{w},q} = \{\mathbf{x} \in \phi^*_{\mathbf{w},q} | \mathbf{x_i} = \mathbf{x_o} + i(\mathbf{w}/q)\}$$

It is easy to see that these critical states correspond to the F_0-invariant torus of rotation vector $\mathbf{w}/q$. *This correspondance is in fact a diffeomorphism if one looks at $\underline{G}^0_{\mathbf{w},q}$ in $\phi_{\mathbf{w},q}$.*

(b) In (5.24), we introduced, following Bernstein-Katok, the coordinates $(\mathbf{v}, \mathbf{t})$ for $\psi_{\mathbf{w},q}, \phi^*_{\mathbf{w},q}, \phi_{\mathbf{w},q}$. $\mathbf{v}$ represents the torus component in both $\phi^*_{\mathbf{w},q}$

and $\phi_{\mathbf{w},q}$, since the identifications only take place there. The retraction map will be given in these coordinates by:

$$(6.5)\qquad \begin{aligned} r : \phi^*_{\mathbf{w},q} &\longrightarrow G^o_{\mathbf{w},q} \\ (\mathbf{v},\mathbf{t}) &\longrightarrow (\mathbf{v},0) \end{aligned}$$

(It is easy to see that $\mathbf{t} = 0$ is indeed the set $G^o_{\mathbf{w},q}$ in these coordinates). This map is obviously continuous and $r\Big|_{G^o_{\mathbf{w},q}} = Id_{G^o_{\mathbf{w},q}}$. In fact, it is a strong deformation retract and it commutes with the action of T. Hence r gives rise to a retraction:

$$(6.6)\qquad \underline{r} : \phi_{\mathbf{w},q} \longrightarrow \underline{G}^o_{\mathbf{w},q}$$

$$\underline{r}([\mathbf{v}],\mathbf{t}) \longrightarrow ([\mathbf{v}],0)$$

(c) We have to study the linearization $\mathbf{T}(\nabla L_0)$ of ∇L_0 on the tangent space $\mathbf{T}(\phi^*_{\mathbf{w},q})\Big|_{G^o_{\mathbf{w},q}}$. To do so, it is convenient to split $\mathbf{T}(\phi^*_{\mathbf{w},q})$ according to the eigenspaces of A, as we did in lemma 5.5. Say

$$(6.7)\qquad \mathbf{T}(\phi^*_{\mathbf{w},q}) = \oplus_{k=1}^n E_k,$$

where E_k corresponds to the eigenvalue λ_k of A. This is an orthogonal splitting, since A is symmetric. $\mathbf{T}(\nabla L_0)$ obviously preserves this splitting and its expression in each E_k is the $q \times q$ matrix:

$$(6.8)\qquad \mathbf{T}(\nabla L_0)\Big|_{E_k} = \lambda_k \begin{pmatrix} 2 & -1 & & & -1 \\ -1 & \ddots & \ddots & & \\ & \ddots & \ddots & \ddots & \\ & & \ddots & \ddots & -1 \\ -1 & & & -1 & 2 \end{pmatrix}$$

(blank spaces are 0's) We get this directly from (12.7).

One recognizes the "discrete Laplacian" on the space of q-periodic sequences in $\mathbf{R}^{\mathbf{Z}}$. To find its spectrum is an elementary exercise in the difference equation:

$$(6.9)\qquad \begin{cases} \mu a_n &= 2a_n - a_{n-1} - a_{n+1} \\ a_{n+q} &= a_n \end{cases}$$

that one solves by setting $a_n = \rho^n$, which imposes the set of eigensequences and eigenvalues:

$$\begin{aligned} V_j &= (\rho_j, \rho_j^2 \dots \rho_j^q) \quad \text{with} \\ \rho_j &= e^{i\frac{2\pi kj}{q}} \quad j \in \{a, \dots, q\} \\ \mu_j &= 2(1 - \cos\frac{2\pi j}{q}) \end{aligned} \tag{6.10}$$

In particular V_0 is the vector $(1, \dots, 1)$ and corresponds to the only 0 eigenvalue. The other eigenvalues are all strictly positive.

Putting it all together, we have found an orthogonal ($\mathbf{T}(\nabla L_0)$ is symmetric) basis of eigenvectors for

$$\mathbf{T}(\nabla L_0) : (V_0^1 \dots, V_0^n, V_1^1 \dots V_1^n \dots V_{q-1}^1 \dots V_{q-1}^n)$$

where the space spanned by $(V_0^1 \dots V_0^n)$ is just the tangent space to $G^o_{\mathbf{w},q}$, which corresponds to the 0 eigenvalue (Harmonic sequences!). As for V_i^k, when $i \neq 0$, it has eigenvalue $\lambda_k \mu_i$ which is non zero and of the sign of λ_k. This proves the normal hyperbolicity of $G^o_{\mathbf{w},q}$.

Since the covering map π is a local diffeomorphism, the above result also holds for $\underline{G}^\lambda_{\mathbf{w},q}(C)$. □

This lemma will enable us to prove the main theorems of this section, of which theorem 4.5 becomes a corollary: the topology of $G^o_{\mathbf{w},q}$ (i.e. $\mathbf{T^n}$) continues in $G^\lambda_{\mathbf{w},q}$ for all values of λ. For this reason, we call $G^\lambda_{\mathbf{w},q}$ a *ghost-torus*.

6.11 THEOREM (Local continuation).

Given $\mathbf{w}/q$, there is an open set $U(o) \subset \Lambda$ such that $G^\lambda_{\mathbf{w},q}$ contains a torus which is a graph over $G^o_{\mathbf{w},q}$. If S_λ is C^r, the torus will be C^{m-1} and C^{m-1} diffeomorphic to $G^\lambda_{\mathbf{w},q}$. This theorem also holds for $\underline{G}^\lambda_{\mathbf{w},q}$.

PROOF: This is an immediate application of classical theorems of perturbation of normally hyperbolic invariant manifolds. The version that we are using here is Fenichel's ([**Fe**]) theorem (see also [**H**,P,S]), who worked out carefully the smoothness of the perturbed invariant manifold. In our case, the ghost-torus will be as smooth as ∇L_λ, due to the fact that $\nabla L_0 = 0$ on $G^o_{\mathbf{w},\lambda}$. □

6.12 THEOREM (Global continuation).

For all λ *in* Λ,

$$\left(r\Big|_{G^{\lambda}_{\mathbf{w},q}}\right)^{*} : H^{*}(G^{o}_{\mathbf{w},q}) \longrightarrow H^{*}(G^{\lambda}_{\mathbf{w},q})$$

is injective.

(hence $G^{\lambda}_{\mathbf{w},q}$ *"contains " the topology of* $\mathbf{T^n}$*).*

This theorem also holds for $\underline{G}^{o}_{\mathbf{w},q}, \underline{G}^{\lambda}_{\mathbf{w},q}$ *and* $\underline{r}$*.*

PROOF: This is an immediate consequence of lemma 6.2, the fact that the $G^{\lambda}_{\mathbf{w},q}$'s are related by continuation and Theorem 2 of Floer [**Fl1**]. Indeed, we have collected here the three ingredients sufficient for his theorem of global continuation to apply, namely:

(1) Normal hyperbolicity of $G^{o}_{\mathbf{w},q}$ (resp. $\underline{G}^{\lambda}_{\mathbf{w},q}(C)$)for ∇L_0 (resp. $\nabla \underline{L}_0$.)
(2) $G^{o}_{\mathbf{w},q}$ (resp. $\underline{G}^{o}_{\mathbf{w},q}$)is a retract of the whole space $\phi^{*}_{\mathbf{w},q}$ (resp. $\phi_{\mathbf{w},q}$)
(3) $G^{\lambda}_{\mathbf{w},q}$ is related to $G^{o}_{\mathbf{w},q}$ by continuation (resp. $\underline{G}^{\lambda}_{\mathbf{w},q}$ with $\underline{G}^{o}_{\mathbf{w},q}$). □

(His theorem fits more general situations).

Finally, we can conclude the proof of theorem 4.5: By theorem 6.12, $l(\underline{G}^{\lambda}_{\mathbf{w},q}) \geq n+1$ and $SB(\underline{G}^{\lambda}_{\mathbf{w},q}) \geq 2^n$. The existence of $n+1$ or 2^n $(\mathbf{w},q)$-periodic orbits follows from Conley-Zenhder's Morse inequalities (valid for index pairs and thus a fortiori for isolated invariant sets) and the fact that we are dealing with a gradiant flow (see [**C-Z**,1 and 2]).

§7. Some extension of Theorem 4.5 and remarks about quasiperiodicity.

In [**Go**,Thm. 5.1], we prove a theorem of existence of periodic orbits for maps that are not known to continue to a completely integrable one.*

We now state this theorem and outline its proof.

*Let us point to the fact that a theorem of suspension of monotone maps by a (time dependent) hamiltonian flow satisfying $Det\frac{\partial^2 \mathbf{H}}{\partial \mathbf{r}^2} \neq 0$ would make theorem 7.1 a corollary of theorem 4.5 in the present paper. Such a theorem of suspension was proven for $n=1$ (Twist maps) by Moser [**Mo**]with the boundary condition 4.4a. In higher dimension, one could follow the suggestion of Albert Fathi and find a clever homotopy of a given global generating function to a quadratic one. The hard part is to keep each function in the homotopy a *globally* generating one.

7.1 THEOREM.

Let $F = F_1 \circ \ldots \circ F_s$ where each F_i is a monotone map of $\mathbb{A}^n$ generated by a global generating function S_i. Suppose that there is a set $\{A_i\}_{i=1}^{s}$ of symmetric, nondegenerate matrices such that

$$(7.2) \qquad [A_i, A_j] = 0 \ \ \forall i, j \in \{1, \ldots, s\},$$

and such that all the A_i's are positive definite (resp. negative definite) in the same subspaces of dimension k_0 (resp. $n - k_0$). Suppose that we have the following condition:

$$(7.3) \qquad \lim_{\|\tilde{\Theta} - \tilde{\theta}\| \to \infty} \frac{1}{\|\tilde{\boldsymbol{\Theta}} - \tilde{\theta}\|} \|\partial_\alpha S_i(\tilde{\boldsymbol{\theta}}, \tilde{\boldsymbol{\Theta}}) - (-1)^\alpha A_i(\tilde{\boldsymbol{\Theta}} - \tilde{\boldsymbol{\theta}})\| = 0, \ \ \alpha = 1 \text{ or } 2$$

uniformly in $\|\tilde{\boldsymbol{\Theta}} - \tilde{\theta}\|$. Then:

(i) F has at least $n+1$ geometrically distinct $(\mathbf{w}, m)$-periodic orbits for each $(\mathbf{w}, m) \in \mathbb{Z}^n \times \mathbb{Z}$ such that (w_i, m) are relatively prime, for at least one i.

Suppose in the following that all (prime) $(\mathbf{w}, m)$-periodic orbits are non-degenerate. Then:

(ii) If either m is odd, s is even, or k_0 is even or equal to n or 0, then F has at least 2^n $(\mathbf{w}, m)$-periodic orbits. Otherwise F has at least 2^{n-1} of them.

Sketch of the Proof:

As for theorem 4.5, we use the existence of an isolating block ($B(C)$ in this case) which looks like figure 5.8.

Conley and Zehnder [**C-Z**, 2] were able to prove, through a thorough study of the cohomology $H(B, B^-)$ of the index pair, together with the use of their generalised Morse inequalities, the existence of at least $n+1$ (2^n non degenerate) critical points for a gradiant like flow with such an isolating block. The problem arises when one tries to count *orbits* and not points in our situation. For this, we have to quotient by the shift map T defined in (5.21), which becomes a deck transformation of a cyclic covering space. Using a crucial part of Conley and Zehnder's proof [**C-Z**, 2], we are able to prove that the cup length $l(\underline{G}_{\mathbf{w},q} \geq l(G_{\mathbf{w},q})$, thus yielding part (*(i)* of the theorem. To prove statement (*ii*) of the theorem is harder: one has to prove the theorem 7.5 for equivariant gradiant flows (see below). This theorem implies that Morse inequalities of Conley-Zehnder apply equally to the pairs (B, B^-) and $(\underline{B}, \underline{B}^-)$, at least in the case where T^* is the identity. This occurs in the cases described in (ii). In the other cases, we are only able to say that $(T^*)^2$ is the identity: we then have to apply theorem 7.5 to T^2 which explains the loss of information as one passes to the quotient in these cases.

7.5 THEOREM. * *Let* $\pi : \mathbf{M} \to \underline{\mathbf{M}}$ *be an q-fold* C^k *covering map between two* C^k *manifolds, with deck transformation group* $\{T^i\}_{i=1}^q$ *such that* $T^m = Id$.

Let $\underline{L} : \underline{\mathbf{M}} \to \mathbf{R}$ *be a Morse function and* $L = \underline{L} \circ \pi$

Let (B, B^-) *be an index pair for* L, *invariant under* T *and such that:*

$$T^* : H^*(B, B^-) \to H^*(B, B^-) \text{ is the identity}$$

Denote $(\underline{B}, \underline{B}^-) = (\pi(B), \pi(\underline{B}^-))$. *Then* $(\underline{B}, \underline{B}^-)$ *is an index pair for* $\underline{L}$ *and*

$$H^*(\underline{B}, \underline{B}^-) = H^*(B, B^-). \tag{7.6}$$

An other extension of the result presented in this paper concerns a more detailed study of the topological structure of the ghost tori $G_{\mathbf{w},q}$. One would hope that, at least in the non degenerate case, they actually contain a torus. If moreover this torus forms a graph over the $\mathbf{x}_0$ axis, the critical points it contains would be natural candidates for the notion of well ordered, or Birkhoff orbits in higher dimensions . In that spirit, we are able to prove ([**G2**]):

7.7 PROPOSITION (GHOST CIRCLES). *Let* $n = 1 =$, *i.e. let* F *be a twist map of the annulus. The isolating block* P *(or* B*) contains an invariant subset for the flow* $-\zeta^t$, *solution of* $\dot{\mathbf{x}} = -\nabla L(\mathbf{x})$, *which is homeomorphic to a homotopically non trivial circle . This circle is generically* C^1 *and projects to the annulus as a Lipshitz graph over the* x*-axis, whose image by the map is also a graph. In fact such a "ghost circle" can be found passing through any Aubry-Mather set of any given rotation number.*

The last statement of the proposition derives from the work of Angenent [**An**] on this gradiant flow: he proves that it is order preserving. Note

*This theorem appears to be a special case of a recent result of C. MacCord and K. Mishaikov where they develop thoroughly an equivariant version of Conley's theory for actions of compact groups [**McC**, M].

that if we could eliminate the possibility of cusps, the circle found in the proposition would be a smooth graph over the $\mathbf{x}_0$ axis.

We end by making some remarks about possible uses of the Aubry setting in the search for quasiperiodic orbits. Let us remind the reader that McKay, Meiss and Kook [**McK**, M, K] have found such orbits for a nontrivial class of monotone maps and that they use the discrete variational approach to do so.

One way to attack the problem would be to consider the flow $\dot{\mathbf{x}} = -\nabla L(\mathbf{x}) = \partial_1 S(\mathbf{x_i}, \mathbf{x_{i+1}}) + \partial_2 S(\mathbf{x_{i-1}}, \mathbf{x_i})$, on the space of bi-infinite sequences of $\mathbf{R}^n$. This flow turns out to be defined even though L is not. Within this big space one can look at the space

$$\mathbf{Y} = \bigcup_{\omega \in \mathbf{R}^n} Y_\omega = \bigcup_{\omega \in \mathbf{R}^n} \{(\mathbf{x}) \in (\mathbf{R}^n)^{\mathbf{Z}} / |\mathbf{x_j} - \mathbf{j}\omega| < \infty\}$$

each slice of which is invariant under ∇L_λ (see (2.12)), and diffeomorphic to l^∞. One could hope that rational $G^\lambda_{\mathbf{w},q}$'s would continue onto some invariant set G^λ_ω in Y_ω, in the sense of Conley. Critical points in G^λ_ω would automatically have rotation vector ω, as elements of Y_ω. There the main stumbling block is: even though $\mathbf{Y} \subset (\mathbf{R}^n)$ can be endowed with the product topology, for which ∇L_λ is continuous, the "slices" Y_ω are not closed in that topology and hence do not constitute a local product parametrization for the flow ∇L_λ (with parameter $\omega \in \mathbf{R}^n$: for a definition of local product parametrization, i.e. the setting in which Conley's continuation is defined, see [**C**],[**Sa**]). The other problem, supposing one could prove the existence of a ghost-torus in each Y_ω, is that the flow is not automaticaly gradiant like on this set, making it hard to use the Conley-Zehnder Morse inequalities to find critical points. Note that Y_ω contains a diffeomorphic image of the KAM torus of rotation vector ω when it exists, as a compact invariant set for the variational flow. Proposition 7.7 shows in particular that G_ω exists for all ω when $n = 1$, and that in this case it contains a "real" circle.

Finally, we state a limiting theorem ([**G1**], thm. 13.2). $((\mathbb{R})^n)^{\mathbf{Z}}$ can be given the coordinates $(\ldots, \mathbf{a}_{-1}, \mathbf{x}_0, a_1, \ldots)$ where $\mathbf{a}_i = \mathbf{x}_i - \mathbf{x}_{i-1}$. When we take the quotient by the action of $\mathbf{Z}^n$, these gives us coordinates for $\mathbf{T}^n \times (\mathbf{R}^n)^{\mathbf{Z}}$. In these coordinates, we can define a retraction map:

$$\mathcal{R} : \mathbf{T}^n \times (\mathbf{R}^n)^{\mathbf{Z}} \to \mathbf{T}^n \times \{(0)\} \tag{7.8}$$

$$(\mathbf{x}_0, \mathbf{a}) \to (\mathbf{x}_0, 0)$$

which is continuous since projection mappings are continuous in the product topology. Also define $K(C)$ to be the set $\mathbb{T}^n \times$ Ball of radius C. It is not hard to see that, given any compact range in the parameter set Λ we can find an appropriate C for which all ghost-tori in a bounded set of rotation vector belong to $K(C)$. We can now state our result:

THEOREM 7.9. *Assume that S_λ are quadratic outside a bounded set (condition (4.4a)), then:*

(a) Any sequence $G^\lambda_{\mathbf{w}_k,q_k}$ has a converging subsequence in K, for the hausdorff metric on compact sets.

(b) If $\mathcal{G} = \lim_{k\to\infty} G^\lambda_{\mathbf{w}_k,q_k}$, then

$$(\mathcal{R}\Big|_{\mathcal{G}})^* : H^*(\mathbb{T}^n \times \{(0)\}) \to H^*(\mathcal{G})$$

is injective.

The proof uses the property of contination of the Alexander cohomology. We cannot claim at this point that $\mathcal{G}$ (or any part of it) is included in Y_ω, where $\omega = lim(\mathbf{w}_k/q_k)$.

REFERENCES

[An] S.B. Angenent, *The Periodic orbits of an area preserving twist map*, Comm. in Math. Physics **115,** no. 3 (1988).

[Ar1] V.I. Arnold, "Mathematical Methods of Classical Mechanics (App. 9)," Springer-Verlag, 1978.

[Ar2] V.I. Arnold, *Sur une propriété topologique des applications globalement canonique de la méchanique classique*, C.R. Acad. Sci. Paris, t. 261, Groupe 1 (novembre 1965), 3719–3722.

[A-LeD] S. Aubry and P.Y. LeDaeron, *The discrete Frenkel-Kontarova model and its extensions I. Exact results for ground states*, Physica **8D** (1983), 381-422.

[B-K] D. Bernstein and A.B. Katok, *Birkhoff periodic orbits for small perturbations of completely integrable Hamiltonian systems with convex Hamiltonians*, Invent. Math. **88** (1987), 225–241.

[B-H] P.L. Boyland and G.R. Hall, *Invariant circles and the order structure of periodic orbits in monotone twist maps*, Topology **25** (1986), 1–15.

[Cha] M. Chaperon, *Quelques questions de Géometrie Symplectique*, Séminaire Bourbaki no.610 (82/83).

[Che] A. Chenciner, *La dynamique au voisinage d'un point fixe elliptique conservatif*, Séminaire Bourbaki no. 622 (1983-84).

[C] Conley, C.C., *Isolated invariant sets and the Morse index*, CBMS, Regional Conf. Series in Math. **38** (1978).

[C-Z,1] C.C. Conley, E. Zehnder, *Morse type index theory for Hamiltonian equations*, Comm. Pure and Appl. Math. **XXXVII** (1984), 207-253.

[C-Z,2] C.C. Conley, E. Zehnder, *The Birkhoff-Lewis fixed point theorem and a conjecture of V.I. Arnold*, Invent. Math. (1983).

[Fe] N. Fenichel, *Persistence and smoothness of invariant manifolds for flows*, Ind. Univ. Math. J. **21,** No. 3 (1971).

[Fl1] A. Floer, *A refinement of conley index and an application to the stability of hyperbolic invariant sets*, Ergod.th. and Dyn. Sys. **7** (1987).

[Fl2] A. Floer, *Witten's complex for arbitrary coefficients and an application to Lagrangian intersections*, Courant Institute, preprint (1987).

[Frs] R. Franzosa, *Index filtrations and connections matrices for partially ordered Morse decompositions*, Ph.D. dissertation. U. of Wisconsin-Madison (1984).

[G1] C. Golé, *Periodic orbits for symplectomorphisms of* $\mathbf{T}^n \times \mathbf{R}^n$, Boston University, Thesis (1989).

[G2] C. Golé, *Ghost circles for twist maps, to appear*, Jour. Diff. Equ..

[Ha] G.R. Hall, *A topological version of a theorem of Mather on twist maps*, Ergod. Th. and Dynam. Sys. **4** (1984), 585–603.

[He] M.R. Herman, *Existence et non existence de tore invariants par des diffeomorphismes symplectiques*, preprint (1988), Ecole polytechnique.

[H,P,S] M.W. Hirsh, C.C. Pugh, M. Shub, *Invariant Manifolds*, Lecture Notes in Math. **583** (1977), Springer-Verlag.

[J] F.W. Josellis, *Forced oscillations on* $\mathbf{T}^n \times \mathbf{R}^n$ *having a prescribed rotation vector*, preprint (1987), Aachen.

[K1] A. Katok, *Some remarks on Birkhoff and Mather twist map theorems*, Ergod. Th. & Dynam. Sys. **2** (1982), 185–194.

[K2] A. Katok, *Minimal orbits for small perturbations of completely integrable Hamiltonian systems*, preprint (1989), Caltech.

[K-M] H. Kook and J. Meiss, *Periodic orbits for Reversible, Symplectic Mappings*, Physica D35 (1989), p. 65-86.

[McC-M] C. McCord and K. Mischaikow. personal communication.

[McK-M] R.S. McKay and J. Meiss, *Linear stability of periodic orbits in Lagrangian systems*, Phys. Lett. **98A** (1983), p. 92.

[Ma1] J. Mather, *Existence of quasi-periodic orbits for twist homeomorphisms*, Topology **21** (1982), 457-467.

[Ma2] J. Mather, *Amount of rotation about a point and the Morse index*, Comm. Math. Phys. **94** (1984), 141-153.

[Mi,2] J. Milnor, "Morse Theory," Princeton University Press.

[Mo] J. Moser, *Monotone twist mappings and the calculus of variations*, Ergod. Th. and Dyn. Sys. **6** (1986), p. 401-413.

[Sa] D. Salamon, *Connected simple systems and the Conley index of isolated invariant sets*, Trans. Amer. Math. Soc. **291,** no. 1 (1985).

[Sp] E. Spanier, "Algebraic topology," McGraw Hill, NY, 1966.

University of Minnesota, Department of Mathematics
127 Vincent Hall, 206 Church Street S.E.
Minneapolis, MN 55455

A Lower Bound for the Number of Fixed Points of Orientation Reversing Homeomorphisms

KRYSTYNA KUPERBERG

Abstract. Let h be an orientation reversing homeomorphism of the plane onto itself. Let X be a plane continuum, invariant under h. If X has at least 2^k invariant bounded complementary domains, then h has at least $k+2$ fixed points in X.

§1. Introduction.

A continuum is a nonempty, compact and connected metric space. The plane is denoted by R^2. If X is a continuum embedded in the plane, then the components of $R^2 - X$ are called the complementary domains of X. In 1951 M. L. Cartwright and J. C. Littlewood proved (see [**5**]) that if h is an orientation preserving homeomorphism of the plane and X is a non-separating plane continuum invariant under h, then h has a fixed point in X (see also [**4**] and [**6**]). H. Bell proved the Cartwright-Littlewood theorem for an arbitrary homeomorphism of the plane, see [**1**], [**2**], and [**3**]. Using Bell's theorem, it has been proved in [**8**] that if $X \subset R^2$ is a continuum invariant under an orientation reversing homeomorphism h of the plane, and X has at least one bounded complementary domain which is invariant under h, then h has at least two fixed points in X.

Throughout this paper h is an orientation reversing homeomorphism of the plane R^2 onto itself, and X is a continuum in R^2 invariant under h.

The results of this paper are the following: If X has at least 2^k bounded complementary domains which are invariant under h, then h has at least $k+2$ fixed points in X. If infinitely many complementary domains of X are invariant under h, then h has infinitely many fixed points in X.

§2. Fixed Points.

LEMMA 1. *Let $\{U_1, U_2, \ldots, U_m\}$ be a collection of invariant, bounded complementary domains of X. Then there exists an orientation reversing homeomorphism $g : R^2 \to R^2$ such that $g|X = h|X$, and g has a fixed point in every U_i for $1 \le i \le m$.*

PROOF: Let u_i be a point in U_i. There exists a closed disk D_i in U_i containing u_i and $h(u_i)$ in its interior. For $i = 1, \ldots, m$, there is a homeomorphism g_i of R^2 such that $g_i \circ h(u_i) = u_i$ and $g_i(x) = x$ outside D_i. Set $g = g_m \circ \cdots \circ g_1 \circ h$.

LEMMA 2. *Suppose that U is a bounded complementary domain of X invariant under h. Let V be the unbounded complementary domain of X. Then, there exists an annulus A with one boundary component in U and the other boundary component in V, and there exists an orientation reversing homeomorphism $G : R^2 \to R^2$ such that $G|[R^2 - (U \cup V)] = h|[R^2 - (U \cup V)]$ and $G(A) = A$.*

PROOF: There exists a closed disk D_1 in U and a homeomorphism G_1 of R^2 such that $G_1 \circ h(D_1) = D_1$ and $G_1(x) = x$ outside U. Similarly, assuming that the outside of a disk is a neighborhood of the added point of a one-point compactification of R^2, we can construct a homeomorphism G_2 of the plane such that $G_2(x) = x$ for $x \notin V$, and $G_2 \circ h(D_2) = D_2$ for some closed disk D_2 containing X in its interior. Let $G = G_2 \circ G_1 \circ h$. Notice that $G|[R^2 - (U \cup V)] = h|[R^2 - (U \cup V)]$, G is an orientation reversing homeomorphism of the plane, and the closure of $D_2 - D_1$ is the required annulus A.

THEOREM. *If X has at least 2^k, $k \geq 0$, bounded complementary domains which are invariant under h, then h has at least $k+2$ fixed points in X.*

PROOF: Let $U_1, \ldots, U_n$, where $n \geq 2^k$, be bounded complementary domains of X such that each U_i is invariant under h. Let V be the unbounded complementary domain of X. By Lemmas 1 and 2, we may assume that h has a fixed point $u_i \in U_i$ for $i = 1, \ldots, n-1$, and X is contained in an annulus A, which is invariant under h, and whose boundary components C_1 and C_2 are contained in U_n and V respectively.

Assume that $A = \{(r, \theta) \in R^2 : 1 \leq r \leq 2, 0 \leq \theta < 2\pi\}$, where (r, θ) denote the polar coordinates of a point in the plane. Let $\widetilde{A} = \{(x, y) \in R^2 : 1 \leq y \leq 2\}$, where (x, y) denote the Cartesian coordinates of a point in the plane. Let $\tau : \widetilde{A} \to A$, defined by $\tau(x, y) = (y, 2\pi x \pmod{2\pi})$, be the covering map; $\widetilde{A}$ is the universal cover of A. (For elements of the theory of covering spaces see for instance [7].) There is a homeomorphism $\widetilde{h}_1 : \widetilde{A} \to \widetilde{A}$ such that the diagram

$$\begin{array}{ccc} \widetilde{A} & \xrightarrow{\widetilde{h}_1} & \widetilde{A} \\ {\scriptstyle \tau}\big\downarrow & & \big\downarrow{\scriptstyle \tau} \\ A & \xrightarrow{h|A} & A \end{array}$$

commutes.

Notice that p is a fixed point of h in A iff the fiber $\tau^{-1}(p)$ is invariant under $\widetilde{h}_1$. Since $h(C_i) = C_i$ $(i = 1,2)$ and h is orientation reversing on C_i, $\widetilde{h}_1(\tau^{-1}(C_i)) = \tau^{-1}(C_i)$ and $\widetilde{h}_1$ is orientation reversing on $\tau^{-1}(C_i)$. For every $p = (r, \theta) \in A$, $\tau^{-1}(p) = \{(\frac{\theta}{2\pi} + n, r) : n = 0, \pm 1, \pm 2, \dots\}$. If $\tau^{-1}(p)$ is invariant under $\widetilde{h}_1$, then there is an integer m_p such that $\widetilde{h}_1(\frac{\theta}{2\pi} + n, r) = (\frac{\theta}{2\pi} - n + m_p, r)$ for every $(\frac{\theta}{2\pi} + n, r) \in \tau^{-1}(p)$, and $\widetilde{h}_1$ has a fixed point in $\tau^{-1}(p)$ iff m_p is even. Let $\widetilde{h}_2 : \widetilde{A} \to \widetilde{A}$ be such that $\widetilde{h}_2(x, y) = \widetilde{h}_1(x + 1, y)$. Let B be a two point compactification of $\widetilde{A}$. Denote the added points by b_0 and b_1. Let $\mu : B \to R^2$ be an embedding. Set $Y = \mu(\tau^{-1}(X) \cup \{b_0, b_1\})$. It has been shown in [**8**], Lemma P4, that there are orientation reversing homeomorphisms $f_i : R^2 \to R^2$ $(i = 1,2)$ such that $f_i(p) = \mu \circ \widetilde{h}_i \circ \mu^{-1}(p)$ for $p \in \mu(\widetilde{A})$, $f_i(\mu(b_0)) = \mu(b_1)$, and $f_i(\mu(b_1)) = \mu(b_0)$. Since Y is the intersection of a nested sequence of disks with holes, Y is a continuum (see [**8**], Lemma P1). Clearly, Y is invariant under each f_i.

Denote by Q the set of fixed points of h in X. Let P_i be the set of fixed points of $\widetilde{h}_i$ in $\tau^{-1}(X)$ for $i = 1, 2$. By the above, the sets P_1 and P_2 are disjoint and $\tau(P_1 \cup P_2) = Q$. We will show that neither of the two sets is empty. Suppose that $P_i = \emptyset$. Then $\widetilde{h}_i$ has no fixed points in $\tau^{-1}(X)$. Consequently, f_i has no fixed points in Y, which contradicts the result in [**8**] : If an orientation reversing homeomorphism of the plane leaves a continuum invariant, then it has a fixed point the continuum. Let R_i be the set of fixed points of $\widetilde{h}_i$ in $\bigcup_{j=1}^{n-1} \tau^{-1}(u_j)$. The cardinality of one of the sets, say R_1, is at least 2^{k-1}. The continuum Y has at least 2^{k-1} bounded complementary domains which are invariant under f_1. The cardinality of the set of fixed points of f_1 in Y equals the cardinality of P_1 which is less than the cardinality of Q.

Proceeding inductively, we obtain a finite sequence of continua in the plane, $Y_k = X, Y_{k-1} = Y, Y_{k-2}, \dots, Y_0$ and a sequence of homeomorphisms $g_i : R^2 \to R^2$, $i = 0, \dots, k$, such that for each i, we have

(1) $g_i(Y_i) = Y_i$,

(2) g_{i-1} has fewer fixed points in Y_{i-1} than g_i has in Y_i,

(3) Y_i has at least 2^i bounded complementary domains, invariant under g_i.

By [8], g_0 has at least two fixed points in Y_0. Hence $g_k = h$ has at least $k+2$ fixed points in $Y_k = X$.

COROLLARY. *If X has infinitely many complementary domains which are invariant under h, then h has infinitely many fixed points in X.*

PROOF: Let k be an integer. In the above Theorem set $n = 2^k$. Hence the number of fixed points of h in X exceeds any given integer k.

PROBLEM 1. *Suppose that X has n bounded complementary domains which are invariant under h. Must h have $n+1$ fixed points in X?*

PROBLEM 2. *Under what conditions must h have a Cantor set of fixed points in X?*

REFERENCES

1. Harold Bell, *A fixed point theorem for planar homeomorphisms*, Bull. Amer. Math. Soc. **82** (1976), 778–780.
2. Harold Bell, *A fixed point theorem for plane homeomorphisms*, Fund. Math. **100** (1978), 119–128.
3. Beverly Brechner, *Prime ends, indecomposable continua, and the fixed point property*, Top. Proc. **(1)4** (1979), 227–234.
4. Morton Brown, *A short short proof of the Cartwright-Littlewood fixed point theorem*, Proc. Amer. Math. Soc. **65** (1977), p. 372.
5. M.L. Cartwright and J.C. Littlewood, *Some fixed point theorems*, Ann. of Math. **54** (1951), 1–37.
6. O.H. Hamilton, *A short proof of the Cartwright-Littlewood fixed point theorem*, Canad. J. Math. **6** (1954), 522–524.
7. Sze-Tsen Hu, "Homotopy Theory,". Academic Press, 1959.
8. K. Kuperberg, *Fixed points of orientation reversing homeomorphisms of the plane*, Proc. Amer. Math. Soc. (to appear).

Keywords. Fixed point
1980 *Mathematics subject classifications*: primary 55M20, secondary 54F15, 54H25, 58C30

Department of Foundations, Analysis and Topology, Auburn University, Alabama 36849-5310

Invariant Tori and Cylinders for a Class of Perturbed Hamiltonian Systems

ERNESTO A. LACOMBA, JAUME LLIBRE AND ANA NUNES

Abstract. We start with a relativistic model of the Kepler Problem, which is an isoenergetically non-degenerate central force problem in 2 dimensions. Then we prove the persistence of invariant cylinders and tori for a class of non Hamiltonian perturbations of this system.

§1. Introduction.

Consider the Hamiltonian $\bar{H}_\varepsilon : \mathbf{R}^+ \times S^1 \times \mathbf{R}^2 \to \mathbf{R}$ defined by

$$\bar{H}_\varepsilon(r, \theta, p_r, p_\theta) = \frac{p_r^2}{2} + \frac{p_\theta^2 - 2\varepsilon}{2r^2} - \frac{1}{r}, \tag{1}$$

where $\varepsilon \in \mathbf{R}^+$. Notice that if $\varepsilon = 0$ then $\bar{H}_\varepsilon$ is the Kepler Hamiltonian, and if $\varepsilon \neq 0$ then it is the correction given by special relativity or by a first order approximation in general relativity to the Kepler problem. These Hamiltonians have the general form

$$\frac{p_r^2}{2} + \frac{p_\theta^2}{2r^2} - (\frac{\alpha}{r} + p_0)^2 ,$$

where p_0 is a positive constant of motion and $\alpha > 0$. If $\alpha < 0$, they describe the motion of a particle in the relativistic coulombian electric field, produced by a charged particle of the same sign (see [**T**]).

For the special relativity correction, the coefficients of the terms in r^{-1} and r^{-2} are

$$2\alpha_s p_{0s} = (1 + E/(mc_0^2))\gamma \quad , \quad \alpha_s^2 = \gamma^2/(2mc_0^2)$$

respectively. Here c_0 is the velocity of light, m is the mass of the particle, γ is the universal gravitational constant times the mass of the central body and E is the constant of total energy. A first order approximation to general relativity (Schwarzschild field) considered in [**P**], gives the values

$$2\alpha_g p_{0g} = (1 + 4E/(mc_0^2))\gamma \quad , \quad \alpha_g^2 = 6\alpha_s^2$$

for the above coefficients. Perturbation computations to estimate the precesion of Mercury show that it depends only on the coefficient α_s^2 or α_g^2

respectively. Hence, the correct value from the general relativity model is 6 times the value predicted by special relativity.

A normalization of coordinates permits to write $2\alpha p_0 = 1$ in both cases, and we get Hamiltonian (1). Even the special relativity Hamiltonian has a historical importance because Sommerfeld obtained correct results for the hydrogen atom. However, it does not work in general since we have to take into account the spin of the electron and relativistic quantum mechanical behavior (see [**B**], [**T**]). These two effects cancel each other in the case of the hydrogen atom.

The exact solution to the Scharzschild field requires to correct the Newtonian potential γ/r to $\gamma/r + \gamma p_\theta^2 c_0^{-2}/r^3$. But the perihelion precesion has to be computed by perturbation methods anyway. For the case of Mercury the result agrees with the one obtained from the above first order approximation.

After this digression, we consider again Hamiltonian (1). Its associated equations of motion are

$$
\begin{aligned}
\dot{r} &= p_r, \\
\dot{\theta} &= p_\theta\, r^{-2}, \\
\dot{p}_r &= (p_\theta^2 - 2\varepsilon)\, r^{-3} - r^{-2}, \\
\dot{p}_\theta &= 0.
\end{aligned}
\tag{2}
$$

Let $\bar{H}_1$: $\mathbf{R}^+ \times S^1 \times \mathbf{R}^2 \times (-\mu^*, \mu^*) \to \mathbf{R}$ be an analytic function in the variables $(r, \theta, p_r, p_\theta, \mu)$ with $\mu^* \neq 0$, and consider a perturbation of (2) of the form

$$
\begin{aligned}
\dot{r} &= p_r + \mu \frac{\partial \bar{H}_1}{\partial p_r}, \\
\dot{\theta} &= p_\theta r^{-2} + \frac{\mu}{1+r^2} \frac{\partial \bar{H}_1}{\partial p_\theta}, \\
\dot{p}_r &= (p_\theta^2 - 2\varepsilon) r^{-3} - r^{-2} - \mu \frac{\partial \bar{H}_1}{\partial r}, \\
\dot{p}_\theta &= -\frac{\mu}{1+r^2} \frac{\partial \bar{H}_1}{\partial \theta}.
\end{aligned}
\tag{3}
$$

System (3) possesses the first integral

$$
\bar{H}(r, \theta, p_r, p_\theta, \varepsilon, \mu) = \bar{H}_\varepsilon(r, \theta, p_r, p_\theta) + \frac{p_\theta^2 - 2\varepsilon}{2} + \mu \bar{H}_1(r, \theta, p_r, p_\theta, \mu),
\tag{4}
$$

but it is not necessarily Hamiltonian. In fact, if system (3) has Hamiltonian form, then it is integrable, (see the appendix).

For $\mu = 0$, (3) reduces to (2), which is an integrable Hamiltonian system, with first integrals the Hamiltonian $\bar{H}_\varepsilon$ and the angular momentum $C(r,\theta,p_r,p_\theta) = p_\theta$. Moreover, the flow associated to the Hamiltonian C, φ_c^t, defines an action of the group $\mathbf{R}$ on the phase space such that the isotropy subgroup of $\mathbf{R}$ at any point of $\mathbf{R}^+ \times S^1 \times \mathbf{R}^2$ is $\mathbf{Z}$. So the phase space of $\bar{H}_\varepsilon$ is foliated by invariant tori and cylinders (see [**AKN**]). The purpose of this paper is to show that, *even when (3) is not Hamiltonian, some of these invariant tori and cylinders persist for $|\mu|$ small enough.*

The proof is based on extending analytically to an augmented phase space the perturbed system (3) giving a new system, which is Hamiltonian (see section 3). In section 4 we see that an invariant submanifold of each energy level of the unperturbed extended system is foliated by tori, and each of these tori either coincides with an invariant torus of $\bar{H}_\varepsilon$ or contains an invariant cylinder of $\bar{H}_\varepsilon$. Applying then the Kolmogorov-Arnold-Moser theorem, we conclude that most of these tori persist under perturbation. Going back through the changes of variables, which allow us to extend the flow, we obtain that system (3) still has invariant tori and cylinders close to the unperturbed ones for $|\mu|$ small enough (see section 5).

The techniques employed here may be applied to other perturbed central force type problems.

The authors wish to thank Alain Chenciner for many helpful comments on the subject of this paper.

The first author was on sabatical leave at Barcelona University during the elaboration of this paper. The other authors are partially supported by CICYT under grant PB 86–0351. The last author was also on leave of absence from Universidade de Lisboa, Departamento de Física, partially supported by Fundaçao Calonste Gulbenkian under grant 321851B.

§2. Summary of results on Hamiltonian systems.

In this section, we summarize the results on Hamiltonian systems that we shall need.

The differential equations of mechanics may be written in Hamiltonian form

$$\text{(5)} \qquad \dot{x}_k = \frac{\partial H}{\partial y_k}\,, \dot{y}_k = -\frac{\partial H}{\partial x_k} \text{ for } k = 1, \ldots, n,$$

where $x = (x_1, \dots x_n) \in \mathbf{R}^n$ and $y = (y_1, \dots, y_n) \in \mathbf{R}^n$ are coordinates in the phase space U, an open subset of $\mathbf{R}^{2n}$, and $H : U \to \mathbf{R}$ is an analytic function.

Recall that a collection of functions $F_1, \dots, F_r$ from U to $\mathbf{R}$ is said to be *in involution* if $\{F_k, F_j\} = 0$ for $k, j = 1, \dots, r$, where $\{\cdot, \cdot\}$ denotes the *Poisson bracket*. The collection $F_1, \dots, F_r$ is said to be *independent* if the set $\sigma(F)$ of critical points of $F = (F_1, \dots F_r)$ has zero measure in U. Recall that

$$\sigma(F) = \{p \in U : dF_1(p), \dots, dF_r(p) \text{ are linearly dependent } \}.$$

A Hamiltonian system (5) defined in an open domain U of $\mathbf{R}^{2n}$ is called *integrable* if there exist n independent first integrals $F_1, \dots F_n$ in involution.

The structure of integrable Hamiltonian systems is particularly simple, as the following theorems show.

THEOREM 1. *(Liouville* [**L**]*). An integrable Hamiltonian system is integrable by quadratures.*

THEOREM 2. *(Arnold, see* [**A2**] *and* [**AM**]*). Assume that the Hamiltonian system (5) is integrable with the first integrals $F_1 = H, F_2, \dots, F_n$, and let $I_{c_1,c_2\dots c_n}$ be the differentiable manifold defined by $F_1 = c_1$, $F_2 = c_2, \dots, F_n = c_n$, where $F^{-1}(c_1, c_2, \dots, c_n) \cap \sigma(F) = \phi$. Then, the following statements hold.*

(a) *Each connected component of $I_{c_1,c_2\dots c_n}$ is invariant for the flow of (5).*
(b) *Each compact connected component of $I_{c_1,c_2\dots c_n}$ is diffeomorphic to the torus T^n. If on a non-compact connected component of $I_{c_1,c_2\dots c_n}$ the Hamiltonian vector fields X_{F_k} are complete for $k = 1, \dots, n$, then this component is diffeomorphic to the cylinder $\mathbf{R}^e \times T^{n-e}$ with $e \in \{1, \dots, n\}$.*
(c) *Under the hypotheses of (b), the flow of (5) on a connected component of $I_{c_1,c_2\dots c_n}$ is differentiably conjugate to a translation type flow on $\mathbf{R}^e \times T^{n-e}$ (for $e = 0$ the flow is quasi-periodic).*

From Theorem 2, if a Hamiltonian system (5) is integrable, then we can introduce on a family of invariant manifolds diffeomorphic to T^n coordinates $p = (p_1, \dots, p_n)$, $q = (q_1, \dots, q_n)$ via a canonical transformation $(x, y) = T(p, q)$ such that T has period 1 in the variables $q_1, \dots, q_n$ and such that the new Hamiltonian $H = H(p)$ is independent of q. In these coordinates, the system takes the simple form

$$\dot{q}_k = \frac{\partial H}{\partial p_k}, \; \dot{p}_k = -\frac{\partial H}{\partial q_k} = 0 \; \text{ for } k = 1, \ldots, n \tag{6}$$

and the solutions are

$$q_k = \frac{\partial H}{\partial p_k}(c_1, \ldots, c_n)t + q_k(0)\,, \; p_k = c_k \; \text{ for } k = 1, \ldots, n.$$

Thus, the p_k are constant along each solution, i.e., the p_k are n integrals of the Hamiltonian system (6). The canonical coordinates (q, p) are called *action-angle variables*.

Now we consider a small perturbation of an integrable Hamiltonian system. Let (x, y) be the action-angle coordinates for the unperturbed Hamiltonian given by the analytic function $H_0(y)$. Assume that the perturbed system is of the form $H(x, y, \mu) = H_0(y) + \mu\, H_1(x, y, \mu)$ where μ is a small parameter and H_1 is analytic in its $2n + 1$ variables and of period 1 in $x_1, \ldots, x_n$. Then, for $\mu = 0$ the $2n$-dimensional phase-space is foliated into a n-parameter family of invariant tori $y_k = c_k$ for $k = 1, \ldots, n$, on which the flow is given by $\dot{x}_k = \frac{\partial H_0}{\partial y_k}(c_1, \ldots, c_n)$. If the Hessian

$$\det\left(\frac{\partial^2 H_0}{\partial y_j \partial y_e}\right) \neq 0 \tag{7}$$

we say that H_0 is *non-degenerate*. In this case, the invariant tori of the foliation may be parametrized by the frequencies $w_k = \frac{\partial H_0}{\partial y_k}$, $k = 1, \ldots, n$.

There is a similar but independent non-degeneracy condition. We say that the unperturbed Hamiltonian $H_0(y)$ is *isoenergetically non-degenerate* if the determinant

$$\begin{vmatrix} \frac{\partial^2 H_0}{\partial y_j \partial y_e} & \frac{\partial H_0}{\partial y_j} \\ \frac{\partial H_0}{\partial y_e} & 0 \end{vmatrix} \neq 0\,. \tag{8}$$

Here the invariant tori of the foliation on each energy level may be parametrized by the frequency ratios $(w_1 : w_2 : \ldots : w_n)$. Actually, only condition (8) is important for this paper, since (7) is not valid in our case.

The next result tells us what happens to this foliations for small values of μ.

THEOREM 3. *(Kolmogorov-Arnold, see* [**A1**], [**A2**]). *Let $H(x,y,\mu)$ be an analytic Hamiltonian in the variables x,y,μ for all x,y in an open subset of $\mathbf{R}^{2n}$ and with μ near zero. Assume that H has period 1 in $x_1,\dots,x_n$ and that $H_0 = H(x,y,0)$ does not depend on x and is non-degenerate (respectively, isoenergetically non-degenerate) . Let $c = (c_1,\dots,c_n)$ be chosen so that the frequencies $w_k = \frac{\partial H_0}{\partial y_k}(c)$ satisfy with some positive constants γ,τ the inequalities*

$$|\Sigma_{k=1}^{n} j_k w_k| \geq \gamma |j|^{-\tau}$$

for all integers j_k with $|j| = \Sigma|j_k| \geq 1$. Then, for sufficiently small $|\mu|$, there exists a torus $x = \theta + u(\theta,\mu)$, $y = c + v(\theta,\mu)$, where $u(\theta,\mu)$ and $v(\theta,\mu)$ are analytic functions of period 1 in $\theta_1,\dots,\theta_n$ and vanishing for $\mu = 0$, which is invariant for the flow of H. The flow on this torus is of translation type and it is given by $\dot{\theta}_k = w_k$ for $k = 1,\dots,n$.

The version of Kolmogorov-Arnold theorem for Hamiltonian systems can be translated to diffeomorphisms of the annulus. Such diffeomorphisms have been studied by Moser, Rüssman and Herman, who showed that C^3 differenciability is sufficient to obtain invariant curves. For more details see [**Mo**] and [**H**].

§3. Recovery of Hamiltonian structure and extension of the phase space.

Consider again system (3), and the change of variables $F : \mathbf{R}^+ \times S^1 \times \mathbf{R}^2 \to (0,\pi/2) \times S^1 \times \mathbf{R}^2$ defined by

$$F(r,\theta,p_r,p_\theta) = (x = \arctan r, \theta, p_x = p_r, p_\theta), \tag{9a}$$

along with the change in time scale given by

$$\frac{dt}{d\tau} = \cos^{-2} x. \tag{9b}$$

In terms of the new variables, system (3) becomes

$$\begin{aligned} x' &= p_x + \mu \frac{\partial H_1}{\partial p_x}, \\ \theta' &= p_\theta \sin^{-2} x + \mu \frac{\partial H_1}{\partial p_\theta}, \\ p'_x &= (p_\theta^2 - 2\varepsilon) \cos x \sin^{-3} x - \sin^{-2} x - \mu \frac{\partial H_1}{\partial x}, \\ p'_\theta &= -\mu \frac{\partial H_1}{\partial \theta}, \end{aligned} \tag{10}$$

where $'$ denotes the derivative with respect to the new time variable τ. System (10) is Hamiltonian with Hamiltonian function

$$(11) \quad H(x,\theta,p_x,p_\theta,\mu) = \frac{p_x^2}{2} + \frac{p_\theta^2 - 2\varepsilon}{2\sin^2 x} - \frac{\cos x}{\sin x} + \mu H_1(x,\theta,p_x,p_\theta,\mu).$$

Moreover, for $\mu = 0$, (10) and (11) may be analytically extended to $(0,\pi)\times S^1 \times \mathbf{R}^2$. Thus, the change of variables defined by F regularizes the $r=\infty$ singularity of (3), and we have the following result.

PROPOSITION 4. *The change of variables defined by equations (9) transforms system (3) into the Hamiltonian system associated to H (see (11)). Moreover, if the perturbing function $H_1(x,\theta,p_x,p_\theta,\mu) = \bar{H}_1(r = \tan x, \theta, p_r = p_x, p_\theta, \mu)$, is analytic in $D = (0,\pi)\times S^1 \times \mathbf{R}^2 \times (-\mu^*, \mu^*)$, then H may be analytically extended to D.*

Notice that since the perturbing function H_1 is arbitrary the Hamiltonian system associated to H will be in general non-integrable, see [**MM**].

§4. Global flow of the unperturbed extended Hamiltonian system.

In this section, we shall study the extended system $H_0 : (0,\pi)\times S^1 \times \mathbf{R}^2 \to \mathbf{R}$ defined by $H_0(x,\theta,p_x,p_\theta) = H(x,\theta,p_x,p_\theta,0)$. We shall see that each energy level of H_0 is foliated by invariant cylinders and tori. Going back through the change of variables (9), some of these tori coincide with invariant tori of the original integrable system $\bar{H}_\varepsilon$, while the others contain invariant cylinders of $\bar{H}_\varepsilon$.

PROPOSITION 5. *For $\mu = 0$, system (10) is integrable with first integrals $H_0 = h$ and $C(x,\theta,p_x,p_\theta) = p_\theta = c$. For every $h < \frac{1}{4\varepsilon} - \varepsilon$, c takes values in the interval*

$$[-\sqrt{h + \sqrt{h^2+1} + 2\varepsilon}, \sqrt{h + \sqrt{h^2+1} + 2\varepsilon}\,],$$

and the invariant surfaces I_{hc} are cylinders (resp. tori, circles) if $0 \le c^2 \le 2\varepsilon$ (resp. $c^2 \in (2\varepsilon, h + \sqrt{h^2+1} + 2\varepsilon)$, $c^2 = h + \sqrt{h^2+1} + 2\varepsilon$). Moreover, I_{hc} coincides with an invariant surface of $\bar{H}_\varepsilon$ (resp. contains an invariant surface of $\bar{H}_\varepsilon$) if $h < \frac{c^2-2\varepsilon}{2}$ (resp. $h \ge \frac{c^2-2\varepsilon}{2}$).

PROOF: Clearly, $C(x,\theta,p_x,p_\theta) = p_\theta$ is a first integral for the equations of motion associated to the Hamiltonian H_0, independent from H_0. Such equations may be reduced to the family of Hamiltonian systems of one degree of freedom given by

$$H_c(x,p_x) = \frac{p_x^2}{2} + V_c(x), \quad V_c(x) = \frac{c^2 - 2\varepsilon}{2\sin^2 x} - \frac{\cos x}{\sin x},$$

along with the complementary equations

$$\theta' = c \sin^{-2} x$$
$$p_\theta = c = \text{ const.}$$

The curves $V_c(x)$ are represented in Figure 1. From this figure the energy level $H = h$ with $h < \frac{1}{4\varepsilon} - \varepsilon$ has two components. One of them , I_h', has its projection in the x-axis contained in $(\pi/2, \pi)$. Since the change of variables F of section 3 maps the interval $r \in (0, \infty)$ into the interval $(0, \pi/2)$, we are not interested in the component I_h'. From now on we shall restrict our study to the other component that we shall denote by I_h.

From Figure 1, $I_h = \bigcup_c I_{hc}$ with $c^2 \in [0, h + \sqrt{h^2+1} + 2\varepsilon]$. The projection of these I_{hc} on the strip (x, p_x) are the curves $H_c(x,p_x) = h$, which are shown in Figure 2. Of course, I_{hc} may be obtained from such curves rotating them around the p_x-axis. This proves that the topology of the I_{hc} is as stated in the proposition.

Let $\bar{h} = h - \frac{c^2 - 2\varepsilon}{2}$ be the constant value of $\bar{H}_\varepsilon$ on I_{hc} and let $\Pi : (0,\pi) \times S^1 \times \mathbf{R}^2 \to (0, \pi/2) \times S^1 \times \mathbf{R}^2$ the natural projection. Then $\Pi(I_{hc}) = I_{\bar{h}c}$, and the proposition follows. □

From Figure 1, we see that if $h < \frac{1}{4\varepsilon} - \varepsilon$. The only singularity of H_0 is the origin $x = 0$. This singularity can be studied blowing it up to a boundary manifold and extending the flow to this manifold (see for instance [**Mc**])

Let then

$$v = p_x \sin x, \tag{12}$$
$$u = p_\theta,$$
$$\frac{d\tau}{d\tau'} = \sin^2 x$$

Equations (10) and (11) become

$$\frac{v^2+u^2}{2} - \varepsilon = (h \sin x + \cos x) \sin x, \tag{13}$$

$$\begin{aligned} \dot{x} &= v \sin x, \\ \dot{\theta} &= u, \\ \dot{v} &= 2h \cos x \sin^2 x + 2 \sin x \cos^2 x - \sin x, \\ \dot{u} &= 0, \end{aligned} \tag{14}$$

where here the dot denotes the derivative with respect to the new time τ'. Equations (13) and (14) are analytic in $x \in [0, \pi]$, and so we may extend the flow over the singularity $x = 0$. According to (13) and (14), this singularity corresponds to the invariant torus (recall that $\varepsilon > 0$):

$$\Lambda = \{(x, \theta, v, u) : x = 0, \theta \in S^1, \ u^2 + v^2 = 2\varepsilon\},$$

and the flow on Λ is as shown in Figure 3.(a).

PROPOSITION 6. *Let $h < \frac{1}{4\varepsilon} - \varepsilon$, and denote by I_h the energy level $H_0 = h$ and by $\bar{I}_h$ the compactified energy level $I_h \cup \Lambda$. Then, $\bar{I}_h$ is a closed solid torus foliated by the invariant surfaces I_{hc} as shown in Figures 3.(b).*

PROOF: It is an immediate consequence of Proposition 5 along with the change of variables (12). □

§5. Invariant tori and cylinders of the perturbed flow.

In this section, we apply Theorem 3 to the invariant tori given by Propositions 5 and 6, and then conclude the persistence under perturbations of the form (3) of the invariant tori and cylinders of the system (2). These results are not valid for $\varepsilon = 0$.

LEMMA 7. *The Hamiltonian H_0 for $\varepsilon > 0$ is isoenergetically non-degenerate.*

PROOF: Let $c^2 \in (2\varepsilon, h + \sqrt{h^2+1} + 2\varepsilon)$ and let us introduce action-angle variables on the tori I_{hc} given by Proposition 5. By definition (see, for instance, [**A1**]), we have (recall that here $S^1 = \mathbf{R}/\mathbf{Z}$):

$$\begin{aligned} J_1 &= \oint p_\theta \, d_\theta = c, \\ J_2 &= \oint p_x \, d_x = 2 \int_A^B \sqrt{2h - (c^2 - 2\varepsilon) \sin^{-2} x + 2 \sin^{-1} x \cos x} \, dx, \end{aligned} \tag{15}$$

where

$$\begin{array}{lll} tg\,A = \frac{-1+\sqrt{1+(c^2-2\varepsilon)a}}{a}, & tg\,B = \frac{-1-\sqrt{1+(c^2-2\varepsilon)a}}{a} & \text{if } a \neq 0, \\ tg\,A = 1/h, & B = \pi/2 & \text{if } a = 0, \end{array}$$

with $a = 2h - (c^2 - 2\varepsilon)$. From (15), we have

$$\frac{\partial J_2}{\partial h} > 0, \tag{16a}$$

$$\frac{\partial J_2}{\partial c} = -2c \int_A^B \frac{dx}{\sin^2 x \sqrt{2h - (c^2 - 2\varepsilon)\sin^{-2} x + 2\cos x \sin^{-1} x}}$$

$$= -\frac{c}{\sqrt{c^2 - 2\varepsilon}}. \tag{16b}$$

Equations (16) imply that

$$h(J_1, J_2) = f(\sqrt{J_1^2 - 2\varepsilon} + J_2),$$

where the derivative f' is always positive. Using the definition (8), the system is isoenergetically non-degenerate if and only if

$$\frac{2\varepsilon[f'(J_2 + \sqrt{J_1^2 - 2\varepsilon})]^3}{(J_1^2 - 2\varepsilon)^{3/2}} > 0,$$

and the result follows. □

Hence, we are in conditions for applying Theorem 3 to the perturbed system (10). Going back through the changes of variables (which are diffeomorphisms in the region $x \in (0, \pi/2)$), we have the following result:

THEOREM 8. *Consider system (3) with* $\varepsilon > 0$ *and denote by* I_h *the invariant manifold defined by* $\bar{H}(r, \theta, p_r, p_\theta, \mu) = h$. *Assume that the extended perturbing function* $H_1(x, \theta, p_x\, p_\theta, \mu)$ *is sufficiently differentiable in* $(0, \pi) \times S^1 \times \mathbb{R}^2 \times [-\mu^*, \mu^*]$. *Then, for* $|\mu|$ *small enough, system (3) has invariant tori and cylinders close to the invariant tori and cylinders that foliate* $I_h \cap \{(r, \theta, p_r, p_\theta) : p_\theta^2 > 2\varepsilon\}$ *for* $\mu = 0$.

Appendix.

Clearly if system (3) has Hamiltonian form its Hamiltonian must be

$$\widetilde{H} = \bar{H}_\varepsilon(r, \theta, p_r, p_\theta) + \mu\, G(r, \theta, p_r, p_\theta, \mu),$$

where G verifies simultaneously

$$G = \frac{1}{1+r^2}\bar{H}_1 + f(r, p_r, \mu),$$
$$G = \bar{H}_1 + g(\theta, p_\theta, \mu),$$

with f and g arbitrary analytical functions. Hence, system (3) has Hamiltonian form if and only if

$$\bar{H}_1 = \frac{1+r^2}{r^2}[f(r, p_r, \mu) - g(\theta, p_\theta, \mu)].$$

In this case the Hamiltonian is

$$\tilde{H} = \bar{H}_\varepsilon(r, \theta, p_r, p_\theta) + \mu[\frac{1+r^2}{r^2} f(r, p_r, \mu) - \frac{1}{r^2} g(\theta, p_\theta, \mu)].$$

Now, let us prove that if system (3) has Hamiltonian form with Hamiltonian $\tilde{H}$, then it is integrable. First, notice that if g is independent on θ then (from (3)) p_θ is a first integral independent from $\tilde{H}$. Otherwise, since $\bar{H}$ is a first integral of system (3) (see (4)) and

$$\bar{H} - \tilde{H} = \frac{1}{2}(p_\theta^2 - 2\varepsilon) - \mu\, g(\theta, p_\theta, \mu),$$

it follows that $\bar{H}$ is independent from $\tilde{H}$.

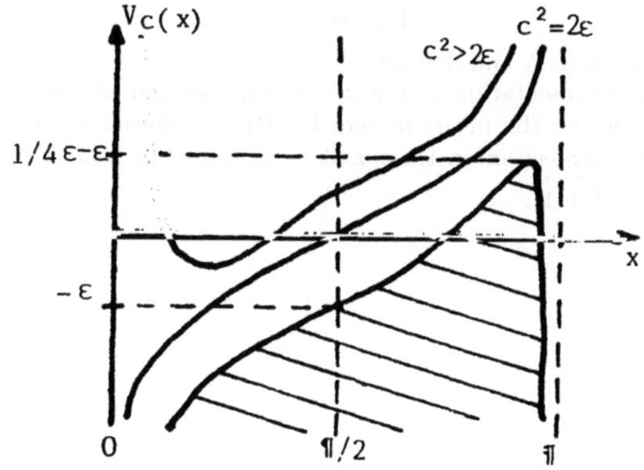

Figure 1. The curves $V_c(x)$. Note that $V_c(\pi/2) = \frac{c^2-2\varepsilon}{2}$, and that when $c^2 > 2\varepsilon$, $\min_{x\in(0,\pi)} V_c(x) = \frac{(c^2-2\varepsilon)^2-1}{2(c^2-2\varepsilon)}$.

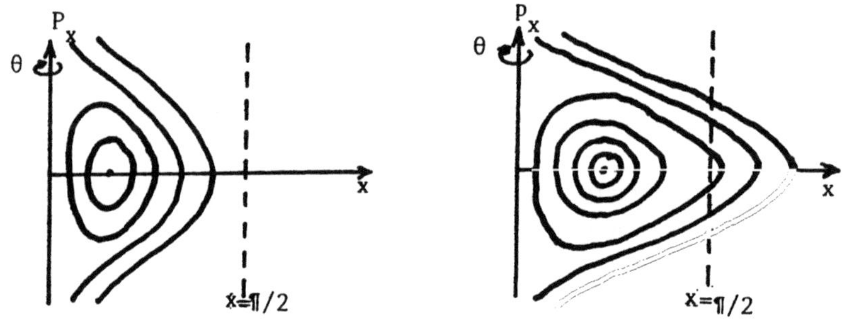

$h < -\varepsilon$ $\qquad$ $h \in (-\varepsilon, \frac{1}{4\varepsilon} - \varepsilon)$

Figure 2. Projection on the (x, p_x) strip of the invariant manifolds I_{hc}. Note that whenever $h \geq \frac{c^2 - 2\varepsilon}{2} = V_c(\pi/2)$ the I_{hc} contain points whose x coordinate belongs to $[\pi/2, \pi]$.

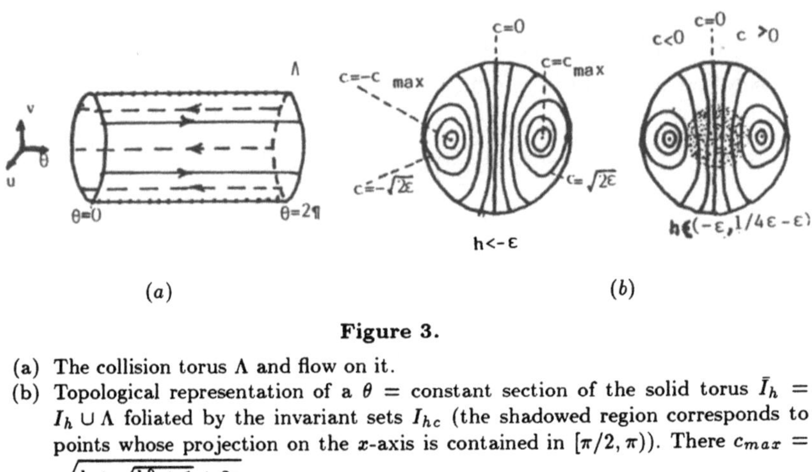

(a) $\qquad$ (b)

Figure 3.

(a) The collision torus Λ and flow on it.
(b) Topological representation of a θ = constant section of the solid torus $\bar{I}_h = I_h \cup \Lambda$ foliated by the invariant sets I_{hc} (the shadowed region corresponds to points whose projection on the x-axis is contained in $[\pi/2, \pi)$). There $c_{max} = \sqrt{h + \sqrt{h^2+1} + 2\varepsilon}$.

REFERENCES

[AM] R. Abraham and J. Marsden, "Foundations of Mechanics," Benjamin, Reading, Mass, 1978.

[A1] V.I. Arnold, *Mathematical Methods of Classical Mechanics*, MIR (russian) (1975), Moscow; (english) (1978), Springer-Verlag, New York.

[A2] V.I. Arnold, *Proof of A.N. Kolmogorov's theorem on the perturbation of quasi-periodic motions under small perturbations of the Hamiltonian*, (russian) Usp. Mat. Nauk SSSR **18** (1963), 13–40; (english) Russian Math. Surveys **18** (1963), 9–36.

[AKN] V.I. Arnold, V.V. Kozlov and A.I. Neishtadt, *Mathematical aspects of classical and Celestial Mechanics*, Encyclopedia of Mathematical Sciences, vol. **3** (1988), Springer-Verlag, Berlin.

[B] P.G. Bergmann, "Introduction to the Theory of Relativity," Dover, New York, 1976.

[H] M.R. Herman, *Sur les courbes invariantes par les difféomorphismes de l'anneau*, Astérisque **144** (1986).

[L] J. Liouville, *Sur l'integration des equations différentielles de la dynamique*, J. de Math. Pures et Appl. **20** (1855), 137–138.

[MM] L. Markus and K.R. Meyer, *Generic Hamiltonian dynamical systems are neither integrable non ergodic*, Memoirs of the Amer. Math. Soc. **144** (1974).

[Mc] R. McGehee, *Triple collision in the collinear three-body problem*, Inventiones Math. **27** (1974), 191–227.

[Mo] J. Moser, "Stable and random motion in dynamical systems," Princeton Univ. Press, Princeton, 1973.

[P] D. Park, *Classical Dynamics and its quantum analogues*, Lecture Notes in Physics **110** (1979), Springer-Verlag, New York.

[T] W. Thirring, "A Course in Mathematical Physics I, Classical Dynamical Systems," Springer Verlag, 1978, New York.

E.A. Lacomba: Departamento de Matemáticas
Universidad Autónoma Metropolitana
Iztapalapa, Apdo. P. 55–534
09340 Mexico D.F.

J. Llibre: Departament de Matemàtiques
Universitat Autònoma de Barcelona
Bellaterra, 08193 Barcelona
Spain

Ana Nunes: Departamento de Física
Universidad de Lisboa
R. Ernesto de Vasconcelos, Ed. C1, Piso 4
1700 Lisboa
Portugal

March, 1989. Revised version, May, 1990

Modified Structures Associated with Poisson Lie Groups

LUEN-CHAU LI AND SERGE PARMENTIER

Abstract. Given a Poisson Lie group, we show how to construct modified (Poisson) structures using suitable linear maps from the dual of the Lie algebra to the Lie algebra. An important class of such maps for which our construction works is given by morphisms of the corresponding Lie bialgebras. The twisted Poisson structures which arose in earlier work of the authors on Lax equations on a lattice are now identified as a special instance of this general construction.

§1. Introduction.

The notion of a Poisson Lie group was introduced by V. G. Drinfel'd in [**1**] as a result of considerations concerning the work of Sklyanin [**10,11**]. Subsequently, this has led to a number of investigations discerning the basic algebraic and geometric aspects, as well as its applications to integrable systems theory [**3,7,8,9**].

By definition, a Lie group G equipped with a Poisson bracket $\{\ ,\ \}_G$ is called a Poisson Lie group if group multiplication is a Poisson map from $G \times G$ (equipped with the product structure) into G. In this paper, we shall consider on a Poisson Lie group $(G, \{\ ,\ \}_G)$ a prescription to obtain modified (Poisson) structures. More precisely, we shall construct on G a family of Poisson brackets $\{\ ,\ \}_\Gamma$ which differ from $\{\ ,\ \}_G$ by anomalous terms associated with linear maps Γ from $\mathfrak{g}^*$ to $\mathfrak{g}$. An important class of Γ's for which our prescription works is given by morphisms of Lie bialgebras. As a matter of fact, an example connected with the Sklyanin bracket was already known to us in our work on Lax systems and higher order structures in the r-matrix approach [**6**].

The paper is organized as follows. In Section 2, we begin by reviewing the basic facts about Poisson Lie groups and r-matrices. In Section 3, we give the abstract theorem describing the modified structures $\{\ ,\ \}_\Gamma$. In the case where Γ is a morphism of Lie bialgebras, a dual pair for the Poisson manifold $(G, \{\ ,\ \}_\Gamma)$ can be constructed using the (local) double Lie group associated with the Poisson Lie group G. The upshot is that we have a description of the symplectic leaves of $(G, \{\ ,\ \}_\Gamma)$ (at least) in a neighborhood of the identity. Section 4 is concerned with Lax systems on a lattice, with most of the results taken from [**6**, Section 3]. They serve to illustrate/motivate the abstract theorem which we believe is of some interest in the theory of Poisson Lie groups. We conclude the paper by giving a tensor form of the (non-ultralocal) twisted Poisson structure in [**6**]. The problem of enumerating all transition matrices which satisfy our fundamental Poisson bracket relations is posed for future work.

§2. Preliminaries.

We begin by reviewing some of the basic facts from the theory of Poisson Lie groups and r-matrices. The main references are [**1,7,8,9**].

DEFINITION 2.1: A Lie group G equipped with a Poisson bracket $\{\ ,\ \}_G$ is called a Poisson Lie group if group multiplication is a Poisson map from $G \times G$ (equipped with the product structure) into G.

For a Lie group G with Lie algebra $\mathfrak{g}$ and dual Lie algebra $\mathfrak{g}^*$, we use $L(\mathfrak{g}^*, \mathfrak{g})$ to denote the space of linear maps from $\mathfrak{g}^*$ to $\mathfrak{g}$.

Given $\varphi \in C^\infty(G)$, its right and left gradients are defined by

$$(2.1)\qquad \begin{gathered} \langle D\varphi(g), X\rangle = \tfrac{d}{dt}|_{t=0}\varphi(e^{tX}g),\ \langle D'\varphi(g), X\rangle = \tfrac{d}{dt}|_{t=0}\varphi(ge^{tX}), \\ D\varphi(g),\ D'\varphi(g) \in \mathfrak{g}^*,\ X \in \mathfrak{g}. \end{gathered}$$

Writing the Poisson bracket on G in the right-invariant frame:

$$(2.2)\qquad \{\varphi, \psi\}_G(g) = \langle D\psi(g), \eta(g) D\varphi(g)\rangle,$$

we have $(G, \{\ ,\ \}_G)$ is a Poisson Lie group if and only if the Hamiltonian operator $\eta : G \to L(\mathfrak{g}^*, \mathfrak{g})$ satisfies the cocycle condition, i.e.

$$(2.3)\qquad \eta(gh) = \eta(g) + \mathrm{Ad}_g \circ \eta(h) \circ \mathrm{Ad}_g^*, \qquad g, h \in G.$$

Related to the notion of a Poisson Lie group is the idea of a Lie bialgebra, also due to Drinfel'd.

DEFINITION 2.2: Let $\mathfrak{h}$ be a Lie algebra, $\mathfrak{h}^*$ its dual. Suppose there is a Lie algebra structure on $\mathfrak{h}^*$ such that the map $\rho : \mathfrak{h} \to \mathfrak{h} \otimes \mathfrak{h} \simeq L(\mathfrak{h}^*, \mathfrak{h})$ dual to the Lie bracket map $\mathfrak{h}^* \wedge \mathfrak{h}^* \to \mathfrak{h}^*$ is a 1-cocycle on $\mathfrak{h}$, i.e.

$$(2.4)\quad \rho([X,Y]) = \mathrm{ad}_X \circ \rho(Y) + \rho(Y) \circ \mathrm{ad}_X^* - \mathrm{ad}_Y \circ \rho(X) - \rho(X) \circ \mathrm{ad}_Y^*,$$

then we say $(\mathfrak{h}, \mathfrak{h}^*)$ is a Lie bialgebra[1].

The Lie bialgebras form a category in which the morphisms are maps $p : \mathfrak{h}_1 \to \mathfrak{h}_2$ such that both p and $p^* : \mathfrak{h}_2^* \to \mathfrak{h}_1^*$ are Lie algebra homomorphisms.

Given a Poisson Lie group $(G, \{\ ,\ \}_G)$, the pair $(\mathfrak{g}, \mathfrak{g}^*)$ is a Lie bialgebra in the above sense. Indeed, the Lie algebra structure on $\mathfrak{g}^*$ is obtained from $\{\ ,\ \}_G$ by the formula

$$(2.5) \qquad [\alpha, \beta] = d\{\varphi, \psi\}_G(e), \ \alpha = d\varphi(e), \ \beta = d\psi(e), \ \varphi, \psi \in C^\infty(G)$$

and (2.4) is a consequence of (2.3). Conversely, given a Lie bialgebra $(\mathfrak{h}, \mathfrak{h}^*)$, it can be shown that the 1-cocycle on $\mathfrak{h}$ can be integrated to a 1-cocycle on the corresponding connected and simply-connected Lie group H [4]. Thus the category of Lie bialgebras is equivalent to the category of connected and simply-connected Poisson Lie groups.

There is still another way to look at Lie bialgebras. This is given by Manin's theorem.

THEOREM 2.3. *Let $\mathfrak{h}$ be a Lie algebra whose dual $\mathfrak{h}^*$ also has a Lie bracket. Then the only Lie algebra structure on $\delta = \mathfrak{h} \oplus \mathfrak{h}^*$ such that* (a) *$\mathfrak{h}$ and $\mathfrak{h}^*$ are subalgebras of δ,* (b) *the pairing $\langle\langle X_1 + \alpha_1, X_2 + \alpha_2 \rangle\rangle = \alpha_1(X_2) + \alpha_2(X_1)$, $X_1, X_2 \in \mathfrak{h}$, $\alpha_1, \alpha_2 \in \mathfrak{h}^*$ on δ is* ad-*invariant, is given by*

$$(2.6) \quad [X_1 + \alpha_1, X_2 + \alpha_2] = [X_1, X_2] + \mathrm{ad}^*_{\alpha_2} X_1 - \mathrm{ad}^*_{\alpha_1} X_2 + [\alpha_1, \alpha_2] + \mathrm{ad}^*_{X_2} \alpha_1 - \mathrm{ad}^*_{X_1} \alpha_2.$$

Moreover, it is a Lie bracket on $\delta \Leftrightarrow (\mathfrak{h}, \mathfrak{h}^*)$ is a Lie bialgebra. The Lie algebra δ equipped with the bracket in (2.6) *will be denoted by $\mathfrak{h} \bowtie \mathfrak{h}^*$.*

Thus, it follows that if $(G, \{\ ,\ \}_G)$ is a Poisson Lie group, with (tangent) Lie bialgebra $(\mathfrak{g}, \mathfrak{g}^*)$, then $(\mathfrak{g}^*, \mathfrak{g})$ is also a Lie bialgebra. Consequently, the connected and simply-connected Lie group G^* corresponding to $\mathfrak{g}^*$ is also a Poisson Lie group, called the dual of G.

The following notion of a double Lie group was employed in [7] to study the geometry of dressing transformations.

DEFINITION 2.4: A double Lie group consists of a triple of Lie groups (H, H_+, H_-) where $H_\pm$ are closed Lie subgroups of H and the map $\alpha : H_+ \times H_- \to H : (h_+, h_-) \mapsto h_+ h_-$ is a diffeomorphism.

[1] Throughout the paper, ad^*_X will denote the dual of ad_X.

There is also the notion of a local double Lie group whose definition should be clear from above. The (local) double Lie group which we shall use in the next section is associated with a Poisson Lie group $(G, \{\ ,\ \}_G)$ and its dual $(G^*, \{\ ,\ \}_{G^*})$. More precisely, if D is the connected and simply-connected Lie group with Lie algebra $\mathfrak{g} \bowtie \mathfrak{g}^*$, then (D, G, G^*) is a local double Lie group. Now, let $G \bowtie G^*$ be the local Lie group corresponding to D, and denote the images of the Lie groups G, G^* under the localization functor by the same symbols. Then we have projection maps

$$\Pi_\pm : G \bowtie G^* \to G(G^*) : g_+g_- \mapsto g_\pm,\ g_+ \in G,\ g_- \in G^* \tag{2.7}$$

and actions

$$\begin{aligned} \Phi^+ &: G^* \times G \to G^* : \Phi^+_{g_+}(g_-) = \Pi_-(g_-g_+), \\ \Phi^- &: G^* \times G \to G : \Phi^-_{g_-}(g_+) = \Pi_+(g_-g_+). \end{aligned} \tag{2.8}$$

The next formula says that these actions are by "twisted automorphisms" [7]

$$\begin{aligned} \Phi^+_{g_+}(g_-h_-) &= \Phi^+_{\Phi^-_{h_-(g_+)}}(g_-)\Phi^+_{g_+}(h_-), \\ \Phi^-_{g_-}(g_+h_+) &= \Phi^-_{g_-}(g_+)\Phi^-_{\Phi^+_{g_+(g_-)}}(h_-). \end{aligned} \tag{2.9}$$

To conclude the section, we recall some ingredients from the theory of r-matrices.

DEFINITION 2.5: A linear operator R on a Lie algebra $\mathfrak{h}$ is called a classical r-matrix if the formula

$$[X, Y]_R = \frac{1}{2}([RX, Y] + [X, RY]), \qquad X, Y \in h \tag{2.10}$$

defines a Lie bracket. The vector space $\mathfrak{h}$ when equipped with $[\ ,\ ]_R$ is denoted by $\mathfrak{h}_R$.

An important sufficient condition for $[\ ,\ ]_R$ to define a Lie bracket is given by the modified Yang–Baxter equation (mYB)

$$[RX, RY] - 2R([X, Y]_R) = -[X, Y]. \tag{2.11}$$

Suppose now that H is a Lie group whose Lie algebra $\mathfrak{h}$ is equipped with a nondegenerate ad-invariant pairing $(\cdot,\cdot)$. If $D\varphi, D'\varphi$ are the right and left gradients of $\varphi \in C^\infty(H)$ defined using $(\cdot,\cdot)$ and R, R' are skew-symmetric solutions of (mYB), we have

THEOREM 2.6. $\{\varphi,\psi\}_{(R,R')} = \frac{1}{2}(R(D\varphi),D\psi) + \frac{1}{2}(R'(D'\varphi),D'\psi)$ *defines a Poisson structure on* H. *If we let* $H_{(R,R')} = (H,\{\ ,\ \}_{(R,R')})$, *then in particular,* $H_{(\pm R,\mp R)}$ *are Poisson Lie groups with tangent Lie bialgebras* $(\mathfrak{h},\mathfrak{h}_{\mp R})$. *On the other hand,* $H_{(R,R)}$ *is symplectic in a neighborhood of the identity if* R *is invertible. The brackets* $\pm\{\ ,\ \}_{(R,-R)}$ *are known as the Sklyanin brackets.*

§3. Modified structures on G.

Let $(G,\{\ ,\ \}_G)$ be a Poisson Lie group, with corresponding tangent Lie bialgebra $(\mathfrak{g},\mathfrak{g}^*)$ and dual Poisson Lie group $(G^*,\{\ ,\ \}_{G^*})$. The following question is motivated by our recent work in [**6**]. Given $\Gamma \in L(\mathfrak{g}^*,\mathfrak{g})$, what is the necessary and sufficient condition on Γ so that the formula

$$(3.1)\quad \{\varphi,\psi\}_\Gamma(g) = \{\varphi,\psi\}_G(g)+\langle D'\psi(g),\Gamma(D\varphi(g))\rangle-\langle D'\varphi(g),\Gamma(D\psi(g))\rangle$$

defines a Poisson bracket on G?

As it turns out, the answer to this can be given in a surprisingly simple manner which does not involve the detailed form of $\{\ ,\ \}_G$. Indeed we have

THEOREM 3.1. (i) *The following conditions are equivalent:*

(a) $\{\ ,\ \}_\Gamma$ *is a Poisson bracket on* G,

(b) $$\langle\alpha_3,\mathrm{Ad}_g(\Gamma[\alpha_1,\alpha_2]-[\Gamma(\alpha_1),\Gamma(\alpha_2)])$$
$$-(\Gamma^*[\mathrm{Ad}_g^*\alpha_1,\mathrm{Ad}_g^*\alpha_2]-[\Gamma^*(\mathrm{Ad}_g^*\alpha_1),\Gamma^*(\mathrm{Ad}_g^*\alpha_2)])\rangle$$
$$+\ c.p.\ in\ (\alpha_1,\alpha_2,\alpha_3)=0\ \forall\, g\in G,\alpha_1,\alpha_2,\alpha_3\in\mathfrak{g}^*.$$

(ii) *If* Γ *satisfies condition* (b) *above and* $\Gamma=\Gamma^*$, *the linearization of* $\{\ ,\ \}_\Gamma$ *at the identity defines a Lie bracket* $[\ ,\ ]_\Gamma$ *on* $\mathfrak{g}^*$, *given by* $[\alpha,\beta]_\Gamma=[\alpha,\beta]+\mathrm{ad}^*_{\Gamma(\beta)}\alpha-\mathrm{ad}^*_{\Gamma(\alpha)}\beta$.

PROOF: (i) Due to its length, we shall give an outline of the calculations in the appendix.

(ii) Let $\eta,\eta_\Gamma : G\to L(\mathfrak{g}^*,\mathfrak{g})$ be the Hamiltonian operators of $\{\ ,\ \}_G$ and $\{\ ,\ \}_\Gamma$ relative to the right-invariant frame. Then we have $\eta_\Gamma(g)=\eta(g)+\mathrm{Ad}_g\circ\Gamma-\Gamma^*\circ\mathrm{Ad}_g^*$. Since $\Gamma=\Gamma^*$, it follows that $\eta_\Gamma(e)=0$ so that the linearization of $\{\ ,\ \}_\Gamma$ defines a Lie bracket $[\ ,\ ]_\Gamma$ on $\mathfrak{g}^*$. The explicit expression now follows from a direct calculation using (2.5). □

The following corollary is important in applications.

COROLLARY 3.2. *If $\Gamma \in L(\mathfrak{g}^*, \mathfrak{g})$ is a morphism of Lie bialgebras, then $\{\ ,\ \}_\Gamma$ is a Poisson bracket on G.*

PROOF: Condition (b) in Theorem 3.1(i) is satified as we have $\Gamma[\alpha,\beta] = [\Gamma(\alpha),\Gamma(\beta)]$ and $\Gamma^*[\alpha,\beta] = [\Gamma^*(\alpha),\Gamma^*(\beta)]$, $\forall\ \alpha,\beta \in \mathfrak{g}^*$. □

In the next proposition, we relate $(\mathfrak{g}^*, [\ ,\ ]_\Gamma)$ to the graph of Γ.

PROPOSITION 3.3. *Suppose $\Gamma = \Gamma^*$ is a Lie algebra homomorphism.*

(a) *The graph of Γ, i.e. $\mathfrak{g}^*_\Gamma = \{\Gamma(\alpha) + \alpha \mid \alpha \in \mathfrak{g}^*\}$ is a Lie subalgebra of $\mathfrak{g} \bowtie \mathfrak{g}^*$.*

(b) *The map $(\mathfrak{g}^*, [\ ,\ ]_\Gamma) \to \mathfrak{g}^*_\Gamma : \alpha \mapsto \Gamma(\alpha) + \alpha$ is an isomorphism of Lie algebras.*

PROOF: The assumption on Γ implies $\mathrm{ad}^*_\alpha\, \Gamma(\beta) = \Gamma \circ \mathrm{ad}^*_{\Gamma(\alpha)}\, \beta$, $\forall\ \alpha,\beta \in \mathfrak{g}^*$ from which the rest is plain. □

For the remainder of the section, we shall consider the important special case where Γ is a morphism of Lie bialgebras. The goal is to construct a dual pair of $(G, \{\ ,\ \}_\Gamma)$ from which the symplectic leaves can be obtained by blowing up points in the double fibering [**12**]. From now on, we shall deal exclusively with local Lie groups and homomorphisms, and the word "local" will be dropped for convenience. So, let $\gamma : G^* \to G$ and $\gamma^* : G^* \to G$ be the group homomorphisms corresponding to $\Gamma, \Gamma^* \in \mathrm{Hom}(\mathfrak{g}^*, \mathfrak{g})$. On the algebra $\mathfrak{g} \bowtie \mathfrak{g}^*$, there is a canonically defined r-matrix:

$$R_\delta = \frac{1}{2}(\Pi_{\mathfrak{g}} - \Pi_{\mathfrak{g}^*}) \tag{3.2}$$

where $\Pi_{\mathfrak{g}}, \Pi_{\mathfrak{g}^*}$ are the projections relative to the splitting $\delta = \mathfrak{g} \oplus \mathfrak{g}^*$. Since R_δ is a skew-symmetric solution of (mYB) which is invertible, it follows from Theorem 2.6 that $(G \bowtie G^*)_{(R_\delta, R_\delta)}$ is a symplectic manifold. We now introduce maps $p, q : G^* \times (G \bowtie G^*) \to G \bowtie G^*$ by

$$\begin{aligned} p(g_-, (h_+, h_-)) &= (\gamma^*(g_-)^{-1}, e)(h_+, h_-)(e, g_-) \\ &= (\gamma^*(g_-)^{-1} h_+, h_- g_-) \end{aligned} \tag{3.3a}$$

$$\begin{aligned} q(g_-, (h_+, h_-)) &= (e, g_-)(h_+, h_-)(\gamma(g_-)^{-1}, e) \\ &= (\Phi^-_{g_-}(h_+ \Phi^-_{h_-}(\gamma(g_-)^{-1})), \Phi^+_{\gamma(g_-)^{-1}}(\Phi^+_{h_+}(g_-) h_-)). \end{aligned} \tag{3.3b}$$

Then p is a right action and q is a left action.

LEMMA 3.4. *The actions p and q are admissible in the sense that the invariant functions on $G \bowtie G^*$ form subalgebras of $C^\infty(G \bowtie G^*)$ equipped with $\{\ ,\ \}_{(R_\delta,R_\delta)}$.*

PROOF: We shall prove the assertion for p. First of all, by combining left and right translations, we get a Poisson Lie group action (see [**10**] for definition):

$$\begin{gathered}(G \bowtie G^*)_{(R_\delta,R_\delta)} \times ((G \bowtie G^*)_{(-R_\delta,R_\delta)} \times (G \bowtie G^*)_{(-R_\delta,R_\delta)}) \to (G \bowtie G^*)_{(R_\delta,R_\delta)} \\ (g_1,g_2) : x \mapsto g_1^{-1} x g_2,\ (g_1,g_2) \in (G \bowtie G^*) \times (G \bowtie G^*), \qquad x \in G \bowtie G^*.\end{gathered}$$

Then we embed G^* into $(G \bowtie G^*)_{(-R_\delta,R_\delta)} \times (G \bowtie G^*)_{(-R_\delta,R_\delta)}$ via $g_- \mapsto ((\gamma^*(g_-),e),(e,g_-))$ and embed $\mathfrak{g}^*$ into $(\mathfrak{g} \bowtie \mathfrak{g}^*)_{R_\delta} \oplus (\mathfrak{g} \bowtie \mathfrak{g}^*)_{R_\delta}$ via the differential of this map. From [**10**], to show p is admissible, it is enough to show $(\mathfrak{g}^*)^\perp \subset (\mathfrak{g} \bowtie \mathfrak{g}^*)_{R_\delta} \oplus (\mathfrak{g} \bowtie \mathfrak{g}^*)_{R_\delta}$ is a Lie subalgebra. But

$$(\mathfrak{g}^*)^\perp = \{(X+\alpha, -\Gamma(\alpha)+\beta) \mid X \in \mathfrak{g},\ \alpha,\beta \in \mathfrak{g}^*\}$$

and $(\mathfrak{g} \bowtie \mathfrak{g}^*)_{R_\delta} = \mathfrak{g} \ominus \mathfrak{g}^*$. Therefore, the result follows from the definition of R_δ and the assumption on Γ. □

From this lemma, it follows that there exists unique Poisson structures on $G \bowtie G^*/G^*$ and $G^*\backslash G \bowtie G^*$ such that the projections $\rho_r : G \bowtie G^* \to G \bowtie G^*/G^*$, $\rho_\ell : G \bowtie G^* \to G^*\backslash G \bowtie G^*$ are Poisson maps. We now identify a left (right) G^*-orbit with its unique intersection with $G \times \{e\} \simeq G$. Then

$$\rho_r(g_+,g_-) = \gamma^*(g_-)g_+ \tag{3.4a}$$

and

$$\rho_\ell(g_+,g_-) = (\Phi^-_{g_-}(g_+^{-1}))^{-1}\gamma(\Phi^+_{g_+^{-1}}(g_-^{-1}))^{-1} \qquad \text{(by (2.9))}. \tag{3.4b}$$

THEOREM 3.5. (a) *The reduced bracket on $G \bowtie G^*/G^*$ coincides with $\{\ ,\ \}_\Gamma$.*

(b) *Left G^*-invariant functions and right G^*-invariant functions Poisson commute in $\{\ ,\ \}_{(R_\delta,R_\delta)}$.*

PROOF: (a) Let $\varphi,\psi \in C^\infty(G)$ and set $\hat\varphi = \varphi \circ \rho_r$, $\hat\psi = \psi \circ \rho_r$. Then $\hat\varphi,\hat\psi$ are right G^*-invariant functions on $G \bowtie G^*$ and we have

$$\begin{aligned}\{\varphi,\psi\}_{\rm red}(g_+) &= \{\hat\varphi,\hat\psi\}_{(R_\delta,R_\delta)}(g_+,e) \\ &= \langle\langle R_\delta(D\hat\varphi(g_+,e)), D\hat\psi(g_+,e)\rangle\rangle + \\ &\qquad \langle\langle R_\delta(D'\hat\varphi(g_+,e)), D'\hat\psi(g_+,e)\rangle\rangle.\end{aligned}$$

Using the basic relations [**7**]

$$\frac{d}{dt}|_{t=0}\Phi^+_{g_+}(e^{t\alpha}) = \mathrm{Ad}^*_{g_+}\alpha, \frac{d}{dt}|_{t=0}\Phi^-_{e^{t\alpha}}(g_+) = -T_eR_{g_+}\eta(g_+)\alpha,$$

we find

$$\begin{aligned} D\hat{\varphi}(g_+,e) &= (\eta(g_+)D\varphi(g_+) + \mathrm{Ad}_{g_+}\Gamma(D\varphi(g_+)), D\varphi(g_+)),\\ D'\hat{\varphi}(g_+,e) &= (\Gamma(D\varphi(g_+)), D'\varphi(g_+)). \end{aligned}$$

Therefore,

$$\begin{aligned} &\langle\langle R_\delta(D\hat{\varphi}(g_+,e)), D\hat{\psi}(g_+,e)\rangle\rangle\\ &\quad= \{\varphi,\psi\}_G(g_+) + \frac{1}{2}\langle D'\psi(g_+),\Gamma(D\varphi(g_+))\rangle - \frac{1}{2}\langle D'\varphi(g_+),\Gamma(D\psi(g_+))\rangle. \end{aligned}$$

Similarly,

$$\begin{aligned} &\langle\langle R_\delta(D'\hat{\varphi}(g_+,e)), D'\hat{\psi}(g_+,e)\rangle\rangle\\ &\quad= \frac{1}{2}\langle D'\psi(g_+),\Gamma(D\varphi(g_+))\rangle - \frac{1}{2}\langle D'\varphi(g_+),\Gamma(D\psi(g_+))\rangle. \end{aligned}$$

Putting the last two expressions together yields the assertion.

(b) Suppose $\varphi \in C^\infty(G \bowtie G^*)$ is right G^*-invariant, $\psi \in C^\infty(G \bowtie G^*)$ is left G^*-invariant. Let $D\varphi(g_+,g_-) = (A,B)$, $D'\varphi(g_+,g_-) = (A',B')$, $D\psi(g_+,g_-) = (X,Y)$, and $D'\psi(g_+,g_-) = (X',Y')$. Then the invariance properties imply $A' = \Gamma(B)$ and $X = \Gamma^*(Y')$. Hence,

$$\begin{aligned} &\{\varphi,\psi\}_{(R_\delta,R_\delta)}(g_+,g_-)\\ &\quad= \frac{1}{2}(-\langle B,X\rangle + \langle Y,A\rangle - \langle B',X'\rangle + \langle Y',A'\rangle)\\ &\quad= -\langle B,X\rangle + \langle Y',A'\rangle \qquad \text{(by ad-invariance of } \langle\langle\ ,\ \rangle\rangle)\\ &\quad= -\langle B,\Gamma^*(Y')\rangle + \langle Y',\Gamma(B)\rangle\\ &\quad= 0. \end{aligned}$$

□

COROLLARY 3.6. *If G is finite dimensional, then $G^*\backslash G \bowtie G^* \xleftarrow{\rho_\ell} (G \bowtie G^*)_{(R_\delta,R_\delta)} \xrightarrow{\rho_r} G \bowtie G^*/G$ is a full dual pair and the symplectic leaf of $(G,\{\ ,\ \}_\Gamma)$ passing through $g_+ \in G$ is given by*

$$\{\gamma^*((x^{-1}g_+\gamma(x))_-)(x^{-1}g_+\gamma(x))_+ \mid x \in G^*\}.$$

PROOF: The first assertion follows from the theorem above and a simple dimension argument, as in [**6**]. For the second assertion, simply note that $\rho_\ell^{-1}(g_+) = \{((x^{-1}g_+\gamma(x))_+, (x^{-1}g_+\gamma(x))_-) \mid x \in G^*\}$. The symplectic leaf passing through $g_+ \in G$ is then computed from $\rho_r(\rho_\ell^{-1}(g_+))$ [**12**].

□

§4. Lax systems on the lattice.

In the study of 'integrable' lattice models (see [**2**], Part II, Chap. 2), an important role is played by the geometric theory of Lax systems [**8,9**]. Recently, based on a class of Poisson structures which we call twisted structures, we have extended the theory to a wider class of r-matrices [**6**]. The first goal of the section is to identify the twisted structures as a special instance of the modified structures. Then we give the Hamiltonian description of the associated Lax systems [**6**]—a question which has led to the present development. We conclude the paper by giving a tensor form of the (non-ultralocal) twisted structure and posing a question of independent interest.

In the following, the Lie algebra $\mathfrak{g}$ (of the Lie group G) is assumed to have a non-degenerate ad-invariant pairing $(\cdot,\cdot)$ and the left and right gradients of functions on G are defined using $(\cdot,\cdot)$.

THEOREM 4.1 [**6**]. *Let $R \in \operatorname{End}\mathfrak{g}$ be a classical r-matrix such that both R and $A \equiv \frac{1}{2}(R-R^*)$ are solutions of (mYB). Let $\tau \in \operatorname{Aut}(G)$ whose induced map on $\mathfrak{g}$ (denoted by the same letter) is orthogonal and commutes with R. Then*

(a) *the formula*

$$\{\varphi,\psi\}_\tau = \frac{1}{2}(A(D'\varphi),D'\psi) - \frac{1}{2}(A(D\varphi),D\psi) + \frac{1}{2}(D'\psi,\tau\circ S(D\varphi))$$
$$- \frac{1}{2}(D'\varphi,\tau\circ S(D\psi)), \varphi,\psi \in C^\infty(G), S \equiv \frac{1}{2}(R+R^*)$$

defines a Poisson bracket on G.

(b) *if φ is invariant under twisted conjugation $g \to hg(\tau(h))^{-1}$, $g,h \in G$, the equation of motion defined by the Hamiltonian φ in the structure $\{\ ,\ \}_\tau$ is given by*

$$\dot{g} = \frac{1}{2}T_eR_g(R(D\varphi(g))) - \frac{1}{2}T_eL_g(\tau\circ R(D\varphi(g)))$$

(c) *functions which are invariant under twisted conjugation commute in $\{\ ,\ \}_\tau$.*

REMARK: For examples of r-matrices which satisfy the hypothesis of the theorem, the reader is referred to [**6**].

PROOF OF THEOREM 4.1: (a) Using the pairing $(\cdot,\cdot)$, we identify $\mathfrak{g}^*$ with $\mathfrak{g}_A$. Then the assumption on the r-matrix and τ implies $\frac{1}{2}\tau\circ S$, $\frac{1}{2}\tau^{-1}\circ S$:

$\mathfrak{g}_A \to \mathfrak{g}$ are Lie algebra homomorphisms (see [**6**] for details). As the first two terms in $\{\ ,\ \}_\tau$ define a Poisson Lie group structure on G whose tangent Lie bialgebra is $(\mathfrak{g}, \mathfrak{g}_A)$, the assertion is now seen to be a direct consequence of Corollary 3.2.

(b),(c) We refer the reader to [**6**] for details. □

For the application which follows, it is enough (see [**2**]) to take G to be a matrix Lie group. Let $G^N = G \times \ldots \times G$ (N copies) and $\mathfrak{g}_N = \oplus_1^N \mathfrak{g}$. Equip $\mathfrak{g}_N$ with the ad-invariant pairing

$$(X,Y) = \sum_{i=1}^{N}(X_i,Y_i),\ X = (X_1,\ldots,X_N),\ Y = (Y_1,\ldots,Y_N) \in \mathfrak{g}_N$$

and let $\tau : G^N \ni (L_1,\ldots,L_N) \mapsto (L_N, L_1,\ldots,L_{N-1})$. Now, extend the operators R, R^* (satisfying the hypothesis of Thm. 4.1) etc. componentwise to $\mathfrak{g}_N$ and denote them by the same symbols. Then obviously, both $R \in \operatorname{End}\mathfrak{g}_N$ and $\tau \in \operatorname{Aut}(G^N)$ satisfy the assumptions of Theorem 4.1 above so that we can equip G^N with the twisted Poisson structure. Thus we obtain

THEOREM 4.2 [**6**]. *If $R \in \operatorname{End}\mathfrak{g}$ satisfies the hypothesis of Theorem 4.1 and τ is the map defined above, then*

(a) *the twisted Poisson structure on G^N takes the form*

$$\begin{aligned}\{\varphi,\psi\}_\tau&(L_1,\ldots,L_N)\\ &= \frac{1}{2}\sum_{j=1}^{N}((A(D'_j\varphi),D'_j\psi) - (A(D_j\varphi),D_j\psi) + (S(D_j\varphi),D'_{j+1}\psi)\\ &\quad - (S(D'_{j+1}\varphi),D_j\psi))\end{aligned}$$

where $D_j\varphi$ (resp. $D'_j\varphi$) denotes the j-th component of $D\varphi$ (resp. $D'\varphi$)

(b) *let φ be a central function on G (i.e. $\varphi(hgh^{-1}) = \varphi(g)$), $\psi_m, T_N : G^N \to G$ be maps defined by $\psi_m(L) = L_{m-1}\cdots L_1$, $T_N(L) = \psi_{N+1}(L)$, $L = (L_1,\ldots,L_N)$. Then the system*

$$\begin{aligned}\dot{L}_j = \frac{1}{2}&R(\psi_{j+1}(L)D\varphi(T_N(L))\psi_{j+1}(L)^{-1})L_j\\ &- \frac{1}{2}L_jR(\psi_j(L)D\varphi(T_N(L))\psi_j(L)^{-1}),\\ j = 1,&\ldots,N\end{aligned}$$

is the Hamilton's equation generated by $H_\varphi(L) = \varphi(T_N(L))$ *in the Poisson structure* $\{\ ,\ \}_r$.

(c) *symmetric functions of the monodromy matrix* $T_N(L)$ *Poisson commute. In* (a) *and* (b), *the subscripts* j *are taken* $\mod N$.

REMARKS: (a) The system of equations in Theorem 4.2(b) is an example of a Lax system, which takes the general form

$$\dot{L}_j = M_{j+1}L_j - L_jM_j, \qquad j \in \mathbb{Z}_N.$$

This is a condition of zero curvature and the matrices L_j which define parallel transport from site j of the lattice to $j+1$ are known as transition matrices (see [**2**], p. 293).

(b) When $R = -R^*$, the Poisson structure in Theorem 4.2(a) becomes the product of the Sklyanin structures on the factors of G^N. This is the case considered by Semenov–Tian–Shansky [**8,9**] and the associated Lax systems are related to many well-known lattice models such as the Toda lattice, the Landau–Lifschitz model etc. (see [**2**], Part II, Chap. 3).

(c) In general, the collection of integrals obtained from the symmetric functions of the monodromy matrix $T_N(L)$ is not necessarily complete. For an example (related to eigenvalue algorithms) where additional integrals can be constructed to prove complete integrability on generic symplectic leaves, the reader is referred to [**5**].

To conclude the paper, we present a tensor form of the bracket in Theorem 4.2(a). Recall that when R is skew-symmetric (see remark (b) above), the fundamental Poisson bracket for lattice models is usually written as [**2**]

$$\{L_n(\lambda) \underset{,}{\otimes} L_m(\mu)\} = [L_n(\lambda) \otimes L_m(\mu), r(\lambda,\mu)]\delta_{nm} \tag{4.1}$$

Here, $r(\lambda,\mu)$ is a skew-symmetric solution of the classical Yang–Baxter (CYB) equation, i.e.

$$P\,r(\lambda,\mu)P = -r(\mu,\lambda) \qquad (P(\xi\otimes\eta) = \eta\otimes\xi) \tag{4.2a}$$

$$[{}^{12}r(\lambda,\mu), {}^{13}r(\lambda,\nu) + {}^{23}r(\mu,\nu)] + [{}^{13}r(\lambda,\nu), {}^{23}r(\mu,\nu)] = 0 \tag{4.2b}$$

where ${}^{12}r(\lambda,\mu) = r(\lambda,\mu)\otimes 1$ etc. In our case, (4.1) is replaced by

$$\begin{aligned} \{L_n(\lambda) \underset{,}{\otimes} L_m(\mu)\} &= [L_n(\lambda) \otimes L_m(\mu), a(\lambda,\mu)]\delta_{nm} + \\ &+ (1\otimes L_m(\mu))s(\lambda,\mu)(L_n(\lambda)\otimes 1)\delta_{m,n+1} \\ &- (L_n(\lambda)\otimes 1)s(\lambda,\mu)(1\otimes L_m(\mu))\delta_{m,n-1}. \end{aligned} \tag{4.3}$$

By a lengthy computation, we find that a sufficient condition for (4.3) to define a Poisson bracket is given by

(4.4a)
$$P\,a(\lambda,\mu)P = -a(\mu,\lambda),\;\; P\,s(\lambda,\mu)P = s(\mu,\lambda),$$

(4.4b)
$$[{}^{12}a(\lambda,\mu), {}^{13}a(\lambda,\nu) + {}^{23}a(\mu,\nu)] + [{}^{13}a(\lambda,\nu), {}^{23}a(\mu,\nu)] = 0$$

(4.4c)
$$[{}^{12}a(\lambda,\mu), {}^{13}s(\lambda,\nu) + {}^{23}s(\mu,\nu)] + [{}^{13}s(\lambda,\nu), {}^{23}s(\mu,\nu)] = 0.$$

In the terminology of Faddeev, we thus have a class of non-ultralocal brackets on the lattice. For given $(a(\lambda,\mu), s(\lambda,\mu))$ satisfying (4.4), the problem of enumerating all L-operators satisfying (4.3) is interesting from the point of view of classifying all possible 'integrable' lattice models associated with (4.3). We hope to return to this and other related questions in future work.

APPENDIX

PROOF OF THEOREM 3.1(i): Let $\eta : G \to L(\mathfrak{g}^*, \mathfrak{g})$ be the Hamiltonian operator of $\{\ ,\ \}_G$ relative to the right-invariant frame and let $\Omega(\varphi, \psi) = \langle D'\psi, \Gamma(D\varphi)\rangle - \langle D'\varphi, \Gamma(D\psi)\rangle$, $\varphi, \psi \in C^\infty(G)$. For $\xi \in \mathfrak{g}$, denote by X_ξ (resp. $\tilde{X}_\xi$) the left-invariant (right-invariant resp.) vector field generated by ξ. We have

$$
\begin{aligned}
(A.1)\quad & \{\Omega(\varphi_1, \varphi_2), \varphi_3\}_G + \text{ c.p.} \\
& = -\langle D\Omega(\varphi_1, \varphi_2), \eta(D\varphi_3)\rangle + \text{ c.p.} \\
& = (\tilde{X}_{\eta(D\varphi_3)} X_{\Gamma(D\varphi_2)}\varphi_1 + \tilde{X}_{\eta(D\varphi_3)} \tilde{X}_{\Gamma^*(D'\varphi_1)}\varphi_2 - (1 \leftrightarrow 2)) + \text{ c.p.}
\end{aligned}
$$

where $(1 \leftrightarrow 2)$ denote terms obtained from previous ones by switching the indices 1 and 2. On the other hand, from the cocycle condition (2.3), we obtain

$$
\frac{d}{dt}|_{t=0}\eta(e^{t\xi}g) = d\eta(e)\xi + \mathrm{ad}_\xi\, \eta(g) + \eta(g)\,\mathrm{ad}^*_\xi,\ g \in G,\ \xi \in \mathfrak{g}
$$

from which it follows that

$$
(A.2)\quad
\begin{aligned}
& \Omega(\{\varphi_1, \varphi_2\}_G, \varphi_3) + \text{ c.p.} \\
& \quad = \langle [D\varphi_1, D\varphi_2], \Gamma^*(D'\varphi_3)\rangle - \langle [D'\varphi_1, D'\varphi_2], \Gamma(D\varphi_3)\rangle \\
& \quad + ((\tilde{X}_{\Gamma^*(D'\varphi_3)} \tilde{X}_{\eta(D\varphi_1)}\varphi_2 + \langle D\varphi_2, [\Gamma^*(D'\varphi_3), \eta(D\varphi_1)]\rangle \\
& \quad + X_{\Gamma(D\varphi_3)} \tilde{X}_{\eta(D\varphi_2)}\varphi_1 - (1 \leftrightarrow 2))) + \text{ c.p.}
\end{aligned}
$$

Adding $(A.1)$ to $(A.2)$ and using the standard commutation relations

$$
(A.3)\qquad [X_\xi, X_\eta] = X_{[\xi,\eta]},\ [\tilde{X}_\xi, \tilde{X}_\eta] = -\tilde{X}_{[\xi,\eta]},\ [X_\xi, \tilde{X}_\eta] = 0,
$$

we find

$$
(A.4)\quad
\begin{aligned}
& \{\Omega(\varphi_1, \varphi_2), \varphi_3\}_G + \Omega(\{\varphi_1, \varphi_2\}_G, \varphi_3) + \text{ c.p.} \\
& \quad = \langle D'\varphi_3, \Gamma[D\varphi_1, D\varphi_2]\rangle - \langle D\varphi_3, \Gamma^*[D'\varphi_1, D'\varphi_2]\rangle + \text{ c.p.}
\end{aligned}
$$

Similarly, by using $(A.3)$,

$$
\begin{aligned}
(A.5)\quad & \Omega(\Omega(\varphi_1, \varphi_2), \varphi_3) + \text{ c.p.} \\
& = -\langle D'\varphi_3, [\Gamma(D\varphi_1), \Gamma(D\varphi_2)]\rangle + \langle D\varphi_3, [\Gamma^*(D'\varphi_1), \Gamma^*(D'\varphi_2)]\rangle + \text{ c.p.}
\end{aligned}
$$

Hence, by adding $(A.4)$ to $(A.5)$, we conclude that $\{\ ,\ \}_\Gamma$ is a Poisson bracket if and only if

$$\langle D'\varphi_3, \Gamma[D\varphi_1, D\varphi_2] - [\Gamma(D\varphi_1), \Gamma(D\varphi_2)]\rangle - \langle D\varphi_3, \Gamma^*[D'\varphi_1, D'\varphi_2] - [\Gamma^*(D'\varphi_1), \Gamma^*(D'\varphi_2)]\rangle + \text{ c.p. } = 0, \ \varphi_i \in C^\infty(G), \qquad i = 1,2,3.$$

ACKNOWLEDGEMENT: The first author would like to thank A. Weinstein for sending the preprint [7] before its publication. Subsequent conversation with him at Phoenix has provided the orientation for this work. Thanks are also due to the organizers of the meetings at Phoenix and at MSRI, Berkeley for their kind invitation.

References

1. Drinfel'd, V.G., *Hamiltonian structure on Lie groups, Lie bialgebras and the geometric meaning of the Yang-Baxter equations*, Sov. Math. Doklady **27** (1983), 69–71.
2. Faddeev, L.D. and Takhtajan, L.A., "Hamiltonian Methods in the Theory of Solitons," Springer-Verlag, Berlin, 1987.
3. Kosmann-Schwarzbach, Y., *Poisson-Drinfel'd groups, Topics in soliton theory and exactly solvable nonlinear equations*, M. Ablowitz, B. Fuchsteiner and M. Kruskal, eds., World Scientific, Singapore (1987), 191–215.
4. Koszul, J., "Introduction to Symplectic Geometry," Science Press, Beijing, 1986. (in Chinese).
5. Li, L.C., *On the complete integrability of some Lax systems on* $GL(n, \mathbf{R}) \times GL(n, \mathbf{R})$, Bull. Amer. Math. Soc. **23** (1990), 487–493.
6. Li, L.C. and Parmentier, S., *A new class of quadratic Poisson structures and the Yang-Baxter equation*, C.R. Acad. Sci. Paris, t. 307, Serie I (1988), 279–281; *Nonlinear Poisson structures and r-matrices*, Commun. Math. Phys. **125** (1989), 545–563.
7. Lu, J.H. and Weinstein, A., *Poisson Lie groups, dressing transformations and Bruhat decompositions*, J. Diff. Geom. **31** (1990), 501–526.
8. Semenov-Tian-Shansky, M.A., *What is a classical r-matrix?*, Funct. Anal. Appl. **17** (1983), 259–272.
9. ——————, *Dressing transformations and Poisson Lie group actions*, Publ. RIMS, Kyoto University **21** (1985), 1237–1260.
10. Sklyanin, E., *On the complete integrability of the Landau-Lifschitz equation*, Preprint LOMI, Leningrad, LOMI (1980).
11. ——————, *Some algebraic structures connected with the Yang-Baxter equation*, Funct. Anal. Appl. **16** (1982), 27–34.
12. Weinstein, A., *The local structure of Poisson manifolds*, J. Diff. Geom. **18** (1983), 523–557.

Luen-Chau Li, Department of Mathematics, Pennsylvania State University, University Park, PA 16802

Serge Parmentier, Department of Physics, Pennsylvania State University, University Park, PA 16802
Current address of S.P.: Max-Planck-Institut für Mathematik, Bonn, Federal Republic of Germany

Research of first author partially supported by NSF Grant DMS-8704097.
Research at MSRI supported in part by NSF Grant DMS-8505550.

Optimal Control of Deformable Bodies and Its Relation to Gauge Theory

R. MONTGOMERY

Abstract. I investigate the question "What is the most efficient way for a deformable body to deform itself so as to achieve a desired reorientation?" I call this the Cat's Problem, since itis the problem faced by the upside-down zero-angular momentum cat in freefall. In order of increasing generality, I show that the Cat's Problem is a special case of problems which occur (1) in the geometry of principal bundles, (2)in subRiemannian geometry, and (3) in optimal control. Some model cases are explicitly solved in which the deformable body consists of a collection of point masses. In one of these models the principal bundle breaks down due to isotropy for the action of the rotation group. Nevertheless we are still able to obtain the general solution.

§1. Introduction.

Consider the

Cat's Problem. What is the most efficient way for a deformable body to deform itself so as to achieve a desired rigid re-orientation?

A cat, dropped from upside-down with no angular momentum, changes her shape in such a way as to land on her feet. In doing so, her initial and final shape are essentially the same, but she has re-oriented herself by a rigid rotation of 180 degrees. In addition, by conservation of angular momentum, her total angular momentum is zero throughout the motion. For a nice analysis of this phenomenon, see Kane and Scher [**1969**]. The cat thus describes a <u>loop</u> in her shape space, with the consequence that in an inertial frame the beginning and final shapes are related by a rigid motion $g \in E(3)$.

Shapere and Wilczek addressed the cat's problem in [**1987**], [**1988**], where they translated it into a problem in gauge theory. Their original paper [**1987**] concerned the motion of microorganisms in water, so perhaps it is not fair to call it the cat's problem. In any case, Shapere and Wilczek observed the following key features of this problem.

(i) The space of inertial configurations, that is, the space of allowable deformations of the body, forms a principal bundle over the shape space of the body. The fiber of this bundle is the group G of rigid motions, an element of which is the desired re-orientation.

(ii) Dynamical constraints define a connection A on this bundle. For the cat in free-fall with no initial angular momentum this constraint is that the angular momentum remains zero. For the micro-organism the constraint is that the viscosity is infinite.

(iii) With the constraint in force, the holonomy (parallel translation operator with respect to the connection A) of a given sequence of shape changes is the net re-orientation.

(iv) Efficiency can be measured by a metric on shape space.

Consequently, the zero angular momentum cat problem becomes the **Isoholonomic problem.** Find the shortest loop in shape space (based at a given shape) with a given holonomy.

The cat's problem is more general because the cat may be spinning. That is, its angular momentum may be constrained to a non-zero constant μ. The main result of our paper, Theorem 1 in §3, is a characterization of the solutions to the cat's problem and hence to the isoholonomic problem. The characterization says that all solutions are obtained by solving Hamilton's differential equations for a certain Hamiltonian H_μ on the cotangent bundle of shape space. H_0 is the horizontal kinetic energy.

Our paper is based on the following two observations.

(I) The kinetic energy k, which is a Riemannian metric on the inertial configuration space Q, together with the action of the group G on Q, contain all the information in observations (i), (ii), and (iii).

(II) The isoholonomic problem is a special case of the problem of finding sub-Riemannian geodesics.

We elaborate on (I). (i) Shape space S is the quotient space Q/G. (ii) The connection A is determined by declaring the A-horizontal directions HOR to be those directions perpendicular to the group orbits. (iii) The metric on shape space is determined by declaring that the projection of the tangent space to Q onto the tangent space to shape space is an isometry when restricted to the horizontal space HOR. Concerning (II): a *sub-Riemannian metric* is the restriction of a Riemannian metric k to a distribution $HOR \subset TQ$. Sub-Riemannian metrics are also known as *Carnot-Caratheodory metrics, non-holonomic Riemannian metrics*, or *singular Riemannian metrics*. In sub-Riemannian geometry one only considers horizontal curves $c(t)$, that is, curves whose derivatives $\dot{c}(t)$ are in HOR

(when they exist). The sub-Riemannian distance between points $p, q \in Q$ is

$$d(p,q) = \inf\{ \text{ length } (c)\text{: } c \text{ a piecewise smooth horizontal curve joining } p \text{ to } q\}$$

Here length (c) is the length of the curve c with respect to the Riemannian metric k. (If there are no horizontal paths joining p to q, set $d(p,q) = \infty$.) This is independent of how k is defined in the non-horizontal directions, since c is horizontal. It is now clear that the isoholonomic problem is a special case of the

Sub-Riemannian geodesic problem. Find the horizontal curve joining p to q whose length is $d(p,q)$.

The contents of this paper is as follows. We begin by describing the configuration space of a deformable body and its geometry. Then we express the cat's problem successively as a problem in:

(A) Riemannian geometry;
(B) gauge theory;
(C) optimal control theory and sub-Riemannian geometry.

Our main result, Theorem 1 in §3, characterizes the solutions $q(t)$ to the problem of the cat with angular momentum μ as the cotangent projections to Q of solutions $(q(t), p(t))$ to a Hamilton's differential equations on T^*Q. We give a formula for the corresponding Hamiltonian function H_μ on T^*Q. H_0 is the horizontal kinetic energy.

This extends the author's previous work Montgomery [**1990**] in two ways. First, it allows for non-zero momentum. Second, it gives a formula for the Hamiltonian in terms of physical data pertinent to the problem, specifically the "locked inertia tensor".

Previously the author [**1984**] showed that this horizontal kinetic energy generates the motion of a particle under the influence of the Yang-Mills potential A. This is the content of Theorem 2 of §4, which is essentially a gauge-theoretic restatement of part of Theorem 1. Theorem 2 is contained in Montgomery [**1990**], with one error: its converse is false.

In §5 we restate the problem in the language of optimal control.

In §6 we discuss sub-Riemannian metrics and quote a theorem (Theorem 4) which enables us to prove Theorems 1 and 3 immediately.

In §7 we present an example: N point particles in space. The case $N = 3$ can be exactly solved. The differential equations are the same as those of a single charged particle under the influence of a magnetic monopole. Interesting singularities occur at the collinear configurations, where the assumption of freeness breaks down. Here the dimension of the space of zero-angular momentum vectors jumps.

Recent History. Guichardet [**1984**] observed that angular momentum define a connection. He applied his observation to molecular dynamics. See also Iwai [**1987**].

The isoholonomic problem was posed to the author by Alex Pines' in connection with some problems in nuclear magnetic resonance. The relevant bundles for Pines are the Stieffel varieties. These are the bundles of unitary k-frames over the Grassmannians of k-planes in $\mathbb{C}^{n+k}$. This problem was dealt with by the author in [**1988**].

For infinitesimal deformations of shape, Shapere and Wilczek reduced (a slight variant of) the isoholonomic problem to that of solving a second order linear o.d.e. defined on the tangent space to shape space (at the given shape). The author [**1988**] found the corresponding nonlinear o.d.e. in the case of finite deformations. This equation is Wong's equation [**1970**] for the motion of a classical spinless particle with color-charge under the influence of the Yang-Mills potential A. In the case of the planar cat G is $U(1)$ and these are the Lorentz equations for a charged particle in a magnetic field.

Recently Hamenstädt [**1986,1988**], Bär [**1988**], and the author [**1991**] made important contributions to the sub-Riemannian geodesic problem.

Future Work. Currently Zexiang Li and the author are working on applications of the ideas presented here to robot gymnastics. In future work the author plans to analyze the optimal control of the model cat of Kane and Scher [**1969**]. This model consists of two identical axially symmetric rigid bodies joined by a ball-and-socket joint at their axes.

Acknowledgements. It is a pleasure to thank Malcolm Adams, Judith Arms, Shankar Sastry, Zexiang Li, Krishnaprasad, Tom Kane, Juan Simo, Jerrold Marsden, Tudor Ratiu, Jair Koiller, Ralf Spatzier, Tadeusz Januszkiewicz, and Ursula Hamensta dt for useful conversations, remarks, and directions to the literature. I would like to thank Eugene Lerman for helping me translate some of the encyclopaedia article of Vershik and Gershkovich, Richard Cushman for critical readings, and Gorky, Claudine

Swickard's cat, for his compliance. This work was funded by the NSF Postdoctoral grant #DMS–8807219.

§2. Configuration Space and Metric for Deformable Bodies.

Let B denote a reference shape for our deformable body. For example, B might be the initial shape. The configuration space of a deformable body is a submanifold Q of the space Emb of embeddings of B into Euclidean 3-space, $\mathbf{R}^3(\mathbf{R}^2$ if the body is planar). So,

$$Q \subset \mathrm{Map}(B, \mathbf{R}^3),$$

and a point q of Q is a map

$$q\colon B \to \mathbf{R}^3; \quad x = q(X) \in \mathbf{R}^3, \quad X \in B.$$

The group of orientation preserving rigid motions of space $SE(3)$ acts on Q by rigidly rotating and translating the body. In other words the action is given by $gq = g \circ q$, for $g \in SE(3), q \in Q$. The shape space of our deformable body is $S = Q/SE(3)$.

For example, if B consists of two rigid bodies joined together by a ball-and-socket joint, then Q is isomorphic to $\mathbf{R}^3 \times SO(3) \times SO(3)$. The components of an element $q = (g_1, g_2, \mathbf{c})$ of Q represent the orientations of the first body, and of the second body with respect to a fixed inertial frame, and the position of the joint (or, alternatively, of the center of mass). Specifically,

$$\begin{aligned} q(X) = &\mathbf{c} \text{ if } X \text{ is the joint} \\ &\mathbf{c} + g_1\mathbf{X} \text{ if } X \text{ is in body 1} \\ &\mathbf{c} + g_2\mathbf{X} \text{ if } X \text{ is in the second body} \end{aligned}$$

Shape space for this problem is $SO(3)$. The projection map $Q \to S$ is $\pi(\mathbf{c}, g_1, g_2) = {g_1}^{-1} g_2$. This matrix represents the orientation of body 2 with respect to a frame attached to body 1, that is, their *relative orientation.*

B will be endowed with a mass density $dm(X), X \in B$. This allows us to define B's total mass, the center of mass of $q(B)$, the kinetic energy of an infinitesimal deformation of $q \in Q$, etc. in the usual way. For example, the kinetic energy of an infinitesimal deformation

$$\delta q : B \to \mathbf{R}^3$$

of q is $k_q(\delta q, \delta q) = \frac{1}{2}\int < \delta q(X), \delta q(X) >_{\mathbf{R}^3} dm(X)$.

Here $\langle \cdot, \cdot \rangle_{\mathbf{R}^3}$ is the standard inner product on $\mathbf{R}^3$. The integral defines a Riemannian metric k on Q. The group $E(3)$ of rigid motions acts by isometries with respect to this metric.

We may want to ignore the translational part of our motion, since we cannot affect it by altering our shape as we fall. (Shapere and Wilczek's parmecium can affect their translation since strong friction is present.) This act of ignorance is performed by going to center of mass coordinates, that is, by setting $\int q(X) dm(X) = 0$. This defines a codimension $3, SO(3)$ invariant, totally geodesic submanifold $Q_0 \subset Q$. In symbols

$$Q_0 \cong Q/\mathbf{R}^3; \quad Q \cong \mathbf{R}^3 \times Q_0; \quad \text{Shape space} = S = Q_0/SO(3).$$

The main property of Q_0 (or Q) which we will use is that it is a Riemannian manifold on which the Lie group G acts by isometries.

§3. The Geometrical Cat.

§3.1.

We now work in the more abstract context just alluded to. We are given a Riemannian manifold Q (previously Q_0), metric k, and a Lie group G (previously $SO(3)$) which acts on Q on the left by isometries.

A vector v in T_qQ which is tangent to the group orbit $G \cdot q$ through $q \in Q$ will be called *vertical (at q)*. Vertical vectors represent infinitesimally rigid rotations of our deformable body $q(B)$. A vector v in T_qQ will be called *horizontal* if it is orthogonal to the group orbit through q. As we will soon see, (fact 1 below) v is horizontal if and only if its *angular momentum* $\mathbf{M}(q, v)$ is zero. The set of vertical vectors will be denoted V, and of horizontal vectors HOR. Thus

$$TQ = V \oplus HOR$$

the direct sum of two vector sub-bundles, assuming the rank of V is constant.

If "maximum efficiency" means minimum length, then the problem of the zero angular momentum cat is the same as the

Geometrical Cat's Problem. Find the shortest horizontal path in Q which join q_0 to gq_0. Here $q_0 \in Q$ and $g \in G$ are fixed.

This is a constrained variational problem. The function to be extremized is the length of the path. In control theory this function is called the *objective function.* It is well-known in Riemannian geometry that the same extremals are obtained if one uses the integrated kinetic energy

$$E[q] = \int k_{q(t)} \left(\frac{dq}{dt}(t), \frac{dq}{dt}(t) \right) dt \tag{3.0}$$

as the objective function, instead of the length. This is also true here. Consequently, we will use the objective function E of [3.0].

Why choose this objective function? This is an important question, to which we do not have a very satisfactory physical answer. In some circumstances it might not be the correct objective function. One justification for this choice is to suppose that frictional torques are present at the joints which are proportional to the moments of inertia of the pieces of the body. Then the integrated kinetic energy represents (energy loss) x (time), which is a desirable quantity to minimize.

More generally, we might suppose that there is some metric on shape space, perhaps empirically determined, whose integrated kinetic energy represents power expenditure. This defines a metric on HOR, so the problem still makes sense. See §6.

The Lie algebra of G will be written $\mathfrak{g}$. For q in Q, let

$$\sigma_q \colon \mathfrak{g} \to T_q Q \text{ defined by } \sigma_q(\xi) = \frac{d}{d\varepsilon}\Big|_{\varepsilon=0} \exp \varepsilon\xi q, \text{ for } \xi \text{ in } \mathfrak{g}.$$

denote the *infinitesimal action.* For example, in the case of our deformable body, $\mathfrak{g} = so(3) =$ skew symmetric 3×3 matrices $\cong \mathbf{R}^3$, and

$$(\sigma_q \omega)(X) = \omega \times \mathbf{q}(X), \quad X \in B, \mathbf{q} \in Q \quad \omega \in \mathbf{R}^3 \cong so(3).$$

The 3-vector ω is the angular velocity of rigid rotation.

The *momentum map* J for the G-action is the dual, $\sigma_q^* \colon T_q^* Q \to \mathfrak{g}^*$, of the infinitesimal generator σ_q, namely,

$$J \colon T^*Q \to \mathfrak{g}^* \text{ is given by } J(q,p) = \sigma_q^* p, \text{ where } p \in T_q^* Q \tag{3.1a}$$

For information and definitions regarding momentum maps, see the appendix of our paper, or Abraham and Marsden [**1978**], ch. 4.

We will use the term *angular momentum*, denoted M, for the momentum map viewed as a map from the tangent bundle TQ. We use the Riemannian metric k to identify T_qQ with T_q^*Q. Set

$$\sigma_q{}^t = \sigma_q{}^* \circ k_q \colon T_qQ \to \mathfrak{g}^*.$$

(The "t" denotes transpose.) Then

$$M(q,v) = \sigma_q^t \cdot v \text{ where } v \in T_qQ. \tag{3.1b}$$

Since M is fiberwise linear, we can view it as a one-form on Q with values in $\mathfrak{g}^*$.

We have the following basic facts concerning M.

Fact 1. $M(q,v) = \int \mathbf{q}(X) \times \mathbf{v}(X) dm(X)$, in case Q is the configuration space of a deformable body. This is the usual expression for the total angular momentum of the system. (In writing M with its values in $\mathbf{R}^3$ we identified $so(3)^*$ with $\mathbf{R}^3$ via a choice of orthonormal basis for $so(3)$.)

Fact 2. $HOR_q = (im\sigma_q)^\perp = \ker(\sigma_q{}^t) = \ker(M(q,\cdot)$.

The first equality is by definition.

Fact 3. Noether theorem: M is a constant along solutions to the Euler-Lagrange equantions for any Lagrangian $L = K - V$ on TQ with V a G-invariant function on Q.

§3.2 The Spinning Cat.

We can now formulate the

Generalized, or Spinning Cat's Problem: Find the shortest path c in Q which joins q_0 to q_1 and has constant angular momentum μ.

Our goal is to give a simple characterization of the solutions to this problem. As in Riemannian geometry, it is easier to characterize those curves which minimize length locally in the arc parameter.

Definition. A curve $c : [0,T] \to Q$ is a *local solution* to the spinning cat's problem if there exists a positive number ϵ such that for all subintervals $[a,b]$ of $[0,T]$ of length less than ϵ the restriction of c to $[a,b]$ is a solution to the spinning cat's problem with endpoints $c(a)$, $c(b)$.

Theorem 1 is a partial characterization of these local solutions. In order to to state it we must first define several functions.

The first of these functions is the "locked inertia tensor" I:

$$I_q(\xi_1, \xi_2) = k_q(\sigma_q \xi_1, \sigma_q \xi_2) \text{ for } \xi \in \mathfrak{g}.$$

for each $q \in Q$. In other words, I is the pull-back of the metric on Q to $\mathfrak{g}$. We call I *the locked inertia tensor* because I_q is the inertia tensor of the rigid body formed by locking all of the joints of the configuration q.

For each q, I_q is a symmetric nonnegative bilinear form on $\mathfrak{g}$ which is positive definite if and only if the action is *locally free* at q. (Locally free means that the isotropy group at q is discrete; equivalently $ker(\sigma) = 0$.) This is true if and only if the body $q(B)$ *is not contained within any single line*. We have the well-known formula

$$I_q = (tr\Psi_q)1 - \Psi_q$$

where

$$(\Psi_q)^{ij} = \int \mathbf{q}(X)^i \mathbf{q}(X)^j dm(X)$$

Here we have identified I_q as a symmetric 3×3 matrix by using the isomorphism between the Lie algebra $\mathfrak{g}$ and $\mathbf{R}^3$. We can also write I_q as a map from $\mathfrak{g}$ to $\mathfrak{g}^*$:

$$I_q = {\sigma_q}^t \sigma_q \tag{3.2}$$

So that, by abuse of notation, $I_q(\xi_1, \xi_2) = I_q(\xi_1)(\xi_2)$.

The other function we must define is the "optimal control Hamiltonian" for the spinning cat problem. This is the real-valued function

$$H_\mu(q, p) = K - \frac{1}{2} I^{-1}(J - \mu, J - \mu) \tag{3.3}$$

on T^*Q. In this formula K denotes the usual kinetic energy

$$K(q, p) = \frac{1}{2} k_q^{-1}(q, p).$$

k_q^{-1} and I_q^{-1} denote the inner products on T_q^*Q and $\mathfrak{g}^*$ which are induced by the inner products k_q and I_q. $I^{-1}(J - \mu, J - \mu)$ is the function $(q, p) \mapsto I_q^{-1}(J(q, p) - \mu, J(q, p) - \mu)$.

THEOREM 1. *Let $q : [0,T] \to Q$ be the cotangent projection of a solution $(q,p) : [0,T] \to Q$ to Hamilton's equations for the above Hamiltonian H_μ. And suppose that the image $q(t)(B), 0 \leq t \leq T$ of each configuration is never contained inside a single line. Then q is a local solution to the spinning cat's problem with angular momentum μ.*

REMARKS CONCERNING THE ZERO-ANGULAR MOMENTUM CASE.: (1) J is a conserved quantity for the $\mu = 0$ Hamiltonian: $\{H_0, J\} = 0$. The value of this constant $J(q(t), p(t))$ can be any element of $\mathfrak{g}^*$ even though *every* solution $(q(t), p(t))$ of the corresponding $\mu = 0$ Hamiltonian system satisfies $M(q(t), \dot{q}(t)) = 0$.

(2) Fix the value of this constant: $J = \alpha = const$. Then we can view the optimal control flow as the motion of a particle in the field of the "effective potential" defined by the second term of the Hamiltonian, $-\frac{1}{2}I_q^{-1}(\alpha, \alpha)$.

(3) H_0 is the *horizontal kinetic energy* as defined by orthogonal direct sum decomposition $T^*Q = V^* \oplus HOR^*$. In other words, if $\mathsf{P}_{HOR} : T_q^*Q \to HOR_q^*$ denotes the corresponding projection then

$$H_0(q,p) = \frac{1}{2}{k_q}^{-1}\left(\mathsf{P}_{HOR^*}(q) \cdot p, \mathsf{P}_{HOR^*}(q) \cdot p\right) \tag{3.4}$$

This can be seen by noting that the effective potential of the previous remark is the vertical kinetic energy.

(4) §7.2.3 discusses an example where the deformable body is allowed to become collinear so that the hypothesis of Theorem 1 fails. The optimal curves are then concatenations of solutions to these Hamilton's equations with the concatenation points occuring at the collinearities. The corresponding curves have derivative discontinuities at these points.

§4. The Gauge-theoretic Cat.

As long as our deformable body is never contained within a single line in 3-space the G action is *free*. This, together with the compactness of G, implies that $Q \to S = Q/G$ forms a principal G-bundle and that HOR defines a connection on this principal bundle.

Guichardet [**1984**], and later Shapere and Wilczek [**1987,1989**] give a formula for the corresponding connection one-form (gauge field) A on Q:

$$A_q = I_q^{-1} M(q, \,\cdot\,) \colon T_q Q \to \mathfrak{g}. \tag{4.1a}$$

Equivalently (see equations 3.1 and 3.2)

$$A = (\sigma^t \sigma)^{-1} \sigma^t. \tag{4.1b}$$

It is immediate from fact 2 of §3 that A satisfies the desired property:

$$\ker A = HOR = \text{ set of deformations with zero angular momentum}$$

It also satisfies

$$\mu \cdot A_q \in J^{-1}(\mu) \subset T^*Q$$

for every $\mu \in \mathfrak{g}^*$ (Any connection satisfies this last property.) The connection A is called nowadays the "natural mechanical connection" or "master gauge". In words, "the natural mechanical connection on shape space is the inverse inertia tensor times the angular momentum".

The physical meaning of this connection is elucidated by considering the following procedure. Let $s(t)$ be a path in shape space and $q(0) \in Q$ be an initial configuration of the body. What is the full motion $q(t)$ of the body as it deforms through space? Suppose we also know that the initial total angular momentum M of the body is zero. It remains zero by conservation of angular momentum. Thus $q(t)$ is a horizontal path which projects onto $s(t)$. It follows that $q(t)$ is recovered from $s(t)$ by *parallel translating* $q(0)$ along $s(t)$ with respect to the connection A.

The curve q described above is called the horizontal lift of s. If $s = \pi \circ q$ is a closed curve then there is a unique group element g such that $q(1) = gq(0)$. This element is called the *holonomy* of s (based at $q(0)$).

The metric k on Q induces a metric k_S on shape space by declaring the restriction of $d\pi_q \colon T_q Q \to T_s S$ to HOR_q to be an isometry. This makes $Q \to S$ into a *Riemannian submersion*.

We can now rephrase the geometric (= zero angular momentum) cat's problem as the **isoholonomic problem: Among all loops in shape space based at an initial shape s_0 find the shortest one whose holonomy is g.** More generally we have the **isoparallel problem: Fix an initial and final shape s_0 and s_1 and also a G-equivariant map $g : \pi^{-1}(s_0) \to \pi^{-1}(s_0)$. Among all curves in shape space which join the given shapes s_0, s_1 find the shortest one whose parallel transport operator is g.**

In [**1990**] I computed the formal Euler-Lagrange equations for this problem and showed that it reduces to the differential equation which governs the motion of a particle traveling through the Riemannian manifold S while under the influence of the gauge field A. These are the so-called "Wong equations" [**Wong**] or "Kerner equations" [**Kerner**]. They are equations for a curve $e(t)$ in the co-adjoint bundle $\mathfrak{g}^*(Q)$ which is a vector bundle over S with typical fiber $\mathfrak{g}^*$, the dual of the Lie algebra of our group G. This bundle is an associated bundle to Q and can be defined by the formula $\mathfrak{g}^*(Q) = Q \times_{Ad^*} \mathfrak{g}^*$. It is naturally isomorphic to V^*/G where V^* denotes the dual of the vertical bundle $V = \ker d\pi$ over Q.

To describe the equations, write $s(t) = \pi(e(t)) \in S$ and $\dot{s} = \frac{ds}{dt} \in TS$. (We occasionally abuse notation and denote any projection by "π".) Let D denote the connection on $g^*(Q)$ induced by the connection A on Q. In coordinates, $De = de + ad^*(A)e$. Let ∇ be the <u>Riemannian</u> (Levi-Civita) connection on S induced by the metric k_S. Let F denotes the curvature of A which we can think of as a two-form on S with values in the adjoint bundle $g(Q) = Q \times_{Ad} \mathfrak{g} \cong V/G$. Then $e \cdot F(\dot{\gamma}, \cdot)$ is a one-form along s (a force) and $k_S \cdot e \cdot F(\dot{\gamma},)$ is a vector field along s. Wong's equations are

$$\nabla_{\dot{s}}\dot{s} = k_S \cdot e \cdot F(\dot{s}, \cdot) \tag{4.3a}$$

$$\frac{De}{dt} = 0. \tag{4.3b}$$

Wong's equations are second order in s and first order in the fiber. They can be written in first order form by adding the equation

$$\dot{s} = k_S(s)^{-1} y, \tag{4.3c}$$

where $y \in T_s^*S$, the cotangent bundle to shape space. Now use this to rewrite the previous differential equations in terms of y and $\dot{y}$. The result is a system of first order differential equations for a curve (s, y, e) in the vector bundle $g^*(Q) \oplus T^*S$ over S. We call this vector bundle the "phase space of a (co-adjoint) particle in a Yang-Mills field".

Definition: A curve c in S is a motion of a Yang-Mills particle if it is the projection to S of some solution of the above system of differential equations in $g^*(Q) \oplus T^*S$.

The above equations for a particle in a Yang-Mills field can be written in Hamiltonian form. See Montgomery [**1984**] and references to Sternberg and Weinstein therein. In fact, these equations are obtained by reducing the equations on T^*Q defined by the optimal control Hamiltonian (see Theorem 1) by the action of the group G. To see this, use the connection A to define a G equivariant isomorphism:

$$T^*Q = V^* \oplus HOR^* \cong \mathfrak{g}^* \times \pi^*T^*S$$

This is the dual of the usual vertical-horizontal splitting of TQ. Upon dividing by G we obtain an isomorphism

$$(T^*Q)/G \cong \mathfrak{g}^*(Q) \oplus T^*S. \tag{4.4}$$

Since H_0 is a G-invariant function on T^*Q its Hamiltonian vector field pushes down to define a Hamiltonian vector field on the reduced (Poisson) manifold $g^*(Q) \oplus T^*S$, which is the phase space for a particle in a Yang-Mills field.

There is an alternative, older, viewpoint on the motion of a Yang-Mills particle which is due to Kaluza-Klein. Let β be a fixed adjoint-invariant positive-definite inner product on $\mathfrak{g}$. So, if G is semi-simple β is a multiple of the Killing form and in particular for the case of most interest to us, $G = SO(3)$, β is the standard inner product on $\mathbb{R}^3$. Define a new metric k_{biv} on Q, the "bi-invariant Kaluza-Klein metric", by

$$\begin{aligned} k_{biv}(v_1, v_2) &= k(v_1, v_2) \text{ if } v_1 \text{ and } v_2 \text{ are horizontal} \\ &= \beta(\xi_1, \xi_2) \text{ if } v_1 \text{ and } v_2 \text{ are vertical with } v_i = \sigma\xi_i \\ &= 0 \text{ if } v_1 \text{ is vertical and } v_2 \text{ is horizontal} \end{aligned} \tag{4.5}$$

Definition. A Kaluza-Klein geodesic is a geodesic on Q with respect to this metric.

Let $H_{biv} = \frac{1}{2}k_{biv}{}^{-1}(q)(p,p)$ denote the corresponding kinetic energy , a function on T^*Q. One calculates (see Montgomery [**1984**]) that

$$H_{biv} = H_0 + \frac{1}{2}\beta^{-1}(J, J)$$

Also the Poisson bracket of the second term, $\beta^{-1}(J, J)$ with *any* G-invariant function is zero. Consequently the push-down (to the Yang-Mills phase space) of the Hamiltonian vector field of this second term is zero. It follows that the push-downs for H_0 and H_{biv} are equal. This proves

THEOREM 2. *The following statements regarding a curve c in S are equivalent (1) c is the motion of a particle in a Yang-Mills field. (2) c is the project to S of a solution to Hamilton's equations for the optimal control Hamiltionian H_0 (3) c is the projection to S of a Kaluza-Klein geodesic on Q.*

Theorem 2 was proved in Montgomery [**1984**]. As an immediate corollary to Theorems 1 and 2 we have

THEOREM 3. *(1) If c is a motion of a particle in a Yang-Mills field then it is a local solution to the zero-angular momentum cat's problem. (2) If q is a Kaluza-Klein geodesic then its projection $c = \pi \circ q$ to S is a local solution to the zero angular momentum cat's problem. The corresponding zero-angular momentum curve $\tilde{q}(t)$ in Q is obtained by taking the horizontal lift of the projection c.*

Theorem 3 can be found in Montgomery [**1990**]. There you can also find the following formula for the passage from q to $\tilde{q}$:

$$\tilde{q}(t) = \exp(-t\xi)q(t)$$

where $\xi = A_q \cdot \dot{q}$, a time-independent Lie algebra element.

§5. Optimal Control.

We can view tangent vectors u to shape space as *control variables*. Let

$$h_q\colon T_sS \to HOR_q \subset T_qQ, \text{where } s = \pi(q),$$

denote the operation of horizontal lift. It can be defined as the unique linear operator from T_sS to T_qQ whose image is HOR_q and which is a right inverse to the differential of the projection: $d\pi_q \circ h_q$ = identity on T_sS. Set

$$\begin{aligned} X_\mu(q) &= k_q{}^{-1}(\mu \cdot A_q) \\ &= \sigma_q(I_q^{-1}(\mu)), \end{aligned}$$

a vector field on Q. According to equation [4.2], $M(q, X_\mu(q)) = \mu$. In fact, $X_\mu(q)$ is the shortest vector in the affine subspace $\{v : M(q,v) = \mu\}$ of T_qQ.

Any tangent vector $\dot{q}$ to Q at q which satisfies $\mathbf{M}(q,\dot{q}) = \mu$ can be expressed uniquely in the form $\dot{q} = h_q u + X_\mu(q)$. And in this case $k(\dot{q},\dot{q}) = k_S(u,u) + I_q^{-1}(\mu,\mu)$. It follows that the spinning cat problem is equivalent to the following problem in optimal control :

$$\text{given } \dot{q} = h_q u + X_\mu(q) \tag{5.1}$$

with

$$q(0) = q_0, q(1) = q_1 \tag{5.2}$$

minimize

$$\frac{1}{2}\int k_S(u(t),u(t)) + V_\mu(q)dt. \tag{5.3}$$

where

$$V_\mu(q) = I_q^{-1}(\mu,\mu)$$

This is the standard form of an optimal control problem. 5.1 is called the control law. 5.2 says that the control steers q_0 to q_1. 5.3 is called the cost function or value function. X_μ is called the drift vector field.

This reformulation is important for two reasons. First, it allows us to view the problem as a feedback control problem. This is probably the correct point of view for of the cat; when blindfolded she usually fails to land on her feet (Kane [**1989**], private conversation). Second, it makes sense even if the metric k_S on shape space has no relation the metric on Q. For example , k_S might be an empirically or analytically determined power dissipation law. This is the case for Shapere and Wilczek's microorganisms [**1987**]. And V_μ might be some "potential" which one must keep small.

§6. Sub-Riemannian Geometry.

§6.1 Basics of Sub-Riemannian Geometry.

The geometrical cat's problem of §3 is a special case of the problem of finding sub-Riemannian geodesics. This more general point of view provides a straightforward proof of our Theorem 1. In fact, we simply quote a result of Rayner [**1967**] or Hammenstadt [**1986**] which we have summarized here as Theorem 4.

Definition A sub-Riemannian metric on the manifold Q consists of a (typically nonintegrable!) distribution $HOR \subset TQ$, together with a smoothly varying positive definite product $\kappa(q)$ on HOR_q.

A contravariant object, for example a curve or vector, is called "horizontal" if it is tangent to the given distribution. In general, we only consider horizontal objects.The *length* of the horizontal curve γ is

$$\text{length}\ [\gamma] = \int \sqrt{\kappa(\dot\gamma(t), \dot\gamma(t)}dt \tag{6.1}$$

The **sub-Riemannian geodesic problem** is the problem of finding the shortest horizontal curve joining two fixed endpoints $q_0, q_1 \in Q$.

Definition: A *minimizing sub-Riemannian geodesic* is a rectifiable horizontal curve which is the shortest such curve among all such curves joining its endpoints.

Definition: A locally minimizing geodesic is a rectifiable horizontal curve for which each sufficiently small subarc of is a minimizing sub-Riemannian geodesic. (Compare with the definition of local solution to the spinning cat problem.)

If we take HOR to be as in the previous sections and take κ to be k restricted to the horizontal distribution then the zero-angular momentum cat problem is precisely the problem of finding minimizing sub-Riemannian geodesics.

Remark Sub-Riemannian metrics are also referred to as *Carnot-Caratheodory metrics, non-holonomic Riemannian metrics, or singular Riemannian metrics.*

A sub-Riemannian structure defines, and is defined by, a constant rank " co-metric" g. This is a symmetric contravariant two-tensor whose rank is the dimension of the distribution. We can think of it as a symmetric vector bundle endomorphism $g\colon T^*Q \to TQ$ in which case it is defined by the requirements:

(1) image(g) $= HOR$

(2) if $v = g(q)(p) \in T_qQ$, then $k_q(v, v) = p(g(q)(p))$.

We can also think of $g(q)$ as a bilinear form on T_q^*Q in which case we write:

$$g(q)(p_1, p_2) = p_1(g(q)(p_2)).$$

The fiber-quadratic form

$$H_0(q,p) = \frac{1}{2}g(q)(p,p) \tag{6.2}$$

is called the ***horizontal kinetic energy***, ***sub-Riemannian kinetic energy*** or ***optimal control Hamiltonian***.

THEOREM 4. *[Rayner [**1967**], Hammenstädt [**1990**]] Let $(q(t), p(t))$ be a solution to Hamilton's equations with the Hamiltonian function H_0 of equation [6.2]. In other words suppose $(q(t), p(t))$ satisfies the differential equations*

$$\frac{dq^i}{dt} = \Sigma g^{ij} p_j; \quad \frac{dp_i}{dt} = -\frac{1}{2}\Sigma\left[\frac{\partial(g^{kj})}{\partial q^i}\right] p_k p_j,$$

where (q^i, p_i) are canonical coordinates on T^*Q, and $H_0 = \frac{1}{2}\Sigma g^{kj}(q)p_k p_j$. Then $q(t)$, the cotangent projection of this curve, is a locally minimizing sub-Riemannian geodesic.

This theorem seems to have first been proved by Rayner [**1967**]. It was later proved and strengthened by Hammenstädt. See her [**1990**] paper. We will not reprove this theorem. Instead, we will content ourselves by in §6.3 with calculating the Hamiltonian H_0 (and H_μ) and by showing that in the case of a deformable body that the H_0 of Theorem 4 is equal to the H_0 of Theorem 1. Thus Theorem 1 is a restatement of Theorem 4.

The converse to Theorem 4 is false. Unfortunately, the converse has been stated as a theorem in many papers going back at least as far as Rayner! Bär in his thesis gives examples of a locally minimizing horizontal curve which does not satisfy the geodesic equations. (Sufficiently small subarcs of Bär's curve admits a "cotangent lift" which satisfies Hamilton's equations but these cannot be smoothly spliced together.) Montgomery [**1991**] gives examples of globally minimizing sub-Riemannian geodesics which do not satisfy the geodesic equations. In each of these examples the distribution generates the tangent bundle to Q. That is, it satisfies the conditions of Chow (also called the conditions of Hörmander ; see the next section.) Such "pathologies" do not occur in Riemannian geometry.

Some History: Sub-Riemannian geometry appears in the study of CR and contact structures, hypoelliptic operators, the analysis of rigidity problems for spaces with nonpositive curvature such as complex or quaternionic hyperbolic space, and the collapsing of Riemannian manifolds. There is a fair-sized literature on sub-Riemannian geometry. Among the works that have come to our attention are Hermannn [**1962,1973**] Rayner[**1967**], Hamenstädt [**1986,1988,1990**], Bär[**1989**], Brockett [**1981,1983**], Baillieul [**1975**], Gunther [**1982**], Strichartz [**1983,1989**], and Taylor [**1989**]. Vershik and V. Ya Gershkovich [**1988**] give a kind of review with a summary of facts and some intriguing pictures. The sub-Riemannian geodesic problem is a special case of the problem of Lagrange in the Calculus Of Variations. This is treated in generality by Carathe'odory [final chapter] and Bliss [**1930**].

§6.2. Chow, Ambrose-Singer and Controllability.

There may be no horizontal paths which join the point q_0 to the point q_1. In this case there are, of course, no solutions to the corresponding sub-Riemannian geodesic problem. To avoid this situation we would like conditions on a given distribution which insure that any pair of points $(q_0, q_1) \in Q \times Q$ can be joined by a horizontal path. A distribution, or more generally, a control law, which satisfies such a property is called *controllable*.

Let $E_i, i = 1, 2, \ldots$ be a local frame for the horizontal distribution. Form the iterated Lie brackets $[E_i, E_j], [E_i, [E_j, E_k]], \ldots$ and evaluate these at $q \in Q$.

Definition: We say that Chow's condition holds at q if these vectors, together with the $E_i(q)$, eventually (i.e. after enough iterates of Lie brackets are taken) span all of T_qQ.

THEOREM. *[Chow, Rashevsky] If Q is connected and Chow's condition holds everywhere, then any two points of Q can be joined by a smooth horizontal curve.*

Chow's original theorem is actually slightly stronger than this. Chow's work actually contains no Lie brackets; instead, his original condition is phrased in terms of push-forwards of vector fields and is a localization of the above theorem about a given horizontal curve.

Remark. What we are calling Chow's condition is very often referred to as "Hörmander's condition".

A corollary to Chow's theorem is the Ambrose-Singer [**1953**] Theorem for connections on principal bundles. (See also Kobayashi-Nomizu [**1963**, p. 83-89]). Suppose we are in the previous setting in which $Q \to S$ is a principal G-bundle. Let $X, Y, Z, \ldots$ be horizontal vector field on Q. Recall that the *curvature* F of the connection A at can be defined by the equality: vertical part of $([X,Y](q)) = \sigma_q(F_q(X,Y))$. Similarly, the vertical part of $[Z,[X,Y]](q)$ is equal to $\sigma_q(D_Z F_q(X,Y))$ and analogous statements hold for the higher covariant derivatives of the curvature. Applying Chow's theorem, we obtain the following weak version of the Ambrose-Singer Theorem.

THEOREM. *[Ambrose-Singer] Suppose Q is connected and let q be some point of Q. Let $\Delta(q)$ denote the Lie subalgebra of the Lie algebra of G which is generated by the values of the curvature $F(X,Y)$ at q, together with all of its covariant derivatives $D_Z F(X,Y), D_W D_Z F(X,Y), \ldots$ evaluated at q as $X, Y, Z, \ldots$ range through HOR_q. If $\Delta(q)$ is the entire Lie algebra then any two points of Q can be joined by a smooth horizontal path.*

§6.3 Calculation of the Hamiltonian.

We will calculate the Hamiltonian H_μ from the constrained Lagrangian $L : HOR \to \mathbb{R}$, $L(q,u) = \frac{1}{2}\kappa_q(u,u) + V_\mu(q)$. This does not constitute a proof of any of the theorems, rather it lends to their credibility. (We have proved the theorems merely by quoting the theorems of others.) Our method of calculation is identical to the method used for prescribing the Hamiltonian of the Pontrjagin maximum principle. This in turn is identical to the method of Legendre transform used in classical mechanics once we realize that $\dot{q} = u + X_\mu(q)$ where $u \in HOR_q$. Note also that $L(q,u) = \frac{1}{2}k_q(\dot{q},\dot{q})$. See §5 where we note that $V_\mu(q) = \frac{1}{2}I_q^{-1}(\mu,\mu)$ for the spinning cat.

Define

$$\tilde{H}(q,p,u) = p(u + X_\mu(q)) - L(q,u)$$

where $q \in Q$, $p \in T_q^*Q$,$u \in HOR_q$ and X_μ is the "drift vector field" introduced at the beginning of section 5. If $\mu = 0$ then $X_\mu = 0$ which is the case stated in Theorem 1. The Hamiltonian H is by definition is the Legendre transform of L:

$$H(q,p) = inf_u \tilde{H}(q,p,u)$$

By elementary calculus this infimum is realized by the unique vector $u \in HOR_q$ such that $p = \kappa_q(u,\cdot)$ when restricted to HOR_q. We can reexpress this relationship by

$$u = g(q)(p)$$

Evaluate $\tilde{H}(q,p,g(q)(p))$ to find that

$$H(q,p) = \frac{1}{2}g(q)(p,p) + p(X_\mu(q)) - V_\mu(q) \tag{6.3}$$

which is the desired result.

In the case of our deformable body κ is the restriction of the Riemannian metric k on Q and there is a simple formula for the cometric g. Let

$$\mathsf{P}_{HOR}: TQ \to HOR; \mathsf{P}_V: TQ \to V$$

denote the k-orthogonal projections. Let $k^{-1}: T^*Q \to TQ$ be the k-induced isomorphism. Then

$$g = \mathsf{P}_{HOR} \circ k^{-1} = k^{-1} - \mathsf{P}_V \circ k^{-1}.$$

Moreover

$$\mathsf{P}_V = \sigma \circ A$$

where A is the natural mechanical connection of §4. Combining these formulas with the formula for A and the definition of $\mathbf{M}$, we find that the H_0 of Theorem 1 (equations [3.3 or .4]) equals the H_0 of Theorem 4 (equation [6.2]). Similarly the H above is the H_μ of Theorem 1.

§7. Examples. Point masses.

§7.1 N Point Masses.

The body B consists of N point masses in d-dimensional Euclidean space. We are most interested in the case $d = 3$ but it is no extra work to follow the general case, at least for awhile. The masses are $m_1, \ldots, m_N$ and have positions $x_1, \ldots, x_N \in \mathbf{R}^d$. We assume that we are in the center of mass frame: $\Sigma m_a x_a = 0$. Then Q is a $d(N-1)$-dimensional subspace of $\mathbf{R}^{dN}$. Notice that we do not delete the collision configurations $\{x_a = x_b\}$. To imagine changing the shape of a configuration when $d = 3$ suppose that each mass slides back and forth upon a massless rod and that these rods can be swung about by means of massless joints attached to the center of mass.

An element of the shape space $S = Q/SO(d)$ can be visualized by connecting the N points in order by line segments so as to form an N-gon. For example, if $N = 3$ then S is the space of triangles. Coordinatizing S for large N is an interesting problem in classical invariant theory (H. Weyl **[1938]**)

The metric on Q is $\Sigma m_a \|dx_a\|^2$. $\mathbf{M} = \Sigma m_a x_a \wedge v_a$ is the angular momentum. When $d = 3$ we have $\wedge = \times$, the vector cross product. The inertia tensor $I = \Sigma m_a x_a{}^i x_a{}^j$. Thus $I(\xi, \xi) = \Sigma m_a x_a{}^i x_a{}^j \xi_i^k \xi_j^k$. (For $d = 3$ the isomorphism of $\mathfrak{so}(3)$ with $\mathbf{R}^3$ converts this into the earlier formula for I.) I is nondegenerate as long as the positions x_a of the masses span either a subspace of dimension $d - 1$ or all of $\mathbf{R}^d$. We will say that such configurations are in general position. For instance when $d = 3$ a configuration is in general position provided all of its points do not lie on the same line.

The action of $SO(d)$ is free on the set of points in general position and so $Q \to S$ forms a principal $SO(d)$ bundle upon restriction to the configurations in general position. As in §4 the distribution $\{\mathbf{M} = 0\}$ of zero-angular momentum deformations defines a connection on this principal bundle. Guichardet **[1984]** showed that this connection satisfies the conditions of the Ambrose-Singer Theorem (§6.2) provided $N > d$. Consequently any two generic configurations can be connected by a zero angular momentum path when $N > d$. The condition "can be joined by a zero-angular momentum path" is a closed condition on the set of pairs of points in Q. It follows that any two configurations can be joined by a zero angular momentum path when $N > d$.

When $N = d$ Guichardet has shown that a generic configuration moving with zero-angular momentum must remain on the $d-1$ dimensional subspace which it initially spans *as long as it remains generic*. (Exercise: use Cartan's lemma for two-forms to prove this.) Note that the fact that the center of mass is zero says that the d vectors cannot be linearly independent.

When $N = d$ it is still possible to join any two configurations by a zero-angular momentum path. By the above observation the only way to do this when the two configurations span different hyperplanes in $\mathbf{R}^d$ is to pass through some intermediate degenerate configuration, that is, a configuration whose span has codimension 2 or greater. These degenerate configurations thus act as "switching yards" between the different hyperplanes. When $N = d = 3$ we will show explicitly how to connect triangular configurations spanning different planes by passing through collinear configurations.

The optimal control Hamiltonian for our zero angular momentum "cat" problem here is given by the formula of Theorem 1:

$$H_0 = K - \frac{1}{2} I^{-1}(J, J)$$

Here $K(x,p) = \frac{1}{2}\Sigma\|p_a\|^2/m_a$ is the kinetic energy and $J = \Sigma x_a \wedge p_a$ is the angular momentum written in terms of momentum variables.

An interesting fact about this Hamiltonian is that it is *collective* for the linear symplectic group $Sp(d)$ of $\mathbf{R}^{2d}$. To see this, note that we can write $T^*Q \subset \mathbf{R}^{2d} \otimes \mathbf{R}^N$, the tensor product of the symplectic vector space $\mathbf{R}^2 d$ with the inner product space $\mathbf{R}^N$ where the inner product is defined by the masses. In general, if (Z, ω) and $(E, \langle , \rangle)$ are two such vector spaces then the symplectic form on $Z \otimes W$ is given by $(z_1 \otimes e_1), (z_2 \otimes e_2) \mapsto \omega(z_1, z_2)\langle e_1, e_2 \rangle$. The groups $Sp(Z)$ and $O(E)$ act in a linear symplectic fashion on $Z \otimes E$ and form one of the standard examples of a Howe dual pair. The $Sp(Z)$ momentum map is given by $\Phi(\Sigma z_\mu \otimes e_\mu) = \Sigma z_\mu \langle e_\mu, e_\nu \rangle z_\nu \in \mathfrak{sp}(Z)^* \cong S^2(Z)$. Here we have identified the Lie algebra $\mathfrak{sp}(Z)$ of $Sp(Z)$ with the space $S^2(V^*)$ of homogeneous quadratic polynomials on Z by taking such a polynomial to its corresponding linear Hamiltonian vector field. In our case, let $x^i, v^j \in \mathbf{R}^{2d}$ be symplectic coordinates so that $\omega = \Sigma dx^i \wedge dv^i$. The metric on $\mathbf{R}^N$ is given by $\Sigma m_a (dt_a)^2$. Then we have nonsymplectic coordinates x^i_a, v^j_b on $\mathbf{R}^{2d} \otimes \mathbf{R}^N$. The $Sp(d)$ momentum map Φ takes values in the space of $2d \times 2d$ symmetric matrices. It consists of the four $d \times d$ blocks $\Sigma m_a x^i_a x^j_a$, $\Sigma m_a x^i_a v^j_a$, $\Sigma m_a v^i_a x^j_a$, $\Sigma m_a v^i_a v^j_a$. By inspection H_0 can be written as a smooth function of the entries of Φ. Thus, by definition, H_0 is collective for the group $Sp(d)$. This means that by solving *one* Hamiltonian differential equation on $\mathfrak{sp}(d)^*$ we will have solved our cat's problem *for all* N. Unfortunately, I do not see how to solve this universal system, even for $d = 2$ or 3.

§7.2. N = 3 Point Masses in the Plane: Exact Solutions.

We will use complex coordinates so that $\mathbf{R}^2 = \mathbf{C}$. Up to equation [7.2.10] our formulas can be found in Iwai [**1987a**].

The configuration space is

$$Q = \{m_1 z_1 + m_2 z_2 + m_3 z_3 = 0\} \subset \mathbf{C}^3.$$

By Graham-Schmidt (see Iwai)

$$q_1 = \sqrt{\frac{m_1 m_3}{m_1 + m_3}}(z_1 - z_3) \tag{7.2.1a}$$

$$q_2 = \sqrt{\frac{m_2(m_1 + m_3)}{m_1 + m_2 + m_3}}\left(z_1 - \frac{m_1 z_1 + m_3 z_3}{m_1 + m_3}\right) \tag{7.2.1b}$$

define orthonormal coordinates on Q, namely,

$$k = |dq_1|^2 + |dq_2|^2. \tag{7.2.2}$$

The rotation group acts according to

$$e^{i\Psi}(q_1, q_2) = (e^{i\Psi} q_1, e^{i\Psi} q_2).$$

and this action is free on $Q \setminus \{(0,0,0)\} = \mathbf{C}^2 \setminus \{0\}$. The natural mechanical connection on $Q \setminus \{0\}$ is

$$A = \frac{1}{r}\Im(\bar{q}_1 dq_1 + \bar{q}_2 dq_2), \tag{7.2.3a}$$

where

$$r = |q_1|^2 + |q_2|^2 = I_q \tag{7.2.3b}$$

is the moment of inertia of the configuration (q_1, q_2). (Check that $A = 1$ on the infinisimal generator (iq_1, iq_2) of the action and that A annihilates the perpindicular to the generator.) The projection

$$\pi\colon Q \to S, \text{ restricted to } Q \setminus \{0\}, \tag{7.2a}$$

is equal to the radial extension

$$\pi\colon \mathbf{C}^2 \setminus \{0\} \to \mathbf{R}^3 \setminus \{0\} \tag{7.2.4b}$$

of the well known Hopf fibration $S^3 \to S^2$. Specifically, we can identify S topologically with $\mathbf{R}^3 = \mathbf{C} \times \mathbf{R}$. Then

$$\pi((q_1, q_2)) = (2\bar{q}_1 q_2, |q_1|^2 - |q_2|^2) = (x + iy, z). \tag{7.2.5}$$

The origin $0 = \pi((0,0,0))$ is a distinguished nonsmooth point of S (an orbifold point). It represents the special shape consisting of all three masses sitting at the origin.

We have $r = \sqrt{x^2 + y^2 + z^2} = |q_1|^2 + |q_2|^2$. (Compare [7.2.3b].) It follows that π maps the 3-sphere $S^3(\sqrt{r})$ of radius $\sqrt{r}$ centered at the origin in $Q = \mathbf{C}^2$ to the sphere $S^2(r)$ of radius r in $S = \mathbf{R}^3$. Setting $r = 1$, we obtain the standard Hopf fibration $S^3 \to S^2$.

Warning: the induced (k_S) metric on the sphere $S^2(r)$ is that of a sphere of radius $\frac{1}{2}\sqrt{r}$. See formula [7.2.8].

We have the following facts concerning the suspended Hopf fibration [7.2.4b]. The curvature F of the connection A is that of a point magnetic monopole with strength $\frac{1}{2}$ at the origin 0 of $S = \mathbf{R}^3$. In symbols,

$$F = \frac{1}{2} d\Omega, \tag{7.2.7a}$$

where $d\Omega$ is the usual solid angle two-form

$$d\Omega = \frac{x dy \wedge dz + z dx \wedge dy + y dz \wedge dx}{r^3}$$

If we identify two-forms F and vector fields $\mathbf{F}$ on $\mathbf{R}^3$ in the usual way: $F(v_1, v_2) = < \mathbf{F}, v_1 \times v_2 >$ then

$$\mathbf{F}(\mathbf{x}) = \frac{1}{2}\mathbf{x}/r, \quad \mathbf{x} \in \mathbf{R}^3 \setminus \{0\} \tag{7.2.7b}$$

Warning: The d of $d\Omega$ is conventional; it does not mean the exterior derivative. $d\Omega$ is not an exact differential but it is closed.

The induced metric on $S \setminus \{0\}$ is calculated to be

$$k_S = \frac{1}{4r}(dx^2 + dy^2 + dz^2) \tag{7.2.8}$$

where $dx^2 + dy^2 + dz^2$ is the standard metric on $\mathbf{R}^3$. In particular it is conformal to the standard metric. It follows from [7.2.7b] that the Lorentz force

$$k_S^{-1} eF(\dot{\mathbf{x}}, \cdot) = \lambda \mathbf{x} \times \dot{\mathbf{x}}, \tag{7.2.9}$$

where λ is a scalar. In fact $\lambda = -2e$.

Formula [7.2.7] for the curvature is well-known. Formula [7.2.8] for the shape metric is not as well-known but can be found in Iwai [**1987a**]. For completeness we will derive these formulas. They are somewhat simpler if we use polar coordinates on $\mathbf{C}^2$:

$$(q_1, q_2) = \sqrt{r}(\cos(\theta/2)e^{i\Psi}, \sin(\theta/2)e^{i\phi}e^{i\Psi}). \tag{7.2.10}$$

with $\theta/2, \varphi$, and Ψ being angles. In these coordinates the $SO(2)$ action is $\Psi \to \Psi + \Delta\Psi$. One finds that

$$\pi(q_1, q_2) = r(\sin\theta e^{i\varphi}, \cos\theta).$$

It follows that (r, θ, φ) are the standard polar coordinates on $\mathbf{R}^3$ with θ the angle which (x, y, z) makes with the z-axis. Differentiating [7.2.9] and plugging this in to [7.2.2, .3] yields

$$A = d\Psi + \sin^2(\theta/2)d\varphi \tag{7.2.11}$$

and

$$k = \frac{1}{4r}\{dr^2 + r^2 d\theta^2\} + r\{\sin^2(\theta/2) - \sin^4(\theta/2)\}d\varphi^2 + rA^2.$$

The metric k_S is obtained by setting $A = 0$ in this last equation. Since $\sin^2(\theta/2) - \sin^4(\theta/2) = [\frac{1}{2}\sin\theta]^2$ we obtain

$$\begin{aligned} k_S &= \frac{1}{4r}[dr^2 + r^2(d\theta^2 + \sin^2\theta d\varphi^2)] \\ &= \frac{1}{4r}[dx^2 + dy^2 + dz^2] \end{aligned}$$

which is formula [7.2.8]. To obtain [7.2.7a] take the exterior derivative of [7.2.11] and note that $\sin^2(\theta/2) = \frac{1}{2}(1 - \cos\theta)$. This yields

$$F = dA = \frac{1}{2}\sin\theta(d\theta \wedge d\varphi)$$

which is the expression for $\frac{1}{2}d\Omega$ in polar coordinates.

The Lorentz equations on $\mathbf{S} \setminus \{0\}$

Our A, being an $SO(2)$ gauge field, is just a one-form (vector potential). It follows that the optimal control equations, which are the equations of motion for a particle in a Yang-Mills field (§4, Theorem 3), are just the Lorentz equations of a particle in the magnetic field F except that the metric on S is not the standard metric on $\mathbf{R}^3$ but rather the funny metric k_S. These Lorentz equations are

$$\nabla_{\dot{\mathbf{x}}}\dot{\mathbf{x}} = \lambda\mathbf{x} \times \dot{\mathbf{x}}$$

where

(7.2.12b) $$\lambda = \text{ constant } = -2e.$$

and where ∇ is the covariant derivative for the shape metric k_S.

Symmetry and Momentum Map. The rotation group $SO(3)$ acts on $\mathbf{R} \times T^*S \cong \mathbf{R} \times \mathbf{R}^3 \times \mathbf{R}^3$ by $R(e, \mathbf{x}, \mathbf{v}) = (e, R\mathbf{x}, R\mathbf{v})$. This action is a Hamiltonian action and preserves the Wong Hamiltonian

$$H(e, \mathbf{x}, \mathbf{v}) = 4r\|\mathbf{v}\|^2.$$

By virtue of these facts, the momentum map

$$\mathbf{L}\colon \mathbf{R} \times \mathbf{R}^3 \times \mathbf{R}^3 \to \mathbf{R}^3 \cong so(3)^*$$

for the $SO(3)$ action, which is given by

(7.2.13) $$\mathbf{L}(e, \mathbf{x}, \mathbf{v}) = \mathbf{x} \times \mathbf{v} + \frac{1}{2}e\mathbf{x}/r$$

is constant along Wong trajectories. This formula for the momentum map is well- known. See for example Jackiw and Manton [**1980**]. For completeness we present a derivation of it in the appendix.

Solutions. Let $(e, \mathbf{x}(t), \mathbf{v}(t))$ be a solution to our Wong's equations, and $\mathbf{L}_0$ the value of the momentum map on this solution. Then

$$\mathbf{x} \cdot \mathbf{L}_0 = \mathbf{x} \cdot \mathbf{L}(e, \mathbf{x}, \mathbf{v}) = \frac{e}{2} r. \tag{7.2.14}$$

This is the equation of a cone $C = C(\mathbf{L}_0, e)$ in $\mathbf{R}^3$ since $\mathbf{L}_0$ is a constant vector in $\mathbf{R}^3$. The curve $\mathbf{x}(t)$ is constrained to lie on this cone, and $\dot{\mathbf{x}}$ must be tangent to the cone at $\mathbf{x}$. Since the radial vector $\mathbf{x}$ is also tangent to the cone, the acceleration $\lambda \mathbf{x} \times \dot{\mathbf{x}}$ of [7.2.11] is always normal to this cone. The shapemetric k_S on $\mathbf{R}^3 \setminus \{0\}$ is conformal to the standard one so that this normality also holds in the shape metric. This demonstrates that

Every extremal trajectory is a geodesic on some cone C in $\mathbf{R}^3$. The geodesic equations are those of the shape-induced metric on the cone.

We now solve the geodesic equations on the cone. Without loss of generality, we can suppose that $\mathbf{L}_0$ is parallel to the z-axis: $\mathbf{L}_0 = e_3 \|\mathbf{L}_0\|$. Then the equation for the cone reads $r\|\mathbf{L}_0\| \cos\theta = er$, or

$$\cos\theta = \frac{e}{2} / \|\mathbf{L}_0\|. \tag{7.2.15}$$

(Note $\|\mathbf{L}_0\| \geq \frac{1}{2}|e|$, so that the right hand side of this equation is less than or equal to 1 in magnitude.) The induced metric on C is then

$$k_S = \frac{1}{4r}[dr^2 + r^2 \sin^2\theta d\varphi^2]. \tag{7.2.16a}$$

Set

$$\rho = \sqrt{r}. \tag{7.2.16b}$$

Then

$$k_S = d\rho^2 + c^2 \rho^2 d\varphi^2, \tag{7.2.16c}$$

where c^2 is the constant

$$c^2 = \frac{1}{4} \sin^2\theta = \frac{1}{4}\left(1 - \left[\frac{1}{2} \frac{e}{\|\mathbf{L}_0\|}\right]^2\right). \tag{7.2.16d}$$

This shows that the effect of the nonstandard metric k_S is to make the cone's geometric angle of opening, smaller than its opening angle as measured in the standard metric, $dr^2 + r^2 \sin^2(\theta) d\varphi^2$.

[7.2.16c] is the equation of the metric on a standard cone whose generator makes an angle $\sin^{-1}(c)$ with its axis of symmetry. In particular, the metric is flat. The geodesics can be found by cutting and unfolding the cone along a generator. The cone is isometric to the closed convex wedge

$$C^* = \{0 \leq \varphi \leq 2\pi c\} \text{ in } \mathbf{R}^2 \text{ with its standard metric,}$$

with boundary rays identified (with the generator) and (ρ, φ) being polar coordinates on this $\mathbf{R}^2$.

From here one can very easily obtain explicit coordinate expressions for the extremal trajectories of the three particle system. We will content ourselves with two observations regarding the extremals.

Observation 1. The largest possible value for c is $c = \frac{1}{2}$. This corresponds to $\theta = \pi$, which is the negative z-axis. This corresponds to C^* being a half-plane. All other values of c lead to opening cones C^* with opening angles $2\pi c < \pi$. On these cones every geodesic except the rays through the cone point (that is, the generators) are self-intersecting. These rays represent simple dilations: $(\mathbf{x}_1(t), \mathbf{x}_2(t), \mathbf{x}_3(t)) = r(t)(\mathbf{x}_1(0), \mathbf{x}_2(0), \mathbf{x}_3(0))$. Except for these rays, all extremal curves have points of self-intersection, and thus define extremal loops in shape space.

Observation 2. The moment of inertia of an optimal trajectory

$$I(t) = \rho(t)^2 \text{ is a quadratic function of time.}$$

This can be seen by writing the $\rho^2 = x^2 + y^2$ where x and y are Cartesian coordinates on C^*, and then writing the parametric equation for a line in terms of x and y.

§7.3. Three Point Masses in Space.

The configuration space is

$$Q = \{m_1\mathbf{x}_1 + m_2\mathbf{x}_2 + m_3\mathbf{x}_3 = 0\} \subset \mathbf{R}^{3\times 3}. \tag{7.3.1a}$$

The change of variables [7.1], with $\mathbf{x}_i$ in place of z_i gives

$$Q \cong \mathbf{R}^3 \times \mathbf{R}^3 \text{ with the standard metric } \|d\mathbf{q}_1\|^2 + \|d\mathbf{q}_2\|^2 \tag{7.3.1b}$$

Geometry of Shape Space. The shape space

$$\begin{aligned} S &= \mathbf{R}^3 \times \mathbf{R}^3/SO(3) \\ &\cong \mathbf{R}^2 \times \mathbf{R}^2/O(2), \end{aligned} \tag{7.3.2}$$

where the groups act diagonally. In order to see this second isomorphism identify $\mathbf{R}^2$ with the x-y plane in $\mathbf{R}^3$. For any pair of points $(\mathbf{q}_1, \mathbf{q}_2) \in \mathbf{R}^3 \times \mathbf{R}^3$ we can find a rotation matrix $R \in SO(3)$ such that $(R\mathbf{q}_1, R\mathbf{q}_2) \in \mathbf{R}^2 \times \mathbf{R}^2 \subset \mathbf{R}^3 \times \mathbf{R}^3$, namely find an R which takes span $\{\mathbf{q}_1, \mathbf{q}_2\}$ into $\mathbf{R}^2$. The ambiguity in R is $O(2)$, not $SO(2)$.

The natural projection

$$\mathbf{R}^2 \times \mathbf{R}^2/SO(2) \to \mathbf{R}^2 \times \mathbf{R}^2/O(2) \tag{7.3.3}$$

is 2: 1 except on the orbits of the collinear points, where it is 1: 1. Since $O(2)/S0(2) \cong Z_2$, the two-element group, we have

$$S = S^*/\mathbf{Z}_2 \tag{7.3.4}$$

where $S^* = \mathbf{R}^2 \times \mathbf{R}^2/SO(2)$ is the planar shape space of §7.2. The $\mathbf{Z}_2$ action on $S^* = \mathbf{C} \times \mathbf{R}$ is given by $(w, t) \to (\bar{w}, t)$ where $\bar{w}$ denotes the complex conjugate of w. (This can be seen by noting that the $\mathbf{Z}_2$ action can be realized on $\mathbf{R}^2 \times \mathbf{R}^2 = \mathbf{C} \times \mathbf{C}$ by $(z_1, z_2) \to (\bar{z}_1, \bar{z}))_2)$ and then inducing the action on S^* by the projection [7.2.5a].) Consequently, S has the structure of a closed half-space with a distinguished point on its boundary. A more accurate picture is that S is a closed convex cone.

S is stratified according to symmetry type. There are three strata, the open interior, the boundary cone, and the apex of the cone. Points in the interior of S represent generic triangles. They have non-zero area. Points on the boundary are the collinear triangles. The apex point represents the "point" triangle, in which all three masses are co-incident.

Our three point masses must remain in the plane which they define up until they become collinear, that is, until their shape hits the boundary of S. More precisely, we have

PROPOSITION 7.3.1. *Suppose that three point masses move continuosly and piecewise differentiably in space in such a way that their total angular momentum about their center of mass is zero (whenever it is defined). Suppose that $a \leq t \leq b$ is an interval of time for which they are never* *collinear. Then the plane P_t containing the three points is* constant *(in the center of mass frame) on any time interval $a \le t \le b$ for which the three points are never collinear.*

Proof. The theorem is more easily proved in terms of the coordinates $(\mathbf{q}_1, \mathbf{q}_2) \in \mathbf{R}^3 \times \mathbf{R}^3$. The plane P_t is span $\{\mathbf{q}_1(t), \mathbf{q}_2(t)\} \subset \mathbf{R}^3$. The proposition is equivalent to the implication $\mathbf{q}_1 \times \mathbf{v}_1 + \mathbf{q}_2 \times \mathbf{v}_2 = 0$ and $\mathbf{q}_1 \times \mathbf{q}_2 \neq 0$ implies $\mathbf{v}_1, \mathbf{v}_2 \in P =$ span $\{\mathbf{q}_1, \mathbf{q}\}$.

To prove this, write $\mathbf{e}_3 = \mathbf{q}_1 \times \mathbf{q}_2$. Since $\mathbf{e}_3 \neq 0, \{\mathbf{q}_1 \times \mathbf{e}_3, \mathbf{q}_2 \times \mathbf{e}_3, \mathbf{e}_3\}$ form a basis for $\mathbf{R}^3$. Write

$$\mathbf{v}_1 = \mathbf{w}_1 + a\mathbf{e}_3, \mathbf{v}_2 = \mathbf{w}_2 + b\mathbf{e}_3$$

with $\mathbf{w}_i \in P$. Then the total angular momentum $\mathbf{M}$ is

$$\begin{aligned} \mathbf{M} &= \mathbf{q}_1 \times \mathbf{v}_1 + \mathbf{q}_2 \times \mathbf{v}_2 \\ &= \lambda\mathbf{e}_3 + a\mathbf{q}_1 \times \mathbf{e}_3 + b\mathbf{q}_2 \times \mathbf{e}_3. \end{aligned}$$

Since this is zero, it follows that $a = b = 0$.

Remark. A less direct proof is obtained from Iwai's expression for the curvature and connection of the natural mechanical connection. With respect to a certain local section Ψ, both of these take values in the one-dimensional subalgebra generated by $\mathbf{e}_3 =$ span $(\Psi_1, \Psi_2)^\perp$. The result follows from the Ambrose-Singer Theorem.

What happens if the particles become collinear? Can the plane they define change? Yes. **See figure 1.**

In this figure $\mathbf{q}_1(t), \mathbf{q}_2(t)$ are continuous $_i$ and approach $\mathbf{q}_0$ as $t \to 0$. Both the left and right derivatives, $\dot{\mathbf{q}}_i(0^-)$ and $\dot{\mathbf{q}}_i(0^+)$, of these vector-valued functions exist at $t = 0$ and these derivatives lie in the plane perpindicular to $\mathbf{q}_0$. These right and left derivatives are not equal. The initial plane of motion is spanned by $\mathbf{q}_0$ and $\dot{\mathbf{q}}_1(0^-)(= -\dot{\mathbf{q}}_2(0^-))$. The final plane of motion is spanned by $\mathbf{q}_0$ and $\dot{\mathbf{q}}_1(0^+)$. The total angular momentum is zero throughout the motion. In summary,

The collinear configurations can act as "switching yards" between different planes of motion.

What are the optimal trajectories? As long as the particles are not collinear, they move in a fixed plane according to proposition 7.3.1, and so the problem is identical to the problem solved in §7.2. When they become collinear, the plane of motion can change. In summary, we have

PROPOSITION 7.3.2. *The optimal curves are piecewise smooth concatenations of the planar optimal curves of §7.2. Derivative discontinuities can only occur when the masses become collinear. If the desired re-orientation $R \in SO(3)$ does not preserve the initial plane, $P =$ span $\{\mathbf{q}_1(0), \mathbf{q}_2(0)\}$, (assuming it is a plane) then the optimal curve must have such a derivative discontinuity.*

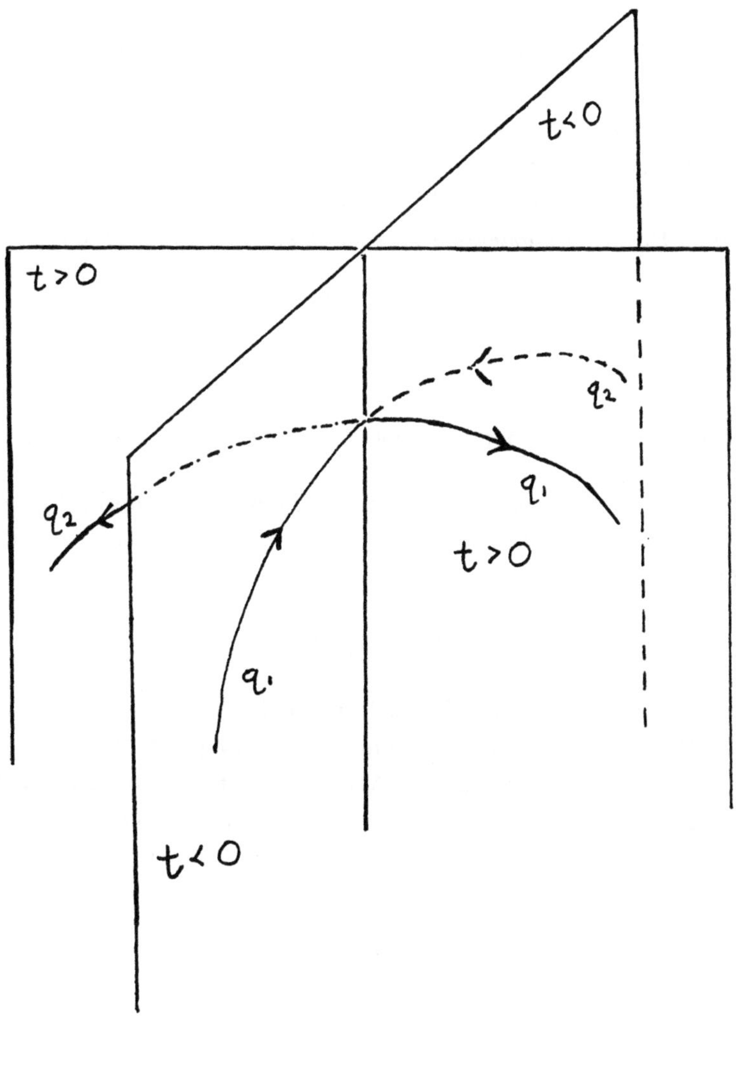

Figure 1

Appendix.

Calculation of the Momentum Map [7.2.13].

We derive the formula [7.2.13]

$$\mathbf{L} = \mathbf{x} \times \mathbf{v} + \frac{1}{2} e\mathbf{x}/r$$

for the $SO(3)$ momentum map $\mathbf{L}$ of §7.2. We will use the symbol "J" instead of "$\mathbf{L}$".

We begin by recalling the definition of a momentum map, and the standard formula for the momentum map associated to an action on configuration space. If a Lie group K acts in a Poisson fashion on the Poisson manifold P, then a *momentum map* for this action is a function

$$J \colon P \to \kappa^* = \text{ dual of Lie algebra of } K$$

which satisfies

$$\{f, J \cdot \xi\} = df \cdot \xi_P \text{ for all smooth functions } f \text{ on } P, \text{ and all } \xi \in \mathfrak{g}.$$

Here $J \cdot \xi$ is the ξ component of J, and ξ_P is the infinitesimal generator on P. If $P = T^*Q$ and the K action is the cotangent lift of an action of K on Q, then

$$(A1) \qquad J(q,p) = p \cdot \xi_Q(q)$$

defines a momentum map. (This is formula [3.1a].)

The Poisson structure on the Wong phase space $\mathfrak{g}^*(Q) \oplus T^*S$ is induced from that on T^*Q so we can use [A1] to calculate the corresponding momentum map on the Wong phase space $\mathfrak{g}^*(Q) \oplus T^*S$. Recall in that set-up (§4) we had $\pi \colon Q \to S$ a principal bundle with structure group G, and connection A. The isomorphism

$$\mathfrak{g}^* \times \pi^* T^* S \to T^* Q$$

is given by

$$(A2) \qquad (\mu, (q, p_S)) \to p = \mu \cdot A_q + h_q^* p_S.$$

Here $q \in Q$ and $p_S \in T^*_{\pi(q)}S$, and $h^*_q\colon T^*_{\pi(q)}S \to T^*_qQ$ is the dual of the horizontal lift operator h. Suppose that K acts by bundle automorphisms of Q, that is, it commutes with the G action. Then by projection K acts on S and $\pi^*\xi_Q = \xi_S$. [A1] reads

$$(A3) \qquad (J\cdot\xi)(q,p) = \mu\cdot A_q\cdot\xi_Q + p_S\cdot\xi_S.$$

(The h^*_q disappears because $\xi_Q = h\cdot\xi_S+$ vertical, and $h^*_q p_S\cdot$ vertical $= 0$.) J is automatically G-invariant, so defines a function on the Wong phase space $\mathfrak{g}^*(Q)\oplus T^*S = (\mathfrak{g}^*\times\pi^*T^*S)/G$. This is the desired momentum map. Note that the second term of [A3] is the ξ component of the standard momentum map [A1] for the K action on T^*S.

We apply formula [A3] to our $SO(3)$ action. First, lift the action to the standard action of $K = SU(2)$ on $\mathbf{C}^2$. Choose the (standard) basis

$$e_1 = \frac{1}{2}\begin{bmatrix} 0 & i \\ i & 0\end{bmatrix} = \frac{i}{2}\sigma_1;$$

$$e_2 = \frac{1}{2}\begin{bmatrix} 0 & 1 \\ -1 & 0\end{bmatrix} = \frac{i}{2}\sigma_2;$$

$$e_3 = \frac{i}{2}\begin{bmatrix} 1 & 0 \\ 0 & -1\end{bmatrix} = \frac{i}{2}\sigma_3.$$

for the Lie algebra of K. (The σ_i are the Pauli matrices.) This gives us an identification $\kappa^* \cong \mathbf{R}^3 \cong so(3)^*$. Moreover, if we write $\omega = \Sigma\omega^i e_i$, then under this identification

$$\omega_S(\mathbf{x}) = \omega\times\mathbf{x}, \text{ where } \mathbf{x}\in S\cong\mathbf{R}^3.$$

It follows that the second term of [A3] is

$$p_S\cdot\omega_S(\mathbf{x}) = (\mathbf{x}\times\mathbf{p})\cdot\omega,$$

which is the ω-component of the standard momemtum map, $\mathbf{x}\times\mathbf{p}$, for the action of $SO(3)$ on T^*S. This yields the first term of [7.2.13].

We will be done if we can show that the first term of [A3] gives the second term of [7.2.13], that is, if

$$(A4) \qquad eA_q\cdot\omega_Q = \frac{e}{2r}\mathbf{x}\cdot\omega$$

Aside: e, the fiber coordinate of $g^*(Q)$ is simply a real number, the same real number appearing in [A4]. The group $G = SO(2)$ is a one-dimensional

Abelian group. Its dual Lie algebra g^* can be identified with $\mathbf{R}$, on which $SO(2)$ acts trivially. [4.4] then reads

$$\mathfrak{g}^*(Q) \oplus T^*S = \mathbf{R} \times T^*S.$$

The $\mathbf{R}$ factor is "central", that is the function e is a Casimir: $\{e, f\} = 0$ for all functions f. But the Poisson structure depends on e. In fact, the symplectic leaves are $\{e\} \times T^*S$ with symplectic form $\omega_0 + eF$, where ω_o is the standard symplectic form on T^*S, and F is the curvature form, pulled back to T^*S by the cotangent projection.

To calculate [A4], one checks that at $(q_1, q_2) \in Q = \mathbf{C}^2$ one has

$$(A5) \qquad (e_1)_Q = \frac{1}{2}(q_2, -q_1); \quad (e_2)_Q = \frac{i}{2}(q_2, q_1); \quad (e_3)_Q = \frac{i}{2}(q_1, -q_2)$$

Recall that the connection is

$$(A6) \qquad A = \frac{1}{r} \text{ im } (\bar{q}_1 dq_1 + \bar{q}_2 dq_2)$$

Plugging [A5] into [A6] , and using formula [7.2.5] for π :

$$\pi((q_1, q_2)) = (2\bar{q}_1 q_2, |q_1|^2 - |q_2|^2) = (x + iy, z),$$

we obtain

$$(A7) \qquad A \cdot (e_1)_Q = \frac{1}{2r} x; \quad A \cdot (e_2)_Q = \frac{1}{2r} y; \quad A \cdot (e_3)_Q = \frac{1}{2r} z.$$

[A4] follows immediately from [A7].

REFERENCES

R. Abraham and J. E. Marsden, [**1978**], "Foundations of Mechanics," Second Edition, Benjamin/Cummings.

W. Ambrose and I. M. Singer, [**1953**], *A theorem on holonomy*, Trans. AMS **75**, 428–453.

V. I. Arnold, V. V. Kozlov, and A. I. Neishtatdt, [**1988**], "Dynamical systems III," vol. 3 in the Springer translation series "Encyclopedia of Mathematical Sciences", Springer-Verlag.

J. B. Baillieu, [**1975**], "Some optimization problems in geometric control theory," Phd. Thesis, Applied Math., Harvard University.

J. B. Baillieu, [**1975**], *Geometric methods for nonlinear optimal control problems*, J. of Optimization Th. and Appl. **25:4**, 519–548.

C. Bär, [**1989**], "Geodesics for Carnot-Caratheodory Metrics," Preprint.

G. A. Bliss, [**1930**], *The problem of Lagrange in the calculus of variations*, Am. J. Math **52**, 674–713.

G. A. Bliss, [**1946**], "Lectures on calculus of variations," University of Chicago Press.

R. W. Brockett, [**1983**], *Nonlinear control theory and differential geometry*, in "Proceedings of the International Congress of Mathematicians, Warszawa."

R. W. Brockett, [**1981**], *Control theory and singular Rieemannian geometry*, in "New directions in applied mathematics," P.J. Hilton and G.S. Young, editors, Springer-Verlag.

C. Carathe'odory, [**1967**], "Calculus of variations and partial differential equations of the first order," vol. 2, Holden-Day, San Francisco, CA.

M. P. do Carmo, [**1976**], "Differential geometry of curves and surfaces," Prentice-Hall.

L. Cesari, [**1983**], "Optimization-theory and applications," Springer-Verlag.

W. L. Chow, [**1939**], *Über Systeme von Linearen partiellen Differentialgleichungen erster Ordnung*, Math. Ann. **117**, 98–105.

L. E. Faibusovich, [**1988**], *Explicitly solvable nonlinear optimal controls*, Int'l. J. Control **48:6**, 2507–26.

C. Frohlich, [**1979**], *Do springboard divers violate angular momentum conservation?*, Am. J. Phys. **47**, 583–92.

A. Guichardet, [**1984**], *On rotation and vibration motions of molecules*, Ann. Inst. H. Poincare, Phys. Theor. **40,** no. 3, 329–342.

V. Guillemin and S. Sternberg, [**1980**], *The moment map and collective motion*, Ann. Phys. **127**, 220–53.

N. L. Gunther, [**1982**], "Hamiltonian mechanics and optimal control," Phd. Thesis, Harvard University.

U. Hamenstädt, [**1986**], "Sur Theorie von Carnot-Caratheodory-Metriken und ihren Anwendungen," Doktorarbeit, Bonn.

U. Hamenstädt, [**1988**], "Some regularity theorems for Carnot-Caratheodory metrics," Preprint, Cal. Tech.

R. Hermann, [**1962**], *Some differential geometric aspects of the Lagrange variational Problem*, Indiana Math. J., 634–673.

R. Hermann, [**1973**], *Geodesics of singular Riemannian metrics*, Bull. AMS **79:4**, 780–782.

T. Iwai, [**1987a**], *A gauge theory for the quantum planar three-body system*, J. Math. Phys. **28**, 964–974.

T. Iwai, [**1987b**], *A geometric setting for internal motions of the quantum three-body system*, J. Math. Phys. **28**, 1315–1326.

T. Iwai, [**1987c**], *A geometric setting for classical molecular dynamics*, Ann. Inst. Henri Poincaire, Phys. Th. **47:2**, 199–219.

R. Jackiw and N. Manton, [**1980**], *Symmetries and conservation laws in gauge theories*, Ann. Phys. **127**, 257–273.

T. R. Kane and M. P. Scher, [**1969**], *A dynamical explanation of the falling cat phenomenon*, Int'l. J. Solids and Structures **5**, 663–670.

R. Kerner, [**1968**], *Generalization of the Kaluza-Klein theory for an arbitrary non-Abelian gauge group*, Ann. Inst. Henri Poincare **9, no. 2**, 143–152.

S. Kobayashi and K. Nomizu, [**1963**], "Foundations of differential geometry," vol. 1, Interscience.

R. Montgomery, [**1991**], *Counterexamples in sub-Riemannian geometry*, submitted to J. Diff. Geom.

R. Montgomery, [**1988**], "Shortest loops with a fixed holonomy," Preprint, Mathematical Sciences Research Institute Preprint Series.

R. Montgomery, [**1984**], *Canonical formulations of a particle in a Yang-Mills field*, Lett. Math. Phys. **8**, 59–67.

Y. G. Oh, P. S. Krishnaprasad, and J. E. Marsden, [**1988**], *The dynamics of coupled planar rigid bodies, Part I*, Dynamics and Stab. of Sys. **3:1,2**, 25–49.

C. B. Rayner, [**1967**], *The exponential map for the Lagrange problem on differentiable manifolds*, Phil. Proc. of the Royal Soc., ser. A **262**, 299–344.

A. Shapere and F. Wilczek, [**1987**], *Self-propulsion at low Reynolds number*, Phys. Rev. Lett. **58**, 2051–54.

A. Shapere, [**1989**], "Gauge mechanics of deformable bodies," Phd. Thesis, Physics, Princeton University.

S. Smale, [**1970**], *Topology of mechanical systems* I (II), Inv. Math. **10, (11)**, 305–331, (45–64).

N. Sreenath, [**1987**], "Modelling and control of multibody systems," Phd. Thesis, Electrical Engineering Dept., University of Maryland.

R. Strichartz, [**1983**], *Sub-Riemannian geometry*, J. Diff. Geom. **24**, 221–263.

T. J. S. Taylor, [**1989**], *Some aspects of differential geometry associated with hypoelliptic second order operators*, Pac. J. Math. **136:2**, 355–378.

A. M. Vershik and V. Ya Gershkovich, [**1988**], *Non-holonomic Riemannian manifolds*, in "Dynamical systems," vol. 7, part of the new Mathematical Encyclopaedia series, vol. 16 *in Russian*, MIR publ., translation to be published by Springer-Verlag.

A. Weinstein, [**1984**], *The local structure of Poisson manifolds*, J. Diff. Geom..

F. Wilczek, [**1988**], "Gauge theory of deformable bodies," Preprint, Inst. for Adv. Studies, Preprint no. 88/41.

S. K. Wong [**1970**], *Field and particle equations for the classical Yang-Mills field and particles with isotopic spin*, Nuovo Cimento **65a**, 689–693.

Mathematical Sciences Research Institute, Berkeley CA 94720
Research at MSRI supported in part by NSF Grant DMS-.

The Augmented Divisor and Isospectral Pairs

RANDOLPH JAMES SCHILLING

Abstract. The Neumann system was shown by J. Moser (1979) to be isospectral with respect to a rank 2 perturbation of a diagonal matrix. In 1982 we showed that the Neumann system is isospectral with respect to a perturbation of a nilpotent 2×2 matrix. In this paper Baker functions are used to derive these matrices and to relate their eigenfunctions. I. M. Krichever used Baker functions to exhibit a homomorphism from a commutative ring of differential operators into a ring of functions. Moser's matrix is derived in terms of a translation in the Jacobian Variety of the spectral curve and the 2×2 matrix is described using a rational extension of the Krichever homomorphism. The *isospectral pair* has three interesting geometric properties which are discussed below. They are related to the geometry of momentum mappings, to Moser's work on the geometry of quadrics and to the Lie algebraic Euler equations.

§0. Introduction.

The Neumann system may be described as a system of $g+1$ harmonic oscillators constrained to the tangent space TS^g to the g-dimensional sphere. Let $\mathbf{a} = \text{diag}(a_1, \dots, a_{g+1})$ be a diagonal matrix with distinct entries. The Neumann equations of motion are given by these formulas:

$$(0.1) \qquad X^* : \begin{cases} \dot{x}_1 = x_2 \quad \text{and} \quad \dot{x}_2 = (\mathbf{a} + \sigma(\mathbf{m}))x_1 \\ \text{where } \cdot = \partial \doteq \frac{d}{d\tau}, \\ \mathbf{m} = (x_1, x_2) \in TS^g, \\ x_1,\ x_2 \in \mathbf{R}^{g+1} : \ x_1 \cdot x_1 = 1, \quad x_1 \cdot x_2 = 0, \\ \text{and } \sigma = -\mathbf{a}x_1 \cdot x_1 - x_2 \cdot x_2. \end{cases}$$

The symplectic form $\omega = dx_1 \wedge dx_2$, restricted to TS^g is nondegenerate; hence, (TS^g, ω) is a symplectic manifold. The vectorfield X^* is Hamiltonian with respect to (TS^g, ω) [**6**].

A Hamiltonian system is said to be *completely integrable* if it has a maximal set of independent involutive constants of motion. The following is a classical theorem in Hamiltonian mechanics. *A (real) level set of a maximal set of involutive constants of motion is diffeomorphic to a product of lines and spheres; if the level set is compact it is a torus.* The theorem gives a foliation of the phase space of an integrable system by Abelian groups. An *algebraic completely integrable system* [**3**] is a completely integrable system whose solutions are expressible in terms of theta functions. The Neumann system is an example of an *algebraic completely integrable* Hamiltonian system. The proof of this statement uses several interesting areas of mathematics. The purpose of this introduction is to outline the proof while explaining how these areas are related to the Neumann system.

GENERAL PROBLEM: It is known that the Korteweg-de Vries equation, $q_t = q_{xxx} - 6qq_x$, has finite dimensional invariant manifolds. The Jacobian variety of a hyperelliptic curve is an invariant manifold of the Korteweg-de Vries equation. One approach to these invariant manifolds uses commutative rings of differential operators. This approach has several advantages but it is in some ways cumbersome. The phase space is a differential algebra so the Hamiltonian mechanics is nonclassical. It is usually not possible to write Hamilton's equations and constants of motion explicitly. The Neumann system is another approach to these invariant submanifolds. Hamilton's equations are obviously explicit and, as we shall soon see, there are explicit formulas for its constants of motion. A level set of the constants of motion is **(i)** invariant under the Korteweg-de Vries flow and **(ii)** isomorphic to an open affine subset of the Jacobian variety of a hyperelliptic curve. In this way we obtain a description of the invariant manifold as a level set of the constants of motion of the Neumann system. This amounts to a system of equations for the Jacobien.

On the other hand the isospectral deformation theory of the Korteweg-de Vries equation plays a key role in the development of several aspects of the Neumann system; e.g., the complete integrabiltiy of the Neumann system. This has led to the following conjecture concerning the nature of a completely integrable system. *The level surfaces of the constants of motion are invariant surfaces of an integrable partial differential equation.* Such a theorem might provide a systematic way of finding constants of motion for a finite dimensional Hamiltonian system because the techniques for dealing with integrable partial differential equations are very well developed.

THE SOLUTION: Let $x^r = (x_1^r, x_2^r)^T$, $u^r = (-x_2^r, x_1^r)^T$ where $1 \leq r, s \leq g+1$ and let $x^r \otimes u^s$ be the 2×2 matrix whose entry (i,j) is $x_i^r u_j^s$. Let $\mathbf{m}$ as in (0.1). Let us consider the initial value problem for the Neumann system; namely, **(IVP):** $\dot{\mathbf{m}} = X^*(\mathbf{m})$, $\mathbf{m}(\tau) \in TS^g$ where $\mathbf{m}(0)$ is prescribed. The method of solution is based on this formula:

$$\text{(0.2)} \qquad \dot{Z} = [B, Z] \quad \text{where} \quad \begin{cases} Z \doteq e_{2,1} + \displaystyle\sum_{r=1}^{g+1} \frac{x^r \otimes u^r}{\lambda - a_r} \\ \text{and} \\ B \doteq (\lambda + \sigma(\mathbf{m}))e_{2,1} + e_{1,2}. \end{cases}$$

The Lax equation in (0.2) implies that the characteristic polynomial of Z, $g(\lambda, z) \doteq |Z(\mathbf{m}, \lambda) - zI|$, is constant along any integral curve of the vector field X^* on TS^g. The characteristic polynomial of Z is given by these formulas:

(0.3*a*)
$$g(\lambda, z) = z^2 - \sum_{r=1}^{g+1} \frac{G_r}{\lambda - a_r}$$

where

(0.3*b*)
$$G_r = (x_1^r)^2 - \sum_{s \neq r} \frac{(x_1^r x_2^s - x_1^s x_2^r)^2}{a_r - a_s}, \quad (r = 1, \dots g+1).$$

Let R be the algebraic curve: $g(\lambda, z) = 0$; R consists of 2 sheets merging at $2g+2$ branch points. There is a branch point at $\lambda = \infty$ which we denote by ∞. The rational function λ has a double pole at ∞ and z has a simple zero at ∞. The remaining $2g+1$ branch points are the zeros and poles of z away from the point ∞. The poles of z are points $(\lambda, z) = (a_r, \infty)$ which we shall denote simply by a_r. Now z has g zeros other than ∞ which we shall denote by b_j, their λ coordinate. The divisor of z is given by

(0.3*c*)
$$(z) = \infty + \sum_{j=1}^{g} b_j - \sum_{r=1}^{g+1} a_r.$$

The genus of R is g.

(0.3D) REMARK: Ulenbeck (1975) and Devaney (1978) (see [**9**], page 186) found these constants of motion evidently through an exhaustive search. Th matrix Z, from which the G_r were just derived, was found in [**10.II**] in a systematic way using the Riemann-Roch theorem for Baker functions [**10.I**]. The first use of Baker functions goes back to Krichever and Novikov (e.g. [**8**]). A Baker function is a function on a Riemann surface with prescribed poles and essential singularities. Krichever's Riemann-Roch theorem gives the dimension of certain linear spaces of such functions.

Here is a restatement of our purpose. The 2-form $\omega(\mathbf{m})$, $\mathbf{m} \in TS^g$, determines a projection $T_{\mathbf{m}}\mathbb{R}^{2g+2} \to T_{\mathbf{m}}TS^g$. Let G_r^* denote the Hamiltonian for the projection to TS^g of the Hamiltonian vector field of G_r. A method for writing down the G_r^* can be found in ([**6**] and [**9**]). Our goal is to show that these Hamiltonians form a set of g independent involutive constants of motion. These facts concerning the Neumann constants may be proved by direct but incredibly tedious computation. We shall give a geometrical proof of their involutivity using a loop algebra. We shall establish a connection between the Neumann system and the Korteweg-de Vries equation and use it to give a proof of their independence.

The matrix spectral problem $Z(\lambda, \mathbf{m})U(p) = z(p)U(p)$ has a unique solution of the form $U(p) = (1, f(p))$ where f is a rational function with a

simple pole at ∞ and g finite poles; say, $\delta = \delta_1 + \cdots \delta_g$. The divisor δ is nonspecial. By our generalized Riemann-Roch theorem, there exists a unique function $\psi(\tau, p)$, *the Baker function*, satisfying these conditions:
(0.4a) Any pole of ψ lies in δ and
(0.4b) $\psi(\tau, p) = (1 + O(\kappa^{-1}))\exp(\kappa\tau)$ at ∞ where $\kappa \doteq \sqrt{\lambda}$.
The Baker function ψ satisfies an operator spectral problem of the form:

$$(0.5a) \qquad L\psi(\tau, p) = \lambda(p)\psi(\tau, p) \quad \text{where} \quad L = (\frac{d}{d\tau})^2 - q(\tau);$$

equivalently,

$$(0.5b) \qquad \dot{U}(\tau, p) = B(\lambda, \mathbf{m})U(\tau, p) \quad \text{where} \quad B = (\lambda + q)e_{2,1} + e_{1,2}$$

and $U(\tau, p) \doteq (\psi, \dot{\psi})^T$. The solution $\mathbf{m}(\tau)$ is obtained by evaluating the Baker function along the polar divisor of z:

$$(0.6) \qquad x_1^r(\tau) = c_r\psi(a_r, \tau) \quad \text{and} \quad x_2^r(\tau) = c_r\dot{\psi}(a_r, \tau)$$

where $c_r = x_1^r(0)$. The proof that this gives the solution to (IVP) follows from a series of residue calculations involving abelian differentials on R. These calculations are analogous to the Deift/Lund/Trubowitz *trace formulas* [**6**]. In our case the *trace formulas* show that the point $\mathbf{m}(\tau)$ defined by (6) lies on TS^g and that

$$(0.7) \qquad q(\tau) = \sigma(\mathbf{m}(\tau)).$$

These facts combined with equation (5b) with $\lambda = a_r$ prove that $\mathbf{m}$ satisfies (IVP).

The Baker function ψ is expressible in terms of the Riemann theta function of the Riemann surface R ([**8**] and [**10.I**]). This establishes the algebraic nature of the solution.

One of the most interesting results of our work was the derivation [**10.II**] of the following spectral problem:

$$(0.8) \qquad Z(\mathbf{m}(\tau), \lambda(p))U(\tau, p) = z(p)U(\tau, p).$$

The Lax equation (0.2) is the integrability condition of this linear equation combined with (0.5b).

APPLICATION: THE GEOMETRY OF QUADRICS: J. Moser [**9**] discovered the following relationship between the Neumann system and the geometry of quadrics. It seems that he was motivated to search for a relationship based on the following 2 facts: **(i)** The solution to the geodesic problem on an ellipsoid is given by hyperelliptic functions and **(ii)** The geodesic problem has a Lax equation that is similar to a Lax equation of the Neumann problem. We shall give another approach using the isospectral matrix Z. It is based solely on isospectrality and so it reaffirms the connection established by Moser between the geometry of quadrics and the Lax idea.

(9) THEOREM. *Let $\mathbf{m}(\tau) = (x_1(\tau), x_2(\tau))$ be a solution to the Neumann problem and let $l(\tau)$ be the line in $\mathbb{R}^{g+1}$ through $x_2(\tau)$ in the direction $x_1(\tau)$:*

$$l(\tau) : x_2(\tau) + s x_1(\tau) \quad \text{where} \quad -\infty < s < \infty.$$

*Then there exists g τ-**independent** confocal quadrics Q_j $(j = 1, \cdots, g)$ in $\mathbb{R}^{g+1}$ such that $l(\tau)$ is tangential to each Q_j.*

PROOF: Let us consider the family of confocal quadrics defined by this formula:

$$Q_\lambda \doteq \sum_{r=1}^{g+1} \frac{Y_r^2}{\lambda - a_r} = 1,$$

and the g finite branch points b_j of R where $z = 0$. Let $Q_j = Q_{b_j}$.
Equation (8) with $p = b_j$ is equivalent to these equations:

(0.9*a*)
$$\sum_{r=1}^{g+1} \frac{x_1^r(x_2^r - \frac{\dot\psi(\tau,b_j)}{\psi(\tau,b_j)} x_1^r)}{b_j - a_r} = 0$$

and

(0.9*b*)
$$\sum_{r=1}^{g+1} \frac{x_2^r(x_2^r - \frac{\dot\psi(\tau,b_j)}{\psi(\tau,b_j)} x_1^r)}{b_j - a_r} = 1.$$

These equations combined imply this one:

(0.9*c*)
$$\sum_{r=1}^{g+1} \frac{(x_2^r - \frac{\dot\psi(\tau,b_j)}{\psi(\tau,b_j)} x_1^r)^2}{b_j - a_r} = 1.$$

Now (0.9c) is equivalent to this formula:

(0.9*d*)
$$Q_j(\xi_j(\tau)) = 1 \quad \text{where} \quad \xi_j(\tau) \doteq x_2(\tau) - t(\tau) x_1(\tau).$$

and $t(\tau) = \partial \ln \psi(\tau, b_j)$. The equation (0.9a) implies that the line $l(\tau)$ is tangential to Q_j at $\xi_j(\tau)$; that is,

(0.9*e*)
$$\nabla Q_j(\xi_j(\tau)) \cdot x_2 = 0.$$

□

(0.9C) REMARK: In addition, Moser showed that $\xi_j(\tau)$, the point of contact, moves in τ along a geodesic on Q_j.

Independence. The formula (0.5a) establishes a relationship between the Neumann system and Krichever's operator theory. We shall develop this theory a bit further to prove that the Hamiltonians G_r^* are independent. There exists for each $j \in \mathbf{N}$ a unique monic differential operator L_j of order j in ∂ such that

$$L_j\psi \cdot \psi^{-1} = \kappa^j + O(\kappa^{-1}). \tag{0.10}$$

It follows then that the operator $[L_j, L]$ has order 0. The Korteweg-de Vries hierarchy is the family of partial differential equations given by these operator Lax equation:

$$\frac{\partial L}{\partial t_j} = -\partial_j q = [L_j, L] \text{ for each } j \in \mathbf{N}. \tag{0.11}$$

The Neumann hierarchy is the family of nonlinear ordinary differential equations in $\mathbf{m}$ given by these formulas:

$$\frac{\partial x_1^r}{\partial t_j} = L_j x_1^r \quad \text{and} \quad \frac{\partial x_2^r}{\partial t_j} = \partial\left(\frac{\partial x_1^r}{\partial t_j}\right). \tag{0.12}$$

The system (0.12) defines a vector field X_j^* which is tangential to TS^g. These vector fields are Hamiltonian with respect to the restriction of ω to TS^g. The Hamiltonian is a linear combination of the G_r^*.

The Baker function ψ may be viewed as a function of $\mathbf{m}$. The action of the vector field X_j^* on ψ is given by $X_j^*\psi = L_j\psi$. A linear dependence relationship among the X_{2j-1}^*, $(j = 1, \ldots, g)$ would correspond to a linear dependence relationship among the $L_{2j-1}\psi$, $(j = 1, \ldots, g)$. This is impossible by (0.10). It follows then that vector fields X_{2j-1}^*, $(j = 1, \ldots, g)$ are linearly independent. This in turn implies that the G_r^*, $(r = 1, \ldots, g)$ are functionally independent.

Involutivity. Let $\mathfrak{g} = gl(2, \mathbf{C})$, the Lie algebra of all 2 *times*2 matrices over $\mathbf{C}$. We shall identify the dual of $\mathfrak{g}$ with $\mathfrak{g}$ itself using the usual trace form $(\xi, \eta) \in \mathfrak{g} \times \mathfrak{g} \to \operatorname{tr}(\xi\eta)$. The *loop algebra*$L(\mathfrak{g})$ in the parameter λ is the infinite dimensional Lie algebra defined by the following formulas:

$$L(\mathfrak{g}) = \{\, \xi(\lambda) = \lambda^N(\sum_{j=0}^{\infty} \xi^{[j]}\lambda^{-j}) : \xi^{[j]} \in \mathfrak{g} \quad \text{and} \quad N \in \mathbf{Z}\} \tag{0.13a}$$

$$[\xi(\lambda), \eta(\lambda)] = \lambda^{N+N'} \sum_{j=0}^{\infty} \left(\sum_{l+k=j} [\xi^{[l]}, \eta^{[k]}] \right) \lambda^{-j}. \tag{0.13b}$$

Let K denote the bilinear form on $L(\mathfrak{g})$ given for $\xi(\lambda)$ and $\eta(\lambda)$ in $L(\mathfrak{g})$ by this formula:

$$K(\xi(\lambda), \eta(\lambda)) = \pi_{-1} \circ \operatorname{tr}(\xi(\lambda)\eta(\lambda)) = -\operatorname{Res}_\infty \operatorname{tr}(\xi(\lambda)\eta(\lambda))d\lambda. \tag{0.13c}$$

Then K is symmetric, nondegenerate and associative. Let us consider the direct sum decomposition of $L(\mathfrak{g})$ into these subalgebras:

(0.14*a*) $$L(\mathfrak{g}) = \mathfrak{N} \oplus \mathfrak{K}$$

where $\mathfrak{N}$ is the subalgebra of polynomial loops, and

(0.14*b*) $$\mathfrak{K} = \mathbb{C}(e_{2,1} + e_{1,2}\lambda^{-1},\ e_{2,1}\lambda^{-1},\ e_{1,1}\lambda^{-1},\ e_{2,2}\lambda^{-1}) \oplus \sum_{j=2}^{\infty} \mathfrak{g}\lambda^{-j}.$$

The projections to $\mathfrak{N}$ and $\mathfrak{K}$ are given by these formula:

(0.14*c*) $$\begin{cases} \pi_{\mathfrak{N}}(\xi(\lambda)) = \pi_{\geq}(\xi(\lambda)) - \xi_{1,2}^{[-1]} e_{2,1}. \\ \pi_{\mathfrak{K}}(\xi(\lambda)) = \pi_{<}(\xi(\lambda)) + \xi_{1,2}^{[-1]} e_{2,1}. \end{cases}$$

Let $\epsilon = 2e_{2,1} \in \mathfrak{N}^0 \cap [\mathfrak{K}, \mathfrak{K}]^0$, an infinitesimal character of $\mathfrak{N}$. The dual of $\mathfrak{N}$ is identified with the annihilator $\mathfrak{K}^0$ of $\mathfrak{K}$ with respect to K; $\mathfrak{N}^* \cong \mathfrak{K}^0$. The annihilator is given by this formula:

(0.14*d*) $$\mathfrak{K}^0 = \mathbb{C}(e_{2,1} - e_{1,2}\lambda^{-1},\ e_{2,1}\lambda^{-1},\ e_{1,1}\lambda^{-1},\ e_{2,2}\lambda^{-1}) \oplus \sum_{j=2}^{\infty} \mathfrak{g}\lambda^{-j}.$$

It is a well known fact that the dual of a Lie algebra is a Poisson manifold with respect to the Kirrillov-Poisson bracket. The space $\epsilon + \mathfrak{K}^0$ is a Poisson manifold under the identification (0.14d). By the Adler/Kostant/Symes Theorem, the following Hamiltonians are in involution with respect to the Kirrillov-Poisson bracket on $\epsilon + \mathfrak{K}^0$:

(0.14*e*) $$h_{i,j}(\epsilon + \xi(\lambda)) \doteq \operatorname{tr}\left(\lambda^i(\epsilon + \xi(\lambda))^j_{-1}\right) \quad \text{if} \quad \xi(\lambda) \in \mathfrak{K}^0.$$

Let $\mathcal{M}$ denote the cotangent bundle to the space of all $n \times \ell$ matrices. Let $\mathbf{x}$ denote an $2 \times \ell$ matrix with columns x^r and rows x_α:

$$\mathbf{x} = (x^1, \dots, x^\ell) = (x_1^T, \dots, x_n^T)^T.$$

An element of $\mathcal{M}$ has the form $\mathbf{m} = (\mathbf{x}, \mathbf{u})$ where $\mathbf{u}$ has rows u^r and columns u_α. The Lie algebra $\mathfrak{N}$ acts on $\mathcal{M}$. The action and its momentum mapping $\Upsilon : \mathcal{M} \to \mathfrak{K}^0$ are given for $\xi(\lambda) \in \mathfrak{N}$ by this formula:

(0.15*a*) $$\xi(\lambda) \cdot \mathbf{m} = (\xi(a_r)x^r, -\xi(a_r)^T u^r)$$

(0.15*b*) $$\Upsilon(\mathbf{m}) = \pi_{\mathfrak{K}^0}(\mathbf{x}(\lambda - \mathbf{a})^{-1}\mathbf{u}^T) = -e_{2,1} + \sum_{r=1}^{\ell} \frac{x^r \otimes u^r}{\lambda - a_r}.$$

The momentum map Υ is equivariant with respect to the infinitesimal action of $\mathfrak{N}$ on $\mathfrak{N}^*$:

$$\begin{array}{ccc} \mathcal{M} & \xrightarrow{\Upsilon} & \mathfrak{N}^* \cong \mathfrak{K}^0 = \mathfrak{K} \\ \xi(\lambda)\Big\downarrow & & \Big\downarrow -\pi_{\mathfrak{K}^0} \circ \mathrm{ad}_{\xi(\lambda)} \\ \mathcal{M} & \xrightarrow{\Upsilon} & \mathfrak{N}^* \cong \mathfrak{K}^0 = \mathfrak{K}. \end{array}$$

Our isospectral matrix Z satisfies

$$Z = \epsilon + \Upsilon(\mathbf{m}) \in \epsilon + \mathfrak{K}^0. \tag{0.16}$$

By a theorem of Guillemin and Sternberg, any momentum mapping is a canonical mapping. It follows then that the Hamiltonians defined by this formula:

$$H_{i,j} \doteq h_{i,j}(\Upsilon(\mathbf{m})) = \mathrm{tr}(\lambda^i Z^j)_{-1} \tag{0.17}$$

are involutive. The G_r being functions of the $H_{i,j}$ are involutive. Since the G_r^* agree with the G_r on TS^g, the G_r^* are involutive.

This completes the proof of the algebraic complete integrability of the Neumann system. A connection with the Korteweg-de Vries hierarchy was established and then used to prove that the Ulenbeck constants of motion are independent. A connection with loop algebras was established and then used to prove that the Ulenbeck constants of motion are in involution.

(0.18) REMARK: This approach to the involutivity of the Ulenbeck constants of motion is based on the geometry of momentum mappings developed in [**2**]; in particular, the infinitesimal $\mathfrak{N}$ action that lead to the interpretation (0.16) of Z in terms of a momentum mapping. The theory in [**2**] does not include the Neumann system. The loop algebra decomposition used there is more directly related to systems of AKNS type [**10.V**].

The obscure point in our construction is the decomposition (0.14a) of $L(\mathfrak{g})$. It might be motivated in the following way. The Hamiltonian vector field D_h on $\epsilon + \mathfrak{K}^0$ associated to the invariant Hamiltonian

$$h(\xi(\lambda)) \doteq \frac{1}{2} K(\lambda\xi(\lambda), \xi(\lambda)) \tag{0.18a}$$

is given by the following formula:

$$D_h(\epsilon + \xi(\lambda)) = [\pi_{\mathfrak{N}}(\lambda(\epsilon + \xi(\lambda)), \epsilon + \xi(\lambda)]. \tag{0.18b}$$

If $\xi(\lambda) = \xi_{1,2}^{[1]} e_{2,1} + \xi^{[1]}\lambda^{-1} + \xi^{[2]}\lambda^{-2} + \cdots$ then

$$\pi_{\mathfrak{N}}\lambda(\epsilon + \xi(\lambda)) = (\lambda + \xi_{1,1}^{[1]} - \xi_{1,2}^{[2]})e_{2,1} + e_{1,2}. \tag{0.18c}$$

The equation (0.18a) with $\epsilon + \xi(\lambda) = Z$ is the Lax equation (0.2a). Our basic linear equation (0.5b) has the form

$$(0.18d) \qquad \partial U(\tau, p) = \pi_{\mathfrak{N}}(\lambda Z)U(\tau, p).$$

The family of differential equations on $\epsilon + \mathfrak{K}^0$ defined by the Hamiltonians $h_{i,j}$ in (0.14e) is equivalent to the Korteweg-de Vries hierarchy.

Generalizations. In this paper isospectral pairs are found under more general circumstances. The Riemann surface is no longer hyperelliptic. It is an n-fold cover of $\mathbb{P}^1$. The basic operator theory is the deformation theory of an nth order scalar differential operator. Let us take a moment to explain the basic idea behind *the Neumann system of a differential operator.* One tries to express the coefficients of an operator in terms of certain eigenfunctions and eigenvalues of the operator itself as in (0.7). The Neumann system is a system of first order equations that is equivalent to the eigenvalue equations satisfied by the special eigenfunctions. It is an algebraic completely integrable Hamiltonian system [**10.V**]. In our algebraic setup the eigenfunctions are determined by the poles of a rational function z. According to Krichever [**8**], a rational function with simple poles is the spectral parameter in a first order matrix spectral problem, *an algebraic AKNS spectral problem.* This secondary spectral problem plays an absolutely essential role in our derivation of Lax pairs and constants of motion for the Neumann system.

The secondary spectral problem associated to the primary problem (0.5a) of the hyperelliptic setup might be described in the following way. The secondary spectral problem is given by these formulas:

$$(0.19b) \qquad \dot{\Psi} = \mathsf{B}\Psi \quad \text{where} \quad \mathsf{B} = \mathbf{a}z + x_1 \otimes u_1 + x_2 \otimes u_2.$$

Its solution is a *vector* Baker function which we denote by Ψ. The polar divisor of Ψ is

$$(0.19a) \qquad \Delta = \delta + (z)_0 - \infty,$$

a divisor of degree $2g$. According to Krichever [**8**], there is a matrix valued function of (τ, p) with eigenvector Ψ and eigenvalue is z. This spectral problem is given by these formulas:

$$(0.19c) \qquad z(p)\Psi(\tau, p) = \mathsf{L}(\mathbf{m}, p)\Psi(\tau, p)$$

where

$$\mathsf{L} = \mathbf{a}z^2 + (x_1 \otimes u_1 + x_2 \otimes u_2)z + x_1 \otimes u_2 = \mathbf{a}z^2 + x(I_2 z - e_{2,1})u^T.$$

It is clear that L is isospectral with respect to the Neumann system. The matrices (Z, L) form what we call an *isospectral pair.*

M.R. Adams, J. Harnad and J. Hurtubise [1], using the geometry of momentum mappings [2], have shown that the coefficients of the characteristic polynomials of the isospectral matrices Z and L generate the same algebra of functions. The term *duality* was used to describe the relationship between the isospectral matrices. The term *dual* is already in use in Baker function theory. We have therefore used the phrase *isospectral pair* in place of *dual pair.*

APPLICATION: **The Euler Equation for a Free Rigid Body.** The Euler equation for an $\ell \times \ell$ matrix E over $\mathbb{C}$ is given by this matrix Lax equation:

$$\partial_t(\mathbf{b}E + E\mathbf{b}) = [\mathbf{b}E + E\mathbf{b}, E] = [\mathbf{b}, E^2] \text{ where } \partial_t = \frac{\partial}{\partial t} \tag{0.20a}$$

and $\mathbf{b}$ is a diagonal matrix with distinct entries b_r. This equation, with real data $E \in so(\ell)$, governs the motion of an ℓ dimensional free rigid body. The solution $E(t)$ moves along a geodesic on so(ℓ) with respect to the left-invariant metric $(X|Y) = -tr(X\mathbf{b}Y)$. Manakov observed that (0.1) is equivalent to the following Lax equation with a parameter z:

$$\partial_t \mathcal{W} = [\mathsf{B}, \mathcal{W}] \text{ where } \mathcal{W} = \mathbf{b}E + E\mathbf{b} + \mathbf{b}^2 z \quad \text{and} \quad \mathsf{B} = -E + \mathbf{b}z. \tag{0.20b}$$

J. Moser ([9], §5d) found that the Euler system contains an integrable subsystem in which $\mathcal{W}$ is the rank 2 perturbation of $\mathbf{b}^2 z$ given for x and y in $\mathbb{R}^\ell$ by

$$\mathcal{W} = x \otimes y - y \otimes x + \mathbf{b}^2 z. \tag{0.20c}$$

The solutions to the subsystem are expressible in terms of hyperelliptic functions. A 2×2 matrix whose spectrum is preserved by the rank two Euler subflow was found in [1]; namely,

$$J = \sum_{r=1}^{\ell} \frac{\begin{pmatrix} x_r \\ y_r \end{pmatrix} \otimes \begin{pmatrix} y_r \\ -x_r \end{pmatrix}}{\lambda - b_r^2}. \tag{0.20d}$$

In [**10.II**] we found a system much like (0.1) involving 4 instead of 2 constraints. The isospectral matrix Z of this system is almost identical to J; Z contains b_r instead of b_r^2 . This suggested that certain solutions to the Euler problem could be found in terms of solutions to the generalized Neumann hierarchies found in [**10.IV**]. This is indeed the case. The second isospectral matrix L in this setup is a solution to the Euler problem. The details are in [**10.VI**]. Moser's example, the motivation behind this work, does not fit into our setup. We are still trying to find the appropriate Baker function setup for it. The paper [**10.VI**] contains an example in which a solution E to the Euler problem is constructed from a solution to Hamilton's equations for a member X^*_E of the Neumann hierarchy associated to a differential operator of order 2. In this way one sees X^*_E as a subsystem of the Euler system. The Korteweg-de Vries flow in Neumann coordinates commutes with X^*_E. Therefore time advance under the Korteweg-de Vries flow preserves the space of solutions to this Euler subsystem.

This Paper. The layout of this paper is as follows. We shall describe one member of the isospectral pair; namely the order n analogue of Z, using a rational extension of Krichever's homomorphism [**8**]. The other isospectral matrix L is found using a translation in the Jacobian of the underlying spectral curve. The translation is the mapping Δ defined as in (0.19a), $\delta \to \Delta \doteq \delta + (z)_0 - \infty$, in terms of the zero divisor of the secondary spectral parameter z. H. Flaschka referred to the divisor Δ as the *augmented divisor*. The key to our approach is a fundamental relationship between the Baker functions of the primary and secondary spectral problems. This relationship was found by H. Flaschka [**7**]. This paper contains a further development of that relationship.

The first section of the paper contains notational preliminaries, a discussion of Baker function analysis and a derivation of Flaschka's formula. The second section contains formulas for our isospectral matrices.

§1. The Augmented Divisor.

Riemann Surface and Rational Functions. Let $\mathcal{R}$ be a Riemann surface of genus $g_{\mathcal{R}}$. Choose a point ∞ of $\mathcal{R}$ and let A_∞ be the ring of all rational functions on $\mathcal{R}$ that are holomorphic in $\mathcal{R} \setminus \infty$. The set of Weierstrass gap numbers of ∞ is given by

$$W = \mathsf{N} \setminus \{-\mathrm{ord}_\infty y | y \in A_\infty\}.$$

Let λ be the unique function in A_∞ whose polar order n at ∞ is minimal. The formula $\lambda = \kappa^n$ defines a function κ whose inverse is a local parameter at ∞. Choose a rational function z on $\mathcal{R}$ that satisfies the following 2 conditions.

(1.1a) If $(z)_\infty = (1) + \cdots + (\ell), (r) \in \mathcal{R}$, denotes the polar divisor of z and $a_r = \lambda(r)$ then we shall assume that the a_r are distinct. It follows then that each (r) is a simple pole of z and z^{-1} is a local paremeter at (r). We shall use the letter $\mathbf{a}$, depending on context, to denote either the divisor $\sum_{r=1}^{\ell} a_r$ over P^1 or the matrix $\mathrm{diag}(a_1, \dots, a_\ell)$.

(1.1b) We shall assume that ∞ is not a pole of z. After replacing z by $z - z(\infty)$, one may assume that $z(\infty) = 0$. Then there exists an integer $m > 0$ and constants z_j such that in a neighborhood of ∞ z is given by

$$z = \kappa^{-m}(1 + \sum_{j=1}^{\infty} z_j \kappa^{-j}).$$

In **[10.IV]**, it was assumed that m and n are relatively prime in order to prove that the Hamiltonian systems defined therein are completely integrable. We shall not need this assumtion to achieve the goals of this paper. **Divisors.** Let δ be positive divisor of degree $g_{\mathcal{R}}$ satisfying this condition:

$$L(\delta - \infty) = \{0\}; \tag{1.2}$$

that is, any function with poles no worse then δ and vanishing at ∞ is identically 0. It follows that $L(\delta) = \mathbb{C}$ and that δ is a nonspecial divisor. Conversely, any nonspecial divisor of degree $g_{\mathcal{R}}$ satisfies (1.2). The set of all such divisors is isomorphic to the affine open subset $\mathrm{Jac}(\mathcal{R}) \setminus \Theta$ of the Jacobian of $\mathcal{R}$ under an appropriately defined Abel mapping; Θ denotes the zero locus of the corresponding Riemann theta function.

We shall associate three divisors with δ, denoted δ^ι, Δ, and Δ^ι in a geometrical way.

(1.3) PROPOSITION. *Let Δ denote the mapping defined by translation by the zero divisor of our rational function z less ∞:*

$$\Delta : Div(\mathcal{R}) \to Div(\mathcal{R}) : \quad \beta^\Delta = \beta + (z)_0 - \infty.$$

This mapping is well defined modulo linear equivalence and its push forward under the Abel map is a translation in the group Jac($\mathcal{R}$). **(1.3a)** *The mapping Δ is, up to linear equivalence, an isomorphism of the positive divisors of degree $g_{\mathcal{R}}$ onto the positive divisors of degree $g_{\mathcal{R}} + \ell - 1$.*

(1.3b) *If δ satisfies (1.2) then its image $\Delta = \delta^\Delta$ satisfies this analogous condition:*

$$L(\Delta - (z)_\infty) = \{0\}. \tag{1.3c}$$

PROOF: Translation in any topolical group is an isomorphism. □

ALTERNATE PROOF: Our statement (a) follows from the Riemann-Roch theorem because the divisor $\beta - (z)_0 + \infty$ has degree $g_{\mathcal{R}}$ for any divisor b of degree $g_{\mathcal{R}} + \ell - 1$. Now to prove (b) one need only note that if f belongs to $L(\Delta - (z)_\infty)$ then zf belongs to $L(\delta - \infty)$. □

Let us consider the mapping corresponding to translation by the zero divisor of our rational function z less ∞,

$$\Delta : \mathrm{Div}(\mathcal{R}) \to \mathrm{Div}(\mathcal{R}) \quad \beta^\Delta = \beta + (z)_0 - \infty.$$

This mapping is well defined modulo linear equivalence; its push forward under the Abel map is simply a translation in the group Jac($\mathcal{R}$).

(1.4) **Proposition.** *Let Δ be a positive divisor of degree $g_{\mathcal{R}} + \ell - 1$ satisfying (1.3c). Then there exists a unique abelian differential Λ such that*

$$(\Lambda) - \Delta + 2(z)_\infty \geq 0 \tag{1.4a}$$

and for each $(r = 1, \dots, \ell)$ one has

$$\Lambda = -z^2(1 + O(z^{-1}))dz^{-1} \text{ at } (r). \tag{1.4b}$$

The proof is an easy consequence of the Riemann-Roch theorem. See [**10.IV**] for details. Since the degree of the divisor of any abelian differential is $2g_{\mathcal{R}} - 2$, there exist a unique positive divisor Δ^ι of degree $g_{\mathcal{R}} + \ell - 1$ such that

$$(\Lambda) = \Delta + \Delta^\iota - 2(z)_\infty. \tag{1.4c}$$

Using the argument in the proof of (1.3b), one may show that Δ^ι satisfies (1.3c). We shall refer to Δ^ι as the *dual* of Δ. It is now clear that the mapping ι is an involution of the set of positive divisors of degree $g_{\mathcal{R}} + \ell - 1$.

Now let δ as in (1.2). By Proposition (1.4) with $(z)_\infty$ replaced by ∞, there exists an abelian differential Ω and a divisor δ^ι of degree $g_{\mathcal{R}}$ such that

$$(\Omega) = \delta + \delta^\iota - 2\infty \text{ and } \Omega = -\kappa^2(1 + O(\kappa^{-1}))d\kappa^{-1} \text{ at } \infty. \tag{1.5a}$$

Let $\Delta = \delta + (z)_0 - \infty$. One may now argue using Proposition (1.4) that

$$\Delta^\iota = \delta^\iota + (z)_0 - \infty \text{ and } \Lambda = z^2\Omega. \tag{1.5b}$$

We have allowed ourselves a slight abuse of notation by using ι for the mapping (1.4) associated with $(z)_\infty$ and the mapping (1.5) associated with ∞. With this in mind, one might say that ι commutes with translation by the divisor $(z)_0 - \infty$.

Baker Functions. Let δ be a divisor as in (1.2) and let Δ be any divisor satisfying (1.3c). Let δ^ι be the divisor dual to δ and let Δ^ι be the divisor dual to Δ. Let

$$\theta(t, p) = \sum\nolimits_W \lambda^k z^\alpha t_{k,\alpha}$$

where the sum is taken over all integer pairs (k, α) such that $nk - m\alpha$ belongs to W. Let $t = (t_{k,\alpha})$, an element of $\mathbb{C}^{g_{\mathcal{R}}}$. The sum covers a proper subset of W unless m and n are relatively prime. Let

$$\partial_s = \partial_{k,\alpha} = \frac{\partial}{\partial_{k,\alpha}} \text{ if } s = nk - m\alpha.$$

Let us consider the linear space consisting of all functions χ satisfying the following 2 conditions:

(1.6a) Any pole of χ lies in δ; $(\chi)+\delta \geq 0$,
(1.6b) The function $\chi\exp(-\kappa\tau-\theta)$ is holomorphic at ∞.

Then according to [**10.I**], this linear space has dimension 1 for all (τ,t) in neighborhood of 0 in $\mathbf{C}^{g_{\mathcal{R}}+1}$. It contains a unique function $\psi = \psi_\delta(\tau, t, p)$ such that

$$\psi = (1+O(\kappa^{-1}))\exp(\kappa\tau+\theta) \text{ at } \infty. \tag{1.6c}$$

We call ψ the Baker function of δ and $\phi(\tau,t,p)\dot{=}\psi_{\delta'}(-\tau,-t,p)$ the *dual* Baker function. Let us consider the linear space consisting of all functions χ satisfying the following 2 conditions:

(1.7a) Any pole of χ lies in Δ; $(\chi)+\Delta \geq 0$,
(1.7b) The function $\chi\exp(\theta)$ is holomorphic at each pole (r) of z.

Then according to [**10.I**], this linear space has dimension ℓ for all t in a neighborhood of 0 in $\mathbf{C}^{g_{\mathcal{R}}+1}$. It contain s a unique function $\Psi^r = \Psi^r_\Delta(t,p)$ such that at the pole (s) of z one has

$$\Psi^r = (\delta_{r,s}+O(z^{-1}))\exp(-\theta). \tag{1.7c}$$

We shall refer to $\Psi\dot{=}(\Psi^1,\ldots,\Psi^\ell)^T$ as the vector Baker functio n of Δ and to $\Phi(t,p)\dot{=}\Psi_{\Delta'}(-t,p)$ as its dual.

We shall now present a rather abstract formula for Ψ. The next subsection contains a much more explicit formula. The condition (1.3c) implies that $L(\Delta)$ contains a basis $f_1,\ldots,f_\ell$ such that $f_r(s)=\delta_{r,s}$. Let us consider these divisors:

$$\delta_r = \Delta + (f_r) - (z)_\infty + (r). \tag{1.8a}$$

One may show that δ_r is a positive divisor of degree $g_{\mathcal{R}}$ that satisfies (1.2). If χ_r is the Baker function associated to δ_r (with (r) replacing ∞ in (1.6) and the appropriate modifications to (1.6b)) then

$$\Psi^r(t,p) = f_r(p)\chi_r(t,p). \tag{1.8b}$$

Relations Between Baker Functions. We shall need the beautiful approach to operator deformation theory due to Cherednik [**4**]. It is contained in the following lemma.

(1.9) LEMMA. *There exists a unique monic differential operator L_j of order j in $\partial = \partial/\partial\tau ='$ such that*

$$L_j\psi \cdot \psi^{-1} = \kappa^j + O(\kappa^{-1}). \tag{1.9a}$$

If functions $\epsilon_{j,r}$ of (τ, t) alone are defined in accordance with (1.9a) by the formula

$$L_j\psi \cdot \psi^{-1} = \kappa^j + \sum_{r=1}^{\infty} \epsilon_{j,r}\kappa^{-r}. \tag{1.9b}$$

then the following formulas hold:

$$L_{j+1} = \partial L_j - \sum_{s=1}^{j-1} \epsilon_{1,s}L_{j-s} - (\epsilon_{j,1} + \epsilon_{1,j}), \tag{1.9c}$$

$$\epsilon_{j+1,r} - \epsilon_{r,j+1} = \epsilon_{1,j+r} + \epsilon'_{j,r} + \sum_{s=1}^{r-1} \epsilon_{j,s}\epsilon_{1,r-s} - \sum_{s=1}^{j-1} \epsilon_{1,s}\epsilon_{j-s,r}, \tag{1.9d}$$

if $j = qn + r$ then

$$L_j = L_r L_n^q - \sum_{s=1}^{j-r} \epsilon_{r,s}L_{j-r-s} \tag{1.9e}$$

and

$$\epsilon_{j,s} = \epsilon_{r,j-r+s} - \sum_{p=1}^{j-r} \epsilon_{r,p}\epsilon_{j-r-p,s}. \tag{1.9f}$$

If the phase in (1.6c) is given by $\theta = \sum_{r\geq 1} \kappa^r t_r$ then

$$\partial_s L_r + L_r L_s = L_{r-s} + \sum_{k=1}^{s} \epsilon_{r,k}L_{s-k} + \sum_{k=1}^{r} \epsilon_{s,k}L_{r-k}; \tag{1.9g}$$

The dual Baker function satisfies

$$L^{\dagger}_{\beta}\phi \cdot \phi^{-1} = \kappa^{\beta} + O(\kappa^{-1}). \tag{1.9h}$$

If we define a series of functions $\epsilon^{\star}_{\beta,j}$ of (τ, t) alone in accordance with (1.9h) by this formula:

$$L^{\dagger}_{\beta}\phi \cdot \phi^{-1} = \kappa^{\beta} + \sum_{j=1}^{\infty} \epsilon^{\star}_{\beta,t}\kappa^{-t}. \tag{1.9i}$$

then the following iterative formula for the $\epsilon^{\star}_{\beta,j}$ *in terms of the* $\epsilon_{\alpha,k}$ *holds:*

$$(1.9j) \quad \epsilon^{\star}_{\beta,j} = \epsilon_{\beta+j-1,1} - \epsilon_{j-1,\beta+1} + \sum_{t=1}^{\beta-1} \epsilon_{j-1,t}\epsilon_{\beta-t,1} + \sum_{t=1}^{j-2} \epsilon^{\star}_{\beta,t}\epsilon_{j-t-1,1}$$

We shall not prove this lemma. There is a nice inductive proof of (1.9a) due to Flaschka [**7.III**]. The formulas, except (1.9h), follow from similar calculations. The paper [**4**] contains a clever proof of (1.9h). We shall need the following notation and technical assumtion for our main theorems.

The formula (1.9b) with $j = n$ implies that ψ satisfies this order n scalar spectral problem: $L\psi = \lambda\psi$ where $L = L_n$. The dual function satisfies the adjoint spectral problem. Let us define this $\mathbb{C}^n$-valued function:

$$\mathcal{U} = (\psi_1, \dots, \psi_n)^T \text{ where } \psi_j = L_{j-1}\psi.$$

Then by (1.9c) $\mathcal{U}$ satisfies this linear system:

$$(1.9k) \qquad \partial\mathcal{U} = B(\mathbf{q}(\tau,t), \lambda(p))\mathcal{U} \text{ where } B = \Lambda + \mathbf{q}, \partial = \partial/\partial\tau =',$$

$$(1.9l) \quad \Lambda = S + \lambda e_{n,1}, \quad \mathbf{q} = \sum_{j=1}^{n-1}(\epsilon_{1,j}S^j + \epsilon_{j,1}e_{j+1,1}) \text{ and } S = \sum_{\alpha=1}^{n-1} e_{\alpha+1,\alpha}.$$

We define another $\mathbb{C}^n$-valued function $\mathcal{V}$ by these formulas:

$$\mathcal{V} = (\phi_1, \dots, \phi_n)^T \text{ where } \phi_n = \phi \text{ and } \partial\mathcal{V} = -B^T\mathcal{V}.$$

We note that $\phi_j \neq L^{\dagger}_{n-j}\phi_n$ if $j \leq n-2$. One might think that a definition of ϕ_β in terms of $L_{n-\beta}$ would be natural. We introduced $\mathcal{V}$ in terms of B becuase the dot product $\mathcal{U}\cdot\mathcal{V}$ shall play a vital role. We note that $\mathcal{U}\cdot\mathcal{V}$ is constant in τ.

The vector functions are evaluated over the polar divisor of z with $\tau = 0$ to make these definitions:

$$(1.9m) \qquad x^r(t) = \rho_r\mathcal{U}(0,t,r) \quad \text{and} \quad u^r(t) = \rho_r\mathcal{V}(0,t,r)$$

where $\rho_r = \sqrt{\text{res}_r z\Omega}$. These vectors form the columns of this $\ell \times n$ matrix:

$$(1.9n) \qquad x = (x^1, \dots, x^\ell)^T = (x_1, \dots, x_n),$$

where the vector x_j contains the component j of each x^r. We define a matrix u in terms of the u^r in the same way. Let $\mathbf{m} = (x, u)$, a point in $\mathbb{C}^{2n\ell}$.

We shall prove the next theorem under the following rather technical but mild assumtion concerning the ring A_∞. We shall assume that the ring A_∞ separates points in the following way: for each pole (r) of z, there exists a y in A_∞ that assumes distinct values at the points of $\mathcal{R}$ over α_r. According to ([**10.II**],formula (1.3.8)), this assum tion implies that the functions $\mathcal{U}(p)\cdot u^r$ and $x^r \cdot \mathcal{V}(p)$ are zero at the points in the set $\lambda^{-1}(\alpha_r)\setminus\{(r)\}$; the order of the zero is equal to that of $(\lambda - a_r)$. It follows then that ∞ and (r) are the only poles of the differential $(\lambda - a_r)^{-1}\mathcal{V}(p)\cdot u^r\phi(p)\Omega$. The formula

$$J_r^1 = x_r \cdot u_r \tag{1.10}$$

follows from the residue theorem and the definition of ρ_r.

(1.11) THEOREM. **(Flaschka [7.II]).** *Let $\Delta = \delta + (z)_0 - \infty$. The vector Baker function Ψ of Δ is given by these formulas:*

$$\Psi^r(t,p) = \mathcal{U}(0,t,p)\cdot u^r(t)\frac{z^{-1}}{\lambda - a_r}\exp(-\theta) \tag{1.11a}$$

or, in vector notation,

$$\Psi(t,p) = (\lambda - \mathbf{a})^{-1}\sum_{\gamma=1}^{n}\psi_\gamma(0,t,p)u_\gamma(t)z^{-1}\exp(-\theta). \tag{1.11b}$$

The dual function is given by this formula:

$$\Phi(t,p) = (\lambda - \mathbf{a})^{-1}\sum_{\gamma=1}^{n}f_\gamma(0,t,p)x_\gamma(t)z^{-1}\exp(\theta). \tag{1.11c}$$

PROOF: In view of the assumtions preceeding (1.10), it is clear that the left hand side (lhs) in (1.11a) is finite at the points of $\mathcal{R}$ over a_r. In a neighborhood of ∞ (lhs) is holomorphic (the exponential in ψ cancels with $\exp(-\theta)$) and

$$\Psi = u_n(t)\kappa^{m-1} + O(\kappa^{m-2}) \text{ at } \infty \tag{1.11d}$$

Now it is clear that (lhs) satisfies (1.7a) and (1.7b). □

The left hand side in (1.11a) does not quite satisfy the condition (1.7c). One must introduce a factor of ρ_r^{-1} into (1.7c). The factor is inconsequential and we shall leave the formula stand as it is. We shall refer to Δ as the **augmented** divisor and to Ψ as the augmented Baker function.

§2. Isospectral Pairs.

(2.1) THEOREM. *The vector function $\mathcal{U}$ satisfies this $n\times n$ matrix spectral problem:*

$$z(p)\mathcal{U}(0,t,p) = Z(\mathbf{m}(t),\lambda(p))\mathcal{U}(0,t,p) \tag{2.1a}$$

where Z is a rank ℓ perturbation

$$Z = Z_0 + \sum_{\alpha=1}^{\ell} \frac{x^r \otimes u^r}{\lambda - a_r} \tag{2.1b}$$

of the nilpotent matrix

$$Z_0 = \sum_{j=0}^{n-m-1} z_j S^{m+j}, \quad S = \sum_{\alpha=1}^{n-1} e_{\alpha+1,\alpha} \tag{2.1c}$$

and $e_{\alpha,\beta}$ is the $n\times n$ matrix with a 1 in entry (α,β) and zeros elsewhere. The dual function satisfies the transposed spectral problem.

PROOF: Let $\tau = 0$ in this proof. The functions $(\psi_1,\dots,\psi_m)\exp(-\theta)$ belong to the linear space (1.7) but $(\psi_{m+1},\dots,\psi_n)\exp(-\theta)$ do not because their polar order at ∞ is too large. Using (1.1b) and (1.9a) one finds that

$$z\psi_{m+s+1}\cdot\psi^{-1} = \kappa^s + z_1\kappa^{s-1} + \cdots + z_s + O(\kappa^{-1}) \text{ at } \infty.$$

This implies that the function $(\psi_{m+s+1} - z^{-1}\sum_{j=0}^{s} z_j\psi_{s+1-j})\exp(-\theta)$ belongs to the linear space (1.7). Its components in the basis Ψ^r are computed using (1.7c) modified by the appropriate factor; one has

$$(\psi_\alpha - z^{-1}(\sum_{j=0}^{\alpha-m-1} z_j\psi_{\alpha-m-j}))\exp(-\theta) = x_\alpha\cdot\Psi;$$

the component r is verified by letting $p\to(r)$. These formulas are equivalent to the matrix equation (2.1a). □

A Rational Extension of the Krichever Homomorphism. Krichever [**8**] constructed a homomorphism $B\colon A_\infty \to A_{op}$ where A_{op} is a commutative ring of differential operators in ∂ in the following way. The Baker function satisfies for each y in A_∞ the spectral problem

$$y(p)\psi(\tau,t,p) = B_y\psi(\tau,t,p).$$

This equation is equivalent to a matrix spectral problem of the form

$$y(p)\mathcal{U}(\tau,t,p) = M_y(\lambda(p))\mathcal{U}(\tau,t,p).$$

The mapping $y\to M_y$ is a homomorphism. If $(\lambda_1,\dots,\lambda_N)$, $\lambda_1 = \lambda$, is a system of generators of A_∞ then z is a rational function of the λ_j and Z is the corresponding rational function of the matrices M_{λ_j}, $(j = 2,\dots,N)$. The mapping $z\in \mathbb{C}(\mathcal{R}) \to Z$ is an extension of Krichever's homomorphism.

(2.2) PROPOSITION. *There exists a unique $\ell \times \ell$ matrix $\mathsf{B}_{k,\alpha}$, a polynomial of degree α in z with leading term $\mathbf{a}^k z^\alpha$, such that the augmented Baker function and its dual satisfy these linear equations:*

$$\begin{cases} \partial_{k,\alpha}\Phi(t,p) = \mathsf{B}_{k,\alpha}(\mathbf{m}(t), z(p))\Phi(t,p), \\ \\ \partial_{k,\alpha}\Psi(t,p) = -\mathsf{B}_{k,\alpha}(\mathbf{m}(t), z(p))^T\Psi(t,p). \end{cases}$$

PROOF: Following Krichever, we let

$$\Phi_\infty = \sum_{j=0}^{\infty} \Phi_j z^{-j} \exp(\Theta)\rho^{-1}$$

where

$$\Theta = \sum\nolimits_W \Lambda^k z^\alpha, \quad \Lambda = \mathbf{a} + \sum_{\beta \geq 1} J_\beta z^{-\beta}, \quad J_\beta = \operatorname{diag}(J_\beta^r),$$

$\rho = \operatorname{diag}(\rho_r)$ and Φ_j is the $\ell \times \ell$ matrix whose column s is defined by the Taylor series of $\Phi\exp(-\theta)$ at (s): $\Phi = \rho_s^{-1}(\sum_{j=0}^{\infty} \Phi_j^{(s)} z^{-j})\exp(\theta)$. We define a series of matrices $\mathsf{B}^{(\eta)}$ by these formulas:

$$\mathsf{B}^{(0)} = \mathbf{a}^k \text{ and } \mathsf{B}^{(\eta)} = T_k^{(\eta)} + \sum_{\beta=0}^{\eta-1}(\Phi_{\eta-\beta}T_k^{(\beta)} - \mathsf{B}^{(\beta)}\Phi_{\eta-\beta})$$

where $\Lambda^k = \sum_{\beta \geq 0} T_k^{(\beta)} z^{-\beta}$. Then the matrix $\mathsf{B} = \sum_{\eta=0}^{\alpha} \mathsf{B}^{(\eta)} z^{\alpha-\eta}$ satisfies

$$(\partial_{k,\alpha}\Phi_\infty - \mathsf{B}\Phi_\infty)\exp(-\Theta) = \sum_{j\geq 0} \Phi_j z^{-j}\Lambda^k z^\alpha - \mathsf{B}\sum_{j \geq 0}\Phi_j z^{-j} + O(z^{-1})$$

$= O(z^{-1})$. This implies that $(\partial_{k,\alpha}\Phi - \mathsf{B}\Phi)\exp(-\theta) = O(z^{-1})$ at each (s). By (1.7c) the matrix $\mathsf{B}_{k,\alpha} = \mathsf{B}$ satisfies the first formula of our proposition. It is clear from our construction that B is unique. According to Cherednik [**4**], the dual function satisfies the adjoint differential equations. □

(2.3) PROPOSITIONS. *The Baker function Φ satisfies the $\ell \times \ell$ matrix spectral problem*

$$\lambda(p)z(p)^{n'}\Phi(t,p) = \mathsf{L}(\mathbf{m}(t), z(p))\Phi(t,p) \quad \textit{where} \quad \mathsf{L} = \mathsf{B}_{1,n'}$$

and n' is the least positive integer such that $mn' \geq n$.

PROOF: One has $\partial_{1,n'}\Phi = \lambda z^{n'}\Phi$ because, according to Krichever [**8**], $\lambda z^{n'}$, being holomorphic at ∞, belongs to the ring of functions holomorphic in $R \setminus (z)_\infty$. □

To prove our next theorem, we shall need to make this definition:

$$\Sigma = Z - Z_0 = \sum_{\alpha=1}^{\ell} \frac{x^r \otimes u^r}{\lambda - a_r}.$$

(2.4) THEOREM. *The $\ell \times \ell$ matrix L is the rank n perturbation of the diagonal matrix $\mathbf{a}$ given by this formula:*

$$\mathsf{L}/z^{n'} = \mathbf{a} + x(I_n - Z_0 z^{-1})^{-1} u^T z^{-1}.$$

PROOF: By Proposition (2.2) there exists matrices Γ_β such that

$$\mathsf{L}/z^{n'} = \mathbf{a} + \sum_{\beta=1}^{n'} \Gamma_\beta z^{-\beta}.$$

In view of (1.11c), the spectral problem (2.3) is equivalent to

$$z^{n'} \sum_{\gamma=1}^{n} \phi_\gamma x_\gamma = z^{n'} \sum_{\beta=1}^{n'} \Gamma_\beta z^{-\beta} (\lambda - \mathbf{a})^{-1} \sum_{\gamma=1}^{n} \phi_\gamma x_\gamma. \tag{2.4a}$$

The component r of the left hand side of (2.4a) is given by this important calculation:

$$\begin{aligned} z^{n'} x^r \cdot \mathcal{V} &= z^{n'-1} x^r \cdot Z^T \mathcal{V} \\ &= z^{n'-1} x^r \cdot Z_0^T \mathcal{V} + z^{n'-1} \Sigma x^r \cdot \mathcal{V} \\ &= \cdots = \sum_{\beta=1}^{n'} \Sigma Z_0^{\beta-1} x^r \cdot \mathcal{V} z^{n'-\beta} \end{aligned}$$

(Here we used $Z_0^{n'} = (S^m)^{n'} = 0$!)

$$= \sum_{\beta=1}^{n'} \left(\sum_{s=1}^{\ell} Z_0^{\beta-1} x^r \cdot u^s \frac{x^s \cdot \mathcal{V}}{\lambda - a_s} \right) z^{n'-\beta}. \tag{2.4b}$$

Comparing the result of (2.4b) to the right hand side of (2.4a), we find that

$$\Gamma_{\beta;r,s} = Z_0^{\beta-1} x^r \cdot u^s \tag{2.4c}$$

and therefore

$$\Gamma_\beta = \sum_{1 \le \alpha, \gamma \le n} (Z_0)_{\alpha,\gamma}^{\beta-1} x_\gamma \otimes u_\alpha = x Z_0^{\beta-1} u^T \tag{2.4d}$$

This formula is equivalent to (2.4a). □

This completes our construction of the isospectral pair (Z, L). One problem remains. The **m**-dependence of our matrices $\mathsf{B}_{k,\alpha}$ in Proposition (2.2) is implicit. This dependence is made explicit by the following theorem.

(2.5) THEOREM. *Let us define a series of matrices $\Gamma_{k,\alpha}$ by this formula:*

(2.5a)
$$(\mathsf{L}/z^{n'})^k = \mathbf{a}^k + \sum_{\alpha \geq 1} \Gamma_{k,\alpha} z^{-\alpha}.$$

Then $\mathsf{B}_{k,\alpha}$ is given by this formula:

(2.5b)
$$\mathsf{B}_{k,\alpha} = z^\alpha(\mathbf{a}^k + \sum_{\beta=1}^{\alpha} \Gamma_{k,\beta} z^{-\beta}) = \pi_+(z^\alpha(\mathsf{L}/z^{n'})^k)$$

where π_+ denotes projection onto the polynomial part of its argument.

PROOF: In the notation used in the proof of Proposition (2.2), one has

$$\Phi_\infty \Lambda = (\mathsf{L}/z^{n'})\Phi_\infty \text{ and } \Phi_\infty \Lambda^k = (\mathsf{L}/z^{n'})^k \Phi_\infty.$$

This formula, taken modulo z^{-1}, is equvalent to the first formula in (2.2) with our formula (2.5b) for $\mathsf{B}_{k,\alpha}$. □

Closing Remarks.

The equations obtained from (1.9k) and (1.9l) by setting $p = (r)$; namely,

$$\partial x^r = B(\mathbf{q}(\mathbf{m}(\tau)), a_r)x^r \text{ and } \partial u^r = -B(\mathbf{q}(\mathbf{m}(\tau)), a_r)^T u^r.$$

make up what we call a *generalized Neumann systems*. An equivalent formulation of these equations is given for $(\alpha = 1, \ldots, n-1)$ and $(\beta = 0, \ldots, n-2)$ by the following formulas:

$$X^* : \begin{cases} \partial x_\alpha = x_{\alpha+1} + \sum_{j=1}^{\alpha-2} \epsilon_{1,j} x_{\alpha-j} + (\epsilon_{\alpha-1,1} + \epsilon_{1,\alpha-1})x_1, \\ \partial x_n = (\mathbf{a} + \epsilon_{1,n-1} + \epsilon_{n-1,1})x_1 + \sum_{j=1}^{n-2} \epsilon_{1,j} x_{n-j}, \\ \partial u_{n-\beta} = -u_{n-\beta-1} - \sum_{j=1}^{\beta} \epsilon_{1,\beta} u_{n-\beta+j}, \\ \partial u_1 = -\mathbf{a} u_n - \sum_{j=1}^{n-1} (\epsilon_{1,j} + \epsilon_{j,1}) u_{j+1}. \end{cases}$$

The curve $(\tau) \to \mathbf{m}(\tau)$ in $\mathbb{C}^{2n\ell}$ defined in (1.9k) and (1.9l) lies on a submanifold M of $\mathbb{C}^{2n\ell}$ given by algebraic constraints. If m and n are coprime then the algebraic constraints define a symplectic submanifold M of $\mathbb{C}^{2n\ell}$

and X^* is a vector field on M; that is, the entries of $\mathbf{q}$ are functions of $\mathbf{m}$ [**10.IV**] and X^* is tangential to M. The vector field X^* is an algebraic completely integrable Hamiltonian system [**10.IV**]. Much less is k nown about the setup when m and n are not coprime. It is not even clear that the entries of $\mathbf{q}$ are expressible in terms of $\mathbf{m}$.

The formulas (2.1) and (2.3) imply that the matrices Z and L are isospectral with respect to the generalized Neumann system. These matrices are the ***isospectral*** pair of the generalized Neumann system. The Lax equation,

$$\partial_{k,\alpha}\mathsf{L} = [\mathsf{B}_{k,\alpha}, \mathsf{L}],$$

is an immediate consequence of Propositions (2.2) and (2.4). We wish to emphasize, as an example would illustrate, that the Lax equation fails in the absence of any constraints.

References

1. M. R. Adams, J. Harnad and J. Hurtubise, preprint (1989).
2. M. R. Adams, J. Harnad and E. Previato, Commun. Math. Phys. **117** (1988), 451–500.
3. M. Adler and P. van Moerbeke, Invent. Math. **67** (1982), 297–331.
4. I. V. Cherednik, Funkts. Anal. Prilozh. **12 (3)** (1978), 45–54.
5. P. Deift, C. Tomei and E. Trubowitz, Comm. Pure and Appl. Math. **XXXV** (1982), 567–628.
6. P. Deift, F. Lund and E. Trubowitz, Comm. Math. Phys. **74** (1980), 141–188.
7. H. Flaschka, World Science Publishing Co. (1981); II. Tohoku Math. J. **36(3)** (1984), 407–426; III. Quart. J. Math. Oxford (2) **34** (1983), 61–65.
8. I. M. Krichever, Anal. Appl. **11** (1977), 12–26.
9. J. Moser, "The Chern Symposium (1979)," Springer-Verlag, 1980, pp. 147–188.
10. R.J. Schilling, Proc. Ams. Vol. **98,** No. 4 (1986), 671-675; II. Comm. Pure and Appl. Math. **XL** (1987), 455–522; III. Comm. Pure and Appl. Math. **XLII** No. 4 (1989), 409–442; IV. Comm. Pure and Appl. Math. (to appear); V. Memoirs A.M.S. (to appear); VI. in preparation.

Acknowledgements This work was inspired by the 1989 workshop at MSRI entitled Geometry of Hamiltonian Systems. I want to thank the organizers and the staff at MSRI for an excellent workshop. I also want to thank Malcolm Adams for several fruitful discussions.

Symplectic Numerical Integration of Hamiltonian Systems

CLINT SCOVEL

Abstract. In this paper I review several techniques to construct Symplectic Integration Algorithms. I also discuss algorithms for systems with other invariants such as Lie-Poisson structures, reversible systems, and volume preserving flows. Numerical results are presented.

§1. Introduction.

The classical integration and transformation theory of Hamiltonian systems always considered canonical transformations, and to generate these transformations a generating function was used. However, with the advent of high speed digital computers it seems the importance of canonical transformations has been forgotten, as the standard numerical technique is to consider the system as a vector field and to integrate it's trajectory by some ODE integrator. In 1956, De Vogalaere [**24**] attempted to publish concerning symplectic methods, and it wasn't until 1983 that Ruth [**66**] and Channell [**15**] developed the notion of a symplectic integration algorithm (SIA). It should be mentioned that it seems well known that the first order integrators

$$Q = q + tP \tag{1}$$

$$P = p - t * \nabla V(q) \tag{2}$$

and

$$Q = q + tp \tag{3}$$

$$P = p - t * \nabla V(Q) \tag{4}$$

for the Hamiltonian $H = p^2/2 + V(q)$ are symplectic. The computational importance of the symplectic condition has been understood in Celestial Mechanics [**22**], [**51**], [**74**], Accelerator Physics [**27**], and recent investigations into the ergodic behavior of the Fermi-Pasta-Ulam system [**10**], [**57**], and Arnol'd diffusion [**75**], [**76**].

Recently, SIA's have recieved much attention [**9**], [**13**], [**16**], [**20**], [**26**], [**30**], [**31**], [**33**], [**34**], [**35**], [**37**], [**39**], [**40**], [**41**], [**48**], [**49**], [**50**], [**52**], [**56**],

[60], [61], [63], [68], [71], [72], and have been shown to produce global phase space structures, commonly found in Hamiltonian mechanics. Since there is a KAM theorem for mappings [5], SIA's can be expected to simply distort all nondegenerate incommensurate invariant tori of the original system, and to possess exponentially slowly deviating adiabatic invariants according to the theory of Nekhoroshev [62]. Equivariant SIA's will enjoy a different class of generic bifurcation phenomena as discussed in Golubitsky et al. [43], and be subject to topological restrictions according to the recent developments of Ekeland and Hofer [28]. Numerical evidence suggests that symplectic integration might reveal hidden symmetries which analytically, at present, are mere fantasy. See Fomenko [32] for example. This, by no means, implies that the symplectic condition is the ultimate invariant, since preserving the energy may have strong compactification properties [52]. Feng and Qin [31] show that energy conservation is possible in a symplectic scheme but that the symplectic structure would have to be altered. In addition, Ge[39] has shown that, for a system which has no other first integrals, a SIA which preserves the energy exactly is a reparameterization, in time, of the real flow. This implies that the energy is the correct monitor of accuracy for SIA's which preserve all invariants but the energy. Indeed, numerical calculations [16] show that SIA's tend to preserve the energy extremely well over long integration times. On the other hand, Jimenez and Vazquez [52] show that SIA's may have blowup where an energy conserving scheme does not, and the latter's flow is volume preserving, and possibly symplectic, in the mean. It is clear that the dual relationship between energy and symplectic structure should be understood.

Many SIA's preserve integral invariants related to symmetries of the system to within roundoff error [16]. In particular, Feng [30] shows that there are SIA's that preserve all quadratic first integrals. Using the relationship between generating functions and Hamilton-Jacobi theory [30], [31], Ge and Marsden [39] have shown that the Hamilton-Jacobi equation reduces when the system has symmetry, and that this reduced Hamilton-Jacobi equation can be used to integrate symplectically, or by Poisson maps,on the reduced spaces. In addition, Ge[40], [41] has demonstrated a generating function for the Poisson map in the setting of Poisson Groupoids [73]. This opens up the possibility of Lie-Poisson integration of systems of Lie-Poisson type: the rigid body [2], [55], Euler fluids[2], the Maxwell-Vlasov and Vlasov-Poisson equations of plasma physics [59], and many more (cf.

Marsden et al. [**59**] and Marsden (these proceedings)). It should be emphasized that a consistent finite dimensional truncation will be required on present finite state machines (cf. Holm et al. [**50**]). Lie-Poisson integration of the rigid body [**17**] and a truncated moment description of the Vlasov equation [**17**] based upon [**21**] show that Lie-Poisson integrators possess the same stability characteristics as the symplectic integrators demonstrated in [**16**]. When the setting is the dual of a Lie algebra [**59**], [**45**], Lie-Poisson integration preserves the coadjoint orbits and defines a symplectic map on the coadjoint orbit with respect to its Kirillov-Kostant-Souriau symplectic structure. Therefore, the stability of these integrators is related to the fact that they keep the dynamics on the correct coadjoint orbit and on it the dynamics is Hamiltonian.

In this paper, I describe some general techniques available for symplectic or Lie-Poisson integration and illustrate the results with some numerical computations. In this spirit, I also discuss reversible integration, equivariant integration, integration of volume preserving flows, and symplectic cellular automata. My intention is not to be exhaustive but to give a representative review.

§2. The Techniques.

2.1 Generating Functions.

A general technique for the construction of symplectic maps is the use of a generating function [3], [42]. Let S be any scalar function. Then the map $(q,p) \to (Q,P)$ determined by solving the implicit system of equations

$$q = \frac{\partial S(Q,p,t)}{\partial p} \tag{5}$$

$$P = \frac{\partial S(Q,p,t)}{\partial Q} \tag{6}$$

is symplectic. Consider a Hamiltonian system

$$\dot{q} = \frac{\partial H(q,p)}{\partial p}$$

$$\dot{p} = -\frac{\partial H(q,p)}{\partial q}.$$

To approximately integrate its trajectories via a generating function we need to determine the generating function in terms of the Hamiltonian.

There is no unique way to do this, so I will illustrate with the following simple example. The actual formulae can be found in [**31**], [**16**]. Expand the solution of the initial value problem in a Taylor series:

$$\begin{aligned} Q &= F_1(q,p,t) \\ P &= F_2(q,p,t). \end{aligned}$$

Inverting the first equation for q and plugging into the second equation gives

$$\begin{aligned} q &= F_1^{-1}(Q,p,t) \\ P &= F_2(F_1^{-1}(Q,p,t),p,t). \end{aligned}$$

which must be of the form of equations (5) and (6):

$$\begin{aligned} F_1^{-1}(Q,p,t) &= \frac{\partial S(Q,p,t)}{\partial p} \\ F_2(F_1^{-1}(Q,p,t),p,t) &= \frac{\partial S(Q,p,t)}{\partial Q} \end{aligned}$$

for some scalar function S.

However, to compute, a truncation of F_1 and F_2 to some approximations f_1 and f_2 will be required giving, generally, a nonsymplectic result for the approximate mapping. Consider the approximate mapping

$$\begin{aligned} Q &= f_1(q,p,t) \\ P &= f_2(q,p,t). \end{aligned}$$

This mapping is assumed to be accurate to within some order. Determine f_1^{-1} to that same order. Then

$$\begin{aligned} f_1^{-1}(Q,p,t) &= \frac{\partial \tilde{S}(Q,p,t)}{\partial p} \\ f_2(f_1^{-1}(Q,p,t),p,t) &= \frac{\partial \tilde{S}(Q,p,t)}{\partial Q} \end{aligned}$$

is approximately correct for some computable function $\tilde{S}$.

Solving

$$\begin{aligned} q &= \frac{\partial \tilde{S}(Q,p,t)}{\partial p} \\ P &= \frac{\partial \tilde{S}(Q,p,t)}{\partial Q} \end{aligned}$$

to within some prescribed tolerance (close to roundoff error) determines a symplectic map, within that tolerance, which approximately integrates the flow. Determination of the timestep by requiring Picard iteration to converge close to roundoff error within 3 to 7 iterations after a nonsymplectic predictor such as Runge-Kutta, seems to work well [**16**], but alternative methods (e.g. Newton's method) have not been explored with respect to additional computational effort being balanced by an increase in possible timestep size.

Classical generating functions are of four types [**42**] depending on mixing old and new variables. Feng [**30**], and Feng and Qin [**31**] generalize the mixing of old and new variables in the setting of Lagrangian submanifolds and construct SIA's which preserve all quadratic integral invariants: those related to linear Lie group actions. Channell and Scovel [**16**] use generating functions which are invariant under symmetry groups and observe that the corresponding integral invariants are preserved to within roundoff error, and show that for linear systems, the general SIA's preserve quadratic integral invariants close to those of the original system. Which generating function to use depends upon the symmetries of the problem, the ease in computing matrix inverses, and the form and size of the Hamiltonian.

The relationship of generating functions to Hamilton-Jacobi theory seems implicitly understood (cf. Goldstein [**42**], p. 279) but is made explicit by Feng and Qin [**31**], and is summarized as follows: the graph of a symplectic map is a Lagrangian submanifold in the product space with the standard product symplectic structure. A Lagrangian submanifold which projects diffeomorphically to the configuration space (as in maps near the identity) is locally the graph of an exact one-form dS. S is called the generating function and the evolution of S as the map evolves by a Hamiltonian system is the Hamilton-Jacobi equation.

The works of Ge and Marsden [**39**] on Lie-Poisson Hamilton-Jacobi theory and Ge [**40**], [**41**] on the generating function for the Poisson map are fundamental in the developement of Lie-Poisson integrators (LPI's) for systems of Lie-Poisson type. For Hamiltonian systems on T^*G, where G is a Lie group, and the Hamiltonian is invariant under G, Ge and Marsden [**39**] reduce the Hamilton-Jacobi equation to one for the reduced Hamiltonian on the dual of the Lie algebra $\mathcal{G}^*$. This reduced Hamilton-Jacobi equation can be used to approximately integrate the Lie-Poisson flow on $\mathcal{G}^*$ by Poisson maps. That is, the LPI constructed preserves the coadjoint orbits and the Poisson bracket. It should be pointed out that these algorithms preserve many additional symmetries that may be present. For example, left reduced systems with right invariances, as in a rigid body with symmetry.

2.2 The Baker-Campbell-Hausdorff Formula.

Consider a Hamiltonian of the form $H(q,p) = F(p) + V(q)$. Each part, F and V, of this Hamiltonian can be solved explicitely: The solution with Hamiltonian $F(p)$ is the map $m_1(t)$ defined by

$$\begin{aligned} Q &= q + t * \nabla F(p) \\ P &= p \end{aligned}$$

and for the Hamiltonian $V(q)$ is the map $m_2(t)$ defined by

$$\begin{aligned} Q &= q \\ P &= p - t * \nabla V(q). \end{aligned}$$

Write e^{tH} as the solution map for the Hamiltonian H. Then, $m_1(t) = e^{tF}$ and $m_2(t) = e^{tV}$. It is easy to see that both $m_1(t) \circ m_2(t)$ and $m_2(t) \circ m_1(t)$ equal e^{tH} plus $O(t^2)$. Therefore they both determine first order SIA's. Note that they are the SIA's mentioned in the introduction.

This is the essence of Ruth's method [**67**]. To obtain higher order integrators, concatenate m_1 and m_2 many times using different time steps to match the coefficients of e^{tH} (cf. [**33**], [**34**], [**63**], [**13**]). For example, as found in Forest and Berz [**34**],

$$m_1(s_1) \circ m_2(d_1) \circ m_1(s_2) \circ m_2(d_2) \circ m_1(s_2) \circ m_2(d_1) \circ m_1(s_1) = e^{tH} + O(t^5),$$

is symplectic. The constants are $\beta = \sqrt[3]{2}$ and

$$\begin{aligned} s_1 &= t/(2 * (2 - \beta)) \\ s_2 &= t * (1 - \beta)/(2 * (2 - \beta)) \\ d_1 &= t/(2 - \beta) \\ d_2 &= -\beta * t/(2 - \beta). \end{aligned}$$

Both m_1 and m_2 are symplectic and explicit and if the system is large with local interaction, as in some molecular dynamics calculations, it is also local. Therefore, SIA's constructed this way are very fast and may be used for very large systems. However, they tend to neglect symmetries of the system which are not simultaneously symmetries of F and V. As mentioned in Neri [**63**], and Forest and Berz [**34**], this method generalizes to an arbitrary Lie group and is applicable when the element of the Lie algebra, which one wants to exponentiate, is a linear combination of elements for which their exponentials can easily be computed, or approximated within the group.

One parameter subgroups are characterized by their linearization at the identity and the fact that they are groups. Therefore, the approximation of such a group can be accomplished by the correct linearization and preserving as much of the group property as possible. One application of this is the following: Suppose $\varphi(t)$ is a one parameter family of elements approximating a one parameter subgroup $\Phi(t)$ which is the identity at $t = 0$ and has the correct linearization there. Φ satisfies $\Phi^{-1}(t) = \Phi(-t)$ but $\varphi(t)$ may not. However, since $\varphi(t)$ is an approximation near $t = 0$, so is $\varphi^{-1}(-t)$ and the composed map $\psi(t) = \varphi^{-1}(-t/2) \circ \varphi(t/2)$ is manifestly the same order approximant but now satisfies $\psi^{-1}(t) = \psi(-t)$. This implies that ψ is of even order accuracy. This provides a way of turning a first order integrator into a second order one, and a third order integrator into a fourth order one, etc. Plus, to integrate backwards, simply change t to $-t$. As an example, consider the first order SIA $m_1(t) \circ m_2(t)$. Since both m_1 and m_2 are one parameter subgroups, a second order SIA is $m_1(t/2) \circ m_2(t) \circ m_1(t/2)$, giving a result of Neri's [**63**].

Lasagni [**56**], and Sanz-Serna [**68**] show that implicit Runge-Kutta, where the quadrature matrices satisfy a quadratic relation, are symplectic. In particular, those with Gauss-Legendre quadrature rules are symplectic [**68**].

2.3 Modified Midpoint Rule.

Feng [**30**], Feng and Qin [**31**], and Stofer [**71**] show that the time centered Euler scheme is symplectic. They generalize to obtain SIA's of arbitrary order. Let $z = (q,p)$ and its image under the map be $Z = (Q,P)$. Using the notation of [**71**], let $g(z) = J\nabla H(z)$ be the Hamiltonian vector field, where J is the symplectic matrix. The time centered Euler or implicit midpoint rule is the map determined by the implicit set of equations

$$Z - z = t * g((z + Z)/2)$$

which determine a second order SIA. To obtain higher order SIA's construct a function $\tilde{g}(z) = J\nabla\tilde{H}(z)$ so that the solution of

$$Z - z = t * \tilde{g}((z + Z)/2)$$

determines a fourth order (for example) SIA. This modified midpoint rule is automatically symplectic since $\tilde{g}$ is a Hamiltonian vector field. Typically, these SIA's require one more derivative of the Hamiltonian for each order of accurracy desired. However, Stofer's ansatz for the form of $\tilde{g}$ provides a fourth order SIA which requires only second derivatives of H . Write

$$\begin{aligned} z^+ &= z + \alpha t * g(z) \\ z^- &= z - \alpha t * g(z) \\ \tilde{g}(z) &= (1 - 2\beta)g(z) + \beta(g(z^+) + g(z^-)) - \alpha\beta t Dg(z)[g(z^+) - g(z^-)], \end{aligned}$$

where $\alpha^2\beta = 1/24$. The new Hamiltonian is

$$\tilde{H}(z) = (1 - 2\beta)H(z) + \beta(H(z^+) + H(z^-)).$$

If g is equivariant under a symmetry group, then the midpoint rule is shown to be equivariant. In addition, Stofer shows that if g is reversible (as discussed in the next section), then the midpoint rule is also reversible. In addition, Stofer shows that the above SIA is equivariant or reversible if g is.

2.4 Reversible Integration.

Birkhoff [**12**] used the fact that the involution $(q,p) \to (q,-p)$ reverses the direction of the flow in the restricted three body problem to prove the existence of and classify periodic orbits. De Vogalaere [**25**] has also utilized this technique. Devaney [**23**] has proved the analogues of the closed orbit, Liapunov, and homoclinic orbit theorems for reversible systems independent of any existing Hamiltonian structure. Feingold, Kadanoff, and Piro [] discuss the implications of reversibility for Liapunoff exponents, and Sevryuk [**69**] has demonstrated a KAM theorem for reversible systems. So it seems that numerical integration of reversible systems should be accomplished by reversible integrators. See also [**8**], [**54**], [**66**]. Bennett [**11**] discusses the reversibility of computation, not exactly in the present sense (to be defined below), with regards to defeating dissipative mechanisms in the computational process. Relegating reversibility to the present context,

Fredkin and Toffoli [36], and Margolus [58], investigate its implications in Physics computations.

Let R be an involution of phase space. A map φ on the phase space is said to be R-reversible if $\varphi^{-1} = R^{-1}\varphi R = R\varphi R$. This implies that the inverse map is obtained by applying the forward map in the changed coordinates determined by R. A vector field χ is said to be R-reversible if $-\chi = R\chi R$.

For an R-reversible dynamical system, the modified midpoint rule produces an R-reversible integration algorithm in addition to being symplectic [71].

There is a projection to R-reversible maps. Let $\varphi(t)$ be an approximation to a reversible system. Then, $R\varphi^{-1}(t)R$ is also an approximate of the same system and the composition $\psi(t) = R\varphi^{-1}(t/2)R\varphi(t/2)$ is R-reversible. Since R is anti-symplectic, ψ is symplectic if φ is. The Hamiltonian system with Hamiltonian $H = p^2/2 + V(q)$ is reversible with respect to the involution $(q,p) \to (q,-p)$. However, the first order SIA

$$\begin{aligned} Q &= q + tP \\ P &= p - t * \nabla V(q) \end{aligned}$$

is not reversible with respect to this involution. The projection discussed above could be used to make it reversible or notice that, as in Kook and Meiss [54], it is reversible with respect to the involution $(q,p) \to (-q, p - t\nabla V(q))$ when V is an even function. It is important to notice that this involution is parametrized by time so that the involution changes depending on the time step. This reversing involution can be explained as follows: Let R be the involution $(q,p) \to (q,-p)$. The symplectic maps m_1 and m_2, defined in section 2.2 are both R-reversible. Consequently, $R \circ m_1$, $R \circ m_2$, $m_1 \circ R$, and $m_2 \circ R$, are all involutions. The set of reversible maps does not form a group, in fact, it is easy to show that the composition of two reversible maps is reversible if and only if the two maps commute. The SIA defined above is $\varphi = m_1 \circ m_2$, and the non-commutativity of m_1 and m_2 explains why their composition φ is not R-reversible. Consider the involution $\tilde{R} = R \circ m_2$. Compute

$$\tilde{R} m_1 m_2 \tilde{R} = R m_2 m_1 m_2 R m_2.$$

Since $Rm_2 = m_2^{-1}R$, this becomes

$$m_2^{-1} R m_1 m_2 m_2^{-1} R = m_2^{-1} R m_1 R = m_2^{-1} m_1^{-1} = (m_1 m_2)^{-1}$$

showing that φ is $\tilde{R}$-reversible. Likewise, it is also Rm_1^{-1}-reversible. The other two involutions can be used for the SIA $m_2 \circ m_1$. In the event that V is an even function, $-I$ commutes with m_1, m_2, and R, and all reversing involutions may be multiplied by $-I$. The reversing involution of Kook and Meiss is $-R \circ m_2$.

2.5 Volume Preserving Integration.

Divergence free vector fields determine volume preserving transformations and it is known that such flows possess a non-canonical Hamiltonian structure which could be used to integrate them **[14]**, **[18]**, **[19]**, **[76]**. However, these structures are singular at fixed points of the vector field, and the Hamiltonian and the canonically conjugate variables might be complicated functions so that it seems desirable to integrate them directly as volume preserving maps. Volume preserving maps preserve the helicity: a topological invariant important in magnetic field line problems **[7]**, **[4]**. Such flows are also important in fluid mechanics **[2]**, **[6]**.

It is interesting to note that the substition of a nongradient term $F(q)$ for $\nabla V(q)$ in (1)(2) and (3)(4) deems the integrator non-symplectic. However, it is exactly volume preserving, and comparison of it with the symplectic version will be reported on in a future publication. Construction of volume preserving integrators(VPI's) for divergence free vector fields is not quite so simple. Following Thyagaraja and Haas **[72]**, mix old and new variables to obtain generating functions for the volume preserving map. Let $v = (v_1, v_2, v_3)$ be a divergence free vector field in R^3. Consider a map $(x, y, z) \to (X, Y, Z)$ represented implicitly by the equations

$$\begin{aligned} X &= f(x, y, Z) \\ Y &= g(x, y, Z) \\ z &= h(x, y, Z). \end{aligned}$$

A simple calculation shows that this map is volume preserving if and only if

$$\frac{\partial h}{\partial Z} = \frac{\partial f}{\partial x}\frac{\partial g}{\partial y} - \frac{\partial f}{\partial y}\frac{\partial g}{\partial x}.$$

This determines h as a partial integral of a function of the two generating functions f and g. The determination of the generating functions f and g in terms of v can be accomplished in the same manner as in section 2.1. I will give the first and second order results. Let

$$f(x,y,Z) = x + f_i(x,y,Z)t^i$$
$$g(x,y,Z) = y + g_i(x,y,Z)t^i$$
$$h(x,y,Z) = Z + h_i(x,y,Z)t^i$$

be the Taylor series of f,g, and h, in t, where the map is determined to be the identity at $t = 0$. A first order VPI is obtained from

$$f_1 = v_1$$
$$g_1 = v_2$$
$$h_1 = -v_3$$
$$h_2 = \int^Z [\frac{\partial v_1}{\partial x}\frac{\partial v_2}{\partial y} - \frac{\partial v_1}{\partial y}\frac{\partial v_2}{\partial x}].$$

and a second order one from

$$f_1 = v_1$$
$$f_2 = 1/2(\frac{\partial v_1}{\partial x}v_1 + \frac{\partial v_1}{\partial y}v_2 - \frac{\partial v_1}{\partial Z}v_3)$$
$$g_1 = v_2$$
$$g_2 = 1/2(\frac{\partial v_2}{\partial x}v_1 + \frac{\partial v_2}{\partial y}v_2 - \frac{\partial v_2}{\partial Z}v_3)$$
$$h_1 = -v_3$$
$$h_2 = 1/2(\frac{\partial v_3}{\partial Z}v_3 - \frac{\partial v_3}{\partial x}v_1 - \frac{\partial v_3}{\partial y}v_2)$$
$$h_3 = \int^Z [\frac{\partial f_1}{\partial x}\frac{\partial g_2}{\partial y} - \frac{\partial f_1}{\partial y}\frac{\partial g_2}{\partial x} + \frac{\partial f_2}{\partial x}\frac{\partial g_1}{\partial y} - \frac{\partial f_2}{\partial y}\frac{\partial g_1}{\partial x}]$$
$$h_4 = \int^Z [\frac{\partial f_2}{\partial x}\frac{\partial g_2}{\partial y} - \frac{\partial f_2}{\partial y}\frac{\partial g_2}{\partial x}].$$

The lower limits of integration have not been specified but if chosen at a value of z for which there are zeros of the vector field, these points will remain fixed under the map. In concrete problems the first order can easily be implemented because the integral can be done by hand. However, the second order may require the use of a symbolic manipulation program such as SMP as discussed in [**16**].

Consider now a divergence free vector field such as the ABC flows [**6**]. They have the special form $v = (v_1, v_2, v_3) = v_1 + v_2 + v_3$, where v_1,v_2, and

v_3 are individually divergence free vector fields which can be integrated explicitly. Specifically,

$$v_1 = v_1(y, z)$$
$$v_2 = v_2(x, z)$$
$$v_3 = v_3(x, y)$$

so that exact integration of the flow corresponding to v_1 is accomplished by the the one parameter subgroup of volume preserving transformations $m_1(t)$ defined by

$$X = x + tv_1(y, z)$$
$$Y = y$$
$$Z = z$$

with corresponding subgroups $m_2(t)$ and $m_3(t)$ for the flows v_2 and v_3.

Thus we can use Ruth's method (section 2.2) to construct VPI's for the flow. For example, composition of m_1, m_2, and m_3, in any order, is a first order VPI. This is the so-called ABC map used by Feingold, Kadanoff,and Piro [**29**]. A second order VPI is given by

$$m_1(t/2)m_2(t/2)m_3(t)m_2(t/2)m_1(t/2)$$

and higher order VPI's can be constructed using the techniques discussed in Ruth [**67**], Forest and Ruth [**33**], Neri [**63**], and Forest and Berz [**34**]. Note that besides being explicit, these VPI's preserve all fixed points of the flow.

2.6 Symplectic Cellular Automata.

Coupled map lattices have been studied with regard to the fundamental question of ergodicity and equipartition of energy in Hamiltonian systems with many degrees of freedom. Most of these studies numerically integrate using the first order SIA (1) and (2), and can be considered symplectic coupled map lattices. However, as of yet no studies have been performed concerning the importance of the symplectic condition, and none with respect to system symmetries and reversibility. I am presently investigating these issues and will report concerning them in a future publication.

Kaneko [**53**] generalizes a technique of Rannou [**65**] to more than one degree of freedom, and shows that finite state symplectic map lattices(cellular automata) can be constructed by requiring a symplectic map lattice to map integer states to integer states and to commute with a mod-q operation. De Vogalaere [**26**] has also developed a version of discrete Hamiltonian mechanics. Cellular automata require integer arithmetic and so have no roundoff error. Therefore, they promise to be important tools in the study of large systems. See also Hannay and Berry [**46**]. Kaneko studies (for mod-3) a complete list of models and classifies them according to their statistical behavior. Their relation to any physical system is yet undertermined but Frisch, Hasslacher, and Pomeau [**38**] show that cellular automata can successfully simulate certain regimes for the Navier-Stokes equation. In addition, Pomeau [**64**] computes an energy-like invariant for reversible cellular automata.

§3. Numerical Calculations.

A SIA for some Hamiltonian system is a symplectic map (or 1-parameter family) which approximates the (symplectic) flow map of the system. This implies that the KAM theorem for mappings [**5**] and Nekhoroshev's theorem [**62**] on adiabatic invariants apply. This immediately implies long term stability results in near integrable, and non-resonance regimes ignoring the important topic of roundoff error. In addition, stochastic layers bounded by regular regions appear to be represented well since the regular regions will be strictly inhibiting to diffusion as they are strictly isolating in systems with two degrees of freedom [**16**]. Symplectic integration of many systems shows that good energy preservation over extremely long runs is typical, and Figure 1. shows the comparison in energy deviation from being constant of a second order SIA with the standard fourth order Runge-Kutta for the Fermi-Pasta-Ulam system as presented in Channell and Scovel [**16**]. The qualitative features of Figure 1 are the same for all systems I have coded except for the three body problem, due to close encounters.

Poor energy conservation has been observed for SIA's by Jimenez and Vazquez [**52**] and Forest [**35**]. A few comments are in order. In Channell and Scovel [**16**], the timestep was determined by several factors. Firstly, the total energy deviation over a fixed long time interval should scale according to the accuracy of the algorithm. This test should be performed at a variety of initial conditions. Secondly, The generating function equations should be solvable in less than 5 to 7 Picard iterations after a Runge-Kutta predictor. The examples of Jimenez and Vazquez **[52]** and Forest **[35]** do not satisfy the first criterion. Numerically induced stochasticity is also investigated in **[9]**, **[37]**, **[48]**, and **[49]**.

For the Hénon-Heiles Hamiltonian at low values of the energy, the system appears completely integrable **[47]**, but comparison of Poincaré sections of an SIA (Figure 2a.) with a RKI (Figure 2b.) for a long run shows the SIA to lie on a closed curve, where the RKI initially travels near this closed curve, but monotonically drifts (similarly to how it drifts in energy in Figure 1.).

Figure 3 shows four views of the results of a single initial condition. Figure 3a is the Poincaré section and Figure 3b is the projection of the three dimensional data onto this plane, while Figures 3c and 3d are the projection onto the other two coordinate planes. Note the almost regular polyhedral figures.

Now consider a system of six Kirchhoff point vortices. The Hamiltonian is

$$H = -1/4\pi \sum_{j \neq k}^{6} \Gamma_j \Gamma_k \ln |z_j - z_k|$$

where Γ_j is the strength of the $j-th$ vortex situated at the point z_j in the complex plane. Three point vortices are completely integrable and four vortices of the same strength are not **[1]**. The system has two integrals of center of mass $\sum \Gamma_j z_j = Q + iP$ and one of angular momentum $\sum \Gamma_j |z_j|^2$. Let the strengths all be 1, and let the position of the vortices be $(1,1), (1+dr, 1), (2,2), (3,3), (4,3), (5,3)$ where the second vortex position is variable. For dr small, the first and second interact with the other vortices as a coherent vortex pair. The other vortices are typically a distance of order 1 from neighboring vortices. What happens to the vortex pair as dr is increased towards 1? I implemented the third order SIA as explained in **[16]**, and monitored the energy of the isolated vortex pair (= minus the log of their separation). The total energy remained constant to one part in 10^{-6}, the integrals remained constant to within roundoff error, and the time step was .03. When dr is less than .4, the pair remain coherent over extremely long runs (t=20,000), but for dr =.46, the pair undergoes separation but then apparently stable reattachment (Figure 4a). For dr=.5, the pair experiences a modulation to slightly larger separation, true separation, reattachment, and then finally, permanent separation (Figure 4b).

For dr=.5005, permanent separation occurs very early (near $t = 1000$), and somewhat earlier for larger values of dr.

My last example is the classical three body problem. I implemented a fourth order SIA for the full non-planar problem as found in [**16**], and all three integrals of center of mass were preserved correctly in terms of the three of center of velocity which remained constant to within roundoff error. The three integrals of angular momentum were also preserved to within roundoff error. However, due to the fact that too many fortran multiplications were needed, it was slow. Energy preservation was typical until a close encounter when the timestep was halved until the generating function equations could be satisfied in less than 7 Picard iterations. For mild encounters this worked well but strong encounters would change the energy drastically or require the time step to be reduced too many times. I terminated the run after 20 time step reductions. Presumably, a reduction in the order of the scheme is appropriate, and/or use of a scheme which requires less derivative information as in Stofer's algorithm in section 2.3, and a symplectic close encounter algorithm, following the two-body regularization procedure as found in Siegel and Moser [**70**], should be implemented.

Consider now the planar problem. As in Graziani and Black [**44**], let the sun have mass 1 and two planets of equal mass m. Initially, the sun is at the origin and the two planets $m1$ and $m2$ are diametrically opposed on the x axis at radii $r1$ and $r2$ such that $r2/r1 = 1.567$ and $(r1+r2)/2 = 10.4$. They are given initial momenta which starts them in almost circular orbits so that they both rotate counterclockwise. When $m = 0$ the planets evolve in perfect circular orbits. Direct comparison with their result's is not possible at the present time since I chose my gravitational constant incorrectly. However, the qualitative phenomena is still of interest. The time step was one and the period of the outer planet $m2$ was approximately 1000 and the inner 500. At $m = .015$ the system survives at least 70,000 orbits of the outer planet, keeping the energy constant to within one part in 10^{-6} over the whole run. At $m = .0155$ the outer planet completes 1200 regular orbits, attempts to switch places with the inner planet, but becomes unstable and ejects. $r1$ is about 20 and $r2$ is about 30 and I called a planet ejected if its position was greater than 1000 from the center of mass. A cleaner version of the switch of the outer planet to the inner and the inner to the outer is shown in Figure 5(a,b,c). This switch terminates (apparently) in a very typical *mode* where the inner planet is undergoing a forced (by the outer planet) two body problem and the outer planet interacts with the inner-sun pair as in a two body problem except on its closest approach, when abrupt switches to other quasi-stable modes are possible. Figure 5d shows the inner planet configuration points at $m = .0188$ where, after a switch similar to Figure 5(a,b,c), an attempt to switch back is made. I refer to a mode as quasi-stable if it remained as in Figure 5(a,b,c) for more than 10 periods of the outer planet. Stability here is a funny thing. When the outer planet is in a highly eccentric elliptical orbit, this should be very unstable. However, the time it takes to travel this orbit is very large. Also, numerical calculations show that modes with highly eccentric outer planet orbits can be fairly stable. In particular, quasi-stable modes with the outer planet reaching 900 were not unusual. At $m = .0167$, the configuration space plots were extremely chaotic. However, Figure 5e., plotting the mutual separation of the two planets, shows that they attach and form a stable satellite system revolving about the sun.

Acknowledgements.

I would like to thank MSRI at Berkeley, ETH-Zürich, and Center for Nonlinear Studies at Los Alamos, for their hospitality and support. Thanks go to David Campbell, Larry Campbell, Etienne Forest, Jack Hills, Darryl Holm, Ge Zhong, Jerry Marsden, Tudor Ratiu, and Alan Weinstein, for many helpful conversations. I would like to express my graditude to Paul Channell for teaching me many things, especially what an algorithm is.

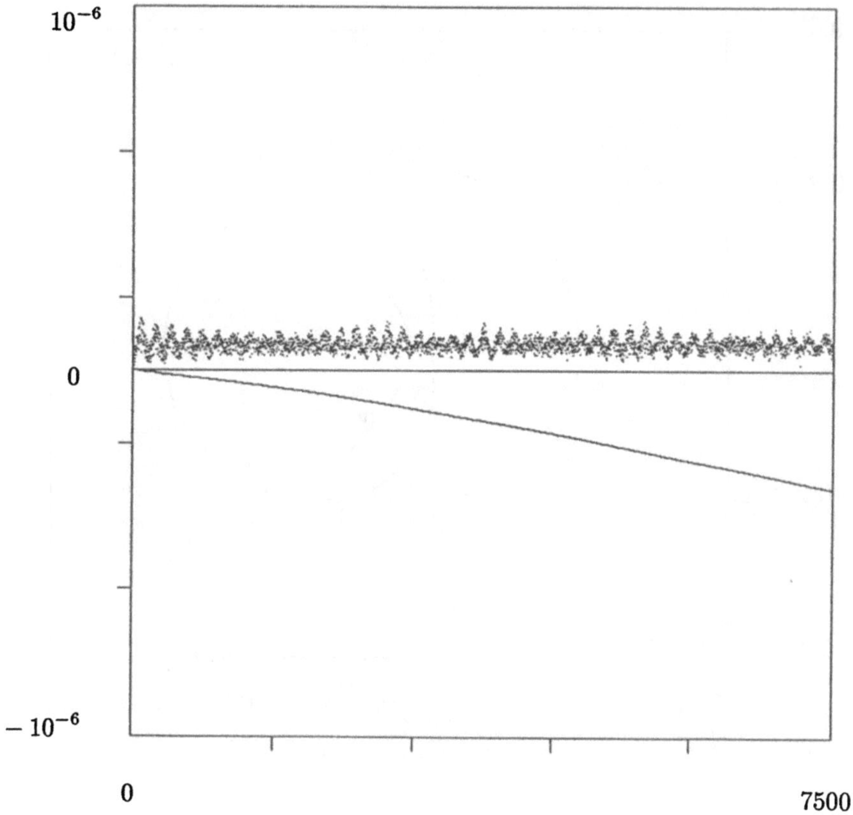

Figure 1
Comparison of relative energy error of a second order SIA and fourth order RKI for the Fermi-Pasta-Ulam system. The time step was .001 for the SIA and .02 for the RKI and the final time was 7500.

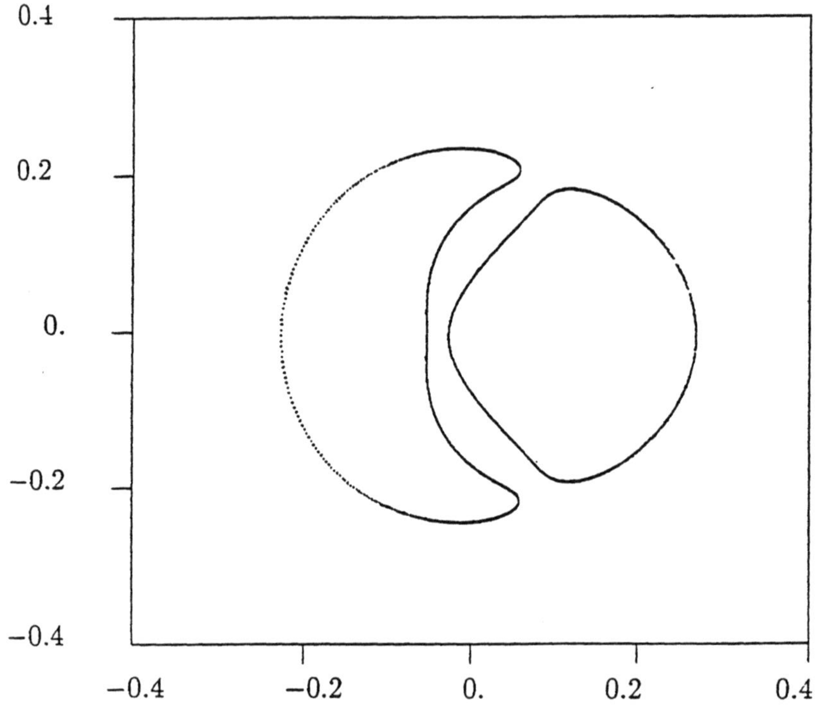

Figure 2a

Comparison of third order SIA (2a) and fourth order RKI (2b) for the Hénon-Heiles system. The initial condition was (.12,.12,.12,.12) with energy .029952. The timestep was 1/6 and 1,200,000 timesteps were computed.

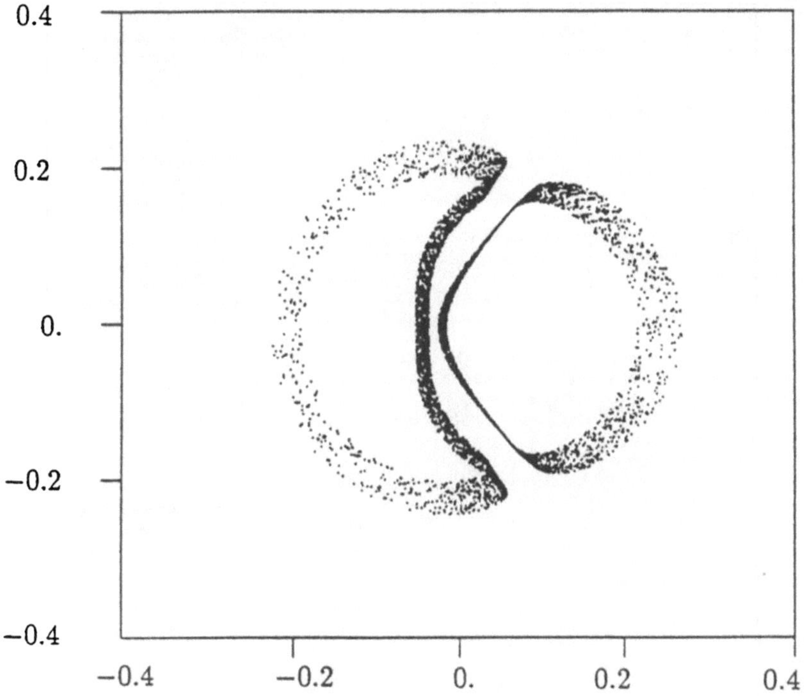

Figure 2b

Comparison of third order SIA (2a) and fourth order RKI (2b) for the Hénon-Heiles system. The initial condition was (.12,.12..12,.12) with energy .029952. The timestep was 1/6 and 1,200.000 timesteps were computed.

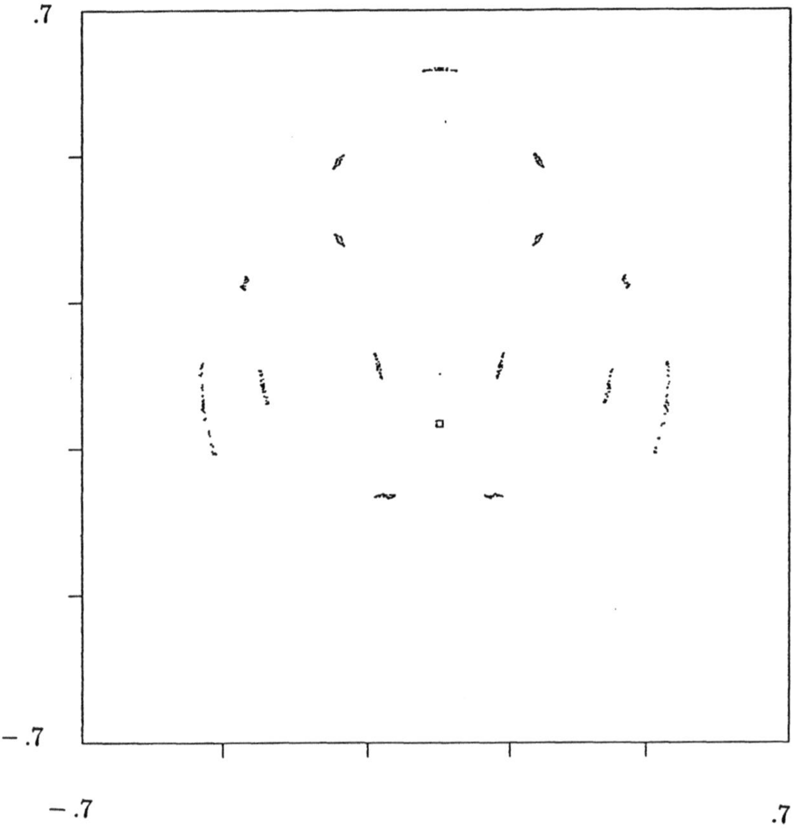

Figure 3a
First of four views of the results from a single initial condition, $(q_1, p_2, q_2) = (0., 0., -0.1)$, with energy 0.117835: Two dimensional slice at the $q_1 = 0$ plane; 14000 timesteps were computed with timestep 0.3.

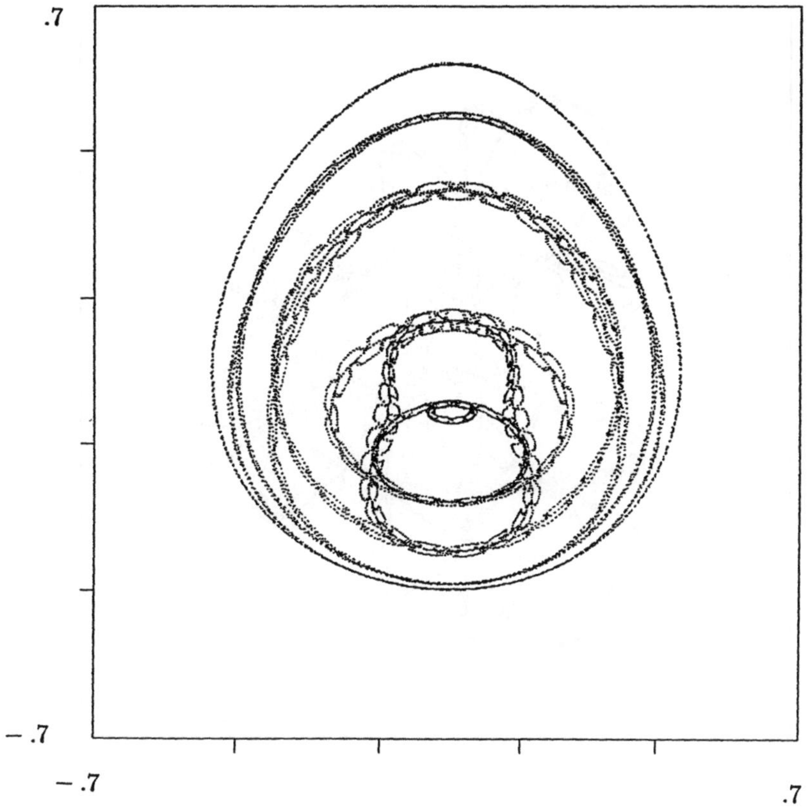

Figure 3b

Second of four views of the results from a single initial condition, $(q_1, p_2, q_2) = (0., 0., -0.1)$, with energy 0.117835: Projection of the three dimensional phase space onto the the $q_1 = 0$ plane; 14000 timesteps were computed with timestep 0.3.

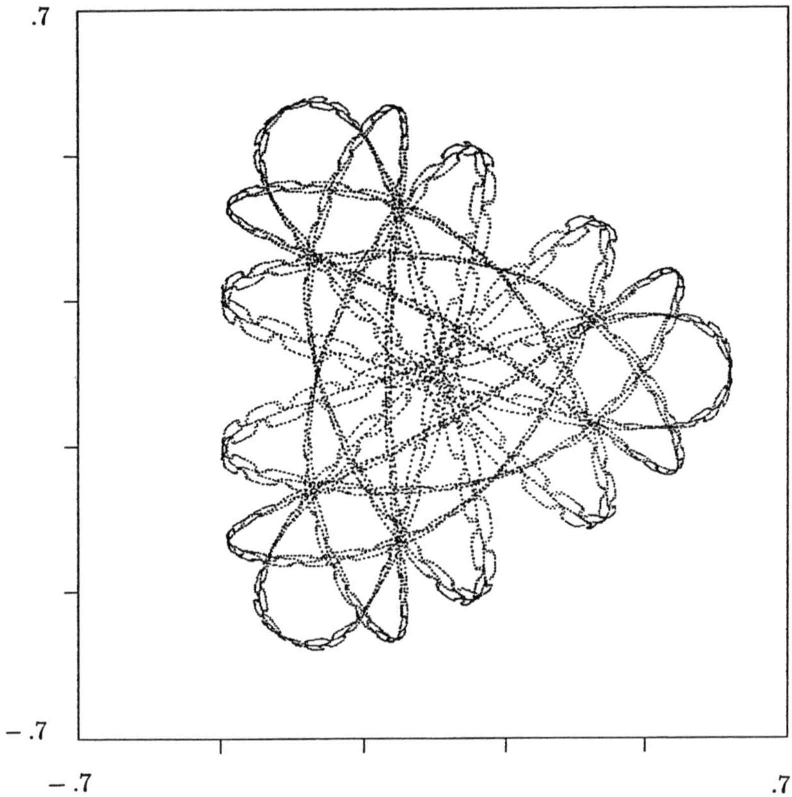

Figure 3c

Third of four views of the results from a single initial condition, $(q_1, p_2, q_2) = (0..0., -0.1)$, with energy 0.117835: Projection of the three dimensional phase space onto the the $p_2 = 0$ plane; 14000 timesteps were computed with timestep 0.3.

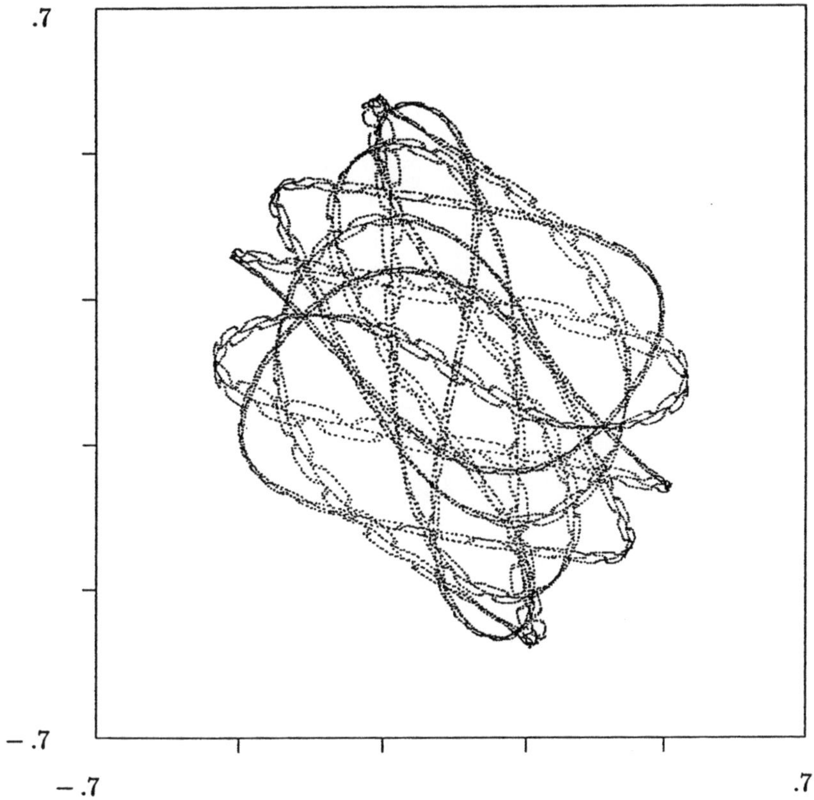

Figure 3d
Fourth of four views of the results from a single initial condition, $(q_1, p_2, q_2) = (0., 0., -0.1)$, with energy 0.117835: Projection of the three dimensional phase space onto the the $p_2 = 0$ plane; 14000 timesteps were computed with timestep 0.3.

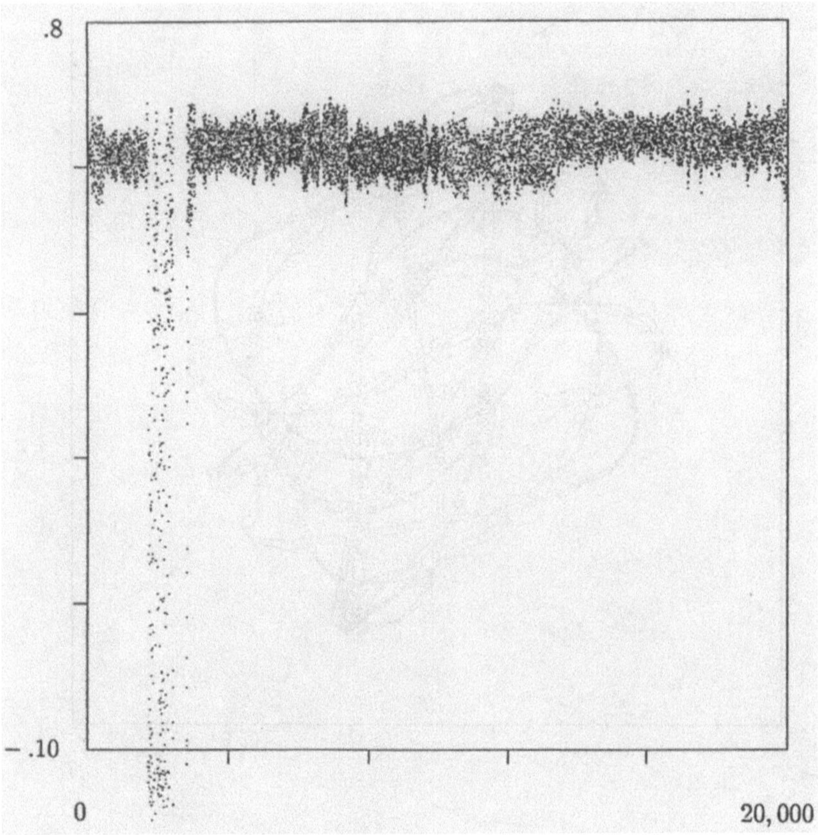

Figure 4a
Plot of energy of vortex pair $= -1/4\pi ln(r)$, where r is their separation. A third order SIA was used with timestep .03; dr=.46.

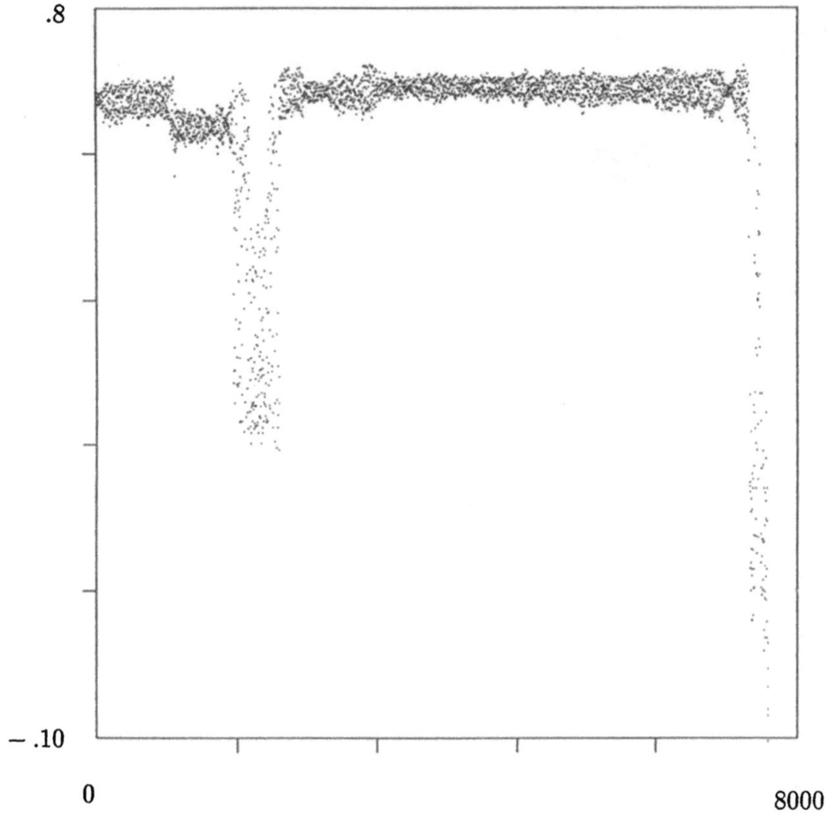

Figure 4b
Plot of energy of vortex pair = $-1/4\pi ln(r)$, where r is their separation. A third order SIA was used with timestep .03; dr=.5.

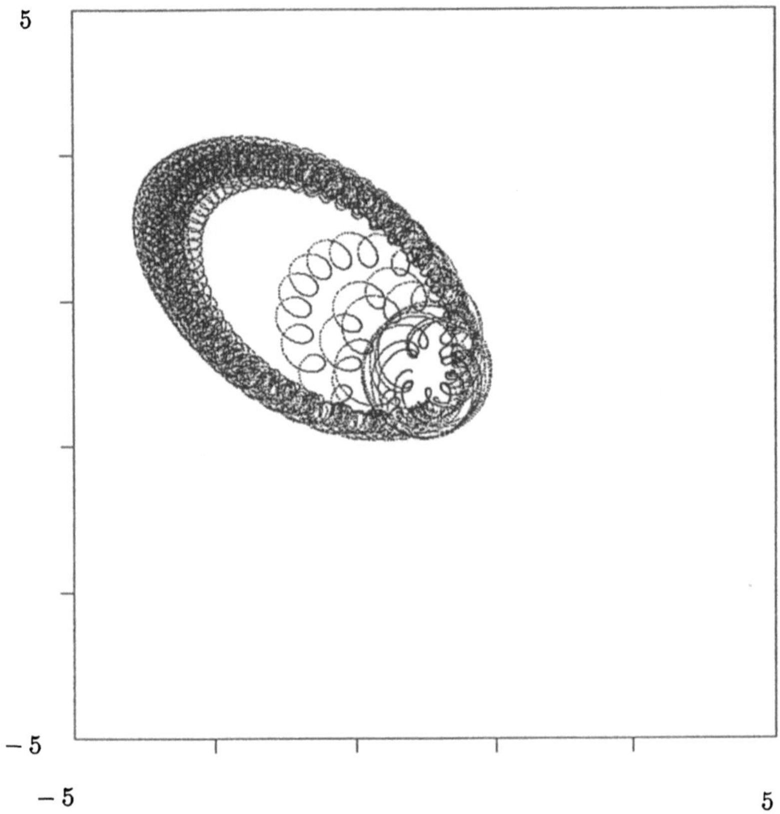

Figure 5a
Configuration space (x,y) plots, with respect to the center of mass, of trajectories computed by a fourth order SIA for the planar three body problem; the sun.

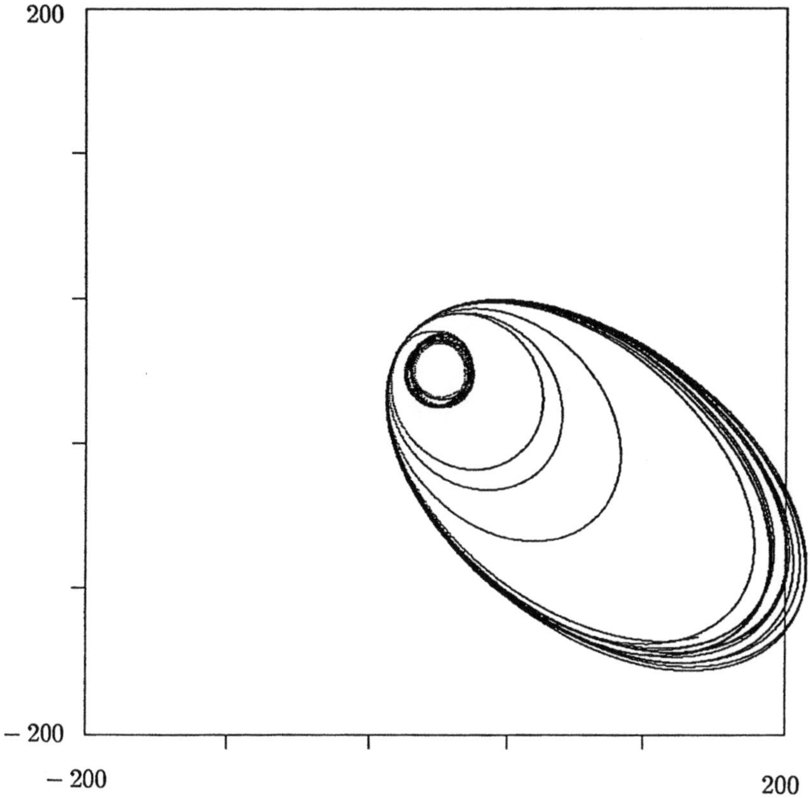

Figure 5b
Configuration space (x,y) plots, with respect to the center of mass, of trajectories computed by a fourth order SIA for the planar three body problem; initially inner planet m_1.

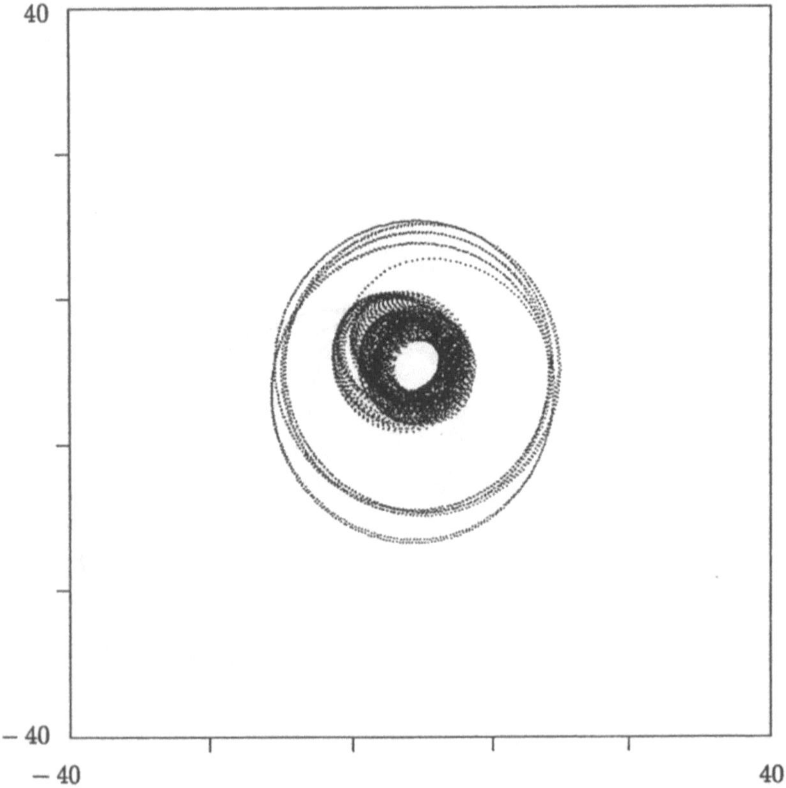

Figure 5c
Configuration space (x,y) plots, with respect to the center of mass, of trajectories computed by a fourth order SIA for the planar three body problem; initially outer planet m_2. Both planets have mass $m = .019$. The timestep was 1 and the final time 133270.

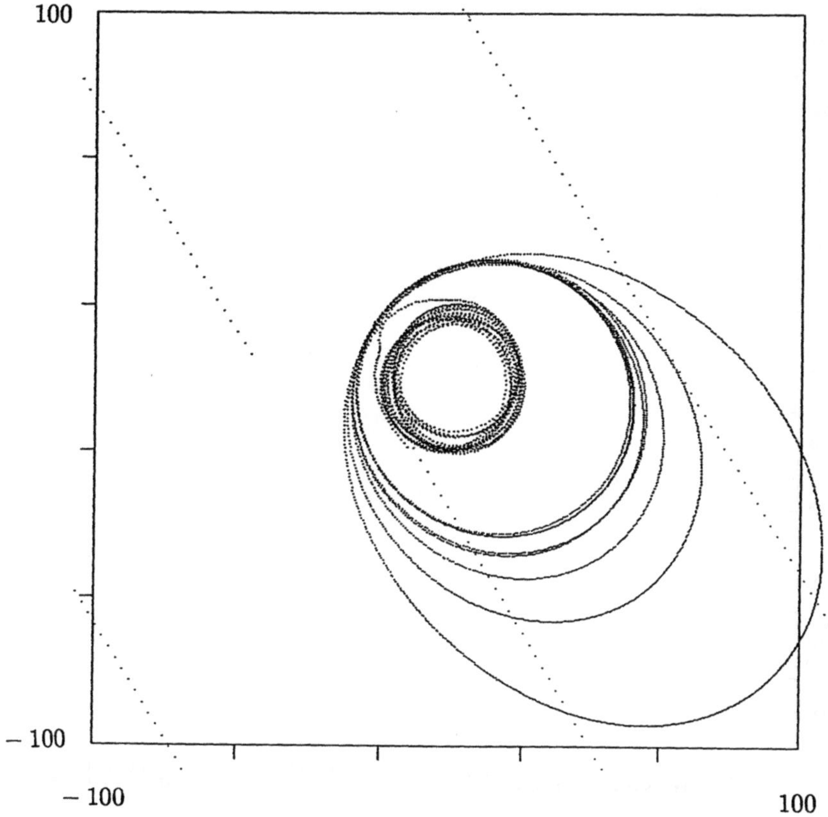

Figure 5d
Configuration space (x,y) plots, with respect to the center of mass, of trajectories computed by a fourth order SIA for the planar three body problem; configuration space plot for m_1 at $m = .0188$.

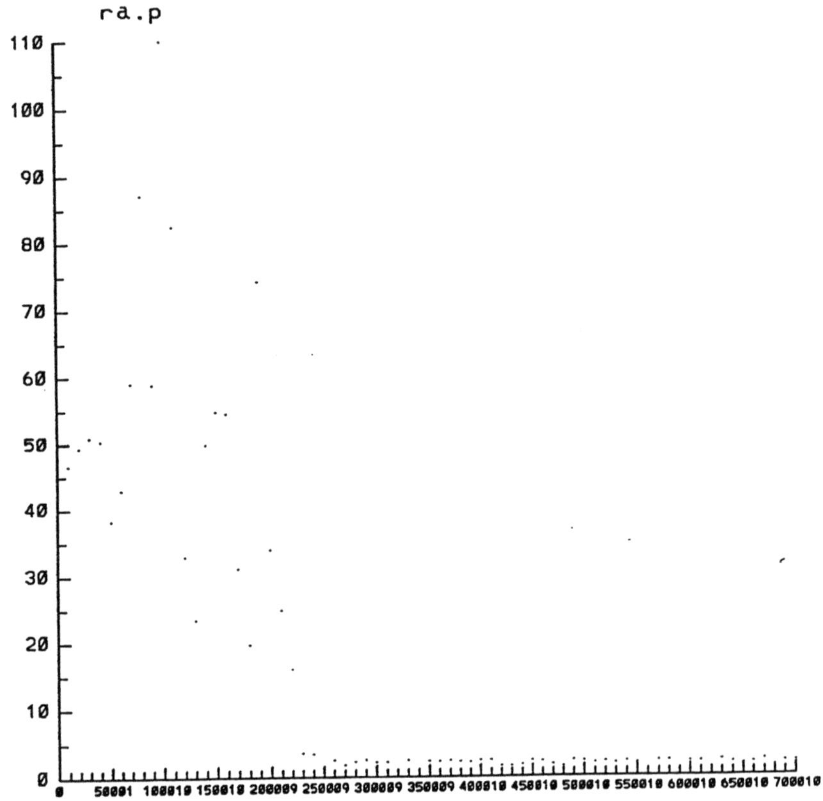

Figure 5e

Configuration space (x,y) plots, with respect to the center of mass, of trajectories computed by a fourth order SIA for the planar three body problem; plot of the separation of the two planets as a function of time at $m = .0167$.

REFERENCES

1. Aref, H., and N. Pomphrey, *Integrable and Chaotic Motions of Four Vortices I. The Case of Identical Vortices*, Proc. R. Soc. London **A 380** (1982), p. 359.
2. Arnol'd, V.I., *Sur la Géometry Differentielle des Groups de Lie de Dimension Infinie et ses Applications a L'hydrodynamique des Fluids Parfaits*, Ann. Inst. Fourier Grenoble **16** (1966), 319–361.
3. Arnol'd, V.I., "Mathematical Methods of Classical Mechanics," Springer-Verlag, New York, 1978.
4. Arnol'd, V.I., *On the Evolution of a Magnetic Field under the Action of Transport and Diffusion*, AMS Transl. **2:137** (198), 119–129.
5. Arnol'd, V.I., and A. Avez, "Ergodic Problems of Classical Mechanics," W. Benjamin, Inc., New York, 1968.
6. Arnol'd, V.I., *Sur la Topologie des Écoulements Stationnaires des Fluides Parfaits*, C.R. Acad. Sc. Paris **261** (1965), 17–20.
7. Arnol'd, V.I., *The Asymptotic Hopf Invariant*, Sel. Math. Sov. **5:4** (1986), 327–345.
8. Arnol'd, V.I., "Reversible Systems, Nonlinear and Turbulent Processes," Acad. Publ., New York, 1984, pp. 1161–1174.
9. Auerbach, S.P., and A. Friedman, *Long-time Behavior of Numerically Computed Orbits: Small Timestep Analysis*, Lawrence Livermore National Laboratory preprint UCRL–97158 (1989).
10. Benettin, G., Galgani, L., and J.M. Strelcyn, *Komolgorov Entropy and Numerical Experiments*, Phys. Rev. A **14:6** (1976), 2338–2345.
11. Bennett, C.H., *Logical Reversibility of Computation*, IBM J. Research and Developement **6** (1973), 525–532.
12. Birkhoff, G., *The Restricted Problem of Three Bodies*, Rend. Circ. Mat. Palermo **39** (1915), 265–334.
13. Candy, J., and W. Rozmus, *A Symplectic Integration Algorithm for Separable Hamiltonian Functions*, preprint (1989).
14. Cary, J.R., and R.G. Littlejohn, *Noncanonical Hamiltonian Mechanics and Its Application to Magnetic Field Line Flow*, Ann. Phys. **151** (1983), 1-34.
15. Channell, P.J., *Symplectic Integration Algorithms*, Los Alamos National Laboratory internal report AT–6:ATN–83–9 (1983).
16. Channell, P.J., and J.C. Scovel, *Symplectic Integration of Hamiltonian Systems*, sub. to Nonlinearity (June 1988).
17. Channell, P.J. personal communication, 1989.
18. Channell, P.J., *The Hamiltonian Structure of Field-Line Equations*, Los Alamos National Laboratory report LA–10048–MS (1984).
19. Channell, P.J., *Hamiltonian Structures Underlying MHD Equilibria*, Los Alamos National Laboratory internal report LA–10460–MS (1985).
20. Channell, P.J., *Explicit Integration of Kick Hamiltonians in Three Degrees of Freedom*, Los Alamos National Laboratory internal report AT–6:ATN–86–6 (1986).
21. Channell, P.J., *The Moment Approach to Charged Particle Beam Dynamics*, IEEE Trans. Nucl. Sci. **NS–30:4** (1983), 2607–2609.
22. Deprit, A., *Canonical Transformations Depending on a Small Parameter*, Celestial Mech. **1** (1969), 12–30.
23. Devaney, R.L., *Reversible Diffeomorphisms and Flows*, Trans. AMS **218** (1976), 89–113.
24. De Vogalaere, R., *Methods of Integration which preserve the Contact Transformation Property of the Hamiltonian Equations*, Dept. of Mathematics, University of Notre Dame report 4 (1956).

25. De Vogalaere, R., *On the Structure of Symmetric Periodic Solutions of Conservative Systems, with Applications*, Contributions to the Theory of Nonlinear Oscillations, vol. IV, Ann. Math. Studies **41** (1958), 53–84, Princeton Univ. Press, Princeton, NJ.
26. De Vogalaere, R., *Symplectic Integration, 32 years later—Finite and p-adic Hamiltonian Mechanics*, report given at the Los Alamos Workshop on Symplectic Integration.
27. Dragt, A.J., and J.M. Finn, *Lie Series and Invariant Functions for Analytic Symplectic Maps*, J. Math. Phys. **17:12** (1976), 2215–2227.
28. Ekeland, I., and H. Hofer, *Symplectic Topology and Hamiltonian Dynamics*, Math. Zeit. **200** (1989), 355–378.
29. Feingold, M., Kadanoff, L.P., and O. Piro, *Passive Scalars, 3D Volume Preserving Maps and Chaos*, J. Stat. Phys. **50** (1988), p. 529.
30. Feng, K., *Difference Schemes for Hamiltonian Formalism and Symplectic Geometry*, J. Comp. Math. **4:3** (1986), 279–289.
31. Feng, K., and M. Qin, *The Symplectic Methods for the Computation of Hamiltonian Equations*, Springer Lect. Notes Math. **1297** (1987), 1–31, Springer-Verlag, Berlin.
32. Fomenko, A.T., *Visual and Hidden Symmetry in Geometry*, Computers Math. Applic. **17:1-3** (1989), 301–320.
33. Forest, E., and R.D. Ruth, *Fourth Order Symplectic Integration*, preprint (1989).
34. Forest, E., and M. Berz, *Canonical Integration and Analysis of Periodic Maps Using Non-standard Analysis and Lie Methods*, Lawrence Berkeley Laboratory internal report LBL–25609, ESG–46 (1989).
35. Forest, E., *The Invariance of the Hamiltonian in Symplectic Integration Schemes*, report given at the Los Alamos workshop on Symplectic Integration (March 1988).
36. Fredkin, E., and T. Toffoli, *Conservative Logic*, Int. J. Theoretical Physics **21:3/4** (1982), 219–253.
37. Friedman, A. and S.P. Auerbach, *Numerically Induced Stochasticity*, Lawrence Livermore National Laboratory, preprint UCRL–96178 (1989).
38. Frisch, U., Hasslacher, B., and Y. Pomeau, *Lattice Gas Automata for the Navier-Stokes Equation*, Phys. Rev. Letters **56:14** (1986), 1505–1508.
39. Ge, Z., and J. Marsden, *Lie-Poisson Hamilton-Jacobi Theory and Lie-Poisson Integrators*, Phys. Letters A **33:3** (1988), 134–139.
40. Ge, Z., *The Generating Function for the Poisson Map*, preprint.
41. Ge, Z., *Generating Functions, Hamilton-Jacobi Equations and Symplectic Groupoids on Poisson Manifolds*, preprint.
42. Goldstein, H., "Classical Mechanics," Addison-Wesley, Reading, 1950.
43. Golubitsky, M., Stewart, I., and J. Marsden, *Generic Bifurcation of Hamiltonian Systems with Symmetry*, Physica D **24** (1987), 391–405.
44. Graziani, F., and D. Black, *Orbital Stability Constraints on the Nature of the Planetary Systems*, Astr. Jrnl. **25** (1981), 337–341.
45. Guillemin, V., and S. Sternberg, *The Moment Map and Collective Motion*, Ann. Phys. **127** (1980), 220–253.
46. Hannay, J.H., and M.V. Berry, *Quantization of Linear Maps on a Torus-Fresnel Diffraction by a Periodic Grating*, Physica D **1** (1980), 267–290.
47. Hénon, M., and C. Heiles, *The Applicability of the Third Integral of Motion: Some Numerical Experiments*, Astr. J. **69** (1964), p. 73.
48. Herbst, B.M., and M.J. Ablowitz, *Numerically Induced Chaos in the Nonlinear Schrödinger Equation*, Phys. Rev. Lett. **62:18** (1989), 2065–2068.
49. Hockett, K., *Chaotic Numerics from an Integrable Hamiltonian System*, preprint.
50. Holm, D.D., Lysenko, W.P., and J.C. Scovel, *Moment Invariants for the Vlasov Equation*, submitted to J. Math. Phys. (1989).

51. Hori, G., *Theory of General Perturbations with Unspecified Canonical Variables*, Publ. Astron. Soc. Japan **18:4** (1966), 287–296.

52. Jiménez, S., and L. Vázquez, *Analysis of Four Numerical Schemes for a Nonlinear Klein-Gordon Equation*, Appl. Math. and Comp. (1989) (to appear).

53. Kaneko, K., *Symplectic Cellular Automata*, Phys. Lett. A **129:1** (1988), 9–16.

54. Kook, H., and J.D. Meiss, *Periodic Orbits for Reversible, Symplectic Mappings*, Physica D **35** (1989), 65–86.

55. Krishnaprasad, P.S., and J.E. Marsden, *Hamiltonian Structure and Stability for Rigid Bodies with Flexible Attachments*, Arch. Rat. Mech. Anal. **98** (1987), 71–93.

56. Lasagni, F.M., *Canonical Runge-Kutta Methods*, ZAMP **39** (1988), 952–953.

57. Livi, R., Pettini, M., Ruffo, S., Sparpaglione, M., and A. Vulpiani, *Equipartition Threshold in Nonlinear Large Hamiltonian Systems: The Fermi-Pasta-Ulam Model*, Phys. Rev. A **31:2** (1985), 1039–1045.

58. Margolus, N., *Physics-Like Models of Computation*, Physica D **10** (1984), 81–95.

59. Marsden, J., Ratiu, T., Schmid, R., Spencer, R., and A. Weinstein, *Hamiltonian Systems with Symmetry, Coadjoint Orbits and Plasma Physics*, Proc. IUTAM-ISIMM Symposium on Modern Developements in Analytical Mechanics, Torino, June 7-11 (1982), 289–340, Atti della Academia della Scienze di Torino, 117.

60. Menyuk, C.R., *Some Properties of the Discrete Hamiltonian Method*, Physica 11D (1984), p. 109.

61. Molzahn, F.H., and T.A. Osborn, *Tree Graphs and the Solution to the Hamilton-Jacobi Equation*, J. Math. Phys. **27:1** (1986), p. 88.

62. Nekhoroshev, N.N., *An Exponential Estimate of the Time of Stability of Nearly-integrable Hamiltonian Systems*, Russ. Math. Surveys **32:1** (1977), p. 65.

63. Neri, F., *Lie Algebras and Canonical Integration*, Univ. of Maryland Dept. of Physics Technical Report (1987).

64. Pomeau, Y., *Invariant in Cellular Automata*, J. Phys. A: Math. Gen. **17** (1984), L415–L418.

65. Rannou, F., *Numerical Study of Discrete Plane Area-preserving Mappings*, Astron. Astrophys. **31** (1974), 289–301.

66. Rod, D.L., and R.C. Churchill, *A Guide to the Hénon-Heiles Hamiltonian*, Singularities and Dynamical Systems, S.N. Pnevmatikos (Ed.), North-Holland (1985), 385–396.

67. Ruth, R.D., *A Canonical Integration Technique*, IEEE Trans. Nuc. Sci. **30:4** (1983), 2669–2671.

68. Sanz-Serna, J.M., *The Numerical Integration of Hamiltonian Systems*, Proc. Conf. on Comp. Diff. Eqns., Imperial College, London, 3-7, July, 1989 (to appear).

69. Sevryuk, M.B., *Reversible Systems*, Lect. Notes. in Math. **1211** (1986), Springer-Verlag, Berlin.

70. Siegel, C.L., and J.K. Moser, "Lectures on Celestial Mechanics," Springer-Verlag, Berlin, 1971.

71. Stofer, D.M., *Some Geometric and Numerical Methods for Perturbed Integrable Systems*, Thesis (1988), ETH-Zürich.

72. Thyagaraja, A., and F.A. Haas, *Representation of Volume-Preserving Maps Induced by Solenoidal Vector Fields*, Phys. Fluids **28:3** (1985), 1005–1007.

73. Weinstein, A., *Coisotropic Calculus and Poisson Groupoids*, J. Math. Soc. Japan **40:4** (1988), 705–727.

74. Wisdom, J., *The Origin of the Kirkwood Gaps: a Mapping for the Asteroidal Motion near the 3/1 Commensurability*, Astron. J. **87** (1982), p. 577.

75. Zaslavskiĭ, G.M., Sagdeev, R.Z., and A.A. Chernikov, *Stochastic Nature of Streamlines in Steady-state Flows*, Sov. Phys. JETP **67:2** (1988), 270–277.

76. Zaslavskiĭ, G.M., Zakharov, M.Yu., Sagdeev, R.Z., Usikov, D.A., and A.A. Chernikov, *Stochastic Web and Diffusion of Particles in a Magnetic Field*, Sov. Phys. JETP **64:2** (1986), 294–303.

Computer Research and Applications, MS–B265
Los Alamos National Laboratory
Los Alamos, New Mexico 87545

Connections Between Critical Points in the Collision Manifold of the Planar 3-Body Problem

C. SIMÓ AND A. SUSÍN

Abstract. We study the relative position of the invariant manifolds of the fixed points on the compactification of the so called non rotating triple collision manifold for the planar general three body problem. The results obtained allow to describe all the possible transition from the approach to triple collision to the escape from it. We also describe how these transitions change as a function of the masses of the three bodies in a domain of the space of masses and are completed by numerical simulations in the remainder set of masses.

§1. Introduction.

We consider the three-body problem in $\mathbf{R}^3$ with bodies of positive masses m_1, m_2, m_3 which interact according to the mutual gravitational attraction. Let $q_i, p_i \in \mathbf{R}^3$, $i = 1,2,3$ the positions and momenta of the masses and denote by $q = (q_1^T, q_2^T, q_3^T)^T$, $p = (p_1^T, p_2^T, p_3^T)^T$ the vectors in $\mathbf{R}^9$ containing all the coordinates. The Hamiltonian of the system is

$$H(q,p) = \frac{1}{2}p^T M^{-1} p - U(q), \tag{1}$$

where $M = diag(m_1, m_1, m_1, m_2, m_2, m_2, m_3, m_3. m_3)$ and

$$U(q) = \sum_{1 \leq i < j \leq 3} \frac{m_i m_j}{\| q_i - q_j \|} \qquad (\| \quad \| = \text{euclidean norm}).$$

Let Δ the set of collisions: $\Delta = \{q \in \mathbf{R}^9 : q_i = q_j \text{ for some } i, j, i \neq j\}$. Then the phase space of (1) is $(\mathbf{R}^9 - \Delta) \times \mathbf{R}^9$. The equations of motion are obtained from (1) as

$$\begin{aligned} \dot{q} &= M^{-1} p. \\ \dot{p} &= \nabla U(q). \end{aligned} \tag{2}$$

They have well known first integrals: the energy integral (H), the center of mass integrals $\left(\sum_{i=1}^{3} m_i q_i, \sum_{i=1}^{3} p_i \right)$ and the angular momentum integral

$\left(\sum_{i=1}^{3} q_i \wedge p_i\right)$. In all there are 10 scalar integrals in $\mathbf{R}^3$, 6 if we restrict to $\mathbf{R}^2$ and 3 if we restrict to $\mathbf{R}$.

Singularities appear in (2) if $q \in \Delta$. When approaching Δ the vector field (2) becomes unbounded. If $i, j, k \in \{1,2,3\}$ denote different indices then if $q_i = q_j \neq q_k$ the singularity is a double collision which is regularizable in any natural sense ([**3**], [**11**], [**15**]). If $q_1 = q_2 = q_3$ we are faced to a triple collision which is not regularizable ([**6**]). Triple collision requires zero angular momentum ([**16**]) and the motion should be planar. Even one can have very close approaches to triple collision in $\mathbf{R}^3$ from now on we restrict the study to the planar case.

If a singularity is regularizable and we consider initial coordinates leading to the singularity in finite (physical) time t_0 any small variation of the initial conditions, not leading itself to singularity, will produce a small change after passage near the singularity for time $t_1 > t_0$. If the singularity is not regularizable, small initial changes lead to big changes after passage close to the singularity. One losses the continuous dependence with respect to initial conditions at time $t_1 > t_0$.

The main objective of this work is to describe the passages near triple collision when the total angular momentum is zero and how do they depend on the masses of the bodies.

In section 2 we write the equations of motion in some useful coordinates. The collision manifold, C and the non-rotating collision manifold, N are introduced. In section 3 we study the critical points of the flow on N and we give a first analysis of their invariant manifolds.

A useful remark is that the planar problem contain as subproblems two cases which have been widely studied: the collinear problem ([**6**], [**12**]) and the isosceles problem if two of the masses are equal ([**1**], [**2**], [**13**], [**14**]). These subproblems have lower dimensionality, they are easier to analyze and the information obtained is very useful for the planar case. Hence in section 4 we present the known connections ([**6**], [**13**]) between the critical points in these subproblems and some quantitative results useful for the planar case. Also the important results of Moeckel ([**8**]) concerning planar connections are given in this section.

Section 5 contains the main new results about the connections in the planar case. The main result is theorem 5.1 showing that for total angular momentum equal to zero, if two of the values of the masses are close enough all the possible passages from approach to triple collision to escape occur except the case of approaching and escaping in equilateral configuration. Some exceptional masses should be skipped. We recall that if the angular momentum is not zero Moeckel showed [**9**] that approaching and escaping in the same type of equilateral configuration is always possible.

Finally in section 6 we show graphycal displays of the evolution of some curves introduced on section 5 and how the space of masses is partitioned by different behaviours. Part of this numerical information was already given in [**17**].

§2. The equations of motion.

The system (2) in the planar case has dimension 12. A classical reduction to dimension 6 can be found in [**20**] but it destroyes the symmetry of the problem and it seems very difficult to use for our purposes. Another more symmetrical reduction was proposed by Waldvogel in [**19**] using ideas of Murnaghan [**10**] and Lemaître [**5**] and combining them with the blow up method of McGehee [**6**]. We summarize the changes of coordinates.

First we consider $q_j \in \mathbf{R}^2$, $j = 1,2,3$ as complex numbers ([**10**]). It is not restrictive to consider that the origin is located at the center of masses of the system. Let

$$q_\ell - q_k = a_j e^{i\varphi_j}, \tag{3}$$

where $a_j \in \mathbf{R}$, $a_j \geq 0$, $\varphi_j \in [0, 2\pi]$ and (j,k,ℓ) is $(1,2,3)$ or a cyclical permutation.

Then let $\varphi = \frac{1}{3}(\varphi_1 + \varphi_2 + \varphi_3)$ and

$$p_j = P_k e^{i\varphi_k} - P_\ell e^{i\varphi_\ell} + \frac{P_\varphi}{3} i \left(\frac{1}{a_k} e^{i\varphi_k} - \frac{1}{a_\ell} e^{i\varphi_\ell} \right), \tag{4}$$

P_φ and P_j, $j = 1,2,3$ being real variables.

The new variables a_1, a_2, a_3 (that is the sides of the triangle formed by the three masses at a given time t), φ (which is related to the rotation of the triangle) and the associated momenta $P_1, P_2. P_3, P_\varphi$, introduced in (3), (4) allow to recover the old ones.

The momentum P_φ is the angular momentum of the system (equal to zero by hypothesis). The associate coordinate φ is a cyclic variable and its time evolution can be recovered from the evolution of the remaining variables (elimination of the node).

The second step is the regularization of the binary collisions ([**5**]). To this end we introduce new variables $\widetilde{\alpha}_j$, π_j, $j = 1,2,3$ and a new time $\widetilde{\tau}$ by

$$
(5) \qquad \begin{aligned}
a_j &= \widetilde{\alpha}_k^2 + \widetilde{\alpha}_\ell^2 , \\
P_j &= \frac{1}{4}\left(-\frac{\pi_j}{\widetilde{\alpha}_j} + \frac{\pi_k}{\widetilde{\alpha}_k} + \frac{\pi_\ell}{\widetilde{\alpha}_\ell} \right) , \\
dt &= a_1 a_2 a_3 d\widetilde{\tau} ,
\end{aligned}
$$

where (j,k,ℓ) is again the triplet $(1,2,3)$ or a cyclical permutation of it.

The final step is a blow up ([**6**]) introducing the moment of inertia of the system, I, new coordinates α_j, $j = 1,2,3$ and a new time τ but keeping the π_j variables:

$$
(6) \qquad \begin{aligned}
r^2 &= I = \Sigma m_j \parallel q_i \parallel^2 , \\
\alpha_j &= \widetilde{\alpha}_j r^{-1/2} , \quad j = 1,2,3 , \\
d\tau &= r^{3/2} d\widetilde{\tau} .
\end{aligned}
$$

From (2) and using the changes (3–6) it is easy to obtain the equations for $\frac{dr}{d\tau}, \frac{d\alpha_j}{d\tau}, \frac{d\pi_j}{d\tau}$, $j = 1,2,3$. These equations are analytical for $r > 0$ and can be extended analytically to $r = 0$. The set $r = 0$ is called the triple collision manifold with regularized binary collisions, C. From (5) it follows that several points in this manifold should be identified. More concretely, each set $(\alpha_1, \alpha_2, \alpha_3, \pi_1, \pi_2, \pi_3)$ has to be identified with the one which one obtains by changing the sign of two of the α's (say α_j, α_k) and, simultaneously, the sign of π_j, π_k.

We remark that for general values of P_φ (not forcely zero) one should scale also this variable introducing $\widetilde{P}_\varphi$ by $P_\varphi = \widetilde{P}_\varphi r^{1/2}$. For $r = 0$ one has $P_\varphi = 0$ but $\widetilde{P}_\varphi$ needs not to be zero, and this is a variable which also appears in C. As the variables α, π satisfy the constraints

$$
(7) \qquad \sum_{1 \le j < k \le 3} m_j m_k (\alpha_j^2 + \alpha_k^2)^2 = \sum_{i=1}^{3} m_i \quad \text{(normalization)}
$$

and

$$
(8) \qquad K(\alpha, \pi) = 0 \quad \text{(energy integral)},
$$

the manifold C is 5-dimensional.

The differential equation for $\widetilde{P}_\varphi$ shows that $\widetilde{P}_\varphi = 0$ is invariant. This defines on C a 4-dimensional invariant submanifold called the non-rotating collision manifold, N already introduced by Waldvogel in **[18]**. If we consider motion in the physical space ($r > 0$) with $P_\varphi = 0$ one has also $\widetilde{P}_\varphi = 0$. Then it is enough to restrict the study to N if we would learn about passages near triple collision with zero angular momentum.

The final equations on N are almost Hamiltonian, given by

$$(9) \qquad \begin{aligned} \frac{d\alpha}{d\tau} &= \frac{\partial}{\partial\pi}K(\alpha,\pi) - \frac{1}{4}z\alpha\,, \\ \frac{d\pi}{d\tau} &= -\frac{\partial}{\partial\alpha}K(\alpha,\pi)\,, \end{aligned}$$

where $\alpha = (\alpha_1, \alpha_2, \alpha_3)^T$, $\pi = (\pi_1, \pi_2, \pi_3)^T$, $z = (\alpha_1^2+\alpha_2^2)(\alpha_2^2+\alpha_3^2)(\alpha_3^2+\alpha_1^2)v$, v being the inner product $v = (\alpha, \pi)$, and K is the function

$$(10) \quad K(\alpha,\pi) = \frac{1}{8}\pi^T B(\alpha)\pi - \sum_{1\le j<k\le 3} m_j m_k (\alpha_{j+1}^2 + \alpha_{j+2}^2)(\alpha_{k+1}^2 + \alpha_{k+2}^2)\,,$$

where $\alpha_s = \alpha_{s-3}$ if $s > 3$. The components of the matrix $B(\alpha)$ appearing in (10) are

$$(11) \qquad \begin{aligned} b_{jj}(\alpha) &= \frac{\alpha_k^2+\alpha_\ell^2}{m_j}(\alpha_1^2+\alpha_2^2+\alpha_3^2) + \frac{\alpha_j^2+\alpha_\ell^2}{m_k}\alpha_\ell^2 + \frac{\alpha_j^2+\alpha_k^2}{m_\ell}\alpha_k^2\,, \\ b_{jk}(\alpha) &= -\frac{\alpha_j^2+\alpha_k^2}{m_\ell}\alpha_j\alpha_k\,. \end{aligned}$$

As before, in (11) the indices (j, k, ℓ) are $(1, 2, 3)$ or a cyclic permutation of it.

On N the variable v is gradient like ($\frac{dv}{d\tau} \ge 0$) and it increases everywhere except at the critical points of (9). We remark that the equations (9) are polynomial in (α, π).

§3. Critical points and the geometry of the collision manifold.

As it is known (**[2]**, **[19]**) the critical points of (9) are related to the central configurations, that is, going back to (2), if the origin is placed at the center of masses, one has $\ddot{q}_j(t) = -\lambda(t)q_j(t)$ for all time. Taking into account the identifications there are 10 equilibrium points, 5 of them with $v > 0$ and the other 5 with $v < 0$. For $v > 0$ (resp. $v < 0$) there are 2 equilateral or Lagrange configurations L_+^s, L_-^s (resp. L_+^i, L_-^i). In the case of L_+^s (resp. L_-^s) the position of m_2 is obtained from the position of m_1 by a rotation of $\frac{2\pi}{3}$ (resp. $-\frac{2\pi}{3}$) with respect to the center of the triangle.

The other 3 configurations for $v > 0$ (resp. $v < 0$) are the so called collinear or Euler configurations E_j^s (resp. E_j^i), with the mass m_j located between the other two masses, for $j = 1, 2, 3$. From (9) it follows that the critical points satisfy $B(\alpha)\pi = z\alpha$ and the related value of v is $v = \pm(8\mathcal{U})^{1/2}$, where $\mathcal{U} = \sum_{1\leq k\leq \ell\leq 3} m_k m_\ell/(\alpha_k^2 + \alpha_\ell^2)$.

The value of v at $L_{+,-}^s$ is less than the value at any of the Euler points. The critical points are hyperbolic ([**2**]). The invariant manifolds dimensions are given in Table I.

	$L_{+,-}^i$	$L_{+,-}^s$	E_j^i	E_j^s	B_j^i	B_j^s
$\dim W^u$	2	2	3	1	4	0
$\dim W^s$	2	2	1	3	0	4

Table I

This table contains also information on some points $B_j^{i,s}$ to be described later.

At $L_{+,-}^i$ the positive eigenvalues are

$$\frac{3}{4} \mid v \mid (\alpha_1^2 + \alpha_2^2)(\alpha_2^2 + \alpha_3^2)(\alpha_3^2 + \alpha_1^2)\sigma_{1,2}\,, \tag{12}$$

where $\sigma_{1,2} = \frac{1}{6}((13 \pm 12(1 - 3\mu)^{1/2})^{1/2} - 1)$. Here μ denotes the ratio $(m_1 m_2 + m_2 m_3 + m_3 m_1)/(m_1 + m_2 + m_3)^2$.

Before going to the connections between the critical points we describe the geometry of N. The configuration space, i.e., the allowed values of $(\alpha_1, \alpha_2, \alpha_3)$, is given by (7) with the additional constrains and the symmetries allow to consider only $\alpha_2 \geq 0$, $\alpha_3 \geq 0$. But then also points like $(\alpha_1, \alpha_2, 0)$ (resp. $(\alpha_1, 0, \alpha_3)$) should be identified to $(-\alpha_1, \alpha_2, 0)$ (resp. $(-\alpha_1, 0, \alpha_3)$).

Therefore we have a two dimensional sphere, were the points on the equator denote the collinear configurations for which at least one of the α's is zero. This includes the binary collisions B_j (for which two α's are zero) and the Euler configurations as special cases. The points on the northern (resp. southern) hemispheres correspond to triangles with direct (resp. retrograde) orientation when one goes from m_1 to m_2 and then to m_3. The poles can be thought as the Lagrange points (see fig. 1 and §5 for the meaning of the curve S). A similar representation can be found in [4].

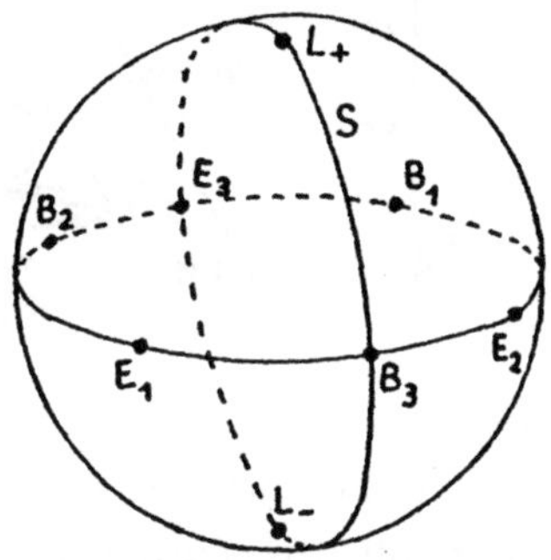

Figure 1

To obtain the non-rotating triple collision manifold we put in each point of the configuration space a fiber which is S^2 unless we are at B_j, $j = 1, 2, 3$ where we put a cylinder. This follows from (8). Then we obtain $S^2 \times S^2$ with 6 holes (after the usual identification). If we compactify the holes by adding a point to each one we get the compact manifold $\overline{N} = S^2 \times S^2$. The holes in the region $v > 0$ (resp. $v < 0$) are denoted by B_j^s (resp. B_j^i), $j = 1, 2, 3$ and they can be considered as "hard" binaries ([7]) where the two bodies forming the binary are really at the same point. It is natural to say that $\dim W^u_{B_j^i} = \dim W^s_{B_j^s} = 4$, as already pointed out in Table I. In $\overline{N}$ the set $v = 0$ is topologically equivalent to S^3.

§4. Connections between critical points. I.

In this section we describe some results already known and some preliminary results to be used in the next section. First we recall that the masses m_1, m_2, m_3, can be normalized in such a way that the sume is equal to 1. Then the values of the positive masses can be thought as the barycentric coordinates of a point of a triangle. We refer loosely to this triangle as the triangle of masses. First we give some results for the subproblems and also for the planar problem.

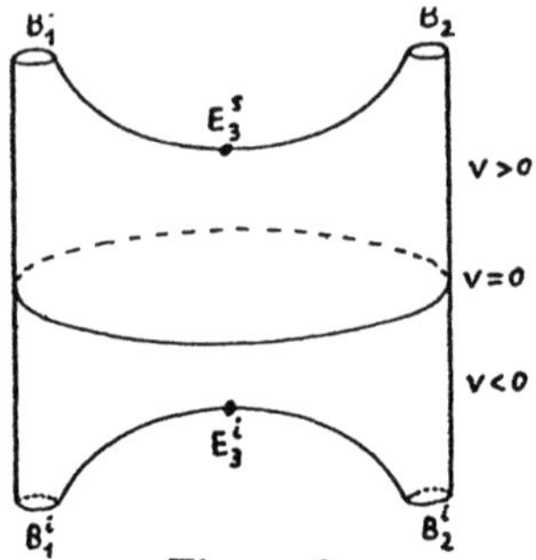

Figure 2

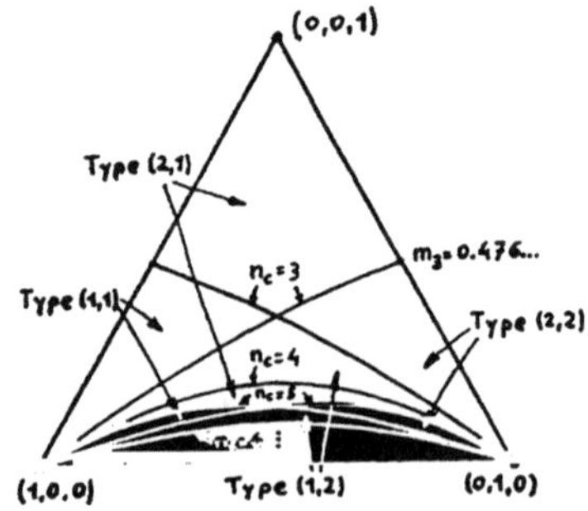

Figure 3

(a) Collinear problem ([**6**]).

We consider the mass m_3 located between the masses m_1 and m_2. In this case $C = N$ and topologically it is equivalent to a sphere with 4 holes. They can be filled up by compactification, what is equivalent to add the hard binaries. In C two critical points appear (one in $v < 0$, the other on $v > 0$) which are related to the Euler equilibrium points $E_3^{i,s}$ (see fig. 2). The following results are known ([**6**], [**12**]):

(i) For all masses one branch of $W^u_{E_3^s}$ goes to B_1^s, the other to B_2^s.

(ii) In an open region (i,j), i and j equal to 1 or 2, of the mass triangle, the right branch of $W^u_{E_3^i}$ goes to B_i^s and the left one to B_j^s according to the qualitative picture given in fig. 3. The boundaries of these regions are lines for which either only one branch of $W^u_{E_3^i}$ coincides with one branch of $W^s_{E_3^s}$ (odd case) or both branches coincide (even case). Even and odd refer to the number, n_c, of binary collisions found along the heteroclinic orbit. The integer n_c ranges from 3 to infinity.

Symmetrical results hold for the branches of $W^s_{E_3^i}$, $W^s_{E_3^s}$.

(b) Isosceles problem ([**1**], [**13**], [**14**] and references therein).

We consider $m_1 = m_2$, $m_3/m_1 = \varepsilon$ and initial positions and velocities such that the form of the triangle spanned by the 3 masses is isosceles for any time. That is, the coordinates of q_1,q_2,q_3 are of the form $(-x(t), -\frac{\varepsilon}{2}y(t))$, $(x(t), -\frac{\varepsilon}{2}y(t))$, $(0, y(t))$, respectively. Now $N = C$ is again a sphere with 4 holes that can be compactified as in (a). But there are 6 critical points (3 for $v < 0$, 3 for $v > 0$) related to the Euler configuration and to the two (direct and retrograde) Lagrange configurations. All of them are hyperbolic.

According to the possible connections between the invariant manifolds of the critical points there are 5 cases (I to V). Cases I, III and V are shown in figure 4.

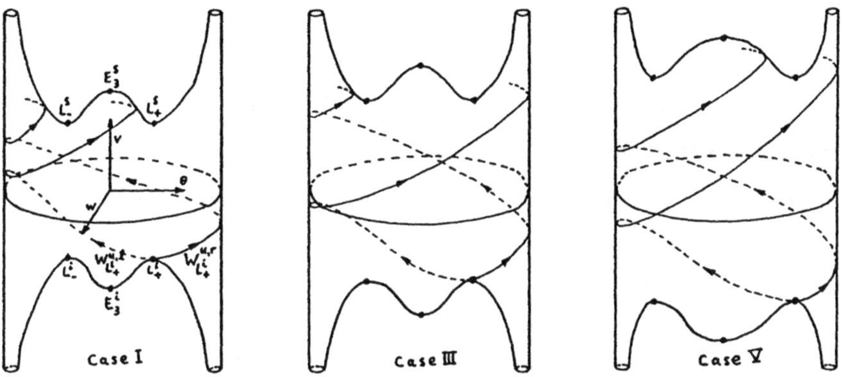

Figure 4

Case II and IV are bifurcation cases. For case II exactly one branch of $W^u_{L^i_+}$ coincides with one of $W^s_{L^s_+}$ (and by symmetry one of $W^u_{L^i_-}$ with one of $W^s_{L^s_-}$) and for case IV exactly one branch of $W^u_{L^i_+}$ coincides with one $W^s_{L^s_-}$ (and one of $W^u_{L^i_-}$ with one of $W^s_{L^s_+}$, again by symmetry).

The existence of values of ε realizing the 5 cases has been proved in [**13**]. There is numerical evidence that cases II and IV occur for only one value of $\varepsilon : \varepsilon = \varepsilon_1 (\simeq 0.378532\dots)$ and $\varepsilon = \varepsilon_2 (\simeq 2.661993\dots)$.

Then, by analizing the limit cases $\varepsilon \searrow 0$ $\varepsilon \nearrow \infty$ one has from this numerical evidence:

(i) Case I occurs for $0 < \varepsilon < \varepsilon_1$.
(ii) Case III occurs for $\varepsilon_1 < \varepsilon < \varepsilon_2$.
(iii) Case V occurs for $\varepsilon > \varepsilon_2$.

It is desirable to have a proof of what happens when $m_1 = m_2 = m_3$, $(\varepsilon = 1)$.

LEMMA 4.1. *If the three masses are equal then we are in case III.*

PROOF: By using the symmetries it is enough to see at what point the branches of $W^u_{L^i_+}$ reach the level $v = 0$. Taking into account the equation of N the differential equation on the triple collision manifold ([**1**], [**13**]) can be put in the form

$$\frac{dv}{d\theta} = \pm\sqrt{\frac{\sqrt{2}}{4\cos\theta} + \frac{2}{(2+4\sin^2\theta)^{1/2}} - \frac{v^2}{4}} \tag{13}$$

where the sign should be changed when one reaches $\theta = \pm\frac{\pi}{2}$. The starting point L^i_+, has coordinates $v = -\sqrt{6}$, $\theta = \frac{\pi}{4}$, one branch of $W^u_{L^i_+}$ starting with positive sign in (13), the other with negative one.

One should show that the right (left) branch reaches $\theta = \frac{\pi}{2}$ $(\theta = -\frac{\pi}{2})$ for some $v < 0$ and then reaches $v = 0$ for some $\theta > 0$ $(\theta < 0\ )$. First we proceed with the right branch (the easiest case). For $\theta \in (\frac{\pi}{4}, \frac{\pi}{2})$ one has from (13)

$$\frac{dv}{d\theta} = \frac{1}{\sqrt{\cos\theta}}\sqrt{\frac{\sqrt{2}}{4} + \frac{2\cos\theta}{(2+4\sin^2\theta)^{1/2}} - \frac{v^2}{4}\cos\theta} \leq \sqrt{\frac{3\sqrt{2}}{4}\frac{1}{\cos\theta}}. \tag{14}$$

By integration of (14) we obtain

$$v(\theta = \frac{\pi}{2}) - (-\sqrt{6}) < \sqrt{\frac{3\sqrt{2}}{4}}\int_{\pi/4}^{\pi/2} \frac{d\theta}{\sqrt{\cos\theta}} =$$

$$= \sqrt{\frac{3\sqrt{2}}{4}}\int_0^{\pi/4} \frac{d\theta}{\sqrt{\sin\theta}} < \sqrt{\frac{3\pi}{8}}\int_0^{\pi/4} \frac{d\theta}{\sqrt{\theta}} = \pi\sqrt{\frac{3}{8}},$$

where we have used $\sin\theta \geq \frac{\sqrt{8}}{\pi}\theta$ in $[0, \pi/4]$. Hence $v(\theta = \frac{\pi}{2}) < \pi\sqrt{\frac{3}{8}} - \sqrt{6} < 0$, as desired.

To finish the proof for the right branch we note first that $v(\theta = \frac{3\pi}{8}) > -\sqrt{6}$. Going from $\theta = \frac{3\pi}{8}$ to $\theta = \frac{\pi}{2}$ we can bound the vector field by

$$\frac{dv}{d\theta} > \sqrt{\frac{\sqrt{2}}{4\cos\theta} - \left(\frac{6}{4} - \frac{2}{(4+\sqrt{2})^{1/2}}\right)}.$$

Hence

$$v(\theta = \frac{\pi}{2}) > -\sqrt{6} + \int_{3\pi/8}^{\pi/2} \frac{dv}{d\theta} > -\sqrt{6} + \int_0^{\pi/8} \sqrt{\frac{\sqrt{2}}{4\theta} - b},$$

where b stands for $\frac{3}{2} - \frac{2}{(4+\sqrt{2})^{1/2}}$. Using

$$\int \sqrt{\frac{a}{x} - b} = \sqrt{x(a-bx)} + \frac{a}{\sqrt{b}} \arctan \sqrt{\frac{bx}{a-bx}},$$

one has

$$v(\theta = \frac{\pi}{2}) > -\sqrt{6} + \sqrt{\frac{\pi}{8}\left(\frac{\sqrt{2}}{4} - b\frac{\pi}{8}\right)} + \frac{\sqrt{2}/4}{\sqrt{b}} \arctan \sqrt{\frac{b\pi/8}{\sqrt{2}/4 - b\pi/8}} > -1.80593\,.$$

We introduce $v_1 = -1.80593$. Now θ goes from $\frac{\pi}{2}$ to the left. As before we have in $\left[\frac{3\pi}{8}, \frac{\pi}{2}\right]$

$$\left|\frac{dv}{d\theta}\right| > \sqrt{\frac{\sqrt{2}}{4\cos\theta} + \left(\frac{2}{(4+\sqrt{2})^{1/2}} - \frac{v_1^2}{4}\right)}\,.$$

Then $v(\theta = \frac{3\pi}{8}) > v_1 + \int_0^{\pi/8} \sqrt{\frac{\sqrt{2}}{4\theta} + b}$ where now b stands for $\frac{2}{(4+\sqrt{2})^{1/2}} - \frac{v_1^2}{4} > 0$. Using

$$\text{(15)} \qquad \int \sqrt{\frac{a}{x} + b} = \sqrt{(a+bx)x} + \frac{a}{\sqrt{b}} \log(\sqrt{b(a+bx)} + b\sqrt{x})\,,$$

one has

$$v(\theta = \frac{3\pi}{8}) >$$

$$v_1 + \sqrt{\left(\frac{\sqrt{2}}{4}b + \frac{\pi}{8}\right)\frac{\pi}{8}} + \frac{\sqrt{2}/4}{\sqrt{b}}\left[\log\left(\sqrt{b\left(\frac{\sqrt{2}}{4} + b\frac{\pi}{8}\right)} + b\sqrt{\frac{\pi}{8}}\right) - \log\sqrt{\frac{\sqrt{2}}{4}b}\right] >$$

$$-1.05466 = v_2\,.$$

Between $\theta = \frac{3\pi}{8}$ and $\theta > 0$ one can use the simple bound $\left|\frac{dv}{d\theta}\right| > \sqrt{\frac{3}{2} - \frac{v^2}{4}}$. Then

$$v(\theta = 0) > \sqrt{6}\sin\left[\frac{3\pi}{16} + \arcsin\frac{v(\theta = 3\pi/8)}{\sqrt{6}}\right] > \sqrt{6}\sin\left(\frac{3\pi}{16} + \arcsin\frac{v_2}{\sqrt{6}}\right) > 0.35134\,.$$

Therefore the right branch of $W^u_{L^i_+}$ reaches $v = 0$ before θ reaches the value 0.

Since for the same value of θ the left branch of $W^u_{L^i_+}$ has a value of v larger than the left branch of $W^u_{L^i_-}$, and using the symmetry, we have that the left branch of $W^u_{L^i_+}$ reaches $v = 0$ for some negative value of θ. The only thing to prove is that it reaches $\theta = -\frac{\pi}{2}$ for some negative value of v.

Write (13) as $\left|\frac{dv}{d\theta}\right| = \sqrt{a(\theta) - \frac{v^2}{4}}$, for $a(\theta) = \frac{\sqrt{2}}{4\cos\theta} + \frac{2}{(2+4\sin^2\theta)^{1/2}}$.

In an interval $J_n = [\theta_n, \theta_{n-1}]$ we define $a_n = \max_{\theta\in J_n} a(\theta)$. Then $\left|\frac{dv}{d\theta}\right| < \sqrt{a_n - \frac{v^2}{4}}$ and therefore

$$(16)\quad v(\theta = \theta_n) < 2\sqrt{a_n}\sin\left(\frac{1}{2}\mid \theta_n - \theta_{n-1}\mid + \arcsin\left(\frac{v(\theta=\theta_{n-1})}{2\sqrt{a_n}}\right)\right).$$

Due to the behaviour of $a(\theta)$ one has $a_n = a(\theta_n)$ (resp. $a_n = a(\theta_{n-1})$) if $J_n \subset [0, \pi/4] \cup [-\pi/2, -\pi/4]$ (resp. $J_n \subset [-\pi/4, 0]$). Starting at $v_0 = -\sqrt{6}$, $\theta_0 = \frac{\pi}{4}$ and using $\theta_n - \theta_{n-1} = -\frac{\pi}{48}$ for $n = 1, \ldots, 32$ the iterated application of (16) gives $v_{32} = v(\theta = -\frac{5\pi}{12}) < -0.6738$. The computations are done with a pocket calculator taking into account the rounding error.

Now we consider the interval $J = \left[-\frac{\pi}{2}, -\frac{5\pi}{12}\right]$. Here we use the bound

$$(17)\quad \left|\frac{dv}{d\theta}\right| < \sqrt{\frac{\sqrt{2}}{4\cos\theta} + \frac{2}{(2+4\sin^2(5\pi/12))^{1/2}}} < $$
$$< \sqrt{\frac{1}{\theta+\pi/2}\,\frac{\sqrt{2}\pi/12}{4\sin(\pi/12)} + \frac{2}{(2+4\sin^2(5\pi/12))^{1/2}}},$$

where we have used $\cos\theta = \sin(\theta+\frac{\pi}{2}) > (\theta+\frac{\pi}{2})\frac{\sin(\pi/12)}{\pi/12}$ in J. The bound (17) is of the form $\sqrt{\frac{\alpha}{x} + \beta}$ with $x = \theta + \frac{\pi}{2}$, $x \in [0, \frac{\pi}{12}]$. The integral has been already given in (15). Suppose $v(\theta = -\frac{\pi}{2}) = 0$. Then, by integration of (16) one obtains $v(\theta = -\frac{5\pi}{12}) > -0.6696$. This is in contradiction with the upper bound of v_{32} found before. Hence $v(\theta = -\frac{\pi}{2}) < 0$. □

c) Planar problem ([**8**]).

The following connections are known to exist:

i) For all masses there are orbits connecting $L^s_{+,-}$ with E^s_i and B^s_i, $i = 1,2,3$. Symmetrical connections occur in the $v < 0$ region. All these connections are transversal.

ii) If there is an orbit connecting E^i_j with L^s_+ transversally, then there are also orbits connecting E^i_j to L^s_- and to E^s_k B^s_k ,$k = 1,2,3$. A similar statement is obtained by symmetry by changing the sign of v.

To prove these results one needs a lemma which will be also used in next section. First we introduce some definitions.

Let $c = v(L^s_{+,-})$, $N_\lambda = \{(\alpha,\pi) \in N : v(\alpha,\pi) \geq \lambda\}$, which is a positively invariant set, and let S_λ the projection of N_λ on the configuration space: $S_\lambda = \text{proj } N_\lambda$. The set $S_{c+\varepsilon}$ is S^2 minus two small neighbourhoods of the poles if ε is small.

LEMMA 4.2. *([8]). Let $\alpha : S^1 \longrightarrow N_{c+\varepsilon}$ be an analytical closed curve such that proj $\alpha : S^1 \longrightarrow S_{c+\varepsilon}$ is not contractible in $S_{c+\varepsilon}$. Then $\alpha(S^1)$ intersects the stable manifolds of E^s_i, B^s_i for $i = 1,2,3$.*

The figure 5 summarizes the known connections in the $v > 0$ region (similar ones exist on $v < 0$). As stated on Table I dim $W^u_{E^s_j} = 1$, $j = 1,2,3$. These manifolds end on B^j_k, $k \neq j$ and are already found in the linear problem.

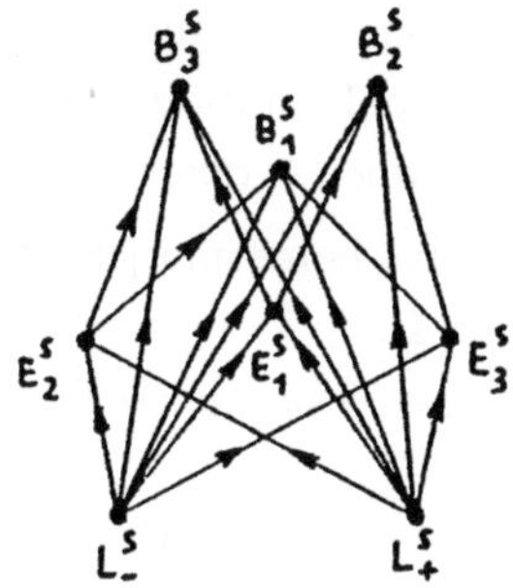

Figure 5

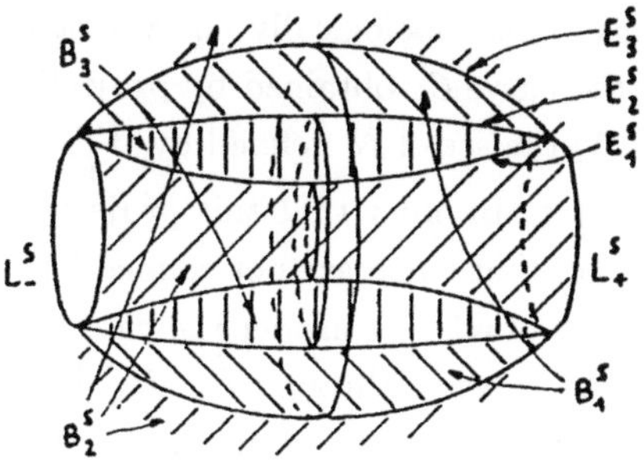

Figure 6

The connections $E^s_j \longrightarrow B^s_k$ consist of just one orbit. It is not known if the connections $L^s_{+,-} \longrightarrow E^s_j$ are made of just one orbit. Assuming that this is true one can obtain the topological structure of $W^s_{P^s} \cap \{v = d\}$, for different values of d where P^s stands for one of the critical points $L^s_{+,-}$, E^s_j or B^s_j. Let $v_2 > \max v(E^s_j)$,$0 < v_1 < \min v(E^s_j)$, $v_0 = 0$. Table II gives the topology of $W^s_{P^s} \cap \{v = d\}$ for the different P^s and $d = v_i$, $i = 0,1,2$.

$d \backslash P^s$	B^s_j	E^s_j	$L^s_{+,-}$
v_2	S^3	$--$	$--$
v_1	$S^2 \times (0,1)$	S^2	$--$
v_0	$E^2 \times S^1$	$S^1 \times (0,1)$	S^1

Table II

In particular, for $v = 0$ one has that $\overline{N} \cap \{v = 0\} = S^3$ has been partitioned in 3 open solid tori going to the hard binaries. Each one of the three two-dimensional tori in the boundary of the three $E^2 \times S^1$ is the union of two S^1 and two open cylinders $S^1 \times (0,1)$. The cylinders go to E^s_j and the circles to $L^s_{+,-}$ (see fig. 6). Similar results are obtained for $W^u_{P^i} \cap \{v = 0\}$. The full description of the possible transitions close to triple collision would be given by the intersections of the type $W^u_{P^i} \cap \{v = 0\} \cap W^s_{Q^s}$, P^i (resp. Q^s) being one of the critical points on $v < 0$ ($v > 0$). The next section gives a partial answer to this question.

§5. Connections between critical points II.

We consider $W^u_{L^i_+} \cap \{v = -c + \varepsilon\}$, $c = v(L^s_+)$, ε positive small enough. This set is topologically equivalent to S^1. Then we study the evolution of the curve C_k obtained by projecting in the configuration space S^2 (see fig. 1) the set of N defined by $W^u_{L^i_+} \cap \{v = k\}$ when k ranges from $-c + \varepsilon$ to $c + \varepsilon$. In order that this projection be always homeomorphic to S^1 we require that there are not connections $L^i_+ \longrightarrow L^s_{+,-}$.

First of all we consider the case $m_1 = m_2$. Let S be the meridian of the figure 1 passing through L_+, B_3, L_-, E_3.

Then, due to the symmetry of (9) C_k is symmetrical with respect to S for any k. For $k = -c + \varepsilon$, C_k is a small circle around L_+. The curve S intersects $C_{-c+\varepsilon}$ in just two points, R (right) and L (left), R between L_+ and B_3, and L between L_+ and E_3. Due to the symmetry when k increases the projections on S^2 of the points obtained by flow transport on N of the preimages r and ℓ of R and L are still on S, and they correspond to the evolution of the right and left branches of the unstable manifold of L^i_+ on the isosceles problem. For this problem one always has $\alpha_1 = \alpha_2$, $\pi_1 = \pi_2$.

Let $t \in [0, 2\pi]$ a parameter of $\mathcal{C}_k$. We denote by $\mathcal{C}_k(t)$ the point of $\mathcal{C}_k$ with parameter t. It can be choosen in such a way that $t = 0$ $(t = \pi)$ corresponds to the point obtained by evolution on $R(L)$ when k increases and points with parameter t and $2\pi - t$ are symmetrical with respect to S for all k. The parameter t is also choosen such that when $k = -c+\varepsilon$ the curve $\mathcal{C}_k(t)$, $t \in [0, 2\pi]$, has index $+1$ around L_+. It is possible (in fact it occurs, see §6) that for some k there are points $\mathcal{C}_k(t_0) \in S$, $t_0 \neq 0, \pi$. However then also $\mathcal{C}_k(2\pi - t_0) \in S$ and, again by the symmetry, the vectors $\frac{d}{dt}\mathcal{C}_k\big|_{t=t_0}$, $\frac{d}{dt}\mathcal{C}_k\big|_{t=2\pi-t_0}$ point towards the same of the two hemispheres defined by S. Of course the vectors $\frac{d}{dt}\mathcal{C}_k\big|_{t=0,\pi}$ are always orthogonal to S, except when the related points coincide with B_3 where these vectors become zero. A local analysis shows that if for some k one has $\mathcal{C}_k(0) = B_3$ then $\frac{d}{dt}\mathcal{C}_{k_1}(0)$ and $\frac{d}{dt}\mathcal{C}_{k_2}(0)$, where $k_1 \in (k-\delta, k)$, $k_2 \in (k, k+\delta)$, point to opposite hemispheres. But this fact will not be used here.

Now we look at the final positions of R and L on S when $k = c + \varepsilon$ in the cases I, III and IV of the isosceles problem (§4, b). The situation is depicted on fig. 7, where the positions of R and L are shown on S.

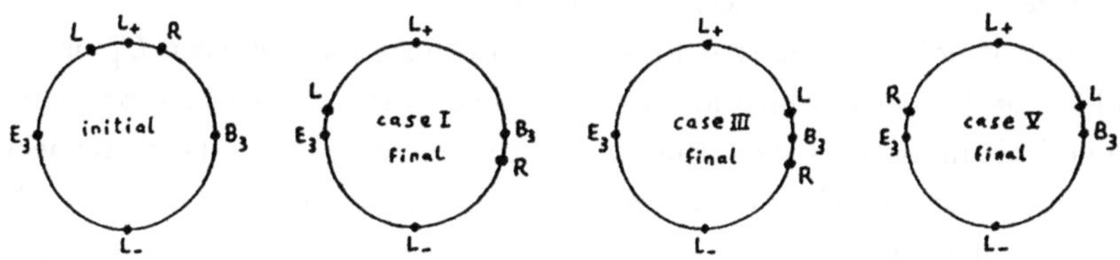

Figure 7

In cases I and V, and taking into account that extra crossings of the curve S at $L_{+,-}$ for some $k \in (-c+\varepsilon, c+\varepsilon)$ and $t_0 \neq 0, \pi$ change the index with respect L_+ or L_- in multiples of 2, the final curve has odd index and therefore is not contractible.

In the case III we use the fact that it contains the case of equal masses (Lemma 4.1). Then 3 isosceles problems (obtained by $\pm\frac{2\pi}{3}$ rotation in the configuration space) are included as subproblems. By symmetry the changes of index should occur in multiples of 3. Hence, also the final curve is not contractible.

This implies that there are connections from L^i_+ to E^s_j, B^s_j, $j = 1,2,3$. If a connection $L^i_+ \longrightarrow L^s_{+,-}$ occurs, we can deformate the initial curve on $v = -c+\varepsilon$ slightly out of $W^u_{L^i_+}$ and the same conclusion follows with the eventual exception that the connection $L^i_+ \longrightarrow E^s_j$ can fail. Similar connections occur starting at L^i_-.

From the previous facts and (ii) in§4, c) it follows that for $m_1 = m_2$ if there are not connections of the type $L^i_{+,-} \longrightarrow L^s_{+,-}$ the connections given in table III occur.

	B^i_k	E^i_k	$L^i_{+,-}$	$L^s_{+,-}$	E^s_k	B^s_k	
B^i_j		$\times$ (*)	$\times$	$\times$	$\times$	$\times$	$j,k = 1,2,3$
E^i_j			$\times$	$\times$	$\times$	$\times$	$(*)\, k \neq j$
$L^i_{+,-}$					$\times$	$\times$	

Table III

Continuing with $m_1 = m_2$, for the masses of case II of the isosceles problem, i.e.,$\varepsilon = \varepsilon_1$ or, equivalently,$m_1 + m_2 + m_3 = 1$, $m_3 = 0.159145\ldots$, there are isosceles connections $L^i_+ \longrightarrow L^s_+$, $L^i_- \longrightarrow L^s_-$. For the masses of case IV of the isosceles problem, i.e., $\varepsilon = \varepsilon_2$ or, equivalently $m_3 = 0.570999\ldots$, there are isosceles connections $L^i_+ \longrightarrow L^s_-$, $L^i_- \longrightarrow L^s_+$. Of course, it is not excluded that for $m_1 = m_2$ connections between L^i_+ or L^i_- and L^s_+ or L^s_- occur which are not contained in the isosceles subproblem. Numerical simulations show that there are two symmetrical (with respect to S say), non isosceles connections of each one of the types $L^i_+ \longrightarrow L^s_+$ and $L^i_- \longrightarrow L^s_-$ for $m_1 + m_2 + m_3 = 1$, $m_3 = 0.632911\ldots$. There is some numerical evidence that these are the only non isosceles connections when two of the masses are equal.

The index of the curve $C_{c+\varepsilon}$ is a locally continuous constant function of the masses. We summarize the result obtained as follows.

THEOREM 5.1. *If two of the values of the masses of the planar three body problem are close enough and there are not connections from the equilateral* *points on $v < 0$ to the equilateral ones on $v > 0$ then the connections of table III occur.*

As a consequence for the masses for which 5.1 holds any type of approach to collision can give rise to any type of ejection, except the approaches and escapes which are simultaneously of equilateral type.

§6. Some numerical simulations.

To show the evolution of $\mathcal{C}_v$ when v changes increasing from $-c+\varepsilon$ we show in figure 8 the projection of the curves contained on S^2 (the configuration space, see fig. 1) into the equilateral plane, for the case of equal masses $m_1 = m_2 = m_3 = 1/3$. In this case $c = 1.2409\ldots$ and figure 8 shows the behaviour for a selection of values of v, the last one with $v > c$.

Each one of the plots contains the position of R and L, the orientation of $\mathcal{C}_v$ and the value of the index (when it is defined). The projection is shown as a continuous curve in the arcs contained in the northern hemisphere (the one which contains L_+) and as a dotted curve in the arcs contained in the southern one. We remark that the final index is -2. In particular this implies that there are at least two connections from L^i_+ to each one of the E^s_j, $j = 1,2,3$.

The complexity of $\mathcal{C}_k$ when k reaches values slightly larger than $v(L^s_{+,-})$ even for the case $m_1 = m_2$ is displayed in figure 9 for $m_1+m_2+m_3 = 1$ and the following values of m_3 : $0.1, 0.2, 0.4, 0.6$ and 0.7 . These curves are given always as dotted curves disregarding whether the corresponding point is in the northern or southern hemisphere. The points R and L as well as the final index are also shown. The numbers along the curves allow to follow how the curve is described when the parameter t (see §5) increases from 0 to 2π. The point R always correspond to the initial and final numbers.

To complete the study for all the masses one should check that there are not masses such that the final index of $\mathcal{C}_k$ (when it is defined) around $L_{+,-}$ is zero. The numerical evidence that we have at this moment using a mesh on the space of masses shows that there are not such masses, but a finer mesh would be desirable and it is being done.

It is also interesting to display the curves on the triangle of masses for which there are connections between equilateral points. Furthermore these

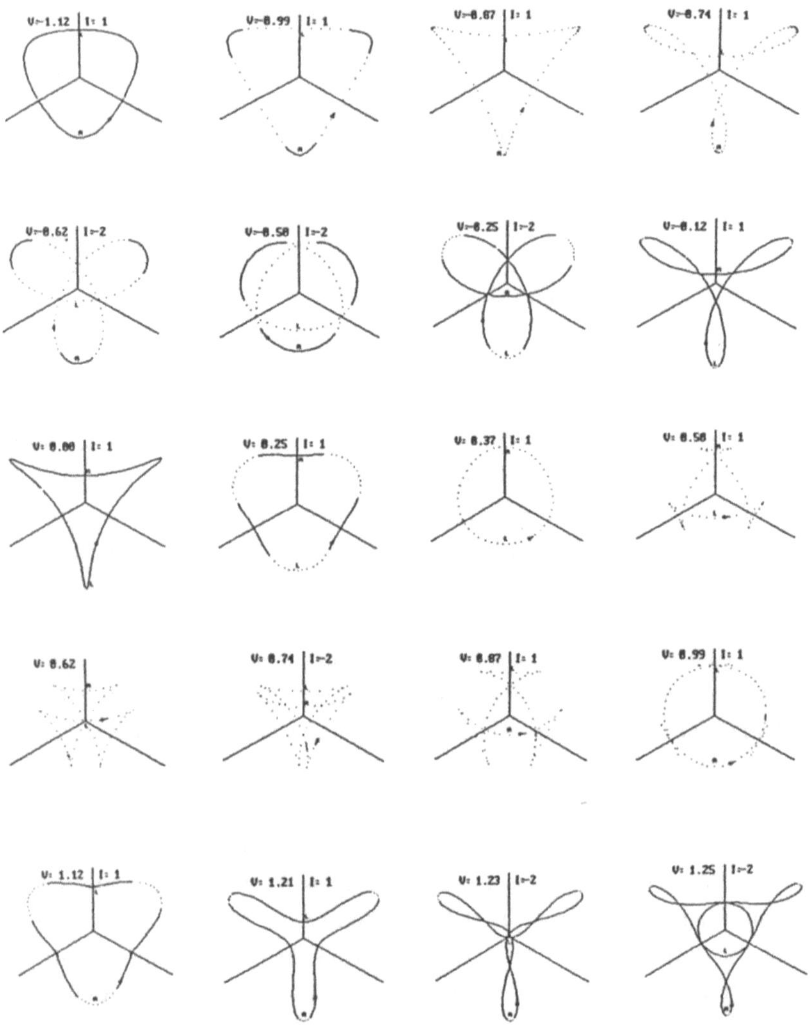
V=-1.12 | I= 1
V=-0.99 | I= 1
V=-0.87 | I= 1
V=-0.74 | I= 1
V=-0.62 | I=-2
V=-0.50 | I=-2
V=-0.25 | I=-2
V=-0.12 | I= 1
V= 0.00 | I= 1
V= 0.25 | I= 1
V= 0.37 | I= 1
V= 0.50 | I= 1
V= 0.62
V= 0.74 | I=-2
V= 0.87 | I= 1
V= 0.99 | I= 1
V= 1.12 | I= 1
V= 1.21 | I= 1
V= 1.23 | I=-2
V= 1.25 | I=-2

Figure 8

curves separate regions with different final index of C_k. Using a continuation method starting at the masses for which these connections occur in the isosceles problem we have obtained so far the results shown in figure 10. In this figure we display only one sixth of the full triangle of masses, the other ones being obtained by symmetry. The masses (m_1, m_2, m_3) associated to each vertex are given and also the final index in each component of the complementary of the curves for which an equilateral connection occurs.

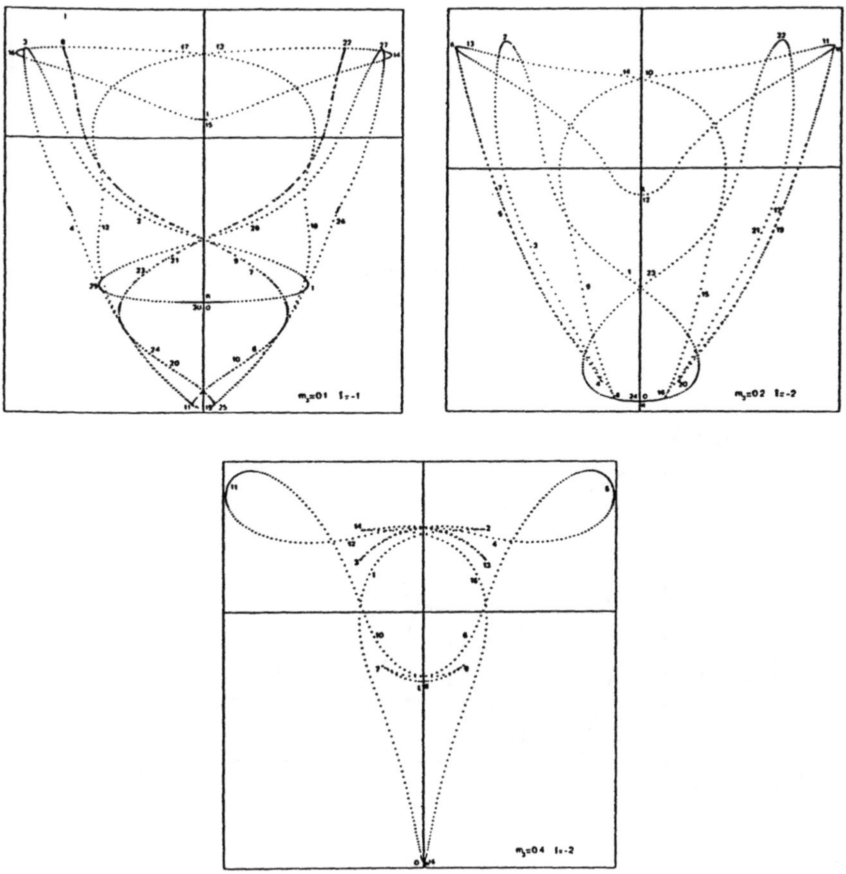

Figure 9

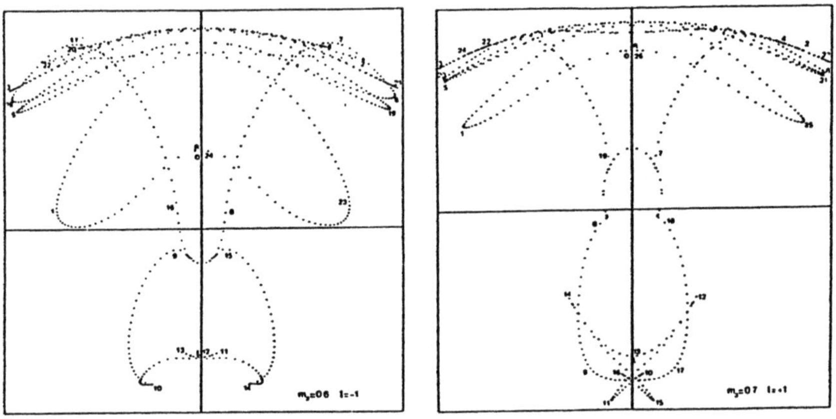

Figure 9 (Continuation)

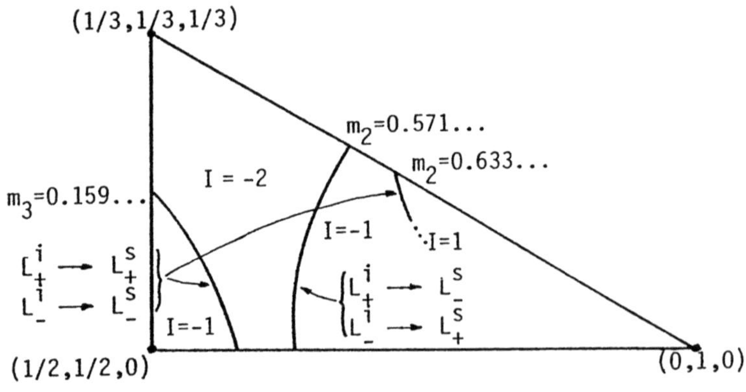

Figure 10

REFERENCES

1. Devaney, R., *Triple collision in the planar isosceles 3-body problem*, Invent. Math. **60** (1980), 249–267.
2. ———, *Singularities in classical mechanical systems*, in "Ergodic theory and Dynamical systems I, Proceedings Special Year," Maryland 1979-80. Ed. A. Katok, Birkhauser, pp. 211–333.
3. Easton, R., *Regularization of vector fields by surgery*, J. Differential Equations **10** (1971), 92–99.
4. Easton, R., *Some Topology of the 3-body problem*, J. Differential Equations **10** (1971), 371–379.
5. Lemaître, G., *Régularisation dans le problème des trois corps*, Bull. Classe. Sci., Acad. Roy. Belg. **40** (1954), 759–767.
6. McGehee, R., *Triple collision in the collinear three body problem*, Invent. Math. **27** (1974), 191–227.
7. Martínez, R., Simó, C., *Blow up of Collapsing Binaries in the Planar Three Body Problem*, Proceed. Colloque Géometrie symplectique et Mecánique, Montpellier, 1988 (to appear).
8. Moeckel, R., *Orbits near triple collision in the three body problem*, Indiana Univ. Math. Jour. **32** (1983), 221–240.
9. Moeckel, R., *Chaotic Dynamics Near Triple Collision*, Arch. Rat. Mechanics and Analysis **107** (1989), 37–70.
10. Murnaghan, F.D., *A Symmetric Reduction of the Planar Three-Body Problem*, Am. J. Math. **58** (1936), 829–832.
11. Siegel, C., Moser, J., "Lectures on Celestial Mechanics," Springer.
12. Simó, C., *Masses for which Triple Collision is Regularizable*, Celestial Mech. **21** (1980), 25–36.
13. ———, *Analysis of triple collision in the isosceles problem*, in "Classical Mechanics and Dynamical Systems," Ed. R.L. Devaney and Z. Nitecki, Marcel Dekker, 1981, pp. 203–224.
14. Simó, C.; Martínez, R., *Qualitative study of the planar isosceles three body problem*, Celestial Mech. **41** (1988), 179–251.
15. Stiefel, E.L.; Scheifele, G., "Linear and Regular Celestial Mechanics," Springer.
16. Sundman, K.F., *Nouvelles recherches sur le problème des trois corps*, Acta Soc. Sci. Fenn. **35** (1909), 1–27.
17. Susín, A., *Passages Near Triple Collision*, in "Long-Term Dynamical Behaviour of Natural and Artificial N-Body Systems," Ed. A.E. Roy. Reidel, 1988, pp. 505–513.
18. Waldvogel, J., *Stable and unstable manifolds in planar triple collision*, in "Instabilities in Dynamical Systems," Ed. V. Szebehely. Reidel, 1979, pp. 263–271.
19. ———, *Symmetric and regularized coordinates on the plane triple collision manifold*, Celestial Mech. **28** (1982), 69–82.
20. Whittaker, E.T., "Analytical Dynamics of Particles and Rigid Bodies," Cambridge Univ. Press.

1980 *Mathematics subject classifications*: 70F07, 58F14

C. SIMÓ: Dept. de Matemàtica Aplicada i Anàlisi
Universitat de Barcelona
Gran Via 585, 08007 Barcelona, Spain
A. SUSÍN: Dept. de Matemàtica Aplicada I
Universitat Politècnica de Catalunya
Diagonal 647, 08028 Barcelona, Spain

Acknowledgements: The research of the first author has been partially supported by CICYT Grant PB86–0527 (Spain). He also thanks the MSRI at Berkeley for giving him the possibility of presenting this work. The interest shown by R. Moeckel concerning this topic was very stimulating.

The Non-collision Singularities of the 5-body Problem

ZHIHONG XIA

Abstract. This paper is a survey on the author's work on the non-collision singularities in the Newtonian n-body problem. The non-collision singularity in the n-body system corresponds to the solution which blow up to infinity in finite time. The question whether there exists such solution was first raised by Painlevé in the last century and since then, it has been open. Here, we show that such solutions do exist in a 5-body problem. The method we use is based on careful analysis of near collisions orbits and McGehee's technique of blowing up collision singularities.

§1. Introduction.

In this paper, we consider the following problem: let $m_1, m_2, \ldots, m_5$ be five particles moving in an Euclidean space $\mathbf{R}^3$ under Newton's gravitation. Let their positions be $q_i \in \mathbf{R}^3$ and their velocities be $\dot{q}_i \in \mathbf{R}^3$. We consider the problem with the following symmetries:

$$
\begin{aligned}
&m_1 = m_2, \qquad m_4 = m_5;\\
&q_1 = (x_1, y_1, z_1), \qquad q_2 = (-x_1, -y_1, z_1);\\
&q_3 = (0, 0, z_3);\\
&q_4 = (x_4, y_4, z_4), \qquad q_5 = (-x_4, -y_4, z_4).
\end{aligned}
$$

These symmetries are preserved under the subsequent motion. In other words, for all the time, m_3 will be on z-axis. Both particles m_1 and m_4 are symmetric to m_2 and m_5 respectively with respect to z axis. Fix the center of masses at the origin. Thus the following equation holds:

$$2m_1 z_1 + 2m_4 z_4 + m_3 z_3 \equiv 0.$$

The resulting system has six degrees of freedom, its uniquely determined by $q_1 = (x_1, y_1, z_1), q_4 = (x_4, y_4, z_4)$ and their derivatives. The z_3 and $\dot{z}_3$ can be expressed in terms of these variables via above equation.

The objective of this note is to show the existence of unbounded solutions in the physical space in finite time for this five body system.

Let T and U be the kinetic and potential energy respectively. With these variables, the functions are:

$$T = m_1(\dot{x}_1^2+\dot{y}_1^2+\dot{z}_1^2)+m_4(\dot{x}_4^2+\dot{y}_4^2+\dot{z}_4^2)+\frac{1}{2}m_3(2m_1\dot{z}_1/m_3+2m_4\dot{z}_4/m_3)^2,$$

$$\begin{aligned}U = &\frac{m_1^2}{2(x_1^2+y_1^2)^{\frac{1}{2}}}+\frac{m_4^2}{2(x_4^2+y_4^2)^{\frac{1}{2}}}+\frac{2m_1m_4}{[(x_1-x_4)^2+(y_1-y_4)^2+(z_1-z_4)^2]^{\frac{1}{2}}}\\ &+\frac{2m_1m_4}{[(x_1+x_4)^2+(y_1+y_4)^2+(z_1-z_4)^2]^{\frac{1}{2}}}+\frac{2m_1m_3}{[x_1^2+y_1^2+(z_1-z_3)^2]^{\frac{1}{2}}}\\ &+\frac{2m_4m_3}{[x_4^2+y_4^2+(z_4-z_3)^2]^{\frac{1}{2}}}.\end{aligned}$$

Note that z_3 appears in the equation for U. However, z_3 is a function of z_1 and z_4 given by the center of mass equation. The Hamiltonian for this system uses the variables

$$\begin{aligned}&Px_1 = 2m_1\dot{x}_1, \qquad Px_4 = 2m_4\dot{x}_4, \qquad Py_1 = 2m_1\dot{y}_1, \qquad Py_4 = 2m_4\dot{y}_4\\ &Pz_1 = 2m_1\dot{z}_1 + 2m_1(2m_1\dot{z}_1/m_3 + 2m_4\dot{z}_4/m_3)\\ &Pz_4 = 2m_4\dot{z}_4 + 2m_4(2m_1\dot{z}_1/m_3 + 2m_4\dot{z}_4/m_3).\end{aligned}$$

Once the kinetic energy T is expressed as a function of $Px_i, Py_i, Pz_i (i = 1,2,3)$, then $x_i, y_i, z_i, Px_i, Py_i, Pz_i (i = 1,2,3)$ satisfy Hamilton's equation with the Hamiltonian

$$H(p,q) = T(p) - U(q).$$

Besides the energy integral $H(p,q) = h$, this system also admits the angular momentum integral $c_{12}+c_{45} = c$, where c_{12} and c_{45} are the angular momentum possessed by m_1, m_2 and m_4, m_5 respectively,

$$\begin{aligned}c_{12} &= 2m_1(x_1\dot{y}_1 - y_1\dot{x}_1) = x_1Py_1 - y_1Px_1,\\ c_{45} &= 2m_4(x_4\dot{y}_4 - y_4\dot{x}_4) = x_4Py_4 - y_4Px_4.\end{aligned}$$

Observe that c_{12}, c_{45} are not constants of motion. Nonetheless their sum, the total angular momentum, is conserved under the motion. We are only interested in the sub-system with zero total angular momentum. From now on, assume $c = 0$, i.e.,

$$c_{12} + c_{45} = 0.$$

The triple collisions between m_1, m_2, m_3 and between m_3, m_4 and m_5 are of special importance to us. There are several different ways for m_1, m_2

and m_3 to reach a triple collision. Classical results show that, as particles approach a collision, they must approach some special configurations, called central configurations. In this special 5-body problem, there are three central configurations for m_1, m_2 and m_3: One is the collinear configuration with m_3 in middle of m_1 and m_2, and the other two are where m_1, m_2 and m_3 form equilateral triangles: one with m_3 below m_1 and m_2, denoted by E_+ and the other one with m_3 above m_1 and m_2, denoted by E_-. Let Σ_1 be the subset of Ω consisting of all the initial conditions such that the corresponding trajectories end in triple collisions of the $1st, 2nd$ and $3rd$ particles with the limiting configuration E_+. We show that Σ_1 is a co-dimension 2 immersed manifold. Similarly, let Σ_4 be the subset of Ω consisting of initial conditions such that the corresponding trajectories end in triple collision of $3rd, 4th$ and $5th$ particles which have a similar limiting configuration E_-. Σ_4 is also a co-dimension 2 immersed submanifold.

§2. Statement of Results.

Our main result of this note is the following two theorems:

THEOREM 1. *There exist positive masses $m_1 = m_2, m_3, m_4 = m_5$, such that the following hold. For $x^* \in \Sigma_1$, let t^* be the time when the trajectory starting from x^* ends. There exist choice of x^* so that $q_4(x^*, t^*) = q_5(x^*, t^*)$, i.e. for the trajectory starting from x^*, when m_1, m_2, m_3 collide at t^*, m_4 and m_5 also have a binary collision at t^*. Then for some three dimensional hyper-surface Π in Ω crossing Σ_1 at x^*, there is an uncountable set Λ of points on Π having the following property: Let $x \in \Lambda$, there exists $\sigma > 0$, $\sigma < \infty$, such that the trajectory with initial condition x is defined for all $0 \le t < \sigma$ (possibly with some binary collisions) and satisfies*

$$z_1(t) = z_2(t) \to \infty, z_4(t) = z_5(t) \to -\infty \text{ as } t \to \sigma.$$

We remark that in the above theorem, the binary collisions are all regularized. Therefore it may be possible that some or all of the solutions given by this theorem involves some binary collisions. However, the following theorem asserts that, when initial values are chosen properly, the solutions given by theorem 1 actually never experience any binary collision at all.

THEOREM 2. *Let x^*, Π, Λ be that of Theorem 2.1. There exist $x^* \in \Sigma_1$, $\Pi \subset \Omega$ such that the following holds. There is a uncountable subset Λ_0*

of Λ*, such that for all* $x \in \Lambda_0$*, we have* $q_1(t) \neq q_2(t)$*,* $q_4(t) \neq q_5(t)$ *for all* $0 \leq t < \sigma$*.*

The solutions given by theorem 2 are examples of non-collision singularities.

The proof of the above theorem is long and fairly complicated, here, we only outline, in an informal fashion, some of the basic ideas involved in the proof. For a complete proof, see Xia [**12**].

To repeat, the objective of this paper is to show the existence of Newtonian motion for the five body problem that is unbounded in physical space in finite time. Moreover, it must be shown that this motion exists without the benefit of an accumulation of infinitely many number of binary collisions. The basic idea behind this dynamical behavior is fairly simple. Consider a five body problem consisting of four particles in two pairs, particles in each pair having the same masses, and a fifth particle of small mass. The four large particles form two binaries. Each binary is in an highly elliptical orbit with a plane of motion parallel to the x-y plane. What differs is that one binary is far above the x-y plane with a rotation in one direction, while the other binary is far below the x-y plane rotating in the opposite direction. This difference in rotation permits the total angular momentum of the system to be zero and also it permits two binaries to have arbitrarily small but non-zero angular momentum.

The fifth particle is restricted to the z-axis. It is the oscillation of this fifth particle that drives the system and creates this unbounded motion. To see the behavior of the system, image the following scenario. Suppose the oscillating particle passes through the plane of motion defined by a binary just when the binary are nearing their closest approach. Because the particles in the binary are almost at their closest point, they also are very close to the fifth particle. This proximity among the particles imposes a considerable force on the fifth particle that is directed back towards the plane of motion of the binary. This forces the small particle to return through the plane of motion of the binary when the binary starts to separate. This separation effect reduces the retaining force on the small particle. In return, this permits the fifth particle to move, at a very fast rate, toward the other binary system. In turn, the action-reaction effect of this three body system causes the original binary to move further away from x-y plane.

The motion, then, is obtained by iterating this scenario. Each time the fifth particle approaches one binary, the timing is such that the close approach of these binaries provides the force to accelerate the fifth particle back toward the other binary. The difficulty is to verify that this scenario actually occurs.

The timing sequence is accomplished with a symbolic dynamic argument. But before the symbolic dynamic argument can be used, several other elements about the dynamics need to be established. In particular, it is clear that if this motion is to become unbounded in finite time, then the acceleration effects on the oscillating particle must become infinitely large. Consequently, this requires the close approaches of each of the binaries to become infinitesimally small. But the only manner in which this can occur is as a by-product of the three body interactions. Therefore a major portion of proof is devoted toward developing a theory to permit us to understand the dynamics of this kind of a three body interaction.

In general, the non-collision singularity in the n-body problem are closely related with the collisions, especially the collisions with three or more particles involved. As we see above, in this special 5-body problem, a full understanding of the nature of the triple collisions is essential in the proof of Theorem 1 and Theorem 2. Our main tool for studying these triple collisions is McGehee's coordinates change to blow up the singularity due to triple collision [**4**].

Our first step is to ignore the presence of m_4 and m_5, thus the five body problem becomes an isosceles three-body problem. This problem has been studied extensively by Devaney [**1**] [**2**] and Moeckel [**6**]. As angular momentum is conserved for the system, Devaney studied the case where angular momentum of the system is zero (co-planar) and Moeckel considered the system with a small fixed angular momentum. Similar to the collinear 3-body problem that McGehee originally considered, a two dimensional triple collision manifold was obtained by Devaney for the co-planar isosceles 3-body problem. From this triple collision manifold, the orbit structure near triple collision can be easily understood by just reading off the flow on this manifold. Moeckel was able to extend Devaney's analysis to the isosceles 3-body problem with a fixed small angular momentum. From his analysis of triple collisions, some chaotic invariant set was found near the collision orbits.

For our case, we do not want to fix the angular momentum as a constant. There are two reasons for this. First of all, we will eventually analyze the five body problem. Although the angular momentum of the total system is a constant of motion fixed at zero, the angular momentum of the subsystem m_1, m_2, m_3 is not conserved under the motion. Besides, if we fix the angular momentum, our discussion will be restricted to an angular momentum hyper-surface. This makes it harder to obtain a global picture of the flow as the angular momentum changes. Naturally, as we allow the angular momentum to vary, we introduce an extra dimension. This makes the analysis a bit more difficult, but the reward is that more results can be obtained.

The basic method we use is again McGehee's coordinates. This time, we use them not only for blowing up triple collision singularities, but also for separating the size of the system from other variables. This allows us to extend Moeckel's results further to be applicable to our problem. The usefulness of the second feature of the McGehee's coordinates can be seen as followings: first, as triple collision manifold corresponds to $r = 0$ (r equal to the square root of the moment of inertia of the system) and flow on this manifold is fictitious, one lacks a general idea about what the flow looks like. However, one can relate this flow to that of the flow of the invariant subsystem with energy being zero. This makes it possible to visualize the flow on the triple collision manifold. Second, by separating r from other variables, one can reduce the dimension of system by one, which sometimes greatly simplifies the problem.

The Cantor set in Theorem 1 is produced by symbolic dynamics. Here we use the full shift on infinitely many symbols. These symbols turn out to be the numbers of complete resolutions of the binaries m_1 and m_2 or m_4 and m_5 made between each close passage of m_3 to these binaries. By making these numbers arbitrarily large, arbitrary large velocity can be produced for each passage. It is this large velocity that makes it possible for the five body system to explode into infinity in finite time.

In the proof of Theorem 1, as we remarked earlier, double collisions are treated as if there are regularized. The infinitesimally close approaches of the binaries creates some worry. Do these binaries collide? They do not if the angular momentum of the binaries is non-zero for all the time. The basic idea for proving Theorem 2 is quite simple. We make the following three simple observations: First, for the solutions constructed in Theorem 1, m_3 stays away from the two binaries for most of times, therefore the two binaries behave very closely to some 2-body problems except for when m_3 is closely encountered. Second, the two binaries are far away from each other, as the presence of m_3 does not affect the angular momentum of each of the binaries, these angular momentum change very little and finally, the binaries are moving in some highly eccentrical elliptical orbits. These three facts make it easy to have an accurate estimate over the angular momentum c_{12} and c_{45}. Since at instant of binary collision $c_{12} = c_{45} = 0$, to show that no binary collision ever exists, one need only to show these quantities never become zero, although their limit has to be zero. To accomplish this, some new variables are introduced and some careful estimate are made.

§3. Historical Comments.

The solutions we show in Theorem 2 are the first examples of a special class of singularities in the n-body problem, called non-collision singularities. The problem of non-collision singularities dates back to the time of Poincaré and Painlevé in last century. Here we give a brief historical account of this problem. For more detailed survey see McGehee [**5**].

Let us consider the classical n-body problem:

$$m_i\ddot{q}_i = \sum \frac{m_i m_j}{|q_i - q_j|}(q_i - q_j) = \frac{\partial U}{\partial q_i},$$

where q_i are positions of n masses m_1, $i = 1, 2, \dots, n$, and

$$U = \sum_{j<i} \frac{m_i m_j}{|q_i - q_j|}.$$

By the existence and uniqueness theory of ordinary differential equations, given $q(0) \in \mathbf{R}^3$ and $q(0) = (\mathbf{R}^3)^n$ with $q_i(0) \neq q_j(0)$, for all $i \neq j$, there exists a unique solution $q(t)$ defined for all $0 \leq t < \sigma$ and where σ is maximum. Here σ may be finite or infinite. In case $\sigma < \infty$, we say that the solution $q(t)$ experiences a singularity at σ. In other words, a solution experiences a singularity at σ if the standard existence and uniqueness theory of ordinary differential equation no longer provide an extension of solution.

An obvious singularity in n-body problem is the collision singularity, i.e., when the solution ends up with two or more particles occupying the same position in the physical space. A question arises naturally: Is there any singularity which is not a collision singularity? i.e., is there any non-collision singularity? This question was first raised by Painlevé in 1985. Painlevé showed that for the three body problem, there is no non-collision singularity.

And since then, it had been an open problem whether there is some non-collision singularity for the n-body system with $n \geq 4$.

An important step toward an answer to Painlevé's question was taken by von Zeipel [**13**] in 1908. He showed that, if the positions of all the particles remain bounded as $t"\sigma$, then the singularity must be due to a collision. In other words, a non-collision singularity can occur only if the system of particles becomes unbounded in finite time. This remarkable makes the existence of the non-collision singularities seemingly impossible, since a particle escaping to infinity in finite amount of time would have to acquire an infinite amount of kinetic energy. However, there is no a priori upper bound on the kinetic energy of a particle.

The subject of the singularities was resurrected by Saari and Pollard [**8**]–[**11**] in the early 1970s. Some important results have been obtained by Moser, Siegel, Saari, Pollard, McGehee, Mather, Devaney, Moeckel and many others. In 1974, McGehee [**4**] introduced a remarkable new set of coordinate for the study of triple collision. McGehee's idea uses "polar coordinate" to "blow up" the singularity set due to collision and to replace it with an invariant boundary called collision manifold. This enables us to readily understand the behavior of solutions near the singularity.

McGehee's coordinate proved to be a useful tool in the study of the collision singularities. By using these coordinate, Mather and McGehee [**3**] constructed an remarkable example of unbounded solutions in finite time in collinear four body problem. However, as binary collisions in the collinear four body problem are inevitable their solution contains infinitely many numbers of binary collisions which have been extended by some elastic bounces. Based on their ideas, Anosov suggested that there might exist the non-collision singularity in the neighborhood of the example constructed by Mather and McGehee, but this approach has not been proved to be successful.

The result we show in this paper can be extended to some system with more than five particles. It will be extremely interesting to see an example of the non-collision singularity in the four body problem, although the author has no doubt on its existence.

ACKNOWLEDGEMENT: The author wishes to take this opportunity to thank Prof. D. Saari for his help and support.

References

[1] R. Devaney, *Triple collision in the planar isosceles three-body problem*, Inv. Math. **60** (1980), 249–267.

[2] R. Devaney, *Singularities in classical mechanical systems*, Ergodic Theory and Dynamical Systems (1981), Birkhauser, Boston.

[3] J. Mather and R. McGehee, *Solutions of the collinear four body problem which become unbounded in finite time*, Lecture Notes in Physics **38** (J. Moser, ed.) (1975), 573–597, Springer-Verlag, Berlin Heidelberg New York.

[4] R. McGehee, *Triple collision in the collinear three-body problem*, Inventiones Math. **27** (1974), 191–227.

[5] R. McGehee, *Von Zeipel's theorem on singularities in celestial mechanics*, Expo. Math. **4** (1986), 335–345.

[6] R. Moeckel, *Heteroclinic phenomena in the isosceles three-body problem*, SIAM J. Math. Analysis **15** (1984), 857–876.

[7] P. Painlevé, "Leçons sur la théorie analytique des èquations différentielles," A. Hemann, Paris, 1897.

[8] H. Pollard and D. Saari, *Singularities of the n-body problem, I*, Arch. Rational Mech. and Math. **30** (1968), 263–269.

[9] H. Pollard and D. Saari, *Singularities of the n-body problem, II*, Inequalities-II (1970), 255–259, Academic Press.

[10] D. Saari, *Singularities and collisions of Newtonian gravitational systems*, Archive for Rational Mechanics and Analysis **49,** no. 4 (1973), 311–320.

[11] D. Saari, *Singularities of Newtonian gravitational systems*, Proceedings of Symposium on Global Analysis, Dynamical Systems and Celestial Mechanics (1971), Brazil, August.

[12] Z. Xia, *The Existence of Non-collision Singularities In Newtonian System*, Thesis (1988), Northwestern University.

[13] H. von Zeipel, *Sur les singularités du problème des n corps*, Arkiv for Matematik, Astronomi och Fysik **4, 32** (1908), 1–4.

Department of Mathematics
Harvard University
Cambridge, MA 02138

Partially supported by an NSF grant and Alfred Sloan Fellowship.

FSC
www.fsc.org

MIX
Papier aus verantwortungsvollen Quellen
Paper from responsible sources
FSC® C105338

If you have any concerns about our products,
you can contact us on
ProductSafety@springernature.com

In case Publisher is established outside the EU,
the EU authorized representative is:
Springer Nature Customer Service Center GmbH
Europaplatz 3, 69115 Heidelberg, Germany

Printed by Libri Plureos GmbH
in Hamburg, Germany